COMPARATIVE ANATOMY
of the Vertebrates

EIGHTH EDITION

COMPARATIVE ANATOMY
of the Vertebrates

GEORGE C. KENT

LARRY MILLER
Gannon University

WCB **Wm. C. Brown Publishers**

Dubuque, IA Bogota Boston Buenos Aires Caracas Chicago Guilford, CT
London Madrid Mexico City Seoul Singapore Sydney Taipei Tokyo Toronto

Project Team

Editor *Marge Kemp*
Developmental Editor *Kathy Loewenberg*
Production Editor *Jane Matthews*
Marketing Manager *Tom Lyon*
Designer *Kaye Farmer*
Art Editor *Kathleen M. Timp*
Photo Editor *Janice Hancock*
Advertising Coordinator *Heather Wagner*
Permissions Coordinator *LouAnn K. Wilson*

 Wm. C. Brown Publishers

President and Chief Executive Officer *Beverly Kolz*
Vice President, Director of Editorial *Kevin Kane*
Vice President, Sales and Market Expansion *Virginia S. Moffat*
Vice President, Director of Production *Colleen A. Yonda*
Director of Marketing *Craig S. Marty*
National Sales Manager *Douglas J. DiNardo*
Executive Editor *Michael D. Lange*
Advertising Manager *Janelle Keeffer*
Production Editorial Manager *Renée Menne*
Publishing Services Manager *Karen J. Slaght*
Royalty/Permissions Manager *Connie Allendorf*

Copyedited by Carol Kromminga

Cover and interior design by Maureen McCutcheon

Cover photos: © 1996 PhotoDisc, Inc., Nature, Wildlife and the Environment and Details of Nature

Printed in the United States of America
2460 Kerper Boulevard, Dubuque, IA 52001

10 9 8 7 6 5 4 3 2 1

C O N T E N T S

16 NERVOUS SYSTEM, 381

17 SENSE ORGANS, 419

his is a textbook of functional and comparative morphology with a developmental and evolutionary perspective. Repeatedly expressed or implied are the concepts that existing structural patterns are modifications of ancestral ones, that the adult is a modification of the embryo, that individual as well as species differences exist, and that development and structure from fishes to human beings are broadly determined by inheritance and adaptively modified by natural selection.

The text is designed for zoology majors, for preprofessional students in medicine, dentistry, and allied health fields, and for liberal arts students interested in the phylogeny, embryonic development, and adult anatomy of vertebrates, including human beings. Discussions of function abound, but the physics and chemistry that drive these are generally left to the province of other courses.

Persons acquainted with earlier editions will find in each chapter the familiar introductory paragraphs that whet student curiosity, the chapter outlines, the widely acclaimed summaries, and the revised lists of selected readings. Current users will also discover deletions and additions in every chapter, each designed to improve readability, update a topic, or otherwise add to the pedagogic value of the discussion.

ORGANIZATION

The first five chapters provide background for the remaining chapters. Chapter 1 describes the basic body plan of vertebrates—notochord, vertebral column, pharyngeal pouches, slits, and arches, dorsal hollow central nervous system, bilateral symmetry, and coelom—then presents thumbnail sketches of the organ systems as an overture to later chapters. Chapter 2 examines concepts of modern vertebrate morphology and phylogeny. Chapter 3 is a brief introduction to protochordates. Chapter 4 provides a glimpse of the diversity of vertebrate life in time and taxa. Chapter 5 describes ontogenesis in amphioxus and vertebrates through gastrulation, and examines the roles of extraembryonic membranes from fishes to mammals. The remaining chapters discuss the vertebrate body systems, highlighting the universal-

ity of basic patterns and the adaptive nature of adult morphology from fishes to human beings. Appendices provide a synoptic classification of chordates and an alphabetical list of prefixes, suffixes, roots, and stems employed in the text. Students should be urged to read the introductory paragraphs to Appendix II, and to avail themselves of this opportunity to enrich their intellectual horizons.

Chapters 5 and 18 are frequently unassigned due to lack of classroom time. However, assignment of one or more *selected* discussions, such as gastrulation in an amphioxus as an introduction to the digestive system, will provide valuable insight into the development of the digestive tract with minimal expenditure of additional study time. Students in premedicine and allied health fields may opt to read about insulin, thyroxine, or gonads as endocrine organs, for example, in chapter 18, provided that the class is simply made aware of the discussions.

Some reviewers of previous editions have expressed opinions that there is "too much" and "too little" detail. We have tried to keep foremost in mind the diversity of the audience, being aware that instructors will omit some discussions and provide enrichment in others in accordance with the needs of their students and their own professional expertise.

A feature of the text is the frequency with which illustrations may be repeatedly cited, including in other chapters. Viewing an illustration more than once, and in different context, reinforces what we try to teach—the functional interdependence of the body systems. It also enhances visual memory and evokes spontaneous recall of earlier subject matter.

NEW TO THIS EDITION

This edition has several new figures. Representatives include a cladogram of postulated fish lineages, the pelvic girdles of dinosaurs, migratory routes of early marsupials, an excretory solenocyte from an amphioxus, bisexual sex cords in developing gonads, and evolution of the cerebral hemispheres. In addition, numerous revisions to existing figures have been made. Among these were changes to figures on middle ear development, pelvic girdles, oral glands, the mammalian kidney, spinal nerve structure, the retina, neuromast organs, and muscle spindles. Readers will also find figure updates on the range and relative abundance of vertebrates in geologic time and a modified family tree of reptiles.

Among the many additions and changes are the following: a revised taxonomy reflects selected traditional and cladistic postulates; the status of hagfishes and lampreys has been examined; biological concepts (homology, adaptation, evolutionary divergence, paedomorphosis, heterochrony, others) include familiar examples; the role of hedgehog genes in vertebrate gastrulation is examined; discussions of the forebrain and its evolution have been extensively revised; and the limbic system has been introduced.

REVIEWERS

Sandra Harrington—*Carthage College*
Stephen Reilly—*Ohio University*
David Shomay—*University of Illinois*
James A. Strauss—*Penn State University*

CHAPTER ONE

INTRODUCTION

This chapter is to the rest of the book as an overture is to an extended musical score. It is a taste of things to come. It commences as an orientation to the taxonomic status of vertebrates in the animal kingdom and outlines their unique and not-so-unique characteristics. Thereafter, it provides thumbnail sketches of the organ systems that will be studied in detail in later chapters. Along with the next chapter, it is a preparation of the mind designed to minimize surprise in the journey ahead. Bon voyage!

OUTLINE

The Phylum Chordata
The Vertebrate Body: General Plan
   Regional Differentiation
   Bilateral Symmetry and Anatomic Planes
   Metamerism
Vertebrate Characteristics: The Big Four
   Notochord and Vertebral Column
   The Vertebrate Pharynx
      Pharyngeal Pouches and Slits
      Pharyngeal Arches
   Dorsal Hollow Central Nervous System
Other Vertebrate Characteristics
   Integument
   Respiratory Mechanisms
   Coelom
   Digestive System
   Urogenital Organs
   Circulatory System
   Skeleton
   Muscles
   Sense Organs

omparative vertebrate anatomy is the study of the structure of vertebrates (descriptive morphology) and of the functional significance of structure (functional morphology). Because structure entails development of the individual (ontogenesis), and individuals have an ancestral history (phylogenesis), the discipline embraces these areas of inquiry as well. Every area of biological inquiry provides relevant data to the discipline. Ecology, genetics, molecular biology, serology, biochemistry, and paleobiology are all sources of valuable data. Geology is an indispensable source. Because of the physical limitations of a single text, the contribution of some of these areas will be mentioned only in passing. The thrust of the chapters will be the organs and systems, their roles in survival, their embryogenesis, and their historical background in geological time. The latter entails consideration of the phylogenesis of vertebrates in general.

To the extent that comparative anatomy is concerned with phylogenesis, it is a study of history and of animals that no longer inhabit the earth and are known to us only by a fossil record. It is interested in the survival value of structure (an adaptation), of the struggle for compatibility with an ever-changing environment, of the invasion of new territory by those best equipped for survival, and of the extinction of species. The history of vertebrates, including humans, is a fascinating story from which is developing a genealogy based on the data just described. Comparative anatomy addresses curiosity about the origin of species, including our own. The generalizations and conclusions arrived at in the discipline add to the enlightenment of the human mind.

THE PHYLUM CHORDATA

It is conventional to think of animals as falling into two categories—those lacking vertebral columns, or invertebrates, and animals with vertebral columns, or vertebrates. Such a dichotomy, although valid, does not recognize a group of small marine animals that are transitional between invertebrates and vertebrates—the protochordates—which we will be studying shortly. Protochordates have no vertebral column, but they share with vertebrates and with no other animals a combination of three other morphological features—a **notochord,** a **dorsal hollow central nervous system,** and a **pharynx with paired pouches and clefts in the embryo stage at least.** These characteristics are so fundamental in the architecture of vertebrates that they are among the first to appear in vertebrate embryos. Indeed, without them no vertebrate could proceed beyond the earliest stages of embryonic development. Because of the primacy of these structures in protochordates and vertebrates alike, these two groups have been incorporated into a single taxon, or classification category, the **phylum Chordata.** The taxonomic relationship of protochordates (two subphyla) and vertebrates (one subphylum) is as follows:

Kingdom Animalia
  Phylum Chordata
    Subphylum Urochordata
    Subphylum Cephalochordata
    Subphylum Vertebrata (Craniata)

A third subphylum of protochordates, of debatable status, is discussed in chapter 3.

Chordates are animals that have a notochord in the embryo stage at least. Vertebrates are chordates with vertebrae. Vertebrae appear during embryonic development after the notochord has formed. Subsequently, they reinforce the notochord or replace it functionally.

THE VERTEBRATE BODY: GENERAL PLAN

All vertebrates conform to a generalized pattern of anatomic structure. This is revealed by dissection and is the result of the expression of similar DNA molecules inherited during the course of vertebrate evolution. Vertebrates also exhibit similar, but not identical, patterns of embryonic development. This, too, is a result of common ancestry. Both morphology and developmental processes have been altered during the passage of time, which, as it lengthens, provides increasing opportunities for genetic changes that result in anatomic diversity. Yet, despite these changes, innumerable primitive structural and developmental similarities still exist. These similarities and diversities will be examined in detail in later chapters. In this chapter we will discuss only the structural highlights that characterize vertebrates in general.

Regional Differentiation

The typical vertebrate body consists of four regional components—head, trunk, postanal tail, and paired pectoral and pelvic appendages. Concentrated on or within the **head** are special sense organs for monitoring the external environment; a brain that is at least large enough to receive and process essential incoming information and to provide appropriate stimuli to the body musculature; jaws for acquiring, retaining, and, in some species, macerating food; and, in fishes, gills for respiration. Expansion of the brain over hundreds of millions of years has resulted in larger braincases that have become increasingly movable independent of the trunk during that time. **Cephalization** has developed to a greater degree in vertebrates than in any other group of animals.

The **trunk** contains a cavity, the **coelom,** that houses most of the viscera (figs. 1.1 and 1.2). Surrounding the coelom is the **body wall** consisting chiefly of muscle, vertebral column, and ribs. The body wall must be opened to expose the viscera. The **neck** is a narrow extension of the trunk of reptiles, birds, and mammals, and lacks a coelom. It consists primarily of vertebrae, muscles, spinal cord, nerves, and elongated tubes—esophagus, blood vessels, lymphatics, trachea—that connect structures of the head with those of the trunk.

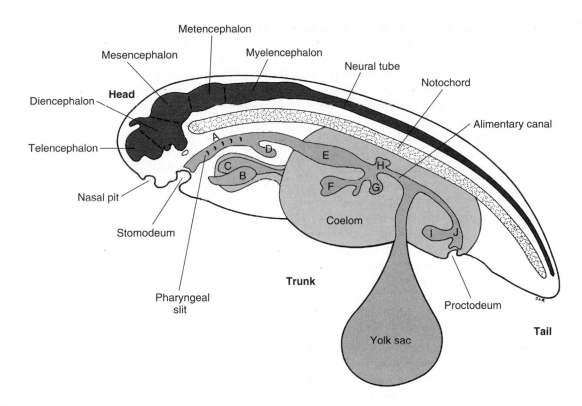

FIGURE 1.1

Sagittal section of a generalized vertebrate embryo. **A**, pharynx (*light red*) with pharyngeal slits; **B** and **C**, ventricle and atrium of heart; **D**, diverticulum that gives rise to the lung in tetrapods and swim bladder in fishes; **E**, stomach; **F**, liver bud and associated gallbladder; **G**, ventral pancreatic bud; **H**, dorsal pancreatic bud; **I**, urinary bladder of tetrapods; **J**, cloaca separated from the proctodeum by a cloacal membrane. The stomodeum is separated from the pharynx by a thin oral plate. The differentiated brain has five major subdivisions: telencephalon and diencephalon (forebrain), mesencephalon (midbrain), and metencephalon and myelencephalon (hindbrain).

FIGURE 1.2

Cross section of a generalized vertebrate trunk. **A**, dorsal aorta, giving off renal artery to kidney; **C**, coelom; **D**, kidney duct; **E**, epaxial muscle; **G**, future gonad (genital ridge); **H**, hypaxial muscle in body wall; **K**, kidney; **R**, rib projecting into a horizontal skeletogenous septum from the transverse process of a vertebra (*black*). **1**, dorsal root of spinal nerve; **2**, ventral root; **3**, dorsal ramus of spinal nerve; **4**, ventral ramus; **5**, parietal peritoneum; **6**, visceral peritoneum; **7**, ventral mesentery. A remnant of the notochord (*dark red*) lies within the centrum of a vertebra. The spinal cord (*light red*) lies above the centrum surrounded by a neural arch (*black*). Body wall muscle is shown in *light red*.

The **tail** commences at the anus or vent; hence it is postanal. It consists almost exclusively of a caudal continuation of body wall muscles, axial skeleton, nerves, and blood vessels. Some adult vertebrates lack a postanal tail, although it is present in all embryos. The swimming larvae of frogs, toads, and wormlike amphibians (caecilians) have a tail, but it is resorbed at metamorphosis. Modern birds have tails reduced to a nubbin, but the first birds had long tails (fig. 4.29). Human beings have a vestigial postanal tail early in embryonic life (fig. 1.10). Remnants remain in adults as the "tail bone" (coccyx).

Two pairs of **appendages**—pectoral and pelvic, supported by an internal skeleton and operated by contributions from the trunk musculature—are characteristic of vertebrates; but again they are sometimes vestigial or have been completely lost. The earliest known vertebrates lacked these appendages, as do living jawless vertebrates (agnathans).

Bilateral Symmetry and Anatomic Planes

Vertebrates have three principal body axes: a longitudinal (anteroposterior) axis, a dorsoventral axis, and a left-right axis. With reference to the first two, structures at one end of the axis are different from those at the other end. The left-right axis terminates in identical structures at each end. Thus the head differs from the tail and the dorsum differs from the venter, but right and left sides are mirror images of each other. An animal with this arrangement of body parts exhibits **bilateral symmetry.**

It is sometimes convenient to discuss parts of the vertebrate body with reference to three **principal anatomic planes.** Two axes define a plane. The transverse plane is established by the left-right and the dorsoventral axes. A cut in this plane is a **cross section** (fig. 1.3, fish). The frontal plane is established by the left-right and longitudinal axes. A cut in this plane is a **frontal section.** The sagittal plane is established by the longitudinal and dorsoventral axes. A cut in this plane is a **sagittal section.** Sections parallel to the sagittal plane are **parasagittal.** Acquainting oneself with these concepts is a simple exercise in anatomy and logic.

Metamerism

Vertebrates exhibit **metamerism,** the serial repetition of structures in the longitudinal axis of the body. It is clearly manifested in vertebrate embryos (fig. 16.6, somite), and is retained in many adult systems. No external evidence is seen in most adult reptiles, birds, and mammals because the skin is not metameric. If, however, the skin is stripped from the trunk and tail of fishes, most amphibians, and some reptiles, one sees a series of

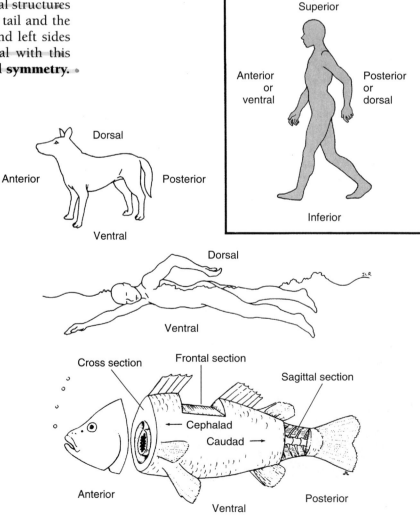

FIGURE 1.3

Terms of direction and position and planes of sectioning of the vertebrate body. Terms employed in human anatomy are shown in the box.

muscle segments that are reflections of the embryonic metamerism (fig. 11.5). In addition, the serial arrangement of vertebrae, ribs, spinal nerves, embryonic kidney tubules, and segmental arteries and veins, which will be studied later, are further expressions of the fundamental metameric nature of vertebrates.

VERTEBRATE CHARACTERISTICS: THE BIG FOUR

Vertebrates exhibit a unique combination of four fundamental morphologic features: (1) a notochord in the embryo stage at least; (2) a pharynx with pouches or slits in its lateral walls in the embryo stage at least; (3) a dorsal, hollow central nervous system; and (4) a vertebral column. These are the "big four" vertebrate characteristics. The first three are also chordate features; the fourth is unique. If a bizarre organism were to be discovered in the abyssal depths of the oceans, these four features, in combination, would admit this creature to the vertebrate hierarchy.

We will now examine these characteristics. The notochord and vertebral column will be discussed together, since they occupy the same site. We will then consider briefly a few other vertebrate features.

Notochord and Vertebral Column

The **notochord** is the first skeletal structure to appear in vertebrate embryos. At its peak of embryonic development it is a rod of living cells located immediately ventral to the central nervous system and dorsal to the alimentary canal extending from the midbrain to the tip of the tail (fig. 1.1). The part of the notochord in the head becomes incorporated in the floor of the skull, and, except in agnathans, the part in the trunk and tail becomes surrounded by cartilaginous or bony **vertebrae.**

These provide more rigid support for the body than does a notochord alone. A typical vertebra (fig. 1.4) consists of a **centrum** that is deposited around the notochord, a **neural arch** that forms over the spinal cord, and various **processes** (fig. 1.4). In the tail a **hemal arch** may surround the caudal artery and vein (fig. 8.1*b, c,* and *e*).

The fate of the notochord in adult vertebrates is variable. In almost all fishes it persists the length of the trunk and tail, although usually constricted within each centrum (fig. 8.2). The same is true in many urodeles and some primitive lizards (fig. 8.3). However, in modern reptiles, birds, and mammals the notochord is almost obliterated during development. A vestige remains in mammals within the intervertebral disks that separate successive centra. The vestige consists of a soft spherical mass of connective tissue called the **pulpy nucleus** (fig. 8.11*d*). Modern reptiles and birds lack even this vestige.

In agnathans the notochord grows along with the animal, and paired **lateral neural cartilages** become perched on the notochord lateral to the spinal cord (fig. 1.5). These cartilages are reminiscent of neural arches, but whether they are primitive vertebrae, vestigial vertebrae from an ancestor that had a typical vertebral column, or entirely different structures is not known.* When a notochord persists as an important part of the adult axial skeleton, it develops a strong outer elastic and inner fibrous **notochord sheath.**

The notochord has been disappearing as an adult structure, but development of a notochord in every vertebrate embryo is a reminder that all vertebrates have a common ancestry.

---

*The backboneless, appendageless, jawless hagfishes and lampreys (agnathans) have been admitted to the vertebrate hierarchy as a concession to logic. They are discussed in chapter 4.

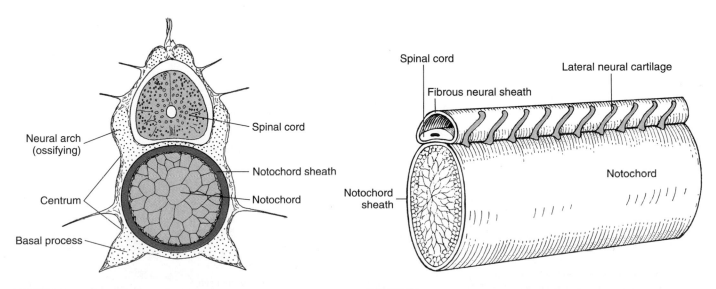

FIGURE 1.4

Transverse section of the vertebral column of a very young trout.

Neural arch (ossifying)

Centrum

Basal process

Spinal cord

Notochord sheath

Notochord

FIGURE 1.5

Lateral neural cartilages (*red*) and notochord of a lamprey.

Spinal cord

Fibrous neural sheath

Lateral neural cartilage

Notochord sheath

Notochord

The Vertebrate Pharynx

The pharynx (pronounced fair-inks) is a vital part of the vertebrate embryo. It produces the gills of fishes, the lungs of tetrapods, the skeleton and musculature of the jaws, the endocrine glands that regulate metabolic rates in every cell of the body and that maintain appropriate calcium levels in the bones, other tissues, and in the circulating blood. It gives rise to the middle ear cavity of tetrapods. Last, but of utmost importance, it provides, in humans, during fetal life and for a short time thereafter, the initial cells of the immune system, which, if destroyed, leads to death. A basic blueprint for the archi-

tecture of the vertebrate pharynx is expressed in all vertebrate embryos from fishes to human beings. We will now examine that blueprint.

Pharyngeal Pouches and Slits

A series of **pharyngeal pouches** arises as diverticula of the endoderm of the foregut and the pouches grow toward the surface of the animal (fig. 1.6). These pouches establish the limits of the pharynx. Simultaneously, an **ectodermal groove** grows toward each pouch (figs. 1.7 and 1.8). Soon, only a thin membrane, the **branchial plate,** separates the groove from the pouch. If, and when, the branchial plate

FIGURE 1.6

Basic pharyngeal architecture as exemplified by a composite vertebrate embryo. The dorsal aorta is paired in the head. The ventral aorta is unpaired. A series of pharyngeal pouches has evaginated from the lateral walls of the digestive tract. Six aortic arches (*red*) connect the heart and ventral aorta with the dorsal aorta. (Typically, the first aortic arch disappears before the sixth one has formed.) Although a lung does not form in all vertebrates, it is an ancient structure and is represented by a swim bladder, in fishes. The anterior end of the dorsal, hollow nervous system is enlarging to form the brain. The notochord commences at the level of the midbrain. All reptile, bird, and mammal embryos exhibit the cephalic flexure at the level of the midbrain.

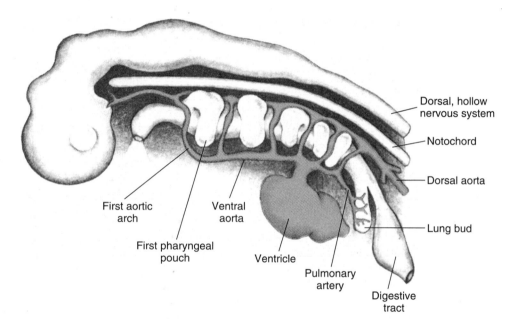

FIGURE 1.7

Early pharyngeal development in embryonic shark, looking down onto pharyngeal floor. (*a*) Formation of pharyngeal pouches and slits, earlier stage in left wall. (*b*) Pharyngeal arches (**I–VII**). **T,** Site of thyroid evagination.

(a)   (b)

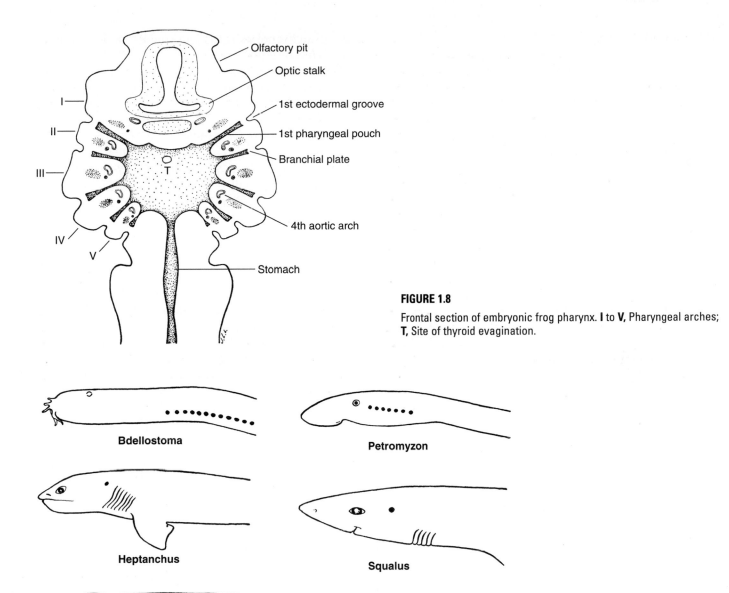

FIGURE 1.8

Frontal section of embryonic frog pharynx. **I** to **V**, Pharyngeal arches; **T**, Site of thyroid evagination.

FIGURE 1.9

Persistent pharyngeal slits in selected agnathans (*top*), sharks (*center*), and tailed amphibians (*bottom*).

ruptures, a passageway is formed between the pharyngeal lumen and the exterior of the animal. This passageway is a pharyngeal slit. The slits may be permanent, as in fishes in which they are exits for respiratory water from the gills, or they are temporary, as in most tetrapods. Eight is the largest number of pouches that develop in any jawed vertebrate, and that number is found only in some primitive sharks. Agnathans have as many as 15 slits (fig. 1.9).

Pharyngeal slits are temporary if the animal is going to live on land. Of the six pharyngeal pouches that form in frog embryos, four give rise to gill slits in tadpoles. These slits close when the tadpole metamorphoses into a frog. In reptiles, birds, and mammals, no gills develop in the pouches, and the slits are transitory. Of the five pouches that develop in chicks, the first three rupture to the exterior and close again. Only one or two of the more anterior pouches of mammals may rupture. Cervical fistulas occasionally seen in human beings are usually the result of the failure of the cervical sinus, housing the third and fourth slits, to close.

Although the pharyngeal pouches of tetrapods rarely give rise to permanent slits, the first one becomes the auditory tube and middle ear cavity of tetrapods, and the second persists as the pouch of the palatine tonsil of mammals. The walls of several pouches give rise to endocrine tissue in all vertebrates.

Pharyngeal Arches

Each embryonic pharyngeal pouch or slit is separated from the next by a column of tissue known as a **pharyngeal arch** (figs. 1.7*b*, and 1.10). Each arch typically contains four basic components, or the blastemas from which these components develop. They are (1) *supportive skeletal elements,* illustrated in an adult shark in fig. 9.1; (2) *striated muscles that operate the arch* (fig. 11.25*a*); (3) *branches of the fifth, seventh, ninth, and tenth cranial nerves,* which innervate the muscles and provide sensory input to the brain (fig. 16.30); and (4) an *aortic arch* that connects the ventral and dorsal aortas (fig. 1.6). These components are found also immediately anterior to the first pouch or slit and, with some omissions, immediately caudal to the last. The boundaries of a pharyngeal arch can be identified from the exterior when there are ectodermal grooves or pharyngeal slits to serve as landmarks, and only then. If the grooves disappear or the slits close, as they do in tetrapods, the boundaries of the arches are lost and the components become reoriented.

The skeleton of the pharyngeal arches, whether in fishes or adult tetrapods, is referred to collectively, for identification purposes, as the **pharyngeal** (or **visceral**) **skeleton**. The muscles of the arches are referred to as **branchiomeric** in recognition of their primitive relationship to gills.

The upper and lower jaws and associated muscles, nerves, and vessels constitute the first, or **mandibular arch**. It develops anterior to the first pouch. Behind the first pouch is the second, or **hyoid arch**. The remaining arches are referred to by number, starting with the third. The modifications that developed in the pharynx as vertebrates shifted from life in water to life on land constitute a fascinating chapter in vertebrate history. Many of the chapters that follow will describe this evolutionary metamorphosis.

Dorsal Hollow Central Nervous System

The central nervous system consists of brain and spinal cord and contains a central lumen, the **neurocoel**. The dorsal location and the lumen result from the fact that the central nervous system typically arises as a longitudinal **neural groove** in what will become the dorsal surface of the embryo (figs. 1.11 and 5.9*c*). The groove closes over or rolls up and sinks beneath the surface to become a hollow **neural tube** located dorsal to the notochord (figs. 1.1 and 1.6). The tube is wider anteriorly, and this part becomes the brain with its ventricles. The process of forming the neural tube is **neurulation**.

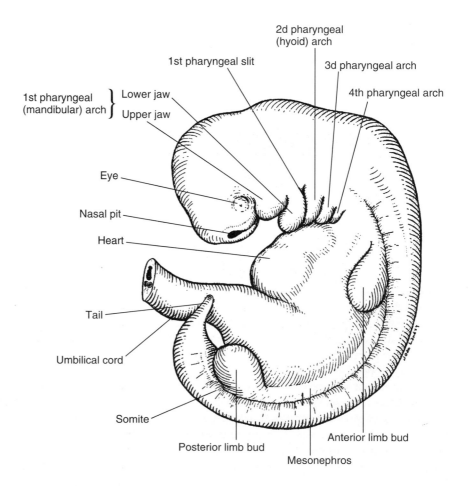

FIGURE 1.10

Human embryo approximately 4½ weeks after fertilization (5 mm stage).   Actual crown-rump length: |

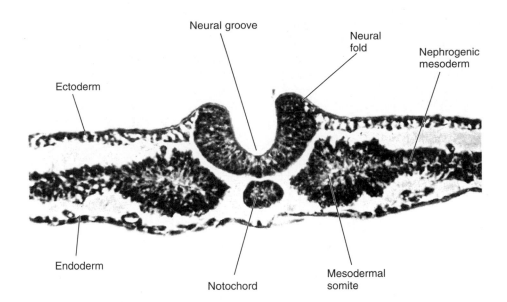

Ectoderm
Neural groove
Neural fold
Nephrogenic mesoderm
Endoderm
Notochord
Mesodermal somite

FIGURE 1.11

Cross section of 24-hour chick embryo. The neural groove and mesodermal somites of a whole chick embryo of the same age are seen in figure 5.9c.

In cyclostomes, teleosts, and ganoid fishes, the basic pattern of neurulation has become slightly modified. Instead of forming a groove, the surface ectoderm dorsal to the notochord proliferates a wedge-shaped **neural keel** (fig. 5.9e). Eventually the keel separates from the surface and a cavity forms within it by rearrangement of the cells in its interior. The result is a typical dorsal hollow central nervous system.

Cranial and spinal nerves connect the central nervous system with the various organs of the body. The nerves, along with associated ganglia and plexuses, constitute the **peripheral nervous system.** The spinal nerves of most vertebrates are metameric (fig. 16.9), arising at the level of each embryonic body segment and passing to the skin and muscles derived from that segment, and to the viscera. Ten cranial nerves arise from the brain of fishes and amphibians and twelve in reptiles, birds, and mammals. The additional two are spinal nerves that became "trapped" within the skull.

OTHER VERTEBRATE CHARACTERISTICS

In the preceding section we noted the four features that collectively characterize vertebrates alone. Other features, not unique to vertebrates or even to chordates, include bilateral symmetry and metamerism, which were also discussed earlier. The paragraphs that follow provide thumbnail sketches of some of the other characteristics of vertebrate animals.

Integument

The skin of vertebrates consists of epidermis and dermis, and the epidermis is multilayered (fig. 6.2). Many varieties of defensive, lubricatory, nutritive, pheromonal, and homeostasis-maintaining glands develop from the epidermis and empty on the surface in one taxon or another. The epidermis of terrestrial vertebrates forms a variety of cornified (horny) appendages such as spines, reptilian scales, feathers, hair, claws, and hooves. Bone in the dermis provides a heavy coat of armor for garfishes and other relicts of the Mesozoic Era and contributes to the dense, sharp, pointed scales of sharks and to the thin, flexible scales of modern bony fishes. Dense fibrous (collagen-bearing) tissue is common in the dermis. In bovines, the dermis is made into leather through a tanning process. The epidermis of terrestrial vertebrates has a surface layer of dead (cornified) cells that protect these species from becoming dehydrated as a result of exposure to air.

Respiratory Mechanisms

Most vertebrates carry on external respiration (exchange of respiratory gases between the animal and the environment) by means of highly vascularized membranes (gills) located on the pharyngeal arches or derived from the pharyngeal floor (lungs, fig. 1.6). In some species respiration takes place through the skin or the lining of the oral and pharyngeal cavities. In embryos that develop within a porous eggshell or within the body of a parent, gaseous exchange usually occurs via special extraembryonic membranes that lie just inside the shell or that are in contact with the lining of the mother's uterus. (Chapter 5 includes discussions of oviparity, viviparity, and extraembryonic membranes.)

Coelom

Like the body of many invertebrates, the vertebrate **trunk** is built like a tube within a tube, the outer tube (body wall) and inner tube (gut) being separated by a cavity, the coelom (fig. 1.2). In fishes, amphibians, and some reptiles, the coelom is partitioned into a **pericardial cavity** housing the heart, and a **pleuroperitoneal cavity**

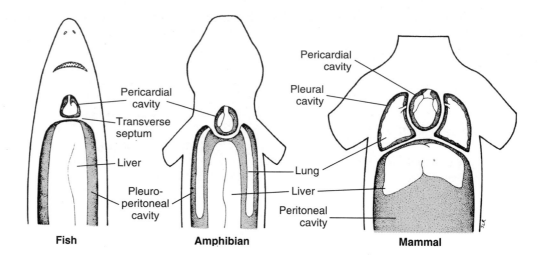

FIGURE 1.12

Chief subdivisions of the coelom of vertebrates.

Fish   Amphibian   Mammal

housing most of the other viscera, including the lungs in tetrapods (fig. 1.12, fish and amphibian). The pericardial and pleuroperitoneal cavities in these vertebrates are separated by a fibrous **transverse septum.** In other reptiles and in birds and mammals, the heart occupies a pericardial cavity, each lung occupies a separate **pleural cavity,** and the digestive tract caudal to the esophagus occupies an **abdominal (peritoneal) cavity** (fig. 1.12, mammal). In most male mammals, a pair of **scrotal cavities** houses the testes. The scrotal cavities arise as evaginations from the abdominal cavity.

The coelom is totally enclosed by a **peritoneal membrane** that lines the body wall and invests the coelomic viscera (**parietal** and **visceral peritoneum,** 5 and 6 in fig. 1.2). These membranes are continuous via **dorsal mesenteries** and, when present, via **ventral mesenteries.** The few visceral organs that do not develop mesenteries—chiefly the kidneys—lie against the dorsal body wall just external to the peritoneum. They are therefore **retroperitoneal.**

Digestive System

The digestive system consists of a digestive tract and several auxiliary (accessory) organs, chiefly a liver, gallbladder, and pancreas. The digestive tracts have specialized regions for the acquisition, processing, temporary storage, digestion, and absorption of food and for elimination of the unabsorbed residue. These include the oral cavity, pharynx, esophagus, stomach, and intestine. The latter is often coiled or has within it a spiral valve, either of which increases the absorptive area without increasing the body length (fig. 12.1, shark). The tract may also exhibit one or more diverticula, or ceca, that play various roles in the digestive process.

The digestive tract in most vertebrates terminates in a **cloaca,** a common chamber that also receives the urinary and reproductive ducts. It opens to the exterior via a **vent.** In modern fishes and a few tetrapods, the cloaca becomes very shallow or nonexistent. In mammals other than monotremes, the embryonic cloaca becomes subdivided into two or three passageways, each with its own exit to the exterior (fig. 15.48*b* and *c*). The exit from the intestine is then called the **anus.**

Urogenital Organs

Kidneys and gonads arise close together in the roof of the coelom (fig. 1.2*g* and *k*), and the two systems share certain passageways. Kidneys (nephroi) are the chief organs for elimination of water in those species in which this is necessary. (As a general rule it is not necessary in marine or desert animals.) They also assist in maintaining an appropriate electrolyte balance in the blood. In the most primitive fishes, fluid wastes accumulate in the coelom and are removed by simple kidney tubules reminiscent of the nephridia of earthworms. In most vertebrates, however, more complex kidney tubules collect fluids directly from blood capillaries. In both instances, the tubules lead to a pair of longitudinal ducts that empty into the cloaca or a urinary bladder or, less commonly, lead directly to the exterior.

Reproductive organs include gonads, ducts, accessory glands, storage chambers, and copulatory mechanisms. Early in development all vertebrate embryos are bisexual, having gonadal and duct primordia for both sexes. If the animal is genetically constituted to become a female, the gonad primordia develop into ovaries, and only the female ducts differentiate. If the animal is to become a male, the gonad primordia become testes, and the male ducts differentiate. The duct system associated with the opposite sex largely disappears. Jawless vertebrates lack reproductive ducts. Their sperm and eggs are shed into the coelom and exit via a urogenital papilla located immediately behind the anus.

Circulatory System

The circulatory system pervades almost every cubic millimeter of the body of the smallest and largest vertebrate, a condition necessitated by logistics. Whole blood, consisting of plasma and formed elements—chiefly red and white blood cells and platelets—is confined to arteries, veins, capillaries, and sinusoids. Sinusoids are broad channels rather than tubes. They are a prominent feature of many invertebrate circulatory systems and are more common in fishes than in tetrapods. Vertebrates also have a system of lymph vessels that collect some of the interstitial tissue fluids and conduct them to large veins. (fig. 14.41).

The heart, which forms immediately ventral to the embryonic pharynx, remains in that location, close to the gills, in fishes. In tetrapods it is displaced somewhat caudad during subsequent development. The heart pumps blood forward into a **ventral aorta,** which is foreshortened in tetrapods, through **aortic arches,** and into a **dorsal aorta,** in which blood courses caudad. Branches from the anteriormost aortic arches carry blood to the head (fig. 1.6).

Fishes have a **single-circuit heart.** Blood passes from the heart to the gills, where oxygen is acquired, then to the body tissues, where oxygen is given up, and then back to the heart to be recirculated. The loss of gills and reliance on lungs resulted eventually in **two-circuit hearts.** In these vertebrates, oxygenated blood in the heart is pumped to all parts of the body where oxygen is released, after which the blood, low in oxygen, returns to the heart. This blood is then pumped to the lungs where it regains oxygen, after which it returns to the heart. Oxygen-rich blood is kept separate from oxygen-depleted blood within the heart by partitions in the chambers. Other adaptations for separation of oxygenated and deoxygenated blood in one vertebrate species or another are described in chapter 14.

Skeleton

Cartilage, bones, and ligaments constitute a jointed framework of rigid or semirigid components that gives the body its shape, protects vital organs, and provides a site for attachment of locomotor and other muscles (figs. 8.24 and 10.32). The framework consists of a longitudinal **axial skeleton,** principally skull and vertebral column; a **pharyngeal skeleton,** best developed in fishes, in which it supports the gills; and an **appendicular skeleton.** As in invertebrates, mineralization of ordinary connective tissues gives skeletal components their rigidity.

Muscles

The axial skeleton is moved principally by trunk and tail muscles that are metameric (fig. 11.5). The appendicular skeleton is operated primitively by budlike extensions of the body wall muscles onto the fins or limbs (fig. 11.18). The pharyngeal skeleton is operated by muscles of the pharyngeal arches (branchiomeric muscles). All of these muscles appear **striated** when viewed by light microscopy (fig. 11.2). **Cardiac muscle**—that of the heart—is a special variety of striated muscle tissue. **Nonstriated (smooth) muscles** are found principally in the walls of hollow viscera other than the heart and in the walls of tubes and vessels. Striated muscles other than cardiac generally can be operated independent of reflex control to the extent that higher brain centers are available for providing stimuli. They are therefore often referred to as voluntary muscles. Cardiac and smooth muscles respond only to reflex control.

Sense Organs

Vertebrates have a wider variety of sense organs (receptors) than any other animals. These monitor the constantly changing external and internal environments. **Exteroceptors** monitor the external environment; they include **mechanoreceptors, chemoreceptors, electroreceptors,** and **thermoreceptors,** and **receptors for radiation** in the visible and infrared spectra. Mechanoreceptors are stimulated chiefly by vibrations of varying amplitudes in the water or air, by barometric pressure, or by direct contact of another object with the skin (touch). With the exception of receptors for radiation, exteroceptors are distributed widely on the head, trunk, and tail of fishes. In tetrapods exteroceptors are generally confined to the head, except those for touch and temperature. **Proprioceptors** monitor the activity of muscles, joints, and tendons. **Visceral receptors** monitor the rest of the internal environment. All sense organs report to the central nervous system. By "keeping in touch," the central nervous system is able to regulate skeletal muscle, visceral muscle, and glandular activity in a manner conducive to survival.

S U M M A R Y

1. Chordates are animals with a notochord, pharyngeal pouches in embryos at least, and a dorsal hollow central nervous system.

2. Vertebrates are chordates with a vertebral column—a skeletal series of bones or cartilages. They exhibit metamerism, marked cephalization, a postanal tail, paired anterior and posterior locomotor appendages, a multilayered epidermis, two to four coelomic chambers, and a closed circulatory system with a single- or double-circuit heart. They respire by gills, lungs, skin, and the lining of the oral cavity or pharynx; as reptile, bird, or mammal embryos, they respire by means of extraembryonic membranes in contact with a porous eggshell or the lining of the maternal uterus.

3. A notochord is the first skeletal structure to appear in vertebrate embryos. In agnathans it is surmounted by lateral neural cartilages. In other vertebrates it becomes surrounded by cartilaginous or bony vertebrae after which the notochord may be reduced to remnants or disappear.

4. Pharyngeal pouches tend to rupture to the exterior to form temporary or permanent pharyngeal slits. These persist as gill slits in fishes and larval amphibians. They close permanently in other vertebrates.

5. A pharyngeal arch lies between each pharyngeal slit, anterior to the first and immediately caudal to the last. The arches support gills when the latter are present. The first two are mandibular and hyoid arches.

6. Pharyngeal arches contain skeletal components, branchiomeric muscles, branches of the fifth, seventh, ninth, or tenth cranial nerves, and an aortic arch.

7. The dorsal hollow central nervous system arises as a neural groove that sinks into the dorsal body wall to become a neural tube and, subsequently, a brain and spinal cord. The peripheral nervous system consists of nerves, ganglia, and plexuses.

8. A wide variety of sense organs monitors the external and internal environment. The information they acquire is employed by the central nervous system to regulate skeletal muscle and visceral muscle and gland activity in a manner conducive to survival.

CHAPTER TWO

CONCEPTS, PREMISES, AND PIONEERS

This chapter provides background for understanding the major concepts and premises upon which the discipline of modern comparative vertebrate anatomy is based. It includes mention of a few of the pioneers in the study of vertebrate anatomy following the Renaissance, and concludes by calling attention to some abstract terms that may be misunderstood. The selected readings at the end provide additional input from easy-to-read articles to in-depth discussions about some of the concepts.

OUTLINE

As stated earlier, one aspect of the discipline of comparative vertebrate anatomy is the study of the structure of living and extinct species. The former reveals similarities and differences among animals of today. The latter reveals what vertebrates were like in the past. Assembling the data in a geological time frame reveals the panorama of changes that have taken place from the distant past to the immediate present. That panorama has led to formulation of certain concepts that we will now examine. Their full import may not be apparent this early in the course; insight will grow as the terms are used in relevant contexts. Some, if not all, may be recalled from previous studies in the biological sciences.

ONTOGENY AND PHYLOGENY

Ontogeny (ontogenesis) is the developmental history of an organism. It begins with embryogenesis, includes postembryonic changes attributable to aging, and ends at death. The primary operants are the genes. In our discussions we will be concerned chiefly with embryogenesis.

Phylogeny (phylogenesis) is the evolutionary history of a taxon. A **taxon** is a related group of organisms that constitute a taxonomic (classification) unit such as a family, order, or class. Phylogeny relates a taxon to ancestral taxa in an evolutionary line. The operant in the establishment of an evolutionary lineage is **speciation,** the emergence of new species from preexisting ones. Ontogeny occupies a single lifetime. Phylogeny requires hundreds of thousands to hundreds of millions of years. *Change* is the universal phenomenon that characterizes both processes.

In the chapters that follow, the morphology—shape, form, and structure—of organisms will be correlated with ontogeny and, to the extent current insight warrants, with phylogeny.

von Baer's Law

All early vertebrate embryos exhibit a similar architectural pattern both of structure and development. This observation was made by the embryologist Karl Ernst von Baer, who said that *features common to all members of a major taxonomic group of animals*—the subphylum Vertebrata, for example—*develop earlier in ontogeny than do features that distinguish subdivisions of the group* (classes, orders, genera, and species). This is known as von Baer's Law. An example of this is the early development of a notochord, dorsal hollow central nervous system, pharyngeal pouches, and aortic arches in all members of the subphylum. These early features become gradually modified in the direction of the species as development progresses.

A corollary of this law evolved following the increasing acceptance of the concept of organic evolution. As it stands today, the corollary is that *features that develop earliest in ontogeny are the oldest phylogenetically, having been inherited from early common ancestors, and features that develop later in ontogeny are of more recent phylogenetic origin.*

HOMOLOGY

In the middle of the sixteenth century, Pierre Belon, a French physician and naturalist, published a woodcut of the skeletons of a bird and a human side by side (fig. 2.1). The caption explains that it represents the bones of a human "put in comparison" with those of a bird, drawn to the same size in order to show "how great is the affinity (kinship) of one to the other." At that time it was believed in Western civilization that similarities in the appearance, for example, of the more than 20,000 species of teleost fishes, or the more than 60 families of perching birds, or the superficial resemblances of salamanders and lizards, horses and deer, monkeys and apes were a manifestation of a divine architectural plan in the mind of the Creator while creating animals from fishes to mammals.

As the concept of organic evolution spread among scholars and enlightened laymen, they realized that *common ancestry* accounts for most similarities, and *adaptive modifications* account for most differences. Consequently, a growing corps of anatomists in universities in western Europe devoted its research to the comparative study of the systems of the vertebrate body with the intent of adding to the knowledge of vertebrate phylogeny. These studies gave rise to the concept of *homology—* homologous structures (**homologues**) in different species.

Initially, structures in the same location with the same function and appearance were said to be homologous. Thus, most of the bones in the bird and mammal illustrated by Belon would be homologues. But we now know, as a result of studies in comparative embryology, that not all look-alikes are homologous, and many homologues bear no resemblance to one another. As phrased by the embryologist Boyden (1943), a homologue is "the same organ in different animals under every variety of form and function." The stapes in the middle ear of mammals and the hyomandibular cartilage that suspends the jaws and hyoid skeleton from the braincase of sharks are homologous, although one would not realize this from their appearance, location, or function. The precaval vein of cats and the right common cardinal vein of lower vertebrates are homologous, as are the intermaxillary bones of human embryos (which become fused with the maxillary bones before birth) and the premaxillary bones of adult apes.

If not all look-alikes are homologous, and homologues may bear no resemblance to one another in location, appearance, or function, how can we establish that two organs in two different taxa are homologous? Comparative embryology can often provide the answer,

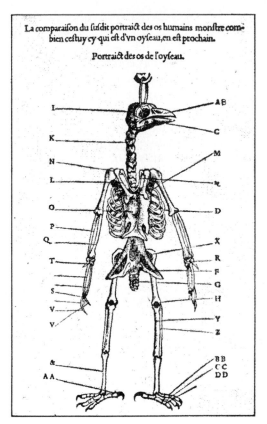

FIGURE 2.1

Comparative anatomical woodcuts by Belon, 1555.

for structures in two different animals are homologous if they come from the same embryonic precursor (i.e., anlage, blastema, fundament). When this criterion is unproductive, other sources of data may be useful. For example, if the two structures are muscles, do they have the same innervation? If confronted with a fossil, useful clues would be the relative location of foramina in bones, or impressions left in a bone by an overlying organ such as a sensory canal. In the instance of the mammalian stapes and the hyomandibular cartilage of sharks, that they are homologous was not difficult to establish once embryos of a critical developmental stage were available. The evidence in that instance is indisputable.

Serial Homology

Since metamerism is characterized by repetition of paired structures in the long axis of the body, it should be pointed out that serial structures—spinal nerves, vertebrae, muscle segments, for instance—are homologous *as a series* with the same series in another species, but the *individual components of the series are not homologues.* For example, a spinal nerve that supplies a muscle of the anterior fin is not homologous with a spinal nerve that supplies a muscle of the posterior fin—not even in the same animal. They do not come from the same embryonic metamere. The term **serial homology,** which is applied occasionally to metameric structures in a single organism, has little relevance in phylogenetic studies.

A related problem arises in comparing the digits of two different tetrapods. *Collectively,* the digits of the manus (hand) of a frog, bird, and horse are homologous, but the first, second, or last digit of one species, if there are more than one digit, may not be homologous with the first, second, or last digit of another. The single digit of the anterior limb of modern horses, for example, appears to be the middle digit of generalized vertebrates and of ancestral horses. The question "Which digit is missing?" is posed with reference to a tailed amphibian in fig. 10.24.

Analogy

The term **analogy,** which refers to coincidental resemblances, is probably familiar since it has application in many disciplines. In zoological usage, *two structures that have the same function but are not homologous are analogous.* Thus the jaw teeth of bony fishes and the horny ectodermal "teeth" of the jawless lamprey eels are analogous, as are the horns of cattle and those of rhinoceros; so, too, are the thymus gland of vertebrates and the bursa of Fabricius of birds (chapter 18). The term, although informative, is employed infrequently in evolutionary literature.

ADAPTATION

The infinitive "to adapt," like the word "analogy," should be familiar, since it is a common concept. Adaptations, whether a modification of a man-made instrument that enables its use in a new, specific context, or a modification having the same effect on a living organism, are analogous phenomena. *A biological **adaptation** is a hereditary modification of a* **phenotype** (the assortment of morphological features of an organism) *that increases the probability of survival.* Since the phenotype is an expression of the **genotype** (the array of genes), adaptations are believed to be the result of environmental pressures that, by natural selection, propagate genetic mutations that have survival value. Natural selection is discussed later in the chapter. The streamlined fishlike body of an ichthyosaur (an extinct marine reptile, fig. 10.29) participated in adapting the reptile to a lifestyle in the sea. It was, therefore, an adaptation. The thick, scaly, water-impervious skin of reptiles (descendants of semiaquatic amphibians) is an adaptation that inhibits water loss (dehydration) in animals living in the air. The vertebrate body from fishes to humans is a complex of adaptations—old and sometimes expendable, or recent (in geologic time) and useful, if not vital.

There is no evidence that adaptive changes occur because of need, present or future. Such a concept is an example of **teleological reasoning.** It attributes natural phenomena to a *preconceived design or purpose.* An example of teleological reasoning is, "Birds have wings *in order that* they may fly." The alternative is, "Birds have wings and *therefore they can* fly." Teleology implies an entelechy, a supernatural intelligence inherent in nature, that guides natural phenomena. Populations that are locked into a changing environment and need a genetic mutation to remain successful and do not acquire one become extinct. Science attributes adaptations to fortuitous genetic changes that are induced by natural laws that govern the behavior of matter.

The term **preadaptation** has been applied to traits that have enabled a phenotype to meet a new environmental challenge before it materializes. There are an abundance of examples of such traits, but they were not adaptations to a future need. They increased the chances of survival in an existing environment. Lungs were present in many fishes long before they were called on to function on land. They enabled fishes to obtain oxygen from the atmosphere when the dissolved oxygen content of the water became too low to sustain life. Subsequently, they enabled the initial short excursions of fishes onto land. Lungs became adapted for a sustained terrestrial existence over millions of years of subsequent evolution. Unlike adaptation, preadaptation is not a biological phenomenon. The term might be misconstrued as having a teleological implication.

SPECIATION

The formation of a new species is preceded in almost every instance, if not invariably, by geographical isolation of a population from other populations of the same species. This could occur, for instance, in a migrant species in which some of the migrants become isolated from the others for one reason or another. This appears to be what happened to the landlocked freshwater lamprey with a marine name, *Petromyzon marinus dorsatus,* which lives in the Great Lakes of North America. Its very close relatives, *Petromyzon marinus marinus,* are marine lampreys that migrate into freshwater streams to spawn, then return to the sea until sexual maturity. There are several possible explanations—all conjectural—for the isolation of these two specific groups, but they are irrelevant in the present discussion. The important point is that further exchange of genes between the two groups has become impossible in the present circumstances. Consequently, chance mutations, recombinations of genes, and random **genetic drift** (spontaneous irregular change in gene frequencies) that characterize all genetic complexes will result ultimately in genetic incompatibility between the two. Thereafter, if brought together again they would no longer be able to produce viable offspring. *Geographical isolation coupled with genetic change is the matrix out of which new species evolve.* Geographical isolation provides the time necessary for the species to drift apart genetically. The length of time required for speciation depends on many factors, but the eventual emergence of new species, given the foregoing conditions, seems inevitable. The result of continuous speciation is phylogenesis, the formation of new taxa.

EVOLUTIONARY CONVERGENCE

When two or more unrelated organisms, such as ichthyosaurs (fig. 10.29) and dolphins (fig. 4.49) occupy similar environments, whether concurrently or millions of years apart, and acquire similar morphological features as adaptations to that environment, the phenomenon is termed **evolutionary convergence.** The streamlined body, dorsal fins, and paddlelike structure of the forelimbs of ichthyosaurs and the same features in dolphins are examples of evolutionary convergence; so, too, are webbed feet in the frog, duck, and platypus. Evolutionary convergence produces look-alike features that are not the result of inheritance from a common ancestor.

Convergence, Homology, and Homoplasy

The terms *convergence, homology,* and *homoplasy* designate specific concepts that are useful to students of phylogeny. Structures in two or more taxa that look alike, whether they are the result of convergence or not, are said to be **homoplastic** or to exhibit **homoplasy.**

However, not all look-alikes are the result of convergence. Some are the result of inheritance from a common ancestor. The paddlelike limbs of ichthyosaurs and dolphins are homoplastic. The paddle *shape* is the result of convergence. The *limbs,* being products of homologous limb buds, are homologues. The abdominal ribs (gastralia) of crocodilians (fig. 8.14) look like ribs, occupy the same position as the ventral segments of thoracic ribs, but unlike ribs, do not develop from preexisting cartilage. Therefore, they do not meet the criteria for homologues. The swim bladders of fishes and the lungs of urodeles (fig. 13.8) are homoplastic. They may or may not be homologous: there are differences of opinion, and proof is lacking. Whether morphological similarities bespeak convergence or whether they are the result of common ancestry is important to systematists who are using morphological data to construct taxonomic classifications designed to reflect the extent of genetic relatedness between two different organisms.

METAMORPHOSIS, NEOTENY, PAEDOMORPHOSIS, AND HETEROCRONY

Naturalists have been interested for centuries in tailed amphibians that become sexually mature and reproduce while retaining one or more larval features, including external gills and gill slits (fig. 4.20, *Necturus* and *Siren*). These species have been characterized as exhibiting "incomplete metamorphosis." The term **neoteny** is applied to this condition in urodeles. Of the eight families of urodeles, every one has at least a few individuals or populations that are neotenic. However, neoteny is not confined to amphibians, or even to vertebrates. It is seen in invertebrate taxa, including larvaceans (fig. 3.4), which, although lacking the vertebrate features of pharyngeal arches, gills, and gill slits, reproduce in a larval morphological state.

Incomplete metamorphosis is an advantage conferred on urodeles that might perish if forced, by loss of gills, to abandon an aquatic environment while the terrestrial environment is unfriendly. Neuroendocrine responses to environmental conditions can temporarily slow down, or speed up, the hormonal component of the control of metamorphosis in any species (fig. 18.5), thereby adapting an organism to the exigencies of the environment. Inheritance plays a significant role in species such as *Necturus,* in which a genetically altered pattern of metamorphosis has been imposed in the course of evolution.

Neoteny is the result of the dissociation of the timing of some of the morphological changes of metamorphosis from the timing of maturation of the germ cells, so that the latter occurs earlier, whether induced by environmental factors, heredity, or both. Among the most visible morphological changes in amphibians are resorption of the gills and closure of the slits. Less visible are the remodeling of the skeleton and muscles of the pharyngeal wall and tongue, alterations of body shape and the integument, and other changes that adapt the organism for survival in a terrestrial environment. Neotenous individuals stop short of completing these morphological changes, temporarily or permanently. The dissociation of sexual maturation from the morphological changes of metamorphosis may become a permanent feature, fixed in the genes, in populations in which geologic or physiologic isolation provides an environment in which natural selection can operate. In those instances, new neotenic species may evolve. These new species exhibit a concept known as paedomorphosis.

Paedomorphosis is defined as ontogenetic changes whereby certain larval or immature features of the immediate ancestor become the end product of metamorphosis in the descendant species. The descendant species is said to be paedomorphic. Knowledge of ancestral morphology is inferred from the morphology of fully metamorphosed species in the same taxon. Most populations of *Cryptobranchus alleghemiensis* come close to completing metamorphosis, ancestral larval features in some populations being limited to retention of a single gill slit. A closely related genus in the same family, *Andrias japonicus,* which is postulated to reflect the adult state of ancestral cryptobranchids, exhibits complete metamorphosis. *C. alleghemiensis* is therefore considered to be a paedomorphic species.

Some populations of axolotls (Family Ambystomatidae) that live in mondane (mountainous) environments (populations of *A. mexicanum* and *A. gracilis,* for example) have complete retention of larval morphology throughout life. Other Ambystomatidae (*A. tigrinum,* for example) exhibit complete metamorphosis. Since the ancestral species exhibited full metamorphosis, as inferred from the morphology of *A. tigrinum* and its closest relatives, the mondane populations of *A. mexicanum* and *A. gracilis* are considered to be paedomorphic. The paedomorphism in *A. mexicanum* is due to insufficient levels of pituitary thyroid stimulating hormone (TSH) and thyroid hormone itself. However, complete metamorphosis can be induced by injection of TSH or thyroid hormone. We may infer from these observations that modern *A. mexicanum* has inherited mutations from ancestors that now limit the output of hormones necessary for metamorphosis.

The red spotted salamander of the woods of the eastern United States, a phase in the life cycle of most populations of *Notophthalmus viridescens*, exhibits a very different pattern of metamorphosis from those described above. In the course of ontogeny, the larvae undergo morphological changes that enable the population to emerge from the ponds and assume a juvenile (sexually immature) existence on land far in advance of maturation of the gonads. These changes include remodeling of the skull, loss of gills and gill slits, development of lungs, and cornification of the skin. Cornification protects the **eft** (see also chapter 4), the terrestrial juvenile, from water loss to the air. Their adoption of a moist location under rocks or rotting logs further promotes water retention. Efts remain on land for one to three years before the onset of sexual maturity, which is signaled by a "water drive" initiated by prolactin. At that time the efts undergo a final metamorphosis. Their anatomy is altered in the direction of an air-breathing aquatic organism, they migrate back to the pond, the gonads mature, and reproduction takes place. Metamorphosis in these populations does not include a neotenous stage.

In some localities, the larvae of *Notophthalmus* remain in the pond, the gonads mature, reproduction occurs, but gill resorption is delayed indefinitely (fig. 4.21*e*). Members of the population remain in the pond throughout life. The population exhibits at least a temporary stage of neoteny, but the species is not paedomorphic.

The term **heterocrony** is applied to changes in the rate and/or timing, hence sometimes the sequence, of metamorphic events in a lineage. Paedomorphosis is attributed to genetic changes resulting from heterocrony in an ancestral species. The concept of heterocrony has inspired a minority theory of the phylogenetic origin of cephalochordates (amphioxuslike protochordates), which derives them from an agnathanlike ancestor, whose ammocoete larvae, through heterocrony, became sexually mature (fig. 3.1*b,c*).

SYSTEMATICS

Systematics is the discipline and practice of ordering organisms into hierarchies (species, orders, classes, and so forth) that reflect one or more preselected criteria: (1) morphologic similarities and differences, (2) phylogeny as inferred from the geologic record, (3) phylogeny as inferred from mathematical analyses of a selected array of morphologic traits (cladistic analysis), and (4) similarities and differences in molecular features such as the sequences of nucleotide bases from ribosomal RNA. The taxonomist eventually constructs a classification **(taxonomy)** based on one or more of the criteria. Current classifications are based on more than one criterion, reflecting, in part, the paucity of data. Verification, validation, and confirmation in phyletic reconstructions is particularly difficult. Therefore, detailed ancestral lineages expressed in any taxonomy are postulates.

A traditional **taxonomy** is based on similarity of structure, evolutionary history (descent) as inferred from the geologic record, rationality, and—as a last resort—intuition. Like all others, traditional classifications are tentative. They reflect new data when warranted, including postulates arrived at by cladistic analysis. Appendix 1 is an example of a current evolutionary taxonomy of chordates.

Cladistic analysis is used to achieve a greater degree of objectivity in the *selection* of morphologic traits to be employed in reconstructing phylogenetic relationships. The data are reflected in cladistic taxonomies. Lampreys and hagfishes, for instance, have an array of morphologic traits characteristic of vertebrates and an array of traits characteristic of certain ostracoderms. They also have many features in common, including no jaws, no paired fins, no vertebrae, no scales, a branchial basket instead of pharyngeal arches, a buccal funnel and muscles for suction, and hundreds of others that are less obvious. The premise of cladistics is, that many morphologic features are irrelevant in ascertaining phylogenetic relatedness, and that the geologic record is not reliable because of possible disruptions of the strata that may have taken place in the past. Cladistic analysis seeks to identify ancestors and subsequent lineages by seeking *arrays of selected characters* that are shared by all members of a group, unique to the group, and derived from a common ancestor—the first to exhibit that array. The group is called a **clade,** and it is a **monophyletic** taxon. (A taxon is monophyletic if it includes the most recent common ancestor of the taxon and all descendants of that ancestor.) For example, in Nelson's tentative cladistic taxonomy of fishes, the Class Sarcopterygii includes crossopterygians (the postulated common ancestors), lungfishes, amphibians, reptiles, birds, and mammals. Cladistics is one method employed in attempts to ferret out phylogenetic lineages, and then to construct taxonomies reflecting the postulated phylogenies. Not all taxa in traditional classifications are monophyletic. (Compare the cladistic class Sarcopterygii with the class Sarcopterygii in the traditional classification in appendix 1. The latter does not include tetrapods.)

Cladistics poses questions such as, "Which group, hagfishes or lampreys, if either, is more closely related to vertebrates than to other agnathans? And which group, if either, is more closely related to some ostracoderm than to one another?" A tentative answer to these questions is then represented in a cladogram (fig. 4.16).

Graphics in Systematics

Family trees (dendrograms) such as those in chapter 4 are designed to provide a graphic impression of postulated lineages. They do not show the distribution of taxa in geologic time, hence, which ones were contemporaries, or their relative abundance in time. Illustrations such as figs. 4.2 and 4.24 show abundance. Family trees and cladograms are graphic representations of hypotheses that are subject to further testing.

ORGANIC EVOLUTION
AND EVOLUTIONARY SELECTION

The concept of organic evolution is a single, simple, unequivocal, yet widely misconstrued premise, which states: *The plants and animals on earth have been changing, and the ones around us today are descendants of those which were here earlier.*

The conclusion that the animals and plants have been changing is based in part on geological evidence. The evidence includes the fossilized remains of plants and animals preserved in sedimentary rock formations. It indicates that, if a human from today could be transported back several hundred million years, the plants and animals would be exotic. Five hundred million years ago one would see mostly water, and on land there would be no plants. No fish could be caught with a fishing pole because the fish would be ostracoderms, which were filter feeders. Three hundred and fifty million years ago fishing would have improved, but there would be no trout, perch, or salmon. The land would be higher and drier and there would be mosses and other simple land plants. Labyrinthodonts would be lumbering in and out of swampy waters, but there would be no frogs or toads. Three hundred million years ago, the land would be still higher and the swamps would be luxuriantly forested with seed ferns and conifers but no flowering plants. Cotylosaurs would be basking in the sun, but there would be no lizards or snakes. One hundred fifty million years ago, there would be birds with teeth; the earth would be heavily populated with reptiles, dinosaurs would be abundant and specialized, and a small minority population of reptilelike mammals, known chiefly from their teeth, jaws, and skulls, would have been struggling to survive. But there would be no cats, rats, or monkeys. It is in these geological findings that the theory of organic evolution has its roots, although not all of them.

The second part of the premise, which says that *the animals and plants around us today are offspring of the animals of yesterday,* is axiomatic. Life comes from preexisting life.

This is the theory of organic evolution, stripped of all satellite theories. It doesn't say that multicellular animals came from protozoa; it doesn't say that humans came from a monkey. It doesn't say *where* humans came from—that is another theory, or several other theories, but they are not the theory of organic evolution. Neither does the theory say what caused, or what causes, species to change. That, too, is another theory. The theory does not state how life began or how the universe began. There are theories about beginnings, but they are not the theory of organic evolution. These theories belong to scientific disciplines beyond the scope of this book.

The theory of organic evolution might also be called the "theory of the mutability of species." There is only one alternative, the "theory of the immutability of species." Proponents of the latter theory must insist that every species on earth today is precisely like it was when it first appeared, and that the first member of the species appeared *de novo,* coming from no previous living organism. To consider the possibility of even a single change ("fishes appeared and from these came all kinds of fishes") is to abandon the theory of the immutability of species in favor of organic evolution. If it is accepted that *one* change can occur, it must be accepted that *two* changes can occur. And once this step in logic has been taken, there is no limit to the number of changes that can occur.

The concept of evolution appeared in early Greek writings, including those of Aristotle. However, under the restraints of scholasticism, the philosophy and theology of the Middle Ages of Western civilization, the concept was not nurtured, and it remained dormant until the Renaissance when inquiry into natural phenomenon revived. **Jean Baptiste de Lamarck** (1744–1829), a French naturalist, was among the first of the Renaissance naturalists to publicly espouse the idea that species are not immutable and that complex species evolved from simpler ones. Lamarck propounded the **Doctrine of Acquired Characteristics** to explain how evolution worked. It postulated that characteristics acquired through the use or disuse of morphological structures are inheritable. Thus when a part is employed in successive generations, it becomes stronger and better adapted for a specific role. Conversely, when a part is neglected in successive generations, it tends to become vestigial. This is how Lamarckism would account for the fact that the olfactory nerves of whales commence development in embryonic life, then regress and become vestigial. A lung-breathing animal could not inhale under water without drowning. Therefore, according to the theory, the lack of use of the olfactory apparatus resulted in its disappearance. Vampire bats have an esophagus so small that they cannot swallow anything other than liquids. According to the Doctrine of Acquired Characteristics, the esophagus would have become too small to swallow solids because the only nourishment that passed through it for generations was mammalian blood. The doctrine as proposed has not been acceptable because there is no known explanation as to how use or disuse of a part in individuals can be translated into alterations of the hereditary code stored in the sperm and eggs. The advent of modern genetics made this aspect of Lamarckism, at least for the present, untenable.

Charles Darwin (1809–1882), born 65 years after Lamarck, had a different explanation to account for evolutionary change. Darwin observed that variation in individual organisms is universal; no two individuals are identical. This is the starting point for Darwin's **Theory of Natural Selection** propounded in 1859. Individual variations, Darwin believed, result in varying degrees of success in competition between individuals and with the physical conditions of life. The resultant "preservation of favorable variations and the rejection of injurious variations I call Natural Selection," said Darwin, who paraphrased the concept as the "struggle for existence" in which the fittest survive. Darwin preceded the advent of genetic theory and thus could not explain natural selection in terms of the molecular units of inheritance. In the age of genetics, natural selection means that random genetic mutations and recombinations that render an organism better able than others to cope with the environment are more likely than other gene complexes (genotypes) to be passed on to another generation. *The concept of natural selection (survival of the fittest) is Darwin's most important contribution to evolutionary theory.* Natural selection is a major route to speciation. Whether or not it is the only route is not known. Current evolutionary research is examining the concept that selection analogous to Darwinian natural selection operates at several levels in addition to the genes (Lieberman and Vrba, 1995). Called a hierarchical theory, it embodies the essence of Darwin's concept, expanded to work at several hierarchical levels (Gould, 1982).

The first paragraph of this discussion cited the *premise* of evolutionary theory. With the advent of genetics the *mechanics* of the process of evolution have gradually become clearer. Our current insight is a synthesis of the basic premise set forth in the first paragraph of this discussion, the concept of natural selection, and the insight gained through subsequent research in genetics, comparative morphology of plants and animals, paleontology, systematics, biochemistry, molecular biology, and other disciplines. Natural selection is a natural, rather than a supernatural, explanation for the variety and number of species on the planet (conservatively estimated at five million [Myers, 1985]), just as the Law of Gravity is a natural explanation for the attraction between the earth and the moon. A natural explanation may gain acceptance from an accumulation of supporting data gathered by independent observers. Conversely, a particular natural explanation will always be subject to rejection pending future observations. By contrast, supernatural explanations are generally not falsifiable and thus fall outside the bounds of scientific inquiry.

ANATOMY FROM GALEN TO DARWIN

Cavemen had some knowledge of the internal organs of mammals, as did the Babylonians and ancient Egyptians, who practiced surgery and embalming. Embalming, which involved removal of most of the internal organs, was an advanced art. Aside from a small number of Egyptian medical papyri dated to about 3000 B.C., the oldest anatomical works in Western civilization were written during the last 400 years B.C. by Greek philosophers and physicians. These works were incomplete, mostly superficial, and often imaginative. Anatomy of

the classical era culminated in the works of **Galen,** a Greek philosopher-physician who practiced in Rome between A.D. 165 and 200. He assembled all available Greek anatomical writings, supplemented them with his own dissections of apes from the Barbary Coast (human dissection was at that time prevented by public opinion and superstition), and, in addition, wrote more than 100 treatises on medicine and human anatomy. Shortly thereafter, Scholasticism took over, and during the next 1,300 years Galen's descriptions were considered infallible and dissent was punishable. Subservience to authority became so accepted that (as has been said, probably partly in jest) if a scholar wanted to know how many teeth horses have, he saddled up and rode 100 miles, if necessary, to the nearest library to see what Galen said. (No doubt the peasants looked into the horse's mouth—an application of the scientific method.)

It was not until the fifteenth century that **Leonardo da Vinci** (1452–1519) and other Italian artists began to make anatomical observations of their own (at the peril of excommunication from their church), and a renaissance in human and animal anatomy began. In 1533 a young Flemish medical student named **Vesalius** attended classes at the University of Paris where Galen's works on human anatomy were read (by a "reader") while the professor tried, often with embarrassing lack of success, to harmonize Galen's descriptions with the dissection. (Recall that many of Galen's descriptions of man were written from dissections of apes.) After three years Vesalius quit Paris, earned a degree at Padua, stayed on to teach, and recorded his own anatomical observations. In 1543 he published *De humani corporis fabrica* ("On the Structure of the Human Body"), ushering in a renewed interest in anatomy of animals in general. His *Fabrica,* however, was influenced to a large degree by Galen, and many of the errors of Galen were perpetuated. Vesalius's illustration of the hyoid and kidney were actually drawn from a dog, which was Vesalius's favorite specimen for dissection. However, he was not interested in mammals other than man beyond the extent to which they might contribute to his knowledge of major organs of the human body. His interest was in descriptive human anatomy, not in species differences.

Eight years after publication of Vesalius's anatomy, **Pierre Belon** (1517–1564), a French naturalist and physician, published an important work on the dissection of cetaceans and other marine animals, invertebrate and vertebrate, including fishes. (Because cetaceans lived in the sea and exhibited adaptations for the marine environment, they were, to Belon, fishes, despite his detailed study of their anatomy, which is otherwise mammalian.) Belon was also interested in birds, saying he never missed an opportunity to dissect one and had examined the internal parts of 200 different species. In 1555 he published his classic illustration of a human and bird skeleton side by side, showing that the skeletons correspond almost bone for bone (fig. 2.1). He published the illustration because he had made the observation, and thought the general public would be enlightened. Like other public figures of his day, he was a teleologist. He attributed the similarity of the skeletons, like the convergent evolutionary similarities that he observed between fishes and cetaceans, to the manifestation of a basic architectural plan, or *architype,* in the mind of the Creator. Today, the term "basic architectural plan" refers to the generalized pattern of development and morphology that has been inherited from the earliest vertebrate ancestors.

The distinction of being considered the founder of comparative anatomy (and also of paleontology) should probably be awarded to Georges Léopold Chrétien Frédéric Dagobert **Cuvier,** Baron de Montbéliard (1769–1832), a French naturalist who studied at Stuttgart. A prolific author in the field of natural history, his works include, among many others, 9 volumes entitled *Leçons d'anatomie comparée* (Paris, 1801–1805), 4 volumes of paleontological observations on fossil skeletons of quadrupeds, a geological description of the environs of Paris, and a work that is now a classic, a 4-volume 1816 abridgement of his course in comparative anatomy at the École Centrale du Pantheon in Paris, which he titled *Le règne animal distribué d'après son organisation pour servir de base à l'histoire naturelle des animaux et d'introduction à l'anatomie comparée* ("The animal kingdom arranged in accordance with its structure for the purpose of serving as a basis for the natural history of animals and as an introduction to comparative anatomy"). In this volume he summed up his lifetime of research on the morphology of fossil and living animals, which he divided, in the role of taxonomist, into four taxa, radiate, articulate, and vertebrate, based on differences in structure of the skeleton and organs. (Note that Cuvier's classification was based solely on phenotypes. Species were considered immutable.) His inspiration throughout life came from a copy of the first edition of Linnaeus's *Systema naturae* (1735), which in Cuvier's early years was his entire scientific library. His final contribution was a 22-volume masterpiece, *Historie naturelle des poissons* ("Natural history of fishes"), published over a four-year span ending in 1832. He died in 1832 in Paris at the age of 56 after an energetic life dedicated to natural history and teaching, including the chancellorship at the University of Paris. Throughout life he steadfastly maintained the absolute *immutability* of species, objecting vehemently to the mutability ideas of Lamarck (a contemporary), of **Geoffroy Saint-Hillarie** (1772–1844), with whom he had collaborated in several published works, and of others who began to envision the possibilities of changes in species with time.

Cuvier would have been a leading Darwin adversary in the great uproar in scientific circles, in academia, among theologians, and, consequently, in the press, that was evoked by the publication of **Charles Darwin's** masterpiece in 1859, *On the origin of species by means of natural selection, or the preservation of favoured races in the struggle for life.* The 1,250 copies sold out on the day of issue. The first four chapters explain the *artificial selection* of domesticated animals and plants by humans, comparing this with *natural selection* as a consequence of the struggle of undomesticated species for existence. The fifth chapter discusses variations and causes of modification other than natural selection. The next five chapters examine the difficulties that stand in the way of a belief in hereditary changes (evolution) in species generally, as well as in natural selection as an operating factor. The last three chapters present the evidence for evolution from the field of paleontology, geographical distribution of species, comparative anatomy, embryology, and vestigial organs. The uproar and almost universal condemnation of Darwin was comparable to that resulting from Copernicus's treatise (1543) proving that the earth revolves around the sun. (Copernicus's treatise was probably completed by 1530, but was circumspectly withheld by him until he was on his deathbed.) Neither Copernicus's nor Darwin's theory conformed to the accounts in the Judeo-Christian Book of Genesis, and both were therefore considered a denial of the authority of God. Darwin's theory was also seen as a denial of the story of Adam and Eve and of creation of the earth in seven days.

With the advent of the study of genetics the mechanism whereby change in phenotypes takes place was elucidated, and mankind was given a natural explanation for the diversity of species on earth.

WORDS TO PONDER

The following words are sometimes employed in discussions of phylogeny. You will read them, hear them, and use some of them. There will be differences of opinion with reference to the connotations of most of them. However, if calling attention to these abstract terms stimulates thoughtful discussion, inclusion of them will have been justified.

Primitive is a relative term. It refers to a beginning or origin. A primitive trait is one that appears in a stem ancestor from which arose an array of subsequent species, some of which may retain the trait. The notochord is primitive, since it occurred in the first chordates. The placoderms were primitive fishes in that they gave rise to an array of later fishes. Ancient insectivorous mammals were primitive *placentals* because they gave rise to an array of later placentals. However, they were not primitive *vertebrates.* Somewhere in phylogeny there was a primitive primate and a primitive species of human. However, one cannot always be certain that a given *structure* is primitive. For example, the lateral neural cartilages of lampreys (fig. 1.5) are primitive only if they reveal an original condition from which typical vertebrae later evolved.

Generalized refers to structural complexes that, at least in some of the descendants, have undergone subsequent adaptation to a variety of conditions. The hand of an insectivore was, and remains, a generalized mammalian hand. It was competent to evolve into the wing of a bat, the hoof of a horse, the flipper of a seal, and the hand of a primate. A generalized group of animals has demonstrated that it was genetically suitable for divergent evolution, that is, evolution in many directions. Labyrinthodonts were generalized tetrapods. The terms "generalized" and "primitive" come into contrast in that generalized connotes a state of potential adaptability and primitive connotes a state of being ancestral.

A **specialized** condition is one that represents an adaptive modification. Vertebrate wings are specializations of anterior limbs, and beaks are specialized upper and lower jaws. Beaks (fig. 2.2) may be needlelike and tubular for extracting nectar from flowers (hummingbirds); chisel-like for drilling (woodpeckers); hooked for piercing and tearing prey (predatory birds, such as hawks and eagles); sturdy and spearlike, for capturing small fish and other aquatic animals in marshes (herons); long, spearlike, but less sturdy, used for nocturnal foraging for grubs and worms (kiwi); or long and upcurved for sweeping through shallow waters and mud in search for small crustaceans, molluscs, and other marsh inhabitants (avocets). Increased specialization connotes increased adaptation. The greater the specialization, the less may be the potential for further adaptive changes.

Derived, or **modified,** connotes any state of change from an ancestral condition; a mutated state. If the presence of bone is a primitive trait, wholly cartilaginous skeletons are modifications of the condition. The modification (loss of the potential to form bone) was a specialization if it adaptively modified the animal. Perhaps not all modifications are adaptive (students of speciation disagree strongly about this); but if they are not, they may portend the demise of the species, since any change is statistically more likely to make the animal less competitive.

Higher and **lower** are used to express the relative position of major taxa on a conventional phylogenetic scale. They are less informative when applied below the class level. Birds and mammals evolved from cotylosaurs and hence are said to be higher than cotylosaurs. In this context the words have some meaning. Sometimes the terms are used to express relative mutational distance of a given taxon from a common ancestor when compared with some other taxon—but are mammals to be considered higher than birds? In this context the term may be misleading. The terms may also be meaningless when used to compare a genus within one taxon with a genus within another taxon, as when comparing a modern frog with a perch, or a human being with a hummingbird. Each would be incapable of competing in the other's environment.

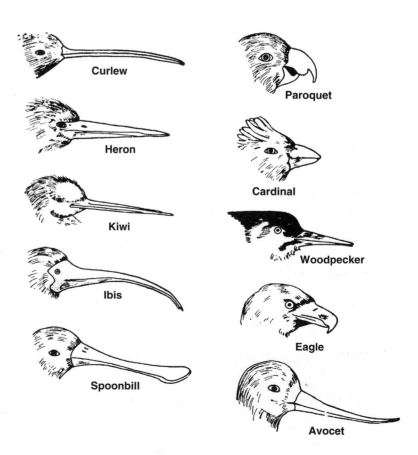

FIGURE 2.2

Adaptations of the beak for feeding in selected birds.

Simple is a relative term connoting a lack of complexity of component parts. A simple state is not necessarily a primitive one. The skull of a human being is simple compared to that of a teleost, but it is not primitive. The primitive may also be far from simple.

The term **advanced** should connote a modification in the direction of further adaptation. Unfortunately the word has overtones connoting progress and hence is subjective and misleading. It is a matter of opinion of one species—humans—whether or not a modification in another species represented or represents progress. The phrase "more recent" or "more specialized" is more informative than the term "advanced."

Degenerate is another value-judgment word. For example, it is sometimes applied to lampreys by those who think lampreys have lost jaw skeletons, paired appendages, bone in the skin, and other characteristics of typical vertebrates. However, the condition of the cyclostomes represents an adaptation to a semiparasitic state and as such may better be characterized as "specialized." Unless lampreys never had jaws, these agnathans may have specialized themselves right into a state of neosimplicity! To call them "degenerate" would seem to discount the value of adaptive modification. Degenerate seems to be a term that should be avoided.

The words **vestigial** and **rudimentary** require explanation. A phylogenetic remnant that was better developed in an ancestor is vestigial. The pelvic girdle of whales is said to be vestigial, since ancestors of whales were tetrapods with functional tetrapod appendages. The yolk sac of the mammalian embryo is vestigial. The term "rudimentary" is used in two different senses, phylogenetic and ontogenetic. In the phylogenetic sense, structures that became more fully exploited in descendants are said to be rudimentary in the phylogenetic precursor. For example, the lagena of the inner ear of fish is sometimes referred to as a rudimentary cochlea, since it evolved into a cochlea in later forms. In the ontogenetic sense, a structure that is undeveloped or not fully developed is said to be rudimentary. The muellerian duct may be considered rudimentary in most male vertebrates. It is not always possible to be certain whether a structure should be called rudimentary or vestigial. The pseudobranch of the shark *Squalus acanthias* is vestigial if it represents a gill that, in ancestral sharks, was a full-fledged functional gill surface.

If the foregoing thoughts trigger discussion, no matter how dissonant, the space employed in presenting them will have been well utilized. It must not be forgotten that words are sounds made by humans to connote a concept. Otherwise, they may be nothing but noise. Or, do you wish to object, on a semantical basis, to the word "noise"?

SUMMARY

Definitions may briefly paraphrase those in the text.

1. Ontogeny is the developmental history of an individual.
2. Phylogeny is the evolutionary history of a taxon.
3. Homology is the manifestation of homologous structures in different species.
4. Homologous structures are derived from the same ancestral anlage. They may be similar or broadly dissimilar morphologically and functionally.
5. Analogous structures have similar functions but they are not homologues.
6. An adaptation is a hereditary modification that increases the chances of survival.
7. Natural selection is the nonrandom differential reproduction of genotypes as a result of the selective effect of the environment.
8. Speciation, the formation of new species, is the result of genetic changes over time in isolated populations of preexisting species.
9. Evolutionary convergence is the evolution of similar structures in unrelated taxa as a result of mutations that are adaptive to similar environments.
10. Homoplasy is the presence in different species of structures that look alike. The structures are said to be homoplastic. They may or may not be homologous.
11. Neoteny is the temporary or permanent retention of larval traits in sexually mature individuals, populations, or species.
12. Paedomorphosis is an evolutionary process wherein certain larval or immature features of the immediate ancestor become the end product of metamorphosis in the descendant species.
13. Heterocrony is a change in the timing and/or rate, hence sometimes the sequence, of the events of metamorphosis in a lineage. Paedomorphosis appears to be a result of heterocrony.
14. Systematics (taxonomy) is the discipline and practice of ordering organisms into hierarchies that reflect their morphological similarities and phylogenetic history.
15. Organic evolution is the changes that have taken place in animals and plants through historical time.
16. Teleology ascribes natural phenomena to purposeful design rather than to natural laws.

17. Darwin's major contribution was the Theory of Natural Selection, an alternative to Lamarck's Doctrine of Acquired Characteristics.

18. Georges Cuvier was the founder of the discipline of comparative anatomy.

SELECTED READINGS

Ayala, F. J.: Teleological explanations in evolutionary biology, Philosophical Society (London) **37**:1, 1970.

Boyden, A.: Homology and analogy: a century after the definitions of "homologue" and "analogue" of Richard Owen, The Quarterly Review of Biology **18**:228, 1943.

Browne, J.: Charles Darwin. Voyaging. Vol. 1 of a Biography, New York, 1995, Knopf.

Cole, J. F.: A history of comparative anatomy, London, 1944, Macmillan.

Davis, J. I.: Phylogenetics, molecular variation, and species concepts, Bioscience **46**(7):502, 1996.

de Beer, G. R.: Homology, an unsolved problem, London, 1971, Oxford University Press.

Gaffney, E. S., Dingus, L., and Smith, M. K.: Why cladistics? Natural History, June, 1995.

Gould, S. J.: Darwinism and the expansion of evolutionary theory, Science **216**:380, 1982.

Gould, S. J.: Ontogeny and phylogeny, Cambridge, Massachusetts, 1985, The Belknap Press of Harvard University.

Hickman, C. P., and Roberts, L. S.: Animal diversity, Dubuque, 1995, Wm. C. Brown. Discussions on traditional and cladistic taxonomies.

Lewis, R. W.: Teaching the theories of evolution, American Biology Teacher **48**:344, 1986. A discussion of Darwin's two major theories—the descriptive theory of descent and the mechanistic theory of natural selection.

Lieberman, B. S., and Vrba, E. S.: Hierarchy theory, selection, and sorting: a phylogenetic perspective, Bioscience **45**:394, 1995.

May, E.: One long argument: Charles Darwin and the genesis of evolutionary thought, Cambridge, 1991, Harvard University Press.

Moore, J. A.: Science as a way of knowing: evolutionary biology, American Zoologist **24**:419, 1984.

Myers, N.: The end of the lines, Natural History **94**(2):2, 1985. A discussion of duration of species and extinction rates.

Nelson, J. S.: Fishes of the world, ed. 3, New York, 1994, John Wiley and Sons.

Reilly, S. M.: Ontogeny of cranial ossification in the eastern newt, *Notophthalmus viridescens* (Caudata, Salamandridae), and its relationship to metamorphosis and neoteny, Journal of Morphology **188**:215, 1986.

Ridley, M.: Evolution, Boston, 1993, Blackwall Scientific Publications.

Shaffer, H. B., and Voss, S. R.: Phylogenetic and mechanistic analysis of a developmentally integrated character complex: alternate life history modes in Ambystomid salamanders, American Zoologist **36**:24, 1996.

Steele, E. J.: Somatic selection and adaptive evolution. On the inheritance of acquired characteristics, ed. 2, Chicago, 1981, University of Chicago Press.

Stock, D. W., and Witt, G. S.: Evidence from 18S ribosomal RNA sequences that lampreys and hagfishes form a natural group, Science **257**:787, 1992.

Thorington, R. W., Jr.: Books, BioScience **44**(10):705, 1994. Cites problems of validation and confirmation in phyletic reconstruction of mammalian orders.

Whiteman, H. H.: Evolution of facultative paedomorphosis in salamanders, Quarterly Review of Biology **69**(2):205, 1994.

CHAPTER THREE

PROTOCHORDATES AND THE ORIGIN OF VERTEBRATES

In this chapter we will speculate concerning the invertebrate origins of vertebrate animals. For background, we will gain some knowledge concerning protochordates—small aquatic animals with a notochord but lacking a vertebral column. We will then hypothesize concerning their affinities with vertebrates. At the end of the chapter we will examine a larval lamprey, for the reason that it bears a remarkable resemblance to protochordates, has been mistaken for a protochordate, yet grows to become a lamprey eel.

OUTLINE

No study of vertebrate anatomy would be complete without providing some knowledge about the closest relatives of vertebrates, the protochordates. A sessile sea squirt squirting water out of its excurrent siphon seems a far cry from even a simple amphioxus, let alone the lowest jawed fishes. Yet, sea squirts, amphioxuses, fishes, and tetrapods exhibit three fundamental structural features in common, a notochord, dorsal hollow central nervous system, and pharyngeal slits (fig. 3.1). Notwithstanding the fact that the notochord becomes encased within vertebrae in vertebrates, the pharyngeal slits close sooner or later in tetrapods, the manifestation of these fundamental structures in all of them during ontogeny indicates that vertebrates are more closely related genetically to protochordates than to any other known animals. Therefore protochordates may provide some clues to the invertebrate origins of animals with backbones.

PROTOCHORDATES

In recognition of the postulate that protochordates and vertebrates are closely related, Ernst Haeckel in 1874 proposed the establishment of the **phylum Chordata** incorporating three subphyla. **Urochordata** (sea squirts and certain other invertebrates), **Cephalochordata** (the amphioxuses), and **Vertebrata.** The two lower subphyla are protochordates, a convenient term, although without taxonomic status. All are marine organisms. Our interest in them stems from the realization that they and the vertebrates probably had common ancestors. Therefore we will examine them briefly, emphasizing the morphologic features that prompt this hypothesis. The place of proto-

(a)

(b)

(c)

(d)

FIGURE 3.1

Chordate larvae showing basic architectural pattern. (*a*) and (*b*) are protochordates; (*c*) and (*d*) are vertebrates. *Dark red,* notochord; *medium red,* dorsal nervous system; *light red,* alimentary canal. In (*a*) a larval sea squirt, part of the left wall of the pharynx has been removed to show the slits in the right wall opening to the atrium.

chordates in the chordate hierarchy as proposed by Haeckel and generally accepted today is as follows:

Phylum Chordata
    Subphylum Urochordata
        Class Ascidiacea. Sea squirts
        Class Larvacea. Larvaceans
        Class Thaliacea. Thaliaceans
    Subphylum Cephalochordata. Lancelets
        (amphioxuses)
    Subphylum Vertebrata (Craniata). Chordates with
        vertebrae

Urochordates

Urochordata is a group of marine chordates in which the notochord is confined to the locomotor tail of the free-living larval stage (fig. 3.1*a*). Of the three classes, only one, the ascidians, undergoes complete metamorphosis, resorbing the tail and the notochord before becoming sexually mature. Larvaceans retain the tail and notochord throughout life and reproduce in the larval state. Most thaliaceans have no larval stage, no notochord, and no tail. Urochordates are enclosed within a delicate, nonliving, and sometimes beautifully colored transparent tunic that accounts for the alternate name, tunicates. They are **filter feeders,** their food consisting of particulate organic matter that is suspended in the seawater and filtered out before the water passes over the gills.

Ascidians

Sea squirts are the best known urochordates (fig. 3.2). There are solitary and colonial populations. The larvae are tiny (0.5 mm to 11.0 mm), they have a fleeting existence—as little as a few minutes and not longer than a few days—and they do not feed. They subsist on nutrients stored in the tissues. The body consists of a trunk containing immature viscera, and a muscular notochord-supported tail the function of which is to provide locomotion during the few hours of its free-swimming existence. The notochord is a stiff rod composed of a small number of epithelial cells arranged in a single layer around a central core of matrix and surrounded by a filamentous sheath. Alongside of the notochord in solitary species there are about 36 uninucleate striated muscle fibers. (Striated muscle fibers of vertebrates are multinucleate.) Other species have as many as 1,600. There are no tendons. The nervous system consists of a dorsal hollow nerve cord, several ganglia, and nerves. A **sensory vesicle** associated with the brain houses an **otolith** that stimulates nerve endings for statoreception. A light-sensitive **ocellus** (a pigment-protected receptor cell) is usually present in its wall. In some species neither definitive blood cells nor a functional heart differentiates until metamorphosis.

    Respiratory water enters the larval pharynx via the mouth, passes over gills lining the pharyngeal slits, and enters a chamber, the **atrium,** surrounding the pharynx.

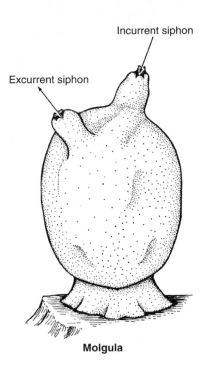

Molgula

FIGURE 3.2

Adult sea squirt.

Oxygen-depleted water is ejected via an **atriopore.** At metamorphosis three **adhesive papillae** with sticky secretions attach the larva to a permanent substrate.

    At metamorphosis the notochord is resorbed and becomes a source of nutrients during metamorphosis. The nervous system is altered in location and structure, and a rearrangement of the viscera takes place (fig. 3.3). The larval mouth becomes an **incurrent siphon,** and the atriopore becomes an **excurrent** one. The forceful discharge of jets of water from the latter whenever the polyp is irritated inspired the name sea squirt. At other times water flows steadily out of the siphon. Food particles are filtered out of the incoming water stream, becoming trapped in mucus in the endostyle, a ciliated glandular groove in the pharyngeal floor directed toward the entrance to the short esophagus. The mucus and entrapped food particles are swept by cilia and esophageal papillae into the stomach, whereas respiratory water passes over the gills and into the atrium, a fluid-filled collecting chamber surrounding the pharynx. Atrial muscles discharge the deoxygenated water to the exterior via the excurrent siphon. The polyp has now become a filter feeder. Arising from each end of the now functioning heart is a blood vessel. Blood is pumped alternately forward for a few beats, then backward, the heart pausing before reversing the flow.

    It wasn't until the middle of the nineteenth century that it was discovered that these polyplike, sometimes colonial, marine animals have a free-living larval stage with a notochord and a dorsal nervous system. From the viewpoint of vertebrate kinships, there is more to be learned from the larvae than from adult sea squirts.

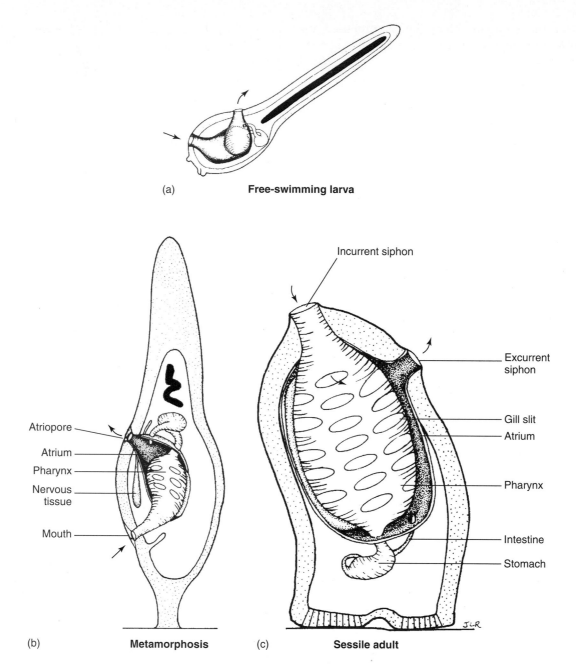

(a)   **Free-swimming larva**

(b)   **Metamorphosis**   (c)   **Sessile adult**

FIGURE 3.3

Metamorphosis in a sea squirt. In (*b*) and (*c*), the atrial wall has been removed to show pharyngeal slits opening into atrium. Arrows indicate direction of water flow. The notochord (*red*) disappears during metamorphosis.

Larvaceans

Larvaceans live so deep in the sea that sunlight penetrates the area only dimly. They have a body and a long, flat locomotor tail supported by a notochord that together measure no more than 8 mm (¼ in) in length (fig. 3.4). The organism is surrounded by a giant mass of fragile transparent mucus that is secreted by the larva. The mucus serves as a filter that excludes coarse particulate matter that is suspended in the seawater and admits to the immediate vicinity of the incurrent aperture only the finest organic particles. These are propelled by a pulsating membrane into the branchial basket (pharynx) along with dissolved oxygen in the water. The tuniclike mass is abandoned at frequent intervals as the filter becomes clogged with debris.

Thaliaceans

Thaliaceans are free-living individuals and colonial individuals in alternate generations. The organisms resemble adult ascidians in many respects. The pharynx, or branchial basket, like that of sea squirts, constitutes most of the body; the digestive tract and other organs are miniscule in comparison. There are considerably fewer slits in the pharyngeal wall. The organisms are cylindri-

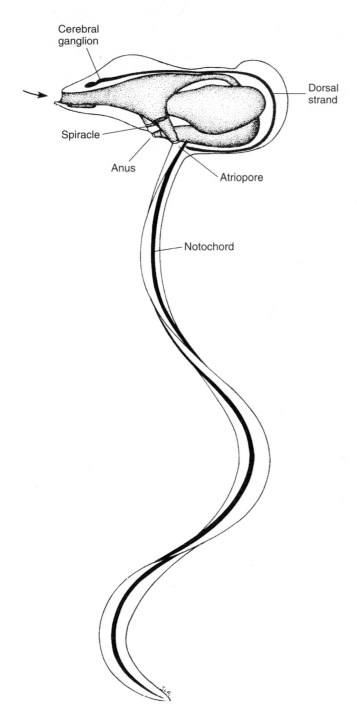

FIGURE 3.4

A larvacean. Red arrow indicates entrance to pharynx. The immense tunic has been removed.

empties into the saccular chamber. Colonies move only with water currents. Only one order of thaliaceans has a free-living larval stage, and this is of very short duration. No notochord has been described.

Cephalochordates

Amphioxus means "sharp at both ends." Any member of the subphylum Cephalochordata may be called an **amphioxus,** or **lancelet** (little spear), but the generic name for the lancelet commonly studied in the laboratory is *Branchiostoma* (fig. 3.5). *Asymmetron* is the only other genus in the subphylum.

Lancelets are marine organisms found a short distance out from sandy beaches throughout most of the globe. They quickly burrow into sand with eel-like movements, make a U turn, and then emerge until only the oral hood area is protruding for filter feeding. Adults vary from less than 2 cm to more than 8 cm in length, the largest being *Branchiostoma californiense.* Off the coast of Southern China, lancelets are collected in quantity and sold as a table delicacy. A living amphioxus is semitransparent but becomes opaque when immersed in preserving fluids.

Locomotor Musculature and Skin

The body is practically all trunk. It exhibits noticeable metamerism because of the muscle segments, or **myomeres,** which lie just under the thin skin, extend the entire length of the organism, and constitute the bulk of the body wall. These, collectively, are the locomotor muscles of an amphioxus. Each myomere is separated from the next by a **myoseptum (myocomma),** a connective tissue partition that is the origin and insertion of the longitudinal muscle bundles that constitute each myomere. Since myomeres are <-shaped, a cross section of the body intersects several myomeres (fig. 3.6). The integument consists of a single layer of epidermal cells and a thin dermis (fig. 6.1). A layer of subcutaneous connective tissue separates the muscles from the skin.

Pharyngeal Slits

The **gill slits** (pharyngeal slits) of an amphioxus do not open directly to the exterior but into a fluid-filled cavity, the **atrium,** which surrounds the pharynx laterally and ventrally (fig. 3.6c). Pharyngeal water, laden with oxygen and relieved of food particles, streams through the slits into the atrium, which expands due to the action of the transverse musculature in the atrial floor. The same muscles expel the oxygen-depleted water to the exterior via a midventral **atriopore** (fig. 3.5). The number of slits varies, but exceeds 60 in adults. Respiration also takes place through the skin.

Notochord

The **notochord** extends from the tip of the rostrum to the tip of the tail (fig. 3.5b). In contrast to all other chordates, it consists of muscular disks arranged like a long roll of coins and separated by fluid-filled spaces. The

cal in shape, and the incurrent and excurrent apertures are at opposite ends of the cylinder. Thaliaceans have no tail, locomotion being by a jet stream from the excurrent aperture. A colony may be linear, or it may consist of a floating water-filled sac the walls of which are composed of organisms that arise as a string of buds proliferated by a sexual individual. The incurrent aperture of a colony member projects into the sea; the excurrent aperture

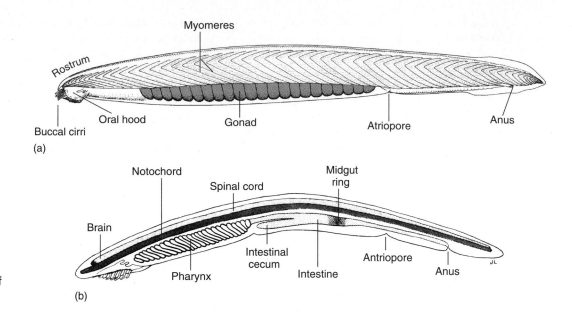

FIGURE 3.5

Branchiostoma. (*a*) Mature adult. (*b*) Internal structure of young specimen.

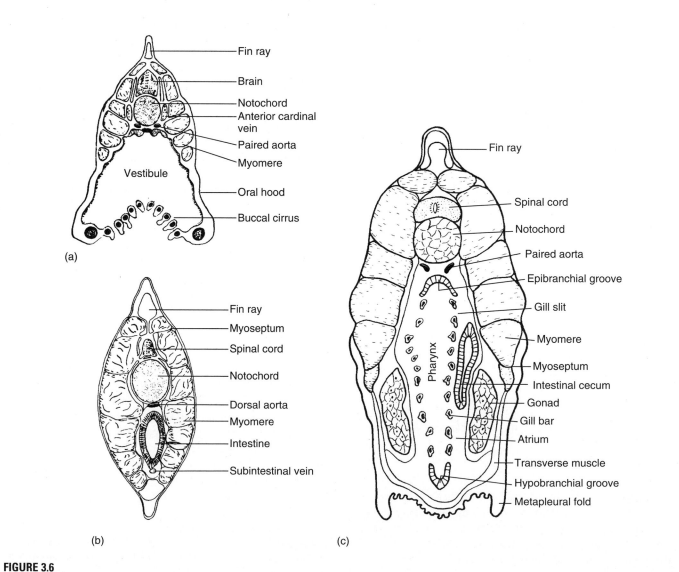

FIGURE 3.6

Cross sections of an amphioxus. (*a*) Anterior to mouth. (*b*) Posterior to atriopore. (*c*) Level of pharynx. The significance of the metapleural folds is unknown.

muscle fibers in each disk are disposed transversely and have cytoplasmic extensions that pass through the tough notochord sheath and end on the surface of the spinal cord. There the extensions receive motor impulses from neurons within the cord. Contraction of the notochordal muscles increases the stiffness of the chord. Continuation of the notochord to the very tip of the rostrum—unlike in any other chordate—may be an adaptation that facilitates burrowing in the sand. The only other skeletal structures are the fibrous rods that support the gill bars, buccal cirri, and the middorsal fin-like ridge.

Nervous System and Sense Organs

An amphioxus, like a vertebrate, has a central nervous system consisting of a hollow brain and a dorsal nerve cord containing a central canal, both lined by a nonnervous supportive membrane, the **ependyma;** and, as in vertebrates, the caudal end of the cord consists of ependyma alone. A single connective tissue membrane, the **leptomeninx** (see appendix 2), surrounds the brain and cord.

Only two brain subdivisions can be identified in contrast to three in vertebrates (fig. 16.13*a*). Attempts at homologizing the parts of the brain with those of vertebrates have not been entirely successful. In an amphioxus the notochord extends anterior to the brain; in vertebrates it stops at the midbrain. Does this indicate the absence of a forebrain? Whether or not the nerves that supply the gills should be considered cranial nerves complicates the problem. If the branchial nerves are omitted, there are 7 pairs of **cranial nerves.** If the branchial and oral nerves are included, there are 39 or more pairs. Vertebrates have 10 or 12 cranial nerves. The absence of semicircular canals, eyes, lateral-line system,

and foramen magnum (there being no skull) deprives us of landmarks that would be helpful. Because of these difficulties, it is not possible to decide where the brain ends and the cord begins.

Spinal nerves in cephalochordates, unlike those of vertebrates, consist solely of dorsal roots. These contain sensory fibers from the integument and sensory and motor fibers from and to the visceral organs. Instead of ventral roots conducting spinal nerve fibers to the muscles of the myomeres, as in vertebrates, tubes in amphioxus conduct cytoplasmic extensions of the muscle fibers to the spinal cord. There they receive their innervation from motor neurons whose nerve fibers remain within the cord! Dorsal roots and ventral tubes are metameric, but not directly opposite one another because left and right myomeres are not exactly opposite one another.

The relatively small size of the brain is correlated with the paucity of special **sense organs.** There are no retinas, semicircular canals, or lateral-line organs. However, chemoreceptors are abundant on the buccal cirri and velar tentacles, where they monitor the incurrent water stream. Chemoreceptors are also scattered on other surfaces of the body, the tail being more sensitive than the trunk. Tactile receptors, which elicit withdrawal, are present in the skin of the entire body.

The most prominent sense organs are the light-sensitive, pigmented **ocelli** embedded within the ventro-lateral walls of the spinal cord throughout its length (figs. 3.7 and 3.8). Each ocellus consists of a receptor cell and a caplike melanocyte. The melanocyte lies between the receptor cell and the incoming light rays and is packed with large melanin granules. A conducting process extends from the base of the receptor cell to the central nervous system.

FIGURE 3.7

Cephalic end of an amphioxus shown in sagittal section. **1,** vestibule bounded by oral hood; **2,** part of the wheel organ projecting into vestibule; **3,** velar tentacle; **4,** velum.

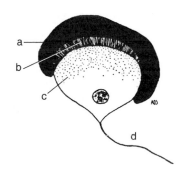

FIGURE 3.8

Ocellus (light receptor) from spinal cord of an amphioxus. **a,** melanocyte; **b,** apical border of receptor cell; **c,** receptor cell; **d,** process for conduction of impulse.

Food Processing

Cephalochordates, like sea squirts, are filter feeders. A collecting chamber for seawater, the **vestibule,** is bounded laterally by an **oral hood** and caudally by a perpendicular membranous **velum** perforated by the mouth, which opens into the pharynx (fig. 3.7). The vestibule is broadly open to the sea ventrally. Cilia on the pharyngeal surface of the gill bars create a steady flow of water through the mouth and into the pharynx. A set of stubby projections in the vestibule, the **wheel organ,** is covered with sticky mucus that retrieves some of the heavier food particles that miss the mouth, and it directs these through the mouth and into the pharynx along with the water stream. **Buccal cirri** partially strain the water as it enters the vestibule and chemoreceptors on their surface monitor it chemically.

Once food is in the pharynx it is processed as follows: In the pharyngeal floor there is a **hypobranchial groove,** or **endostyle** (fig. 3.6c), and in the roof there is an **epibranchial groove.** On the gill bars, ciliated **peripharyngeal bands** connect the two grooves. The cells of these bands and grooves secrete mucus. Food particles trapped in the mucus are incorporated into a stringy food cord that is propelled by cilia dorsally into the epibranchial groove and then caudad into the midgut behind the pharynx. There it is temporarily arrested by the **midgut ring** (fig. 3.5b) and mixed with digestive enzymes. Some of the digesting foodstuffs then pass beyond the ring into the hindgut, and some are propelled forward by cilia into the **intestinal cecum.** This evagination arises in the same manner as the vertebrate liver, but the two are not homologous and they also differ in function. The cecum secretes enzymes for digestion within the lumen of the cecum and intestine, and its lining cells phagocytose the smallest food particles and digest them by *intracellular* digestion. The intestine terminates at the anus.

Coelom

The coelom is not a prominent cavity in adults. It becomes laterally compressed as additional gill bars form in the elongating larvae. Consequently, only traces remain in a mature amphioxus, and they are difficult to find in stained cross sections. Small pockets of coelom persist lateral to the epibranchial groove, in the gill bars, and just internal to the body wall musculature. However, the early larval coelom is prominent and metameric because it forms from metameric outpocketings of the dorsolateral roof of the embryonic archenteron (fig. 5.3).

Circulatory System

Unlike a vertebrate, an amphioxus has no heart, and the blood is a colorless serum that lacks blood cells, platelets, and other formed bodies. A bulbous venous sinus (**sinus venosus,** fig. 3.9), analogous to the first chamber of the vertebrate heart, is the termination of the venous system

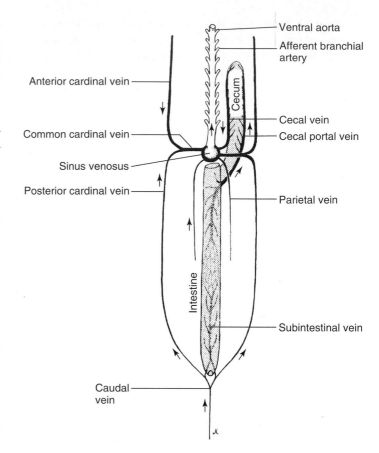

FIGURE 3.9

Basic venous channels and ventral aorta of an amphioxus, dorsal view. The cecum has been rotated 90 degrees around the long axis. The cecal portal vein in life is ventral to the cecum, and the cecal vein is dorsal.

and the beginning of arteries. Deoxygenated blood from all parts of the body empties into this sinus, then continues forward beneath the pharynx in an unpaired vessel, the **ventral aorta.** The ventral aorta gives off arteries that ascend in the gill bars, acquire oxygen, and then become tributaries of the paired **dorsal aortae,** located above the gill bars. This is basically the pattern we find in gnathostome fishes (fig. 14.16). The details differ. The cecal vein, ventral aorta, and bulbous swellings at the base of the afferent branchial arteries participate in pumping blood through the circulatory system. The sinus venosus is not a pump. Muscular contractions during body movements may contribute to the flow.

The paired aortae give off branches to the rostrum, pass caudad above the pharynx (fig. 3.6c), and unite just behind it to form an unpaired dorsal aorta (fig. 3.6b). The aortae distribute blood by paired **parietal arteries** to the body wall and by median vessels (**visceral arteries**) to the viscera. The dorsal aorta continues caudal to the anus as the **caudal artery.**

The venous channels are similar to those of embryonic vertebrates (fig. 3.9). From the capillaries of the tail a **caudal vein** courses forward and divides into right and left posterior cardinal veins. These pass forward in the lateral body wall to a point just behind the pharynx. Here the posterior cardinal veins meet the anterior cardinal veins from the rostrum and pharyngeal wall. The blood then enters a common cardinal vein leading to the sinus venosus. Parietal veins drain the body wall of the trunk. These also terminate in the sinus venosus.

Drainage from the visceral organs is via a median subintestinal vein that receives blood from the caudal vein. The subintestinal vein passes cephalad along the ventral surface of the intestine. There it breaks up into smaller channels, receives tributaries from the intestine, and reconvenes to continue forward as a **portal vein** ending in the capillaries of the cecum. From the cecum the contractile cecal vein pumps blood to the sinus venosus.

Elimination of Metabolic Wastes

Amphioxus has no compact organ known as a kidney. Metabolic wastes are collected by cells known as **solenocytes,** that are clustered below the paired aortae lateral to the epibranchial groove (fig. 3.10). Each solenocyte has (1) footlike projections **(pedicels)** that make contact with an adjacent **glomerulus** (an arterial rete), and (2) a long strand of microvilli that extends across a remnant of the coelom and terminates in the lumen of a **nephridial tubule.** The tubules are tiny ductules that collect fluids from a cluster of solenocytes and

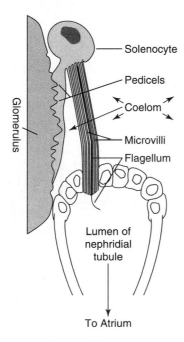

FIGURE 3.10

One of a cluster of amphioxus solenocytes with pedicels in intimate contact with the thin wall of a glomerulus lateral to the epibranchial groove and ventral to the paired aortae.

empty into the atrium via tiny pores. A single long flagellum from the solenocyte lies at the core of the microvillar strand and, like microvilli, projects into the nephridial tubule. Collectively, flagella presumably propel tubule fluids to the atrium. The atrium opens to the exterior via the atriopore.

Solenocytes are common among invertebrates. The details of how they work in an amphioxus are unknown. Contact of the pedicels with a glomerulus suggests that glomeruli are the source of one or more components of what is called urine in vertebrates. The fact that microvillar strands project across a coelomic space—especially one that has been spared when many parts of the larval coelom are greatly reduced during metamorphosis—may or may not be significant; coelomic fluid is the collection site for similar wastes in many multicellular invertebrates and in larval hagfishes.

Gonads

Mature gonads are visible through the body wall (fig. 3.5a) and bulge into the water-filled atrium, into which sperm or eggs are shed. The gametes are then flushed into the sea via the atriopore along with respiratory water and metabolic wastes. The amphioxus is **dioecious;** that is, ovaries and testes do not develop in the same individual.

Amphioxus and the Vertebrates Contrasted

Although an amphioxus resembles a vertebrate in many respects, differences are evident. An amphioxus has almost no cephalization and no paired sense organs; it has a notochord but no vertebral column; it has pharyngeal (gill) slits but in large numbers, emptying into an atrium; it has a dorsal, hollow central nervous system, but the brain lacks the major vertebrate subdivisions; it has a segmented musculature, but the segments extend to the anterior tip of the head; it has a two-layered skin, but the outer layer is only one cell thick; it has arterial and venous channels similar to the basic channels of vertebrates but no muscular heart; it is coelomate, but the coelom is greatly restricted in adults. A hypothetical relationship of the amphioxus to other protochordates and to vertebrates is diagrammed in fig. 3.13.

HEMICHORDATES: INCERTAE SEDIS

The phrase *incertae sedis* means "of uncertain status." It denotes that the taxonomic status of a group of organisms is widely disputed. Among such groups are acorn worms—marine organisms that live in mud under shallow waters and, although fragile when handled, have been known to reach 1.5 meters (5 ft) in length (fig. 3.11a). They have invertebrate and chordate features, and have been classified at one time or another in each category.

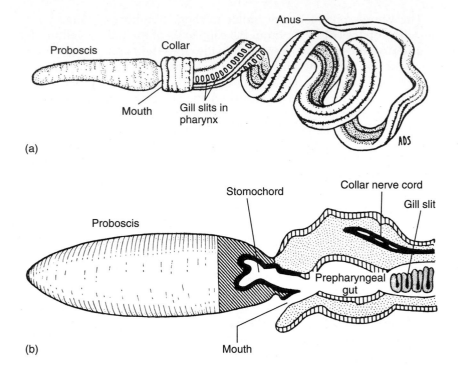

FIGURE 3.11

An acorn worm, *Saccoglossus*. (*a*) Entire worm.
(*b*) Head, sagittal section (except proboscis).   (b)

In 1870 Karl Gegenbaur established a new taxon, **Enteropneusta**, to accommodate acorn worms. Four years later, Ernst Haeckel established the phylum **Chordata** to include all organisms with notochords. After ten more years, William Bateson (1884) added enteropneusts to Haeckel's phylum Chordata, believing them to have a notochord. He named the new group **Hemichordata**. His decision was based on the following considerations.

1. Acorn worms have a strand of nerve cells and fibers in a longitudinal groove of ectoderm on the dorsal surface of the collar and extending to the end of the body. In the collar, the groove sinks beneath the surface to become a **collar nerve cord** with a continuous or discontinuous lumen. This dorsal nerve strand and the tendency to have a lumen in the collar region was believed by Bateson to be an expression of the same phenomenon that induces the formation of a dorsal neural groove and tube in chordates (fig. 1.11).
2. Acorn worms have slits that open to the exterior in the lateral walls of the foregut. Pharyngeal slits are basic features of chordates.
3. Acorn worms have a short diverticulum of the foregut, the **stomochord,** that extends forward into the proboscis (fig. 3.11*a*). Bateson hypothesized that the diverticulum may be homologous with the notochord of chordates. Evidence for this is lacking.

Bateson discounted certain facts that indicate a close kinship with echinoderms. Enteropneusts have bilaterally symmetrical larvae (**tornaria**) so similar to some echinoderm larvae that nineteenth-century biologists frequently confused the two. This suggests an affinity of echinoderms and enteropneusts. But even if this is an erroneous inference, enteropneust larvae do not resemble ascidian or cephalochordate larvae and could not be mistaken for them. Second, similarities between enteropneusts and echinoderms in certain developmental processes, in muscle proteins, and in other traits seem to link enteropneusts closer to echinoderms than to chordates. Third, a detailed examination of the nervous systems of enteropneusts reveals a system quite unlike the hollow central nervous system of larval sea squirts and cephalochordates. In addition to the dorsal nerve strand, a second strand lies in the coelomic floor, as in invertebrates. Dorsal and ventral strands are united by a circum-intestinal nerve ring at the caudal boundary of the collar. But nowhere, collar or trunk, is there a region that can be regarded as a *central* nervous system, that is, a center that has been shown to integrate stimuli and responses. The collar nerve cord is not a brain in either the invertebrate or chordate sense, having no neurohistologic specialization, and neither receiving nor giving off nerves (*bundles* of nerve fibers).

The similarities between enteropneusts and echinoderms cited above coupled with the highly improbable homology of stomochord and notochord, and the enigmatic nature of the enteropneusts, nervous system prompt the current majority view that enteropneusts should be classified, at least for the present, close to echinoderms, close to protochordates, but in a taxon independent of both. The taxon **Hemichordata** includes, in addition to enteropneusts, one or more classes that have little resemblance to one another or to enteropneusts.

FIGURE 3.12

One of many varieties of heterostracan ostracoderms.

THE ORIGIN OF VERTEBRATES

The oldest vertebrates we know of are the ostracoderms (figs. 3.12 and 4.4). These strange little fishes, 2 to 30 cm long and of diverse appearances, had no jaws and no paired fins, and many of the earliest ones were filter feeders. Broad bony plates in the skin formed a protective shield over the head and much of the trunk, inspiring the nickname "armored fishes." Where there were no bony shields smaller bony scales fitted tightly together over the body like tiles. Fossils that can be clearly identified as ostracoderms date back to the beginning of the Ordovician, after which ostracoderms became an abundant and very diversified group. One order is described in chapter 4.

The broad outlines of vertebrate evolution after the Ordovician have been remarkably well ascertained (fig. 4.2). The jawless ostracoderms were succeeded in the seas by jawed fishes, and amphibians eventually became established on land. The perplexing problem is, who were the ancestors of ostracoderms?

Because ostracoderms were chordates, we must look for clues in the protochordates that are with us today and in the fossil record that immediately preceded ostracoderms. What clues to vertebrate ancestry might we find in living protochordates? Earlier in this discussion an amphioxus was contrasted with vertebrates with respect to differences in their morphologic traits. Let us now look at these cephalochordates, emphasizing similarities rather than differences.

Cephalochordates have a notochord, pharyngeal slits that are used in respiration, a dorsal hollow central nervous system with brain and cord, a metameric body wall musculature, a two-layered skin, and arterial and venous channels similar to those of fishes and to the early embryonic vessels of higher vertebrates. Cephalochordates are deuterostomous and coelomate. Finally, cephalochordates are filter feeders, as were many early ostracoderms. These similarities bespeak close genetic ties between the ancestors of cephalochordates and those of vertebrates. But did cephalochordates precede ostracoderms in the earth's waters?

The answer is, "Yes!" Early twentieth-century paleontologist C. D. Walcott unearthed, in a limestone quarry 8,000 feet up in the Canadian Rockies, several dozen specimens of soft-bodied, laterally compressed, ribbon-shaped fossils about 5 cm (2 in) long that resembled an amphioxus. The stratum from which they were recovered was the Middle Cambrian. They had zigzag myomeres and a notochord that extended the length of the body. This recovery places this specific chordate, *Pikaia gracilens,* in the earth's waters considerably earlier than the ostracoderms. Gould (1989) has related the broad lessons in natural history that are implicit in Walcott's discoveries.

Although we know, or believe we know on the basis of current data, that protochordates preceded vertebrates in the course of natural history, we have to speculate concerning the lineages that might have led from prechordate invertebrates to protochordates on the one hand, and to vertebrates on the other. Surely, cephalochordates *as we know them today* were not genetic ancestors of the first vertebrates. We must consider the fact that echinoderms, like vertebrates, have mineralized tissue in their mesoderm (not in their ectoderm, as in molluscs, for example); that echinoderms, like cephalochordates, form their mesoderm and coelom as outpocketings of their archenteron; that echinoderms, enteropneusts, protochordates, cephalochordates, and vertebrates are all deuterostomes, a trait found in only one other invertebrate taxon (Chaetognatha); and that all have larvae in their history. On the basis of these and other morphologic and ontogenetic facts, it is rational to hypothesize (1) that genetic ties unite one or more of these in unknown lineages, (2) that the characters shared by protochordates and vertebrates are derived from a common ancestor, and (3) that the bony dermal armor of ostracoderms must have developed late in the Cambrian in one or more soft-bodied invertebrates that already had a notochord, dorsal hollow nervous system, and pharyngeal slits. Figure 3.13 depicts one postulated phylogeny. A cautionary note should be interjected: morphologic similarities (homoplastic traits) may, in one or several instances, be attributable to evolutionary convergence.

THE AMMOCOETE: A VERTEBRATE LARVA

The ammocoete larvae of lampreys (eel-like jawless fishes) provide compelling support for the concept that vertebrates share a common ancestor with protochordates. These larvae metamorphose very slowly, spending from two to six years before metamorphosis is completed and the gonads mature. They were at one time classified as protochordates with the generic name *Ammocoetes.* Only later was it discovered that they are vertebrate larvae.

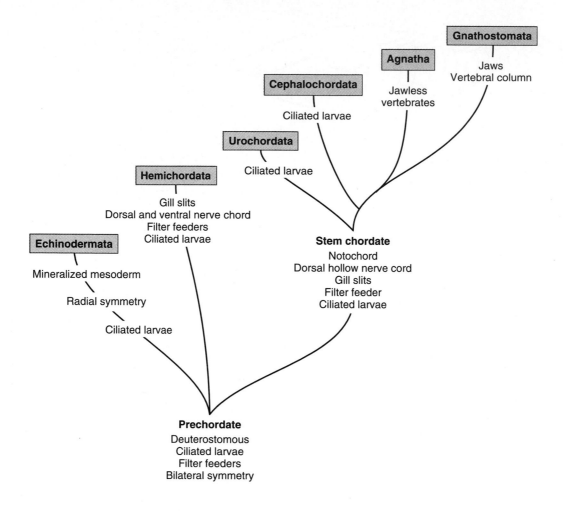

FIGURE 3.13

One of several postulated phylogenetic relationships between invertebrates, protochordates, and vertebrates. Agnathans are classified as vertebrates but have not been shown to have a vertebral column.

Ammocoetes have a notochord that commences at the midbrain and continues to the end of the body (fig. 3.1c). The dorsal hollow nervous system develops in the same manner as that of an amphioxus, but the ammocoete brain has three vesicles instead of two. Seven pairs of pharyngeal slits open to the exterior. The body wall musculature is disposed as overlapping myomeres that provide locomotion.

Ammocoete larvae, like protochordates, are filter feeders. Food particles in the respiratory water stream, chiefly diatoms, are ensnared in mucus produced by the endostyle, which is homologous with the hypobranchial groove of an amphioxus. In ammocoetes, however, the groove eventually sinks into the floor of the pharynx to become a **subpharyngeal gland** (fig. 18.15). Beyond the ammocoete pharynx there is a *solid* liver diverticulum. An amphioxus has a *saccular* diverticulum at the same location, but the two do not have the same functions.

The circulatory system of an ammocoete larva is remarkably similar to that of an amphioxus, except that an amphioxus has no muscular heart. The ammocoete heart consists of a sinus venosus, atrium, and ventricle. Blood in both organisms is pumped forward in the ventral aorta, dorsad into the gills, then to the dorsal aorta. Venous blood returns from the head via anterior cardinal veins and from most of the trunk and tail via posterior cardinals. All converge on paired common cardinals that empty into the sinus venosus (see figs. 3.9 and 3.14). Blood from the intestine returns via a ventral intestinal (subintestinal, portal) vein that ends in the capillaries of the liver of an ammocoete and in the cecal diverticulum of an amphioxus. This blood then continues via a hepatic or cecal vein to the sinus venosus.

Ammocoete larvae exhibit vertebrate features that are not present in an amphioxus, among which are special sense organs and a vertebrate means of eliminating excess tissue fluids and metabolic wastes. Among early developing sense organs are **otic vesicles,** which are precursors of a primitive membranous labyrinth (inner ear, but not for hearing), and an **olfactory sac** for chemore-

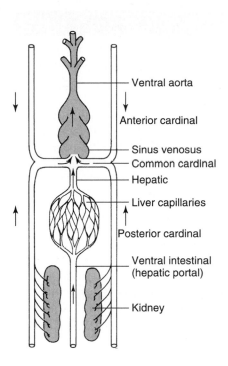

FIGURE 3.14
Venous channels of an ammocoete. See also fig. 3.9.

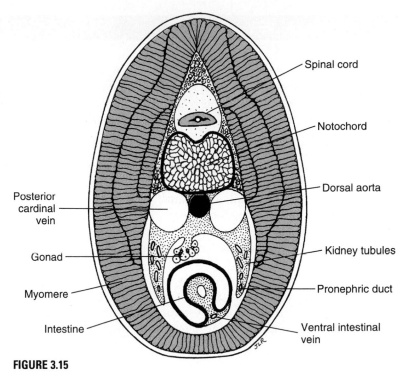

FIGURE 3.15
Relatively old ammocoete in cross section just behind the liver. Several myomeres are cut in a single section because they curve caudad in passing inward from the skin.

ception (associated with feeding). Two middorsal light receptors (**pineal** and **parapineal organs,** also known as median eyes, fig. 17.20) lie in tandem beneath the skin on top of the head. They monitor light rays, but there are no retinas and the lenses cast no image. The function of these light receptors is somewhat analogous to that of protochordate ocelli, but they are complex organs, not mere cells. Paired lateral eyes eventually form, but they remain rudimentary. (Eventually, they lie under the thick skin of adults, and are useless.) Whether or not the paired eyes are vestiges (chapter 2) depends on the phylogenetic history of lampreys. Special neuromast organs, discussed in chapter 17, lie in the skin of the head and open onto the surface via pores.

Ammocoete larvae have a different mechanism for elimination of fluid wastes than do protochordates. The larvae have a primitive vertebrate kidney that, in an early larva, consists of three to six tubules with funnel-like **nephrostomes** that collect coelomic fluids for processing and excretion (figs. 3.15 and 15.1). These fluids are *expressed into the coelom* from glomeruli. The tubules, which increase in number as the larvae elongate, lead to longitudinal ducts that empty to the exterior near the anus. Between the kidneys is an elongated rudimentary gonad that is paired in early larvae,

unpaired later. The epidermis is multilayered as in other vertebrates (fig. 6.5).

DOES AMPHIOXUS HAVE A VERTEBRATE ANCESTOR?

The phenomenon of heterochrony has been proposed to explain the similarity of an amphioxus and an ammocoete larva. If you will reexamine figure 3.1, you will see four chordate larvae. Examples (*a*) and (*b*) are protochordates, (*c*) and (*d*) are vertebrates. If, in an *isolated population* of ammocoetes, a genetic differential in the timing and/or rate of maturation of the gonads had developed so that the larvae became sexually mature before metamorphosis was complete, the eventual result might have been either extinction or a new paedomorphic species that would be classified as a protochordate. In that scenario, amphioxus would be a descendant of an agnathan vertebrate. Objections to the heterochrony hypothesis include, among others, the lack of an atrium that collects water from the gill slits. The hypothesis is untenable if an amphioxus and *Pikaia gracilens* share *derived characters* (see chapter 2). But who can prove they don't? Or that they do?

SUMMARY

1. Protochordates are bilaterally symmetrical, filter-feeding marine chordates with notochord, gill slits, a dorsal hollow central nervous system, and no vertebral column. There are two subphyla, Urochordata and Cephalochordata.
2. Urochordates include sea squirts (ascidians), in which the notochord is confined to the tail of larvae; larvaceans, which are permanent larvae with notochord confined to the tail; and thaliaceans. The latter have no tail, and no notochord.
3. Cephalochordates are the lancelets (amphioxuses). The notochord extends the length of the body to the tip of the postanal tail.
4. Hemichordates are invertebrates with pharyngeal slits, a longitudinal dorsal nerve strand that does not constitute a central nervous system, a ventral nerve strand characteristic of major invertebrate phyla, and no notochord. Best known are the acorn worms (Enteropneusta), which are sometimes included in the phylum Chordata.
5. The basic chordate features common to protochordates and vertebrates and the similarity of the details of the embryonic development of these features bespeak close genetic ties between the two groups. It is rational to postulate that vertebrates are derived from, or share common ancestors with, protochordates.
6. Ammocoetes are the larvae of lamprey eels. They have many features of an amphioxuslike protochordate.
7. A minority hypothesis derives amphioxuslike protochordates from paedomorphic agnathan larvae. However, current geologic data place protochordates in the Middle Cambrian and the earliest known vertebrates in the Late Cambrian or Early Ordovician.

SELECTED READINGS

Barrington, E. J. W., and Jefferies, R. P. S., eds.: Protochordates, Symposium of the Zoological Society of London, no. 36, New York, 1975, Academic Press.

Cloney, R. A.: Ascidian larvae and the events of metamorphosis, American Zoologist **22:**817, 1982.

Flood, P. R.: Fine structure of the notochord of amphioxus, Symposium of the Zoological Society of London **36:**81, 1975.

Gould, S. J.: Wonderful life, New York, 1989, Norton.

Mallatt, J.: The suspension feeding mechanisms of the larval lamprey. *Petromyzon marinus,* Journal of Zoology, London **194:**103, 1981.

CHAPTER FOUR

PARADE OF THE VERTEBRATES IN TIME AND TAXA

This chapter is designed to introduce vertebrates that are referred to in later chapters, relating briefly their natural history and their postulated ancestry, so they will not be total strangers when you meet them again. The chapter also provides a glimpse of the diversity of vertebrate life on our planet today. It may prove rewarding to revisit the chapter should you meet an organism with which you are not acquainted.

OUTLINE

here are approximately 50,000 known species of animals with vertebral columns. Fortunately for Noah, more than half of these are fishes. However, many amphibians and some reptiles and mammals share with fishes the freshwater ponds and streams as their preferred or permanent abode, and a much smaller number share the seas. Amphibians that can tolerate salt water are rare. There are some. They don't live in the sea, but they forage there. Among reptiles a few snakes, turtles, and iguanid lizards are permanently marine, but oviparous females must emerge to deposit their eggs. Among mammals, cetaceans and sirenians never leave the sea, and marine carnivores (pinnipeds) leave the sea only to breed. Although no bird lives in the sea, many are dependent on marine organisms for food. All the foregoing animals other than fishes, however, are descendants of terrestrial ancestors and have returned to water. In the chapters that follow we will examine the adaptations that make these different lifestyles possible.

VERTEBRATE TAXA

The classification scheme in this chapter is a conventional evolutionary taxonomy that divides vertebrates into two superclasses, the jawless **Agnatha**, and the jawed **Gnathostomata**. Figure 4.1 shows the eleven classes of vertebrates that are discussed in this chapter and that are listed in the synoptic classification of vertebrates in appendix 1. Not all taxonomies are identical. Cladistic (phylogenetic) taxonomies, for example, differ in that all taxa must be monophyletic. Not all taxa in a conventional evolutionary classification are monophyletic. You are urged to read the short discussion of systematics in chapter 2 before going farther into this chapter. It will prepare you to understand why all classifications are not identical. It will also provide insight into the problems faced by scholars who are trying to reconstruct phylogenetic lines that began over 400 million years ago.

Taxons consist of species that are designated by a latinized *binomial name*, an invention of the Swedish

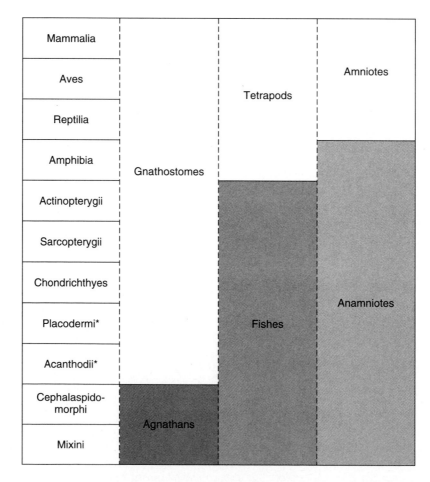

FIGURE 4.1

Major categories of vertebrates. Classes are listed in the first column. Asterisks indicate extinct groups.

naturalist Carl von Linné (Latin, Linnaeus). The Latin name enables zoologists of all nationalities to understand one another without translation when a specific animal is mentioned. For instance, not every zoologist knows what is *un chat, die Katze,* or *el gato.* All these words refer to the domestic cat, but *Felis cattus* is the binomial name for the species. The generic name *Felis* separates certain cats—mountain lions, European wildcat, domestic cat, and others—from lions, tigers, leopards, and so forth, which bear the generic name *Panthera. Felis cattus* separates domestic cats from mountain lions (*F. concolor*) and from European wildcats (*F. sylvestris*).

Figure 4.2 shows the range and relative abundance of some of the major categories of vertebrates in geologic time. Figure 4.3 summarizes graphically the postulated phylogeny of vertebrates that are alive today.

AGNATHA

Agnathans include two contrasting groups: the extinct, heavily armored ostracoderms, and the living hagfishes and lampreys. All other vertebrates are gnathostomes.

Ostracoderms

Ostracoderms is a general term that has been applied to a diverse assembly of the oldest known vertebrates. The term has no taxonomic status. Ostracoderms date back at least to the middle of the Ordovician, if not to the Cambrian. The entire body was covered with bony dermal armor consisting of broad plates and small tilelike scales (fig. 4.4). The plates were largest on the head where they formed a bony shield. Ostracoderms lacked jaws and paired fins, although some had bizarre projections from the body wall. Many were only 2 or 3 cm long

Eras	Duration of periods (millions)	Years ago (millions)	Periods	Fishes	Tetrapods
Cenozoic 65 million years duration	65	65	Recent to Paleocene		
Mesozoic 185 million years duration	80	145	Cretaceous		
	65	210	Jurassic		
	40	250	Triassic		
Paleozoic 350 million years duration	36	286	Permian		
	74	360	Carboniferous		
	48	408	Devonian		
	30	438	Silurian		
	62	500	Ordovician		
	100	600	Cambrian	At the end of the Cambrian, the oceans teemed with invertebrates that left their shells in the sediments.	

FIGURE 4.2

Range and relative abundance of vertebrates through geologic time. Four billion years of earth's existence preceded the Cambrian period. The two pairs of double lines denote times of significant extinctions. The Devonian (*red*) is known as the age of fishes. All figures are approximations.

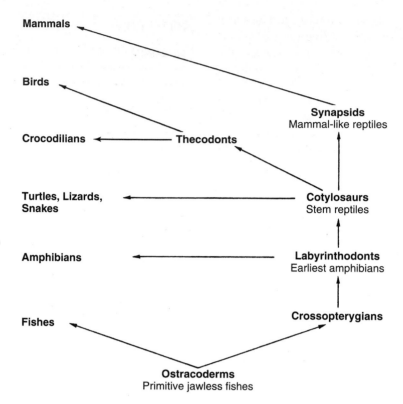

FIGURE 4.3

Mainstreams of vertebrate evolution leading to modern taxa (*left column*).

(1 cm = 0.3937 in); a few attained a length of up to 30 cm (1 ft). What appear to have been the earliest ostracoderms lived largely, if not exclusively, in freshwater. By the later Silurian, there were also marine species.

Knowledge of the anatomy beneath the fossilized head shield of an osteostracean (a cephalaspidomorph ostracoderm [Cephalaspidomorphi], appendix 1) is due largely to the painstaking research of E. A. Stensiö, an early twentieth-century paleontologist at the University of Stockholm. His studies revealed that the head skeleton was a more or less flattened spicule-covered bony shield with four dorsal apertures (fig. 4.5). Two of these accommodated a pair of upward-staring eyes, a third accommodated the median, or pineal, eye, and a small, anterior opening was a single naris from which a nasohypophyseal duct led to an olfactory sac, and beyond. The shield turned under along its lateral edges, where it was then replaced beneath the gills by tilelike scales. Between the anterior edge of the shield and the scales a very small mouth opened into the oropharyngeal chamber, which was lined by gills. A curved row of external gill slits extended from the corners of the mouth to the caudal margins of the head shield. In addition to the shield, the head contained an endoskeleton of endochondral bone and considerable cartilage. Little is known of the postcranial skeleton except that the tail was heterocercal like that of most Paleozoic fishes. (See fig. 10.18a.)

The origin of ostracoderms is unknown. Possible phylogenetic ties with protochordates were discussed in the preceding chapter. Their numbers dwindled and they disappeared at the end of the Devonian, having been replaced by jawed fishes.

Living Agnathans

Hagfishes and lampreys, the living agnathans (fig. 4.6), may not be as closely related to one another as their superficial resemblances suggest. Hagfishes, of which there are 42 living species, have been given class status (Class **Mixini**) in a superclass Agnatha in one cladistic taxonomy of fishes (Nelson), and lampreys have been tentatively included with the ostracoderms in the Class **Cephalaspidomorphi.** (Appendix 2 may be consulted for derivation of the term.) This taxonomy has been employed in appendix 1. It is a tentative postulate. Other methods of studying phylogenetic relationships (comparison of the sequence of nucleotide bases from ribosomal DNA, for example) have yielded other postulates (see Forey and Janvier). Hagfishes and lampreys are generally referred to by the descriptive term **cyclostomes.**

Cyclostomes have a prominent notochord that serves as the sole axial skeleton throughout life. They have no paired fins, no skeleton comparable to that of the vertebrate jaws, no vertebral column comparable to the vertebrate spine, and no bone anywhere in the body—no bony

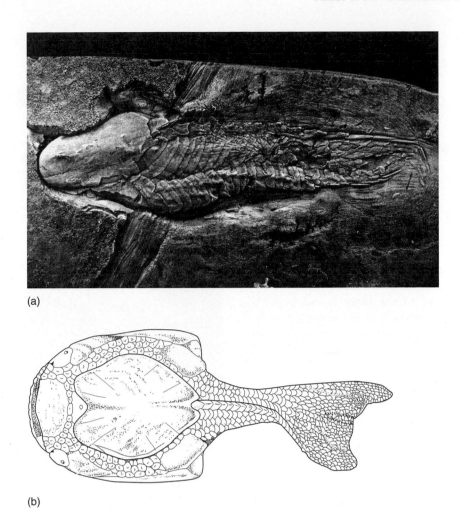

(a)

(b)

FIGURE 4.4

Ostracoderms of the order Pteraspidiformes. (*a*) A fossil from a Silurian-Devonian formation in Canada. (*b*) Dorsal view of a reconstruction of the flat-bodied *Drepanaspis* (× ⅓).

(*a* and *b*) From *Vertebrate Paleontology and Evolution,* by Carroll. Copyright © 1988 by W. H. Freeman and Company. Used with permission.

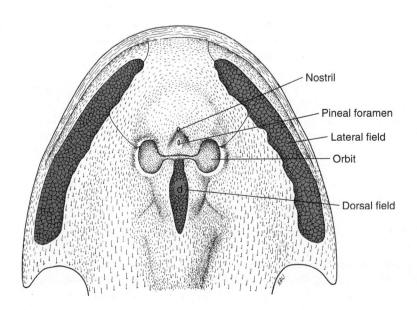

— Nostril

— Pineal foramen

— Lateral field

— Orbit

— Dorsal field

FIGURE 4.5

The bony denticulated head shield of the ostracoderm *Cephalaspis*. The dorsal and lateral fields may have been sites of exteroceptors and electric organs, respectively.

FIGURE 4.6

Lamprey *Petromyzon* above, hagfish *Bdellostoma* below.

skeleton, no integumentary armor or scales, and no bony teeth. They have only one (hagfishes) or two (lampreys) semicircular ducts instead of the three found in jawed vertebrates, and a single nostril that is connected with a single olfactory sac. They exhibit adaptations for parasitism, including a buccal funnel and a rasping "tongue."

Lampreys

Lampreys have not changed appreciably since the Carboniferous (compare lamprey in figs. 4.6 and 4.7). A large buccal funnel lined with horny denticles helps keep the parasitic adult lamprey attached to the host while a tonguelike cartilaginous rod covered with horny teeth rasps the flesh of the victim, leaving only skin and skeleton. A nasohypophyseal duct leads from the median nostril to the olfactory sac and then terminates blindly in a nasohypophyseal sac (fig. 13.1*b*). Seven pairs of gill pouches open separately to the exterior via porelike gill slits that conduct water *into* as well as *out of* the pouches, thus freeing the buccal funnel for feeding on the host.

Petromyzon marinus is a species of **anadromous** lampreys. Adults live in the sea but migrate upstream to lay their eggs. In about three weeks the small, free-living ammocoete larvae emerge. After several years in freshwater, the larvae have fully metamorphosed and, as juveniles, they migrate to the sea. There they attain sexual maturity and become physiologically adapted for the freshwater journey back to the spawning place. A landlocked population of this species inhabits the freshwater Great Lakes of North America. The freshwater brook lampreys are not parasitic. They cease feeding on reaching sexual maturity, spawn, and like all other lampreys, they die shortly thereafter.

Hagfishes

Hagfishes are marine cyclostomes with a shallow buccal funnel that lacks rasping denticles. The funnel is surrounded by a ring of stubby fingerlike papillae. Hagfishes are chiefly bottom-feeding scavengers whose diet includes a variety of small invertebrates and the vis-

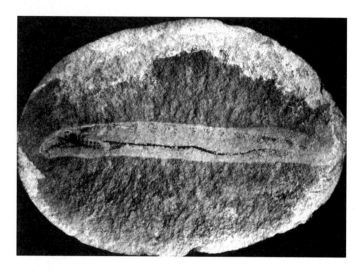

FIGURE 4.7

Fossil lamprey *Mayomyzon* from the Carboniferous.

cera of dead or weakened fishes. A canal leads from the median nostril to the olfactory sac and then continues to the pharyngeal cavity, carrying a respiratory stream of water (fig. 13.1*a*). The eyes, unlike those of lampreys, are vestigial and covered by opaque skin.

Myxine glutinosa, the Atlantic hagfish, has six pairs of porelike gill pouches (occasionally five or seven) that open into a common efferent duct (fig. 13.1*a*). *Bdellostoma stouti*, common off the coast of California, has ten to fifteen pairs of gill pouches that open directly to the exterior. Hagfishes don't enter freshwater, and the larvae stay within the egg membranes until metamorphosis. This may be an adaptation to the saltwater environment in which the eggs are laid.

ACANTHODIANS AND PLACODERMS

Two abundant groups of bizarre jawed fishes, Acanthodians and Placoderms (figs. 4.8 and 4.9), emerged from ostracoderms and preceded the cartilaginous, lobe-finned, and ray-finned fishes.

FIGURE 4.8

An Acanthodian.

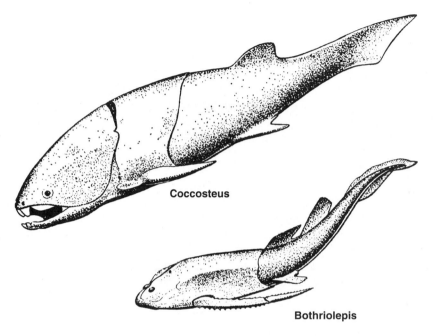

Coccosteus

Bothriolepis

FIGURE 4.9

Two armored Devonian placoderms, each about one-third natural size. *Coccosteus* is an arthrodire; *Bothriolepis* is an antiarch.

Acanthodians (fig. 4.8) were spiny fishes, as the name implies. Stout hollow spines were associated with the median and paired fins, and in some species an additional series of paired spines, probably associated with finlike membranes, extended along the lateral body wall. As in ostracoderms, the body was covered by bony armor consisting of small scales and, on the head, dermal plates. Most acanthodians were only a few inches long, but some reached 5 feet in length. The skeleton consisted of bone and cartilage. Unlike sharks, they had a large operculum overlying the gill slits. Acanthodian spines have been found in the early Silurian; fossils are abundant from the Devonian and Carboniferous, after which acanthodians disappear from the geologic record.

Placoderms, named for their bony dermal plates, had paired fins unlike those of later fishes (fig. 4.9). Among the better known are the **arthrodires** (fig. 4.9, *Coccosteus*). Heavy bony shields covered the head and gill region and another covered part of the trunk. The shields met in movable joints. The remainder of the body had small bony scales or, in later species, was naked.

Arthrodires appear to have been swift predators. **Antiarchs** (fig. 4.9, *Bothriolepis*) were small armored Placoderms with eyes on top of the head and a flattened belly, features that suggest they may have been bottom feeders. Some Placoderms reached a length of 15 feet, but the majority were much shorter. Placoderms flourished throughout the Devonian along with the Acanthodians, and disappeared rapidly thereafter.

The pylogenetic relationships of Acanthodians and Placoderms to later jawed fishes has been a matter of conjecture. Acanthodians may have been the ancestral jawed fishes (fig. 4.2) and Placoderms may have disappeared leaving no descendants.

CHONDRICHTHYES (CARTILAGINOUS FISHES)

Living Chondrichthyes, the cartilaginous fishes, are usually placed in two subclasses, **Elasmobranchii,** which includes sharks, rays, skates, and sawfishes, and

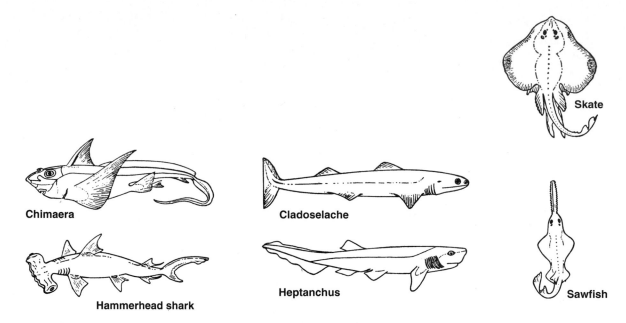

FIGURE 4.10

Representative Chondrichthyes. *Cladoselache* is primitive and extinct.

Holocephali, the chimaeras, or ratfishes (figs. 4.10 and 17.2*b*). They have no bone in their body other than in their scales and teeth. The mouth is on the ventral surface rather than terminal, except in Paleozoic sharks such as *Cladoselache* (fig. 4.10). Their unique dermal scales, called placoid, are composed of a basal plate and a bony spine of dentin that protrudes through the epidermis, giving the skin of sharks the texture of sandpaper. The skeleton of the pelvic fins of males is modified to form claspers used in transferring sperm to the female reproductive tract, fertilization being internal. The eggs are macrolecithal, and in oviparous species they are often encased in a horny or leathery shell with tendrils that entwine around vegetation as holdfasts.

Elasmobranchs

The taxon **Elasmobranchii** consists of many orders of sharks, Paleozoic to modern, and of skates, rays, and sawfishes. The gill slits are exposed (naked) rather than covered by a fleshy or bony operculum, and there are usually five pairs, although the sharks *Hexanchus* and *Heptanchus* have six and seven pairs, respectively. Seven slits may have been the primitive number in sharks, since that was the number in *Cladoselache*. It is the largest number in any jawed fish. Anterior to the first gill slit there is usually a spiracle that bears in its walls a miniature gill-like surface (pseudobranch). The caudal fin of Paleozoic sharks was heterocercal, the predominant type in the Paleozoic, and that of today's sharks, skates, and rays has not changed (fig. 10.18*a*).

Dissection of a shark (usually in the order **Squaliformes**) is a valuable point of departure for study of comparative vertebrate morphology because of its generalized anatomy. If one were seeking a living blueprint of a generalized vertebrate—an architectural pattern that could provide preliminary insight into the systems of the bodies of all vertebrates—it would be difficult to find a better example than a shark. Most frequently studied is the viviparous spiny dogfish of the Atlantic, *Squalus acanthias,* named for the prominent spine associated with each dorsal fin. *Squalus suckleyi* is the Pacific spiny dogfish.

Rays, skates, and sawfishes—**Rajiformes**—are compressed dorsoventrally, and the anterior fins are attached all along the sides of the head and trunk, forming broad, undulating, winglike locomotor organs (fig. 4.10). The five gill slits are ventral (fig. 13.7), but the spiracle, which is the chief incurrent route for respiratory water, remains on the dorsum. Most rays and skates subsist on molluscs scooped up from muddy bottoms, and the location of the spiracle removes it from the mud and debris that is stirred up by bottom feeding. Manta rays, on the other hand, are filter feeders. The tail of Rajiformes has changed from a muscular locomotor organ to a lean, often whiplike, organ of defense and, in some species, of offense. In electric rays (torpedos) it houses an electric organ capable of a high-voltage discharge that immobilizes prey. Sting rays have rows of spines along their tails that inflict wounds and often contain a poison. Sawfishes, not to be confused with sawfish sharks, have the rajiform shape to a lesser degree. The sword is used for impaling small fishes and disturbing the sea bottom in a search for burrowing animals.

Holocephalans

Chimaeras lack scales on most of the surface, have a fleshy operculum that hides their gill slits, and the spiracle is closed. The upper jaw, unlike that of elasmobranchs, is solidly fused with the cartilaginous braincase; and instead of teeth, hard flat bony plates on the jaws crush molluscan shells, the diet of chimaeras being similar to that of rays.

SARCOPTERYGII (LOBE-FINNED FISHES)

Sarcopterygians are bony fishes that have a prominent fleshy lobe at the base of their paired fins. The lobe contains part of the fin skeleton (figs. 4.11 and 10.45*a*). Most lobe-finned fishes also have internal nares that open into the oropharyngeal cavity, and they have a gas-filled swim bladder. The gill slits are covered by a bony operculum that grows caudad from the second pharyngeal arch. There are two major taxa, **Crossopterygii,** which includes the ancestors of amphibians, and **Dipnoi,** the lungfishes. The two were distinct in the early Devonian. Their pre-Devonian ancestors are unknown.

Crossopterygians

All crossopterygians are extinct with one exception, the relict coelacanth, *Latimeria*. The skeletal elements within the fin lobes of crossopterygians correspond closely to the proximal skeletal elements of early tetrapod limbs (fig. 10.45); the skull was similar to that of the first amphibians (fig. 9.8); they had swim bladders that were probably used at least occasionally as lungs; and most had internal nares, although these may not have been used for breathing. Conventional classifications place crossopterygians in several subtaxa, including rhipidistians and coelacanths. **Rhipidistians,** which are freshwater species, are thought to have been the ancestors of amphibians.

Dipnoans

These lobe-finned fishes are sometimes referred to as "true" lungfishes to distinguish them from those few ray-finned fishes that use swim bladders for respiration in times of drought only. There are three living genera, *Protopterus* from Africa, *Lepidosiren* from Brazil, and *Neoceratodus* from Australia. The gills are inefficient, and *Protopterus* and *Lepidosiren* would suffocate if held under water, since they are dependent on air and their swim bladders for oxygen. *Neoceratodus,* on the other hand, utilizes gills except when the oxygen content of the water is low. During the wet season, dipnoans inhabit streams and swamps, but when the rains cease and the tropical sun dries up their environment, the African and Brazilian species dig deep burrows in the mud and spend the dry hot season in a state of aestivation. Their lowered metabolism minimizes water loss and reduces the need for nutrients and oxygen.

Dipnoans resemble amphibians in many respects. In both dipnoans and amphibians, the swim bladder or lungs have ducts leading to the pharynx with blood supplied by a branch from the sixth aortic arch instead of from the dorsal aorta as in crossopterygians. The atrium of the heart is partially divided into two chambers. Both usually have a larval stage with external gills and both have internal nares.

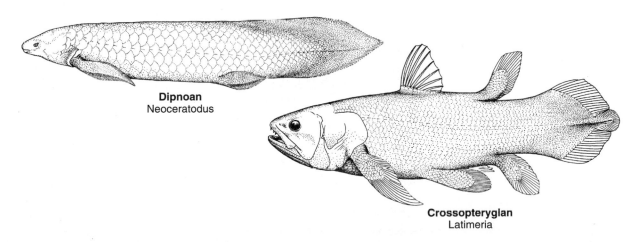

Dipnoan
Neoceratodus

Crossopterygian
Latimeria

FIGURE 4.11

Two sarcopterygians (lobe-finned fishes). A modern dipnoan and a relict crossopterygian.

ACTINOPTERYGII (RAY-FINNED FISHES)

Actinopterygians are ancient and modern bony fishes whose membranous fins are supported by slender fin rays radiating from basal skeletal elements within the body wall (fig. 10.14). Therefore, the fins are not lobed. Their gill slits are covered by a bony operculum, a swim bladder is usually present, and internal nares are lacking. During the Paleozoic their bony dermal armor and scales were overlain with an enameloid called ganoin (fig. 6.29*a–c*), and their caudal fins, like those of Paleozoic sharks, were heterocercal (fig. 4.12, sturgeon). Both traits have all but disappeared in more recent species. Actinopterygians have been placed in two major taxa, **Chondrostei** and **Neopterygii.**

Chondrosteans

Chondrosteans are the oldest known ray-finned fishes. They are chiefly **paleoniscoids**—Paleozoic fishes that flourished from the Devonian until the beginning of the Mesozoic, after which only relicts (remnants) remained (fig. 4.2). These are the African freshwater fishes *Polypterus* and *Calamoichthys,* and the sturgeons and paddlefishes. Representative species of these relicts are seen in fig. 4.12.

Polypterus and *Calamoichthys* have features of the paleoniscoids—large ganoid scales, a well-ossified endoskeleton, and swim bladders that are connected to the pharynx by an air duct that enables aerial respiration. They are sometimes called "African lungfishes," but they shouldn't be confused with dipnoans.

Sturgeons and **paddlefishes** have an endoskeleton that is largely cartilaginous, since most of the embryonic cartilage is not replaced by bone as in other bony fishes. The scales lack ganoin; and the skin of paddlefishes is naked except for small bony scales on the tail. The taxon Chondrostei is a mostly extinct nonmonophyletic assemblage of much structural diversity and the subject of much disagreement.

Neopterygians

The subclass Neopterygii includes the surviving Mesozoic ray-finned fishes—gars (two genera, including *Lepisosteus*) and bowfins (one species, *Amia calva*)—and the vast array of more recent ray-finned fishes, the **teleosts** (figs. 4.13 and 4.14). Gars and bowfins, classified as **Holosteans** in conventional evolutionary taxonomies, are placed in a cladistic order **Semionotiformes.** The latter classification has been adopted in this text. Teleosts are placed in the taxon Teleostei (appendix 1).

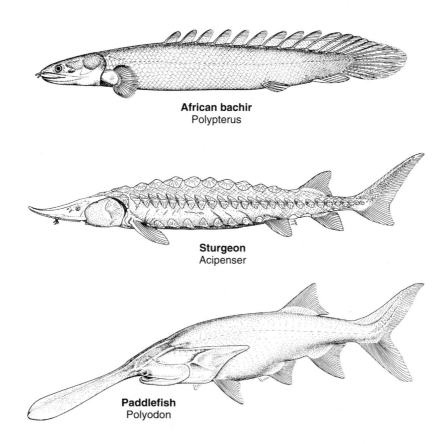

African bachir
Polypterus

Sturgeon
Acipenser

Paddlefish
Polyodon

FIGURE 4.12
Three surviving chondrosteans.

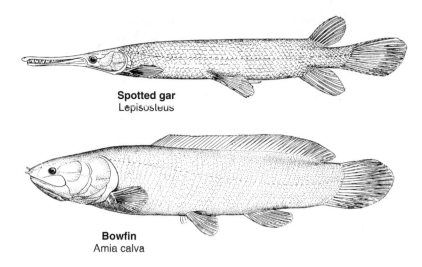

FIGURE 4.13

The two surviving genera of Semionotiformes (holostei).

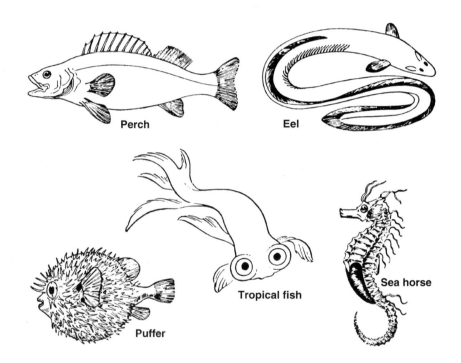

FIGURE 4.14

Group of teleosts.

Gars and **bowfins** are freshwater fishes. The trunk and tail of gars are covered with ganoid scales that are modifications of the ganoid scales of paleoniscoid fishes (see fig. 6.29*b* and *c*). The trunk and tail of *Amia* have modern fish scales. Much of the endoskeleton is ossified in both genera, but the braincase (neurocranium) remains largely cartilaginous throughout life.

Teleosts are the most recently evolved ray-finned fishes, and hence are referred to as "modern." From the perspective of geologic time, they are modern indeed, although some have been around for 65 million years, more or less. Their scales, which are in the dermis, have become very thin and flexible; the dermal bones of the skull are generally thinner and more numerous than in other bony fishes; the jaws and palate have become more independently maneuverable; the pelvic fins in many species are far forward; and the body has become altered in innumerable other ways, resulting in organisms that occupy all aquatic niches on the planet. The number of species has been exploding since the Mesozoic (fig. 4.2), and the more than 23,000 extant species now constitute 96 percent of all living fishes.

There are long, slim teleosts that have lost paired appendages; short, fat ones with sails; transparent ones; fishes that stand on their tails; fishes with both eyes on the same side of the head; fishes that carry lanterns, that climb trees, that carry their unhatched eggs in their mouth, that appear to be smoking pipes, that possess periscopic eyes; and hundreds of other bizarre genera. They inhabit the abyssal depths far out from the continental shelves, cavort in modest brooks, and make nightly sorties onto land. They display a range of coloration, although a relatively small number of pigments assisted by myriads of light-dispersing crystals are responsible for all hues. Figures 4.15 and 4.16 are postulated lineages of fishes.

AMPHIBIA

Amphibians have been placed in three subclasses: **Labyrinthodontia,** the oldest tetrapods, **Lepospondyli,** a diverse group that was at its height in the upper Carboniferous, and **Lissamphibia,** which includes

Triassic and modern amphibians. Successive mutations and natural selection increasingly adapted labyrinthodont descendants for terrestrial life.

Labyrinthodonts

The oldest amphibians were the swamp-dwelling labyrinthodonts (fig. 4.17), so named because the dentin of their teeth was folded in such a manner as to resemble a labyrinth when viewed in cross section. There are three orders: **Ichthyostegalia,** the oldest, which appeared in the Devonian, **Temnospondyli,** a group that was common in the Permian, and **Anthracosauria,** a relatively small Paleozoic group thought to be in a direct line to reptiles.

Labyrinthodonts were a large, widely dispersed, and diverse assemblage, and their kinships are unclear because of a lack of fossil evidence. The structure of labyrinthodont vertebrae has played a prominent role in attempts to reconstruct labyrinthodont lineages and, especially, to identify the labyrinthodont ancestors of reptiles. On the basis of the morphology of labyrinthodont vertebrae, paleontologists have been of the opinion

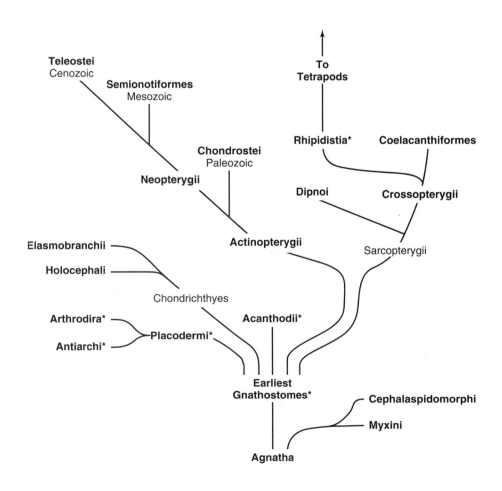

FIGURE 4.15

One postulated ancestral lineage of fishes. Asterisked groups are extinct.

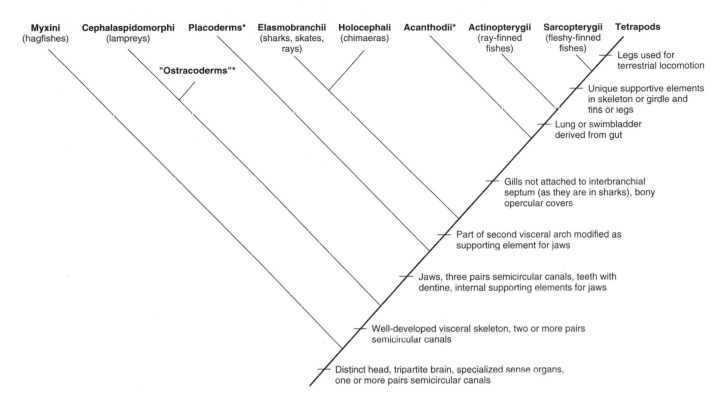

FIGURE 4.16

Cladogram of fishes. Diverging lines are postulated clades. Extinct groups are indicated by asterisks. Some of the shared derived characters that mark the branchings are shown to the right of the branch points.

From Cleveland P. Hickman and Larry S. Roberts, *Animal Diversity.* Copyright © 1995 Times Mirror Higher Education Group, Inc., Dubuque, Iowa. All rights reserved. Reprinted by permission.

FIGURE 4.17

Reconstruction of the early labyrinthodont *Ichthyostega* from the Devonian of Greenland. The hind limbs had seven digits and the forelimbs had at least six, rather than the five that had been found at the time the skeleton was reconstructed.

that stereospondyls and embolomeres were not in the reptilian line (figs. 4.18 and 8.10).

Labyrinthodonts had many features seldom seen in modern amphibians. Among them were minute bony scales in the dermis of the skin, a fishlike tail supported by dermal fin rays, and skulls similar to those of rhipidistian crossopterygians. Grooves in skull bones that lay just under the skin show that labyrinthodonts, like crossopterygians, had a sensory canal system of neuromast organs that monitored the aquatic environment. Today's aquatic amphibians have this system; terrestrial species lose it at metamorphosis. Labyrinthodonts were as small as today's newts and as large as today's largest crocodiles. One or another of them was ancestral to the first reptiles.

Lepospondyls

These were a diverse assortment of small amphibians that became extinct before the last of the labyrinthodonts. Among them were species that lacked limbs, species resembling salamanders, and species with large, bizarre, triangular skulls unlike those of any extant amphibian. These and other features suggest that at least some lepospondyls may have arisen directly from a crossopterygian ancestor independent of other labyrinthodonts, and that today's urodeles and, perhaps, apodans are their descendants. This is the **theory of the diphyletic origin of Amphibians** (fig. 4.18, broken lines).

Lissamphibians

Lissamphibians include the three orders of extant amphibians: **Apoda (Gymnophiona),** the limbless, burrowing caecilians; **Urodela (Caudata),** the tailed amphibians; and **Anura (Salientia),** the frogs, toads, and tree toads. The taxon also includes Triassic and Jurassic anurans whose skeletons resemble those of today's frogs and toads. According to the **theory of the monophyletic origin of Amphibians,** all three orders stem from a common crossopterygian ancestor (fig. 4.18). The ancestry of apodans is purely conjectural because only a single fossil remnant has been found, and it consists solely of vertebrae from the Cenozoic.

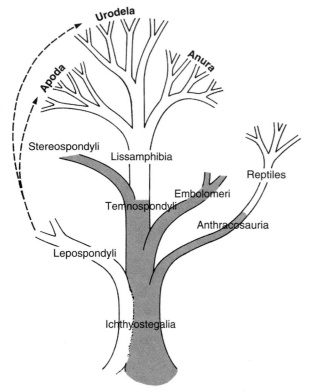

FIGURE 4.18

One of several postulated genealogies of amphibians. Broken lines indicate alternative hypotheses.

TABLE 4.1   Distribution of Gills, Pharyngeal Slits, and Lungs among Adult Urodeles

Family	Representative Genera	Number of Pairs		
		Gills	Slits	Lungs
Proteidae	*Necturus*	3	2	Yes
Amphiumidae	*Amphiuma*	0*	1	Yes
Hynobiidae	*Hynobius*	0*	0	Occasionally
Cryptobranchidae	*Cryptobranchus*	0*	1	Yes
Salamandridae	*Notophthalmus*	0*	0	Yes
Ambystomatidae	*Ambystoma*	0*	0	Yes
Plethodontidae	*Plethodon*	0*	0	No
Sirenidae	*Siren*	3	3 to 1	Yes

*Some individuals are occasionally perennibranchiate.

FIGURE 4.19

An apodan. The annular structure probably aids in burrowing.

Apodans

Apodans (caecilians) are circumtropical, limbless amphibians that, except for aquatic species, live in burrows in swampy locations (fig. 4.19). Their eyes are small and, in some of the 50 or more species, buried beneath the bones of the skull. (Was this the cause, or is it the effect, of a subterranean habitat? The explanation is surely less simplistic.) Some species have minute scales in their dermis. Most caecilians are about 30 cm long, but some attain a length of more than a meter and have as many as 250 vertebrae. The vent is almost at the end of the body, hence the tail is very short. Terrestrial species lay large yolky eggs, and the larval stage is passed in the egg envelopes. Several aquatic genera are viviparous.

Urodeles

There are eight families of urodeles (Table 4.1), six of which are represented in figure 4.20. Two families, **Proteidae** and **Sirenidae** (represented by *Necturus* and *Siren*) are perennibranchiate; that is, they retain larval gills and one or more gill slits throughout life. The other families may have a few perennibranchiate populations in one locality or another, and sometimes in the same pond with fully metamorphosed individuals. In some localities, perennibranchiates undergo complete meta-

morphosis when administered iodinated compounds or thyroid hormone.

Necturus is the only genus of Proteidae in North America. *Necturus* attains sexual maturity in about five years. *Proteus,* its European relative, is blind and lives in dark caves. *Siren* (family Sirenidae) lacks hind limbs and lives in muddy ditches and weed-choked lakes or ponds.

Plethodon, like most other members of its family, loses its gills at metamorphosis but fails to develop lungs. It lives in moist sites and, except in the tropics, lays its eggs under damp logs or in caves. Respiration is via the skin. Larvae hatch with legs already formed and may never enter water.

Amphiuma is an eel-like urodele up to 1 m long, but with tiny appendages. *Amphiuma pholeter* has one digit. *A. means* has subspecies with two or three digits.

Cryptobranchus looks ferocious because of its broad flat head and wrinkled skin, which often conceals a single gill slit, if one is present.

Hynobius, a primitive urodele, and *Ambystoma* are terrestrial genera. However, *A. mexicanum,* the Mexican axolotl, is an aquatic perennibranchiate that can be

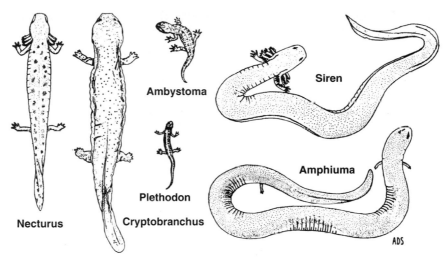

FIGURE 4.20
Representative urodeles.

caused to discard its gills experimentally. (Xolotl was the Aztec god of twins and monsters.)

Notophthalmus and *Salamandra* are true salamanders (family **Salamandridae**). *Salamandra atria* is viviparous. *Notophthalmus viridescens* (fig. 4.21) has an interesting life history. After several months of an aquatic larval existence, the red-spotted newt loses its gills and gill slits, sprouts legs, and, now an **eft,** leaves the pond for life on land. The skin develops a thick cornified layer that blocks the openings of the sensory canal system and skin glands, the body assumes a bright orange-red color, and a series of dorsolateral black-bordered red spots appear. The red eft stage lasts one to three years, depending on the locality.

At the approach of sexual maturity, the eft, stimulated by the hormone prolactin, joins a mass migration to freshwater ponds where mating occurs and eggs are deposited. The tail commences to change from round to laterally compressed with dorsal and ventral keels, the thick cornified layer of epidermis is shed, exposing the openings to mucous glands and the sensory canal system, and the body gradually becomes olive green above and light yellow below, a protective coloration appropriate for life in a freshwater pond. The animal is now a **newt.** In some localities the larvae remain in the pond, mature, and retain vestigial gills throughout life.

Anurans

Anurans are tailless amphibians in which several caudal vertebrae are fused into one elongated urostyle (fig. 8.13). No single morphological trait distinguishes frogs from toads. The most representative frogs are the long-legged, slender-bodied members of the family **Ranidae.** The most representative toads are in the family **Bufonidae.** Toads are more terrestrial than frogs. Tree toads, also called tree frogs, are in the family **Hylidae.** Anurans breathe with lungs and skin and most are

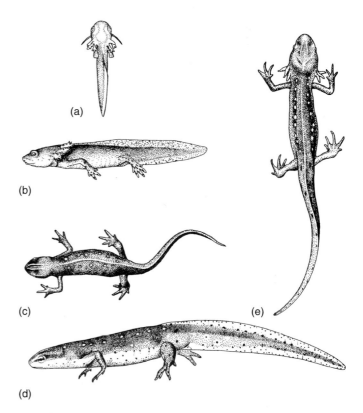

FIGURE 4.21

Life history of *Notophthalmus.* (*a*) Newly hatched larva (7 mm), balancers behind eyes, gills only slightly branched, forelimbs are blunt buds, hind limbs not formed, keeled back and tail. (*b*) Late larva. (*c*) Terrestrial eft. (*d*) Adult male (95 mm). (*e*) Neotenous adult. All to same scale, except (*a*).

amphibious. A few, such as *Rana cancrivora*, the crab-eating frog of Thailand, can tolerate salt water.

Most anurans breed only during or shortly after a rain and deposit their jelly-encapsuled eggs in water-filled ditches or in ponds. Tadpoles emerge from the jelly envelopes and further develop in the water. Robber frogs, however, lay their eggs in rain-filled crevices, in a pile of leafmold, or at the base of a grass hummock away from the water's edge and above flood level. The larval stage therefore lacks an aquatic environment. As an adaptation, the tadpoles remain within the jelly envelopes until metamorphosis. The interval between egg laying and completion of metamorphosis is much shorter than in other anurans. Some species of tree toads also do not lay their eggs in bodies of water. Instead, they carry developing eggs in a brood pouch under the skin of their back and fully metamorphosed tree frogs emerge through an opening in the skin. A similar condition is found in aquatic frogs of the genus *Pipa.* The East African toad *Nectophrynoides vivipara* is, as the name implies, viviparous. As many as 100 young develop in the uterus.

The earliest known anuran, *Triadobatrachus* (= *Protobatrachus*) is from the Triassic. The skull was quite similar to that of today's anurans and the body was similarly shortened as a result of a reduction that had already taken place in the number of trunk vertebrae. The ribs were not as foreshortened, the tibia and fibia were not fused, and the abdomen was covered by bony scales.

REPTILIA

From some group of early labyrinthodonts, perhaps anthracosaurs, there arose during the early Carboniferous a group of tetrapods destined to be named more than 300 million years later, **cotylosaurs** (fig. 4.22a). These were the stem reptiles. From them evolved a varied and abundant assemblage of descendants who established themselves securely on land, conquered the air, and reinvaded the seas, flourishing as a tetrapod majority for 260 million years until the mass extinction of

reptiles that ended the Cretaceous. Remaining are a relatively few descendants, the **turtles,** an ancient group that thus far has doggedly persevered in the struggle for existence; *Sphenodon,* a lizardlike relict; **modern lizards,** recent additions to reptilian society; **snakes and amphisbaenians,** which are lizards deprived of their appendages; and **crocodilians.** These survivors mirror the effects of mutations and of a changing environment on cotylosaurs and their descendants. Birds and mammals represent further mutations of reptilian genes.

Reptiles underwent changes that made them better adapted for a permanent terrestrial existence than were their amphibious ancestors. Perhaps the major one was acquisition of three extraembryonic membranes, the **amnion, chorion,** and **allantois,** which emancipated them from the necessity of returning to water to lay their eggs. The amnion is a fluid-filled membranous sac in which the embryo develops (fig. 5.13b). The fluid is secreted by the lining of the sac. To state the situation figuratively, instead of the mother having to go to water to deposit her eggs, the embryo is provided with its own private pond, and the egg can be laid on land. Thus, reptiles became the first amniotes. The chorion and allantois constitute a vascularized membrane that lies against the porous eggshell, another reptilian innovation, taking the place of larval gills for respiration.

Because of these three extraembryonic membranes, oviparous reptiles were able and, indeed, forced, to lay their eggs on land. The young hatch fully formed, without a larval stage, ready to seek their own food (fig. 4.23). Not only were reptiles liberated from returning to water to lay their eggs, aquatic oviparous species must go onto land to do so, since porous eggs would become waterlogged in an aqueous environment.

Reptiles exhibit other adaptations to terrestrial life. The body surface is covered with a thick layer of cornified epidermal cells that is organized into plaques, shields, or surface scales unlike the bony dermal scales of fishes and

(a)

(b)

(c)

FIGURE 4.22

(a) Stem reptile—an extinct cotylosaur. (b) *Sphenodon*—a "living fossil." (c) Modern iguanid lizard.

FIGURE 4.23

Siamese crocodile emerging from egg.

Courtesy of Metrozoo, Dade County, FL.

early amphibians. The scales are impervious to water, which results in water conservation, a necessity for animals living in air. Reptiles have developed a neck by specialization of several postcranial vertebrae. This modification, combined with their single occipital condyle, enables reptiles to scan the horizon. The pelvic girdle now articulates with two sacral vertebrae, providing a stouter brace for more powerful hind limbs. The digits are supplied with claws, and a new kidney, the metanephros, has come into existence as a modification of the amphibian kidney. The heart is partially or completely divided into right and left chambers, thereby separating the systemic and pulmonary circulations to an extent not achieved in anamniotes.

Living reptiles, like fishes and amphibians, are **ectotherms,** that is, they cannot maintain a more or less constant body temperature in the face of variations in the ambient temperature. *Sphenodon* and some lizards have retained a parapineal organ that is exposed to the environment and aids in thermoregulation by monitoring the duration of exposure of the animal to the sun's rays (fig. 17.19).

These then are the reptiles: scaly, clawed, mostly terrestrial tetrapods lacking feathers and hair, which,

excepting viviparous species, lay macrolecithal, shell-covered (**cleidoic**) eggs on land, the embryos of which develop within an amnion, and the young of which are hatched fully formed. We will look briefly at the three subclasses, **Anapsida, Diapsida,** and **Synapsida,** which have been erected on the basis of the number of fossae in the temporal region of the skull. Anapsids have none, synapsids have one, and diapsids have two, with modifications in some recent species. (You may wish to consult appendix 1 for the derivation of the terms.)

Anapsids

Anapsids have no fossae in the temporal region of the skull (figs. 9.15 and 9.16). This was the primitive condition exhibited in cotylosaurs, all of whom are included in the taxon. The diapsid and synapsid conditions were evolutionary descendants. Turtles are the sole living members of the subclass.

Turtle skulls are enigmatic. There is no evidence of a temporal fossa; but unlike in any other reptile, there is a deep cavitation on each side of the midline at the rear of the skull (fig. 9.14*b*). The phylogenetic history of this condition is unknown. Consequently, in some taxonomies turtles (**Testudinata**) are assigned a separate subclass status.

Diapsids

The subclass Diapsida includes all living reptiles except turtles. It also includes a vast array of extinct reptiles that ruled the land, swam the seas, and flew in the air during the Mesozoic. The subtaxa **Archosauria** and **Lepidosauria** include extant reptiles. Two additional major taxa include **plesiosaurs** (Sauropterygia) and **ichthyosaurs** (fig. 4.24).

Archosaurs

Archosaurs, which include thecodonts, pterosaurs, dinosaurs, and crocodilians, were the dominant land vertebrates during the Mesozoic. Today's descendants are the modern crocodilians and birds (fig. 4.24). **Thecodonts** were the stem archosaurs. They are so named because their teeth were set in sockets. **Pterosaurs**, like birds, had pneumatic bones, but their

FIGURE 4.24

Postulated relationships of selected reptiles. The major taxa (*red*) reflect the architecture of the temporal region of the skull. Gray indicates surviving reptiles.

wings were more like those of bats, being supported by an elongated fourth finger (fig. 10.28*b*). The largest had a wingspread of 10 m (35 ft).

Saurischian and **ornithischian dinosaurs** are differentiated by the structure of their pelvic girdles. Those of ornithischians are structurally similar to those of birds, their ischial and pubic bones being parallel to each other and directed caudad as opposed to those of saurischians (fig. 4.25). Dinosaurs came in many sizes and body builds and many were bipedal. Massive ones did not represent the majority. Many saurischians were swift predatory carnivores, whereas ornithischians were herbivores lacking front teeth, although some had horny beaks. A small unidentified bipedal saurischian is thought to have given rise to birds. The hypothesis is based in part on the similarity of the pelvis in the two groups.

Crocodilians include the subtropical and tropical alligators, crocodiles, caimans, and a single species of gavials. Crocodiles can be distinguished from alligators by the shape of the snout, which is more slender and triangular in crocodiles, broad and rounded in alligators; and by the enlarged fourth tooth of the lower jaw, which in crocodilians fits into a notch on the upper jaw and can be seen when the mouth is closed, whereas in alligators it fits into a pit medial to the upper jaw tooth line and cannot be seen. Caimans and alligators look very much alike. The gavial snout is long and very slender because the left and right halves of the mandible are united by a long symphysis that extends all the way back to the fifteenth tooth.

Lepidosaurs

There are two extant groups of Lepidosaurs: rhynchocephalians and squamates (fig. 4.24). **Rhynchocephalians** are primitive lizardlike reptiles, but with quite different scales, teeth, and internal morphology than modern lizards. **Squamates** are the modern lizards, snakes, and amphisbaenians. Rhynchocephalians retain the diapsid skull of their ancestors. Squamates have a diapsid skull that has been adaptively modified.

Rhynchocephalians   The only surviving rhynchocephalian is *Sphenodon*, commonly known as a tuatara (fig. 4.22*b*). Sphenodons feed on small vertebrates and insects, and attain a length of 0.75 m (2.5 ft). They become sexually mature at 20 years of age, and their normal life span may exceed 60 years. They may face extinction in their only surviving habitats in New Zealand and a few adjacent coastal islands.

Squamates   Of all reptiles, lizards are the most versatile. Because of well-developed appendicular muscles and a suitably constructed skeleton some run agilely on their hind limbs; some are amazing broad jumpers; some, the nocturnal geckos, have suction discs on their toes that adhere to smooth vertical surfaces; and some like *Draco*, the flying dragon, glide through the air on rib-supported extensions of the lateral body wall. A few lizards are either limbless or have only vestiges of limbs. Some lizards are blind, and some have transparent eyelids **(spectacles)**. There is a third eyelid, the **nictitating**

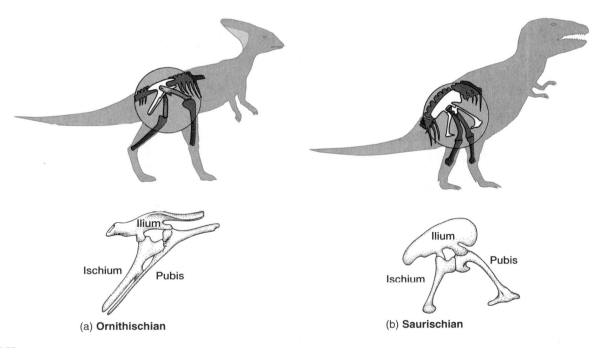

FIGURE 4.25

Lateral view of the pelvis distinguishing two dinosaur groups. (*a*) A birdlike pelvis in which the pubis is parallel to the ischium and points to the rear. (*b*) A lizard pelvis—the pubis points ventrad.

membrane, and the teeth are in sockets. Iguanid lizards are more generalized. *Iguana* is an arboreal, strictly vegetarian species that enters seawater to forage on seaweed. The largest lizard, the Komodo dragon of Indonesia, reaches 2.75 m (9 ft) in length and 115 kg (250 lbs) in weight. The horned toad of the North American desert, *Phrynosoma,* is a lizard.

Snakes evolved from lizards. Having lost limbs, they have acquired other modes of locomotion, including a unique set of muscles that connects the ribs with the large ventral scales, or scutes. Despite the absence of limbs, snakes inhabit mountains, deserts, trees, freshwater, and the sea. Most sea snakes are viviparous, thereby eliminating the necessity to return to land to lay their eggs.

Amphisbaenians are subterranean lizards 30 to 70 cm long (1 to 2 ft or more), mostly limbless, with annulated bodies like those of the burrowing apodans. The eardrum and eyes are covered with opaque skin. There are about 120 species.

Plesiosaurs and Ichthyosaurs

Plesiosaurs **(Sauropterygia)** and ichthyosaurs are marine reptiles. They were contemporaries of the dinosaurs but, unlike the latter, they had a dorsal temporal fossa on each side (a "euryapsid" skull). It has been postulated that the single dorsal fossa is a result of the loss of a lower fossa that was present in a diapsid ancestor. For that reason these reptiles are classified with the diapsids. It is also hypothesized that the euryapsid condition was a result of evolutionary convergence, and that the two groups were not closely related.

Synapsids

Synapsid reptiles were diverging from anapsid ancestors during the Carboniferous. They are the reptiles from which mammals emerged during the Triassic. The synapsid lineage began with **pelycosaurs,** which were transitional between amphibians and cotylosaurs. The pelycosaurs were succeeded by the **therapsids** (fig. 4.26), to which mammals appear to be closely related on the basis of comparative morphologic data. Like mammals, therapsids had two occipital condyles, a secondary palate, and heterodont dentition with incisors, canines, and grinding molars (cheek teeth). The dentary was the largest bone in the lower jaw, presaging the condition in modern mammals (fig. 9.39c). The distribution of selected reptiles in geologic time is shown in fig. 4.27.

FIGURE 4.26

(*a*) Pelycosaur, a Carboniferous synapsid. (*b*) A Triassic therapsid about the size of a large dog. Nothing is known of the skin.

(a)   (b)

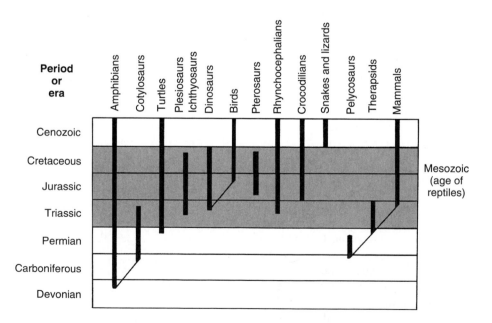

FIGURE 4.27

Range of selected reptiles through time. Connecting lines indicate postulated lineages.

AVES

Birds are endothermic vertebrates with feathers. **Endothermy** is the ability to maintain a relatively stable body temperature despite fluctuations in the ambient temperature. A less precise term is "warm blooded." Birds are descendants of a bipedal archosaur, perhaps an ornithischian dinosaur, if the similarity of their pelvic girdles is valid evidence. Bipedal ornithischians could stand on their hind limbs, which freed the forelimbs for other functions including, ultimately, flight in pterosaurs and birds. Birds have retained reptilian scales on their legs, feet, and beak, a single occipital condyle, and a diapsid skull that has been modified by partial or complete loss of the zygomatic arch.

Feathers are exquisitely structured keratinized integumentary appendages that replace reptilian scales wherever feathers are present. Feathers make flight possible in birds. Whereas pterosaurs and bats have a wing membrane stretched between the forelimbs and trunk, or even the tail, that serves as an airfoil, feathers attached to the bones of the forearm and hand are the only airfoil birds have. Feathers also insulate against seasonal heat and high-altitude cold. (When birds produce excessive heat during flight, it is eliminated in expired air.) Feather pigments facilitate recognition by other members of the species and often provide protective coloration. Plumage coloration can vary by age and sex, thereby signaling the reproductive state of the individual.

In addition to feathers and wings, birds have other adaptations for flight. Body weight has been reduced in several ways. Long bones have become slender, and most bones, including vertebrae, lack the central marrow, which leaves cavities that contain air-filled extensions of air sacs from the lungs. The skull has been lightened by a thinning of the compact layers of the membrane bones, but it remains durable because sutures have been eliminated in adults, and the spongy bone that remains is architecturally strong. The bones of the wrist, palm, and digits have been reduced in number. Teeth do not develop; no urinary bladder develops; the large intestine has been shortened. All these modifications, and others, have reduced the cost of flight in terms of energy expended. There are two generally recognized subclasses of birds, **Archaeornithes** and **Neornithes.**

Archaeornithes

Two fossil birds about the size of a crow were found in Late Jurassic limestone deposits in Bavaria, Germany, in the nineteenth century. These were given the appropriate name *Archaeopteryx*, the first birds (fig. 4.28). The first specimen was complete except for a few cervical vertebrae, the right foot, and lower jaw. The second specimen was even more complete; and five additional specimens, not exactly alike, have been recovered elsewhere. *Archaeopteryx* had a long reptilian tail, thecodont teeth on

FIGURE 4.28

Fossil *Archaeopteryx* embedded in limestone. Wings and their skeleton are near bottom of photograph; the long feathered tail is at top.

Courtesy of Dr. John Ostrom, Yale Peabody Museum. Portrait O. C. Marsh, photograph *Archaeopteryx.*

both jaws, and feathers on the wings and tail that were no different from today's. The skull was more reptilian than avian, the nostrils were far forward, there was no beak, and the braincase had not expanded to accommodate an enlarged brain. The cervical vertebrae were not saddle shaped at the ends as in today's birds, trunk vertebrae were not rigidly fused, the synsacrum was not well developed, and the sternum was small, except in the last specimen recovered. The small sternum, which could not have accommodated strong pectoral muscles for sustained flight, suggests that these birds may have soared more than they flew. Figure 4.29 contrasts some of the more obvious features of the skeleton with that of a pigeon.

In 1986, two crowlike fossil skeletons 75 million years older than *Archaeopteryx* were discovered in a mudstone quarry in Texas. These specimens, for which the genus *Protoavis* was proposed, were more dinosaur-like than is *Archaeopteryx,* and they had smaller wings. Whether or not they should be regarded as birds is in dispute. A *Protoavis* feather would be helpful!

Neornithes

The subclass Neornithes includes three major subgroups, **Odontognathae, Palaeognathae,** and **Neognathae.** Odontognaths were toothed marine birds

(b)

(a)

FIGURE 4.29

(*a*) *Archaeopteryx*, from the Jurassic. (*b*) A carinate bird (pigeon) for comparison.

From E.H. Colbert & M. Morales, *Evolution of the Vertebrates: A History of Backboned Animals Through Time,* 4th edition. Copyright © 1991 John Wiley & Sons, New York, NY. Reprinted by permission of John Wiley & Sons, Inc.

that used the land as a base for marine operations. The only known odontognaths are *Hesperornis* and *Ichthyornis* species. The first *Hesperornis* fossils were recovered from rocks formed 75 million or more years ago at the bottom of what was then a shallow North American sea extending inland northward as far as Kansas. Similar fossils have since been recovered in Europe. *Hesperornis* was covered with small, hairlike feathers. It had vestigial wings, so it couldn't fly; but it had stout legs for wading, and it was a good diver. Its diet consisted of fishes that it caught with its sharp, pointed teeth. *Ichthyornis* was an active flier that was able to go far offshore to feed.

Palaeognaths, known generally as ratites, have small incompetent wings, but they have powerful leg muscles that enable them to run well. They are descendants of active fliers. Many are known as fossils only, having been depleted by human society. Among current survivors are the rheas, ostriches, emus, and cassowaries. The extinct moas of New Zealand were more than 3 m (8 ft) tall and

they laid eggs more than 30 cm (1 ft) in diameter. Ostrich eggs weigh about 3 pounds.

Neognaths are, for the most part, birds that have a large carina to which relatively massive flight muscles attach. They are therefore generally known as carinates. There are about 10,000 species and they include all living birds except the palaeognaths. The largest living carinate is the Andean condor, with a wingspread of 3 m (10 ft) and weight of up to 14 kg (30 lbs). The Giant Teratorn, a vulture that lived in Argentina five million years ago, had a wingspread of more than 7 m (25 ft) and an estimated weight of 23 kg (50 lbs). Penguins have a large carina, but their forelimbs have become flippers, so they cannot fly. However, they are powerful swimmers.

Many carinates are annual migrants, a behavioral adaptation to predictable annual climatic changes. The Arctic tern spends several months above the Arctic Circle and the remaining months in the Antarctic, traveling 22,000 miles round trip each year! During migration, birds move in mass flights, often at night and at an elevation of approximately 600 m (2,000 ft). Migration is physiologically demanding and birds have become able to store and rapidly metabolize large energy reserves. An internally regulated annual cycle, entrained by seasonal environmental cues, coordinates the physiological and behavioral events associated with migration. How birds navigate has long challenged ornithologists. The sun, stars, barometric pressure, polarized light, naturally fluctuating magnetic fields, and low-frequency sound all seem to provide input, probably along with other cues at any time or location.

The number of bird species increased greatly in the Late Mesozoic. The spread of human habitation with the resulting destruction of the foraging and breeding sites is the major threat to birds in the twenty-first century.

MAMMALIA

Mammals succeeded therapsid reptiles at the end of the Triassic. They are amniotes with a synapsid skull, hair, and, except monotremes, mammary glands and nipples. Further distinguishing modern mammals from other vertebrates are the single dentary bone on each side of the lower jaw articulating with the squamosal bone; three bones in the middle ear cavity; a muscular diaphragm separating thoracic and abdominal cavities; sweat glands (in most mammals); absence of an adult cloaca in all but the lowest order; heterodont dentition (except in toothed whales); only two sets of teeth (milk teeth and a permanent set) instead of continual replacement as in lower vertebrates; biconcave, enucleate red blood cells that are circular (except in camels and llamas); loss of the right fourth aortic arch; a sound-collecting lobe (pinna, or auricle), accessory to the outer ear; a more specialized larynx; and extensive development of the cerebral cortex.

Because of many variations in limb structure, mammals have been able to achieve a greater diversity of habitats than any other tetrapod. They burrow in the ground, hop, lumber, or gambol over the plains, scramble over mountain crags, swing through trees, propel themselves through the air in true flight, and inhabit the oceans, each lifestyle made possible by modifications of body structure.

Some taxonomies divide mammals into two major groups, **Prototheria,** which lays eggs and has a cloaca throughout life, and **Theria,** which gives birth to its young (fig. 4.30). Living prototherians are in the class **Monotremata.** Therians are in two groups: those that have a yolk sac placenta (**Metatheria)** and those that have a chorioallantoic placenta (**Eutheria).**

Monotremata

Monotremes and therians are thought to have diverged early in mammalian evolution. Their name mirrors the fact that the cloaca has a single opening to the exterior (fig. 15.48*a*). The sole surviving monotremes are the platypus, or duckbill, *Ornithorhynchus,* and two genera of spiny anteaters, *echidnas,* all from Australia or nearby

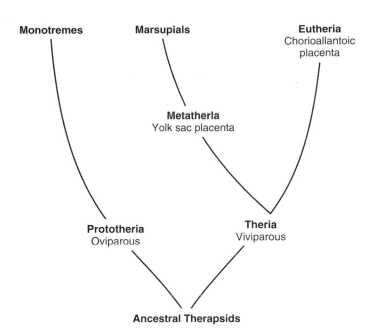

FIGURE 4.30

Major categories of living mammals.

Tasmania and New Guinea (fig. 4.31). Like reptiles, they lay heavily yolked eggs; they have a ventral mesentery that extends the length of the abdominal cavity; the testes are confined to the abdominal cavity; the outer ear has no pinna; and the brain lacks the great transverse fiber tract, the corpus callosum, that connects the two cerebral hemispheres in other mammals, except marsupials. The malleus and incus are larger than in other mammals, resembling the articular and quadrate bones of therapsids. They have no nipples; a milky fluid exudes from modified sweat glands onto tufts of hairs in shallow pits on the abdomen, from which it is licked up by the young. Monotremes are endothermic, but their body temperature is less stable than that of most placental mammals.

Duckbills live in pairs in burrows up to 18 m (60 ft) long, in the banks of creeks and rivers, and spend most of their life in the water. They have webbed feet that enable them to walk on muddy river bottoms, and a soft, rubbery, sensitive, beaklike muzzle with which they detect invertebrates, especially molluscs, in the mud and dredge them. The food is then stored in a cheek pouch until it is crushed later with horny teeth that replace a prenatal set of regular ones. During breeding seasons, the female moves out of the burrow and constructs a new burrow just above the water line, where she lays one to three eggs in a nest she has constructed. The eggs are nearly round, about 2 cm in diameter, and covered with a pliable white shell. She incubates the eggs for 10 to 14 days, after which they hatch.

Echidnas are terrestrial and have long, sticky tongues and stout claws that are used to gather ants and termites. Except on the abdomen they are armed with sharp quills interspersed among coarse hairs, and they roll into a ball

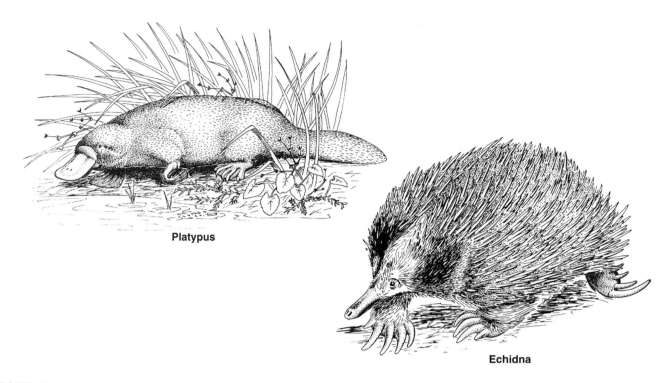

Platypus

Echidna

FIGURE 4.31

The two surviving monotremes, the platypus *Ornithorhynchus,* and a spiny anteater.

for protection. Usually, a single egg, about 4 mm in diameter, is incubated in a temporary pouch that develops as a thin flap of abdominal skin in females. The milk-secreting glands are on the abdominal wall in the pouch, and the young hatch and are carried in the pouch for several weeks until they can forage for food.

Marsupialia

Marsupials (fig. 4.32) are primitive mammals in which the fetal yolk sac (in contact with the chorion) serves as a placenta. The young are born in almost a larval state and are incubated and nursed after birth in a maternal abdominal pouch **(marsupium)** of muscle and skin until they are old enough to be independent. The walls of the pouch are supported by two slender marsupial (epipubic) bones that project forward from the pelvic girdle. In several South American genera, the pouch is incomplete or absent. Newborn young make their way to the pouch by squirming and wriggling and by using the claws on their forelimbs, which are considerably larger than the hind limbs at birth. The lips of neonates are sealed at the angles of the mouth, which therefore is only a small, circular opening. Once the young has taken a nipple into its mouth, the tip of the nipple swells, and the young cannot easily drop off.

Marsupials and monotremes were the only mammals in Australia, with the possible exception of bats, until the arrival of the first human inhabitants. The people brought with them rats and mice, which inhabited the wooden sailing vessels. Australia at one time was part of a huge supercontinent that included the present Antarctica, Africa, and South America. The breakup of this supercontinent into several land masses eventually separated Australia from Antarctica, and these two continents drifted apart. The absence of eutherian fossils in Australia indicates that placentals were not present when that land mass separated from Antarctica.

Marsupials diverged from other therians in the Late Cretaceous. They appear to have arisen in what is now the northwestern part of North America, and they arrived in Australia before it was isolated. Their southbound trail crossed a land bridge to South America, and eventually reached Australia via Antarctica (fig. 4.33). (The present Central American land bridge was formed later.) Metatherian fossils of increasingly later ages have been found in North America along the southbound route.

While Australian marsupials were undergoing an extensive adaptive radiation, those on other continents were facing extinction in competition with evolving placentals. With the exception of opposums and a few other genera in North and South America, marsupials are now found only in Australia, Tasmania, New Guinea, and on a few offshore islands. Marsupial fossils found in Europe attest to their presence there until about 15 million years ago. Continental breakup provided the geographical isolation that fostered marsupial radiation in Australia and placental radiations elsewhere.

Tasmanian wolf

Opossum

Lesser gliding phalanger

Wallaby

FIGURE 4.32

Four marsupials.

Among marsupials are kangaroos, wallabys, Tasmanian wolves, bandicoots, wombats, anteaters, and phalangers. These and other marsupials resemble placental mammals—wolves, foxes, bears, rabbits, mice, cats—in surprising detail. Some phalangers resemble "flying" squirrels, and there are marsupial moles. The term "evolutionary convergence" doesn't apply to this phenomenon, since metatherians and placentals shared ancestral therian genes. The term "parallel evolution" has been applied to this phenomenon. It connotes that similar traits have evolved in isolated taxa whose common ancestor lacked such traits.

Insectivora

The members of this order are primitive, generalized mammals (fig. 4.34). Although at one time they were abundant, they are represented today by relatively few survivors, including moles, shrews, and hedgehogs.

There are numerous insectivorous mammals in other orders, but the latter are thought to have been derived from less primitive ancestors. Most insectivores subsist on a diet of insects, worms, and other small invertebrates.

Among primitive traits of the members of this order are a flat-footed (plantigrade) stance; five toes; smooth cerebral hemispheres; small, sharp, pointed teeth with incisors, canines, and premolars poorly differentiated; a large embryonic allantois and yolk sac; and, in some species, a shallow cloaca. The testes are retained in the abdominal cavity in some genera (a primitive trait), and they never descend fully into scrotal sacs in any.

Moles have short, stout anterior limbs, with forefeet that are broad and more than twice as large as the hind feet, which is an adaptation for digging (fig. 10.22). The neck is short, and the shoulder muscles are so powerful that head and trunk seem to merge. Their tiny eyes are practically useless, but an acute sense of smell locates

FIGURE 4.33
Postulated dispersal of evolving marsupials during the Mesozoic.

distant food, and the sensitive tip of the elongated snout tells them when they have encountered it. Shrews superficially resemble mice. They are shy, busy little fighters with a keen sense of hearing. They have an elongated, bewhiskered, sensitive snout, and their incisor teeth are long and curved.

Xenarthra

Xenarthrans are New World insectivorous mammals that are considerably more specialized than members of the order Insectivora. They are **armadillos** (fig. 4.35*a*), **sloths,** and **South American anteaters.** None have incisor or canine teeth; and cheek teeth, when present, are peglike and lack enamel, although histological evidence of enamel organs has been reported. Anteaters are completely toothless. Enlarged front claws are used for digging into ant nests or mounds and, conveniently for sloths, for hanging from the limbs of trees, which are their major habitats. Armadillos are notable for always giving birth to identical quadruplets from a single fertilized egg. They are also the only mammals that develop a true bony dermal armor.

Tubulidentata

Two species of Central and South American insectivorous **aardvarks** comprise the order Tubulidentata (fig. 4.35*b*). Their elongated snout, long sticky tongue, and strong claws on the front feet facilitate routing out and capturing insects. Aardvark teeth, like those of many other insectivores, are peglike, lack enamel, and have shallow roots.

Pholidota

Another anteater, the **pangolin** of Africa (fig. 4.36), is toothless and peculiarly scaly. It belongs to the order Pholidota. Its scales are made of keratin and appear to be agglutinated hairs. Typical hairs grow between the scales. There is only one genus, *Manis*.

Chiroptera

Bats (fig. 4.37) comprise a large mammalian order that is probably derived from a primitive insectivore. They, pterosaurs, and birds are the only known vertebrates to have achieved true (powered) flight. The wing, or

Gliding lemur

Mole

Solenodon

European hedgehog

FIGURE 4.34

Four members of the order Insectivora.

patagium, is a double membrane of skin stretched along the length of the body between the trunk, forelimbs, and hind limbs, and extending from there to the tail. It incorporates four greatly elongated, clawless fingers. The first digits, or thumbs, project from the anterior margins of the wing membranes and bear claws that aid in clambering about on rough vertical surfaces. All five digits of the hind limbs bear claws, and these are used for hanging upside down, wings folded, from crevices in caves, from small branches, or within hollow trees. The pectoral muscles are strong, and the sternum is keeled, although not as much as in carinate birds. All the bones are slender, but not pneumatic. Nipples, usually a single pair, are limited to the thoracic wall. Bats have large external ear auricles (pinnas), and their facial glands are prominent,

giving the head a bizarre appearance. The development of a wing in pterosaurs, birds, and bats are instances of convergent evolution. The bumblebee bat, weighing less than 2 gm, is the smallest mammal on earth.

Bats are insectivorous, frugivorous (fruiteaters), or sanguinivorous (subsisting on the blood of other mammals). Vampire bats have received attention because of sanguinivorous habits. Incisor teeth occur on the upper jaw only, and there is only one pair. They are razor sharp and point toward one another so they slit the skin of prey. As blood oozes from the wound, the bat licks it up without awakening the sleeping victim, which is usually a domestic animal. Associated with the vampire bat's practice of taking only fluid nourishment is the very small lumen of the esophagus, through which no solid food could possibly pass.

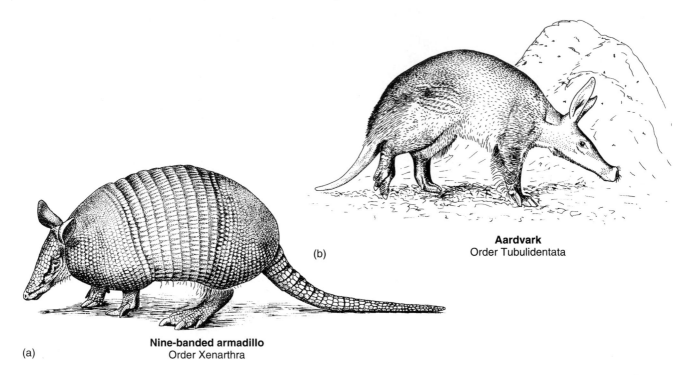

Aardvark
Order Tubulidentata

(b)

Nine-banded armadillo
Order Xenarthra

(a)

FIGURE 4.35

Two anteaters.

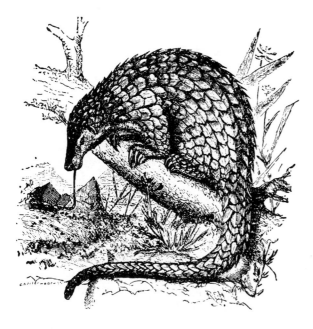

FIGURE 4.36

A pangolin, the scaly anteater.

Primates

Primates (prī-ma′tēz) are primarily arboreal mammals that arose as an offshoot of Cretaceous insectivore stock. One of many different classification schemes divides them into two suborders, **Prosimii** and **Anthropoidea.**

Lemurs, lorises, and tarsiers are prosimians. Monkeys, apes, and humans are anthropoids.

Among primate specializations is the grasping hand, built so that the thumb can touch the tips of the other four fingers of the same hand. The big toe is also opposable in most primates, and at least some digits have nails instead of claws. The cerebral hemispheres are larger than in any other mammal. There is frequently only one pair of nipples. Among primitive features are a flat-footed stance, five digits, a large clavicle, a central carpal bone in the wrist of many primates, and a generalized mammalian dentition.

Prosimians

Prosimians are arboreal and mostly nocturnal primates found in the tropics of the Old World. **Lemurs** (from the Latin word for ghost) receive their name from the habit of swinging silently through the forest at night. Unlike higher primates, they have the long axis of their head in line with the vertebral column (fig. 4.38b). The largest lemurs are the size of a domestic cat. **Lorises,** from India, Sri Lanka, and Southeast Asia, have no tail and the index finger is vestigial. In the same family are the bush babies and pottos.

Tarsiers (fig. 4.38a) resemble anthropoids to a greater degree than do the other prosimians. Their head is more nearly balanced at right angles to the vertebral column, the snout is truncated, and their eyes are close together and directed forward so there is an overlap in the left and

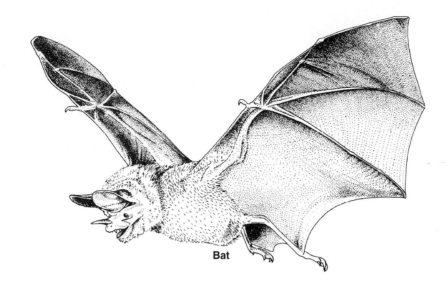

Bat

FIGURE 4.37

A chiropteran.

right visual fields. All fingers have nails, as do all toes except the second and third. Their placenta is deciduate, that is, the fetal membranes become rooted into the wall of the maternal uterus, as in anthropoids.

Anthropoids

There are two groups of anthropoids, platyrrhines and catarrhines. They are differentiated on the basis of the direction that the nostrils open. Those of platyrrhines are separated by a broad internarial septum and open to the side (fig. 4.39). The nostrils of catarrhines are close together and open downward.

Platyrrhines are the New World monkeys and marmosets. They include capuchins (*Cebus*), spider monkeys (*Ateles*), and howler monkeys (*Alouatta*). Howler monkeys are named for the screeching cries they produce from an enlarged larynx and hyoid bone (fig. 13.14).

Catarrhines are Old World monkeys, apes, and humans. The Old World monkeys are in a separate superfamily, **Cercopithecoidea.** Among Old World monkeys are baboons and macaques, or rhesus monkeys. The abbreviation "Rh," used in designating human blood types, was derived from the latter, in which it was first detected. Apes (chimps, gorillas, orangutans) are in the pongid family, and humans are hominids (family **Hominidae,** appendix 1). Catarrhines have no tail. A well-developed tail is present in fetal catarrhines. In humans it disappears during the third month of fetal life.

The skull of all anthropoids is set at right angles to the vertebral column; the eyes are directed forward and are close together; the cerebral hemispheres are highly developed; there are 32 teeth in the permanent set; the pla-

centa is deciduate; and, usually, a pregnancy produces only one infant. Despite any specializations, anthropoids, including humans, are much less specialized than many other mammals—whales and perissodactyls, for example.

Since the discovery of Neanderthal man (*Homo neanderthalensis*), who lived in Europe, the Middle East, and parts of central Asia until about 30,000 years ago, fossil remains of much earlier hominids have been found in considerable numbers, particularly in Africa. With each new discovery, the problem arises of whether to include the new member in an existing species or to erect a new one, or even a new genus. The criteria for classifying other vertebrates are applied to the classification of hominids, and are subject to the same contingencies. To date, the oldest fossil considered to be a hominid is *Ardipithicus,* about 4.5 million years old. The oldest *Australopithicus* is about 4 million years old. Sufficient *Australopithicus* bones have been recovered to estimate that these individuals were about 5 ft tall and had a humanlike skeleton, facial bones of an ape, and an ape-sized cranial cavity. They made crude tools of stone or bone. *Homo erectus,* 1.5 million years old, is thought to be the immediate ancestor of both *Homo neanderthalensis* and *Homo sapiens* (modern humans). *Homo sapiens,* about 100,000 years old, lived in Europe at the same time as the Neanderthals and eventually replaced them on that continent. The history of humans includes continual intraspecific warfare. It is possible this behavior brought about the extinction of the Neanderthals.

With the emergence of humans from an earlier anthropoid, many changes occurred. An **S**-shaped curve in the vertebral column permitted an erect

(a)

(b)

FIGURE 4.38

Two Prosimians: (*a*) *Tarsius,* (*b*) *Lemur.*

human posture; the facial angle became less acute (fig. 4.40); the teeth, especially the canines, became smaller; the frontal lobes of the cerebral hemispheres enlarged, resulting in a larger braincase and a more prominent forehead; the eyebrow ridges became reduced; the nose became more prominent; the tail became confined to embryonic stages; the arms became shorter; a metatarsal arch developed in the otherwise flat feet; the big toe moved in line with the other toes and ceased to be opposable; articulate speech became structurally possible; and many other changes occurred.

It is to the development of the massive frontal lobes of the cerebral hemispheres, to the opposable thumb, and to articulate speech that humans owe their present dominance in the animal kingdom. With fingers they can construct instruments of offense and defense and equipment to lighten burdens, and with them they write symbols that convey experiences and techniques

Platyrrhine   **Catarrhine**

FIGURE 4.39

Two anthropoids. The platyrrhine is a scarlet-faced red ouakari, a small South American monkey with a nonprehensile tail. The catarrhine is an orangutan.

to the corners of the earth and to generations unborn. With voices they communicate with contemporary fellow creatures and exchange ideas with delicate shades of meaning. With larger brains, they associate sensory stimuli that are currently received with those recalled from earlier experiences, and after meditation elect a mode of action that they call "intelligent." With their brains, too, they can enjoy the esthetic beauty of the universe, search for ultimate truth, and dream of Utopia, which, but for their residual animal nature, might be their heritage.

Lagomorpha

There are only two families of lagomorphs, the pikas (fig. 4.41), and the hares and rabbits, all of which are herbivores. Lagomorphs differ from rodents in having *two* pairs of incisors on the upper jaw, a small pair lying immediately behind, not alongside of, a much larger pair. The large front pair are rodentlike and grow throughout life. The smaller pair lack cutting edges.

Rabbits differ from hares in being born in a fur-lined nest, blind, virtually hairless, and helpless. Hares are born without benefit of a nest, fully formed, with eyes that open within an hour. They are able to hop around before they are a day old. Some hares of the genus *Lepus* acquire white fur in winter (fig. 4.41*a*); no rabbits do. Jack rabbits of western North America and snowshoe rabbits of the Arctic are hares, but the so-called Belgian hare is a rabbit! Rabbits and hares have a split upper lip, which inspired the term "harelip" for the human anomaly.

Pikas live above the timber line in western North America and northern Asia. They differ from rabbits and hares in having a smaller body, shorter ears, and fore and hind limbs of about equal length.

Rodentia

Rodents comprise the largest mammalian order and are distributed worldwide. They have a *single* pair of long, curved incisor teeth on each jaw that are used for gnawing. These teeth have enamel on their outer surface only, which provides a chisel-like edge as the softer dentin wears away. The teeth grow throughout life. Canine teeth are absent, so there is a diastema, or a stretch of toothless jaw, between the incisors and the first grinding teeth. Rodents are cellulose eaters, and at the beginning of the large intestine there is a long, coiled cecum that houses commensal cellulose-digesting microorganisms.

There are four suborders. **Sciuromorpha** includes squirrels, chipmunks, woodchucks, gophers, beaver, and others. **Myomorpha** includes micelike rodents such as rats, voles, hamsters, and lemmings. **Caviomorpha** includes cavys, nutria (coypus), and chinchillas. **Hystricomorpha** includes porcupines (fig. 4.42).

Carnivora (Fissipedia)

Carnivora is a large, diverse terrestrial group some species of which have powerful jaws and sharp upper canine teeth capable of spearing and tearing flesh. The saber-toothed tiger, now extinct due to human avarice,

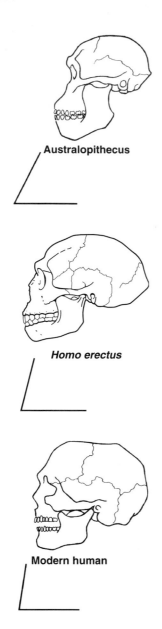

FIGURE 4.40

Facial angles of three hominids. The angles are shown beneath the skulls.

(a)   *Lepus*

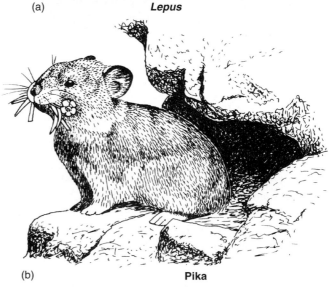

(b)   Pika

FIGURE 4.41

Two Lagomorphs. The hare, *Lepus*, is shown in its brown summer coat and white winter coat.

(*a*) © Leonard Lee Rue III.

was an extreme example. There are typically 42 permanent teeth. (Human beings have 32.) Premolars tend to be tricuspid; molars have four cusps. The clavicles are reduced, vestigial, or missing altogether. The cerebral cortex is unusually convoluted compared to most mammals. A combination of these and other morphologic traits—not a meat-eating, or carnivorous, diet—define the order. Most terrestrial carnivores have five toes (a few have four) with sharp retractable claws.

There are currently seven families of carnivores: **cats** (domestic cats, panthers, tigers, lynx, others), **civets, hyenas, canines, bears, racoons** and **pandas,** and **mustelids** (mink, otter, skunks, ferrets, badgers, others), all with large scent glands near the anus (fig. 4.43). Not

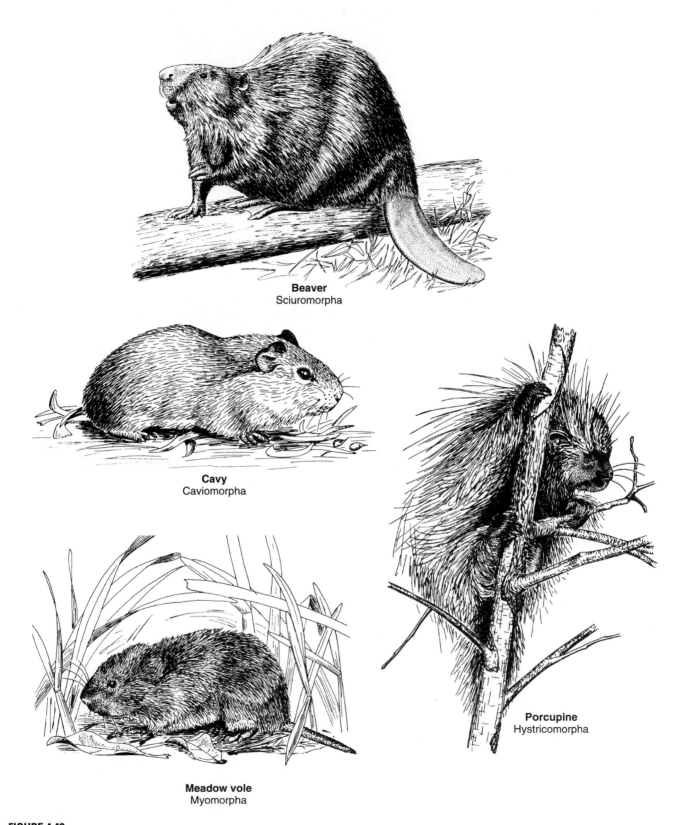

Beaver
Sciuromorpha

Cavy
Caviomorpha

Meadow vole
Myomorpha

Porcupine
Hystricomorpha

FIGURE 4.42

The three suborders of rodents. The cavy is a feral guinea pig.

FIGURE 4.43

Representatives of four families of terrestrial carnivores.

all members of the order are carnivorous, or even predominantly so. For example, bears other than polar bears (fish eaters) are partly herbivorous, and pandas eat only nuts, roots, bamboo shoots, and other succulent vegetation.

Pinnipedia

These are carnivores that are morphologically and physiologically adapted for an aquatic life. They are closely related to the terrestrial carnivores, but have been separated for a considerable time. They feed on fish, squid, molluscs, and crustaceans. Pinnipeds are the earless (wriggling) seals, eared (fur) seals, sea lions, and walruses (fig. 4.44). Earless seals lack the pinna of the outer ear characteristic of most mammals, but they have excellent hearing, nonetheless. All pinnipeds leave the water at least once a year, at breeding time. Females give birth a year after mating, and the young are born on land, unable to swim. Among pinniped adaptations for life in the water are webbed paddlelike limbs known as flippers. The digits are mostly enclosed within the distal end of the flipper and, unlike those of land-dwelling carnivores, they usually lack claws.

Seals and sea lions seek their food deep in the oceans, including beneath the dense Antarctic ice pack. Northern elephant seals have been recorded at depths of over 900 m (3,000 ft), where they may remain for as long as an hour. The time spent at the surface between dives for replenishing their oxygen supply is only a few minutes, during which time the oxygen in the inhaled atmospheric air is transferred from the lungs to the bloodstream, to be released as needed. The animals then exhale before going back down, and the lungs collapse at 40 m (130 ft). Walruses, on the other hand, are shallow divers.

Ungulates and Subungulates

Ungulates are mammals that walk on the tips of their toes, protected by hoofs. There are two orders: Perissodactyla and Artiodactyla. They have no more than four toes on each foot and some, such as horses, have only one. Ancestral ungulates had five, the generalized tetrapod number. Reduction in the number of toes is documented by extensive studies of the evolution of horses, whose small Eocene ancestors had four toes in front and three behind.

Ungulates are herbivores. Their molar teeth are high crowned and deeply grooved for grinding vegetation, and there is little morphologic difference between premolars and molars. Ungulates have no clavicle. Its absence facilitates grazing. Ungulates are the only mammals with horns.

"Subungulate" is a term that has been applied to elephants (Proboscidea), hyraxes (Hyracoidea), and manatees (Sirenia), which are discussed shortly. The taxonomic status of these groups, like that of many mammals, is debatable.

FIGURE 4.44

Harbor seals (marine carnivores).

Perissodactyla

There are three families of perissodactyls, **horses** and **horselike mammals, tapirs,** and **rhinoceros** (fig. 4.45). They walk on the hoofed tips of one, three, or occasionally four toes and are distinguished by the fact that most of the body weight is borne on a single digit (figs. 10.34, rhinoceros and horse, and 10.35). This is a **mesaxonic foot.** Perissodactyls are usually called odd-toed ungulates, but tapirs and some rhinos have four toes on the forefeet.

Artiodactyla

Artiodactyls are ungulates in which the weight of the body is borne on two toes (fig. 4.46). This is a **paraxonic foot** (fig. 10.34). Living artiodactyls have an even number of toes, but at least one extinct artiodactyl had five toes on the forelimb. Artiodactyls include pigs, hippopotamuses, peccaries, cattle, camels, deer, antelope, and giraffes. With the exception of pigs, hippopota-muses, peccaries, and camels, they have stomachs divided into not fewer than three chambers, sometimes four. They bolt their food, which then passes to the rumen, the first segment of the stomach. Such animals are **ruminants.** At their leisure, they force undigested food balls back up the esophagus and masticate this cud more thoroughly.

Hyracoidea

The order consists of three genera of **hyraxes** (fig. 4.47), with four digits on the forefeet and three on the hind feet. Although hyraxes are flat-footed (plantigrade), all digits except one end in small flat hoofs. The teeth of grazing members are high crowned like those of ungulates. Because of these and other features, some systematists include hyraxes with the perissodactyls. The upper lip is split (harelip), and the incisor teeth grow continuously, as in rodents.

Indian rhinoceros

Brazilian tapir

FIGURE 4.45

Two perissodactyls.

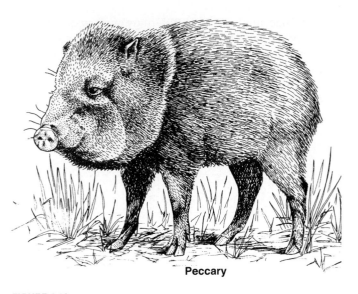

Peccary

FIGURE 4.46
An artiodactyl.

Hyrax

FIGURE 4.47
A rock hyrax, order Hyracoidea.

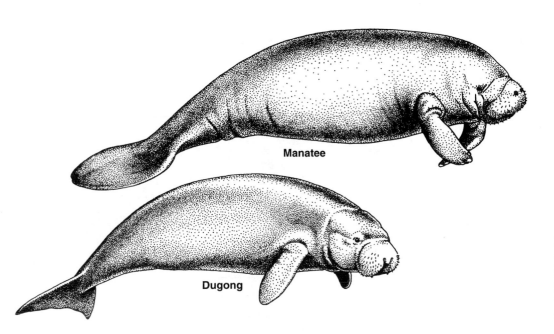

Manatee

Dugong

FIGURE 4.48
Representatives of the order Sirenia. Notice the notch in the fluke of the dugong, and vestigial nails on the flippers of the manatee.

Proboscidea

The order Proboscidea includes elephants, mastodons, and their relatives. They have a proboscis, scanty hair, and thick, wrinkled skin. The incisor teeth of one or both jaws are elongated to form tusks, canine teeth are absent, and the molars are large grinders, as in ungulates. Proboscideans are bulky animals, and the limbs are almost vertical pillars of bone and muscle. They have five toes that end in thick, hooflike nails. An elastic pad on the back of each toe bears much of the body weight. The pad is not present in ungulates.

Because of the structure of the teeth and feet, proboscideans have been considered subungulates. But applying gene sequence data of nuclear and mitochondrial genes leads to the postulate that elephants are no more closely related to ungulates than they are to other orders of mammals.

Sirenia

Manatees and dugongs, known also as sea cows, constitute the order Sirenia (fig. 4.48). Like ungulates, they are strictly vegetarian and are thought to be descendants of primitive ungulate stock. The forelimbs are paddles, but within them the tetrapod complement of bones is intact. The hind limbs are lost, but there are internal skeletal vestiges of them and of the pelvic girdle. The snout is covered with coarse bristles, and the rest of the body is naked except for a few scattered hairs. The flukelike tail is a result of evolutionary convergence. Sirenians and cetaceans are the only completely adapted aquatic mammals. All others must return to rookeries to breed.

Cetacea

Today's cetaceans—whales, dolphins, and porpoises—are permanently marine mammals (fig. 4.49). Their tail consists of two horizontally oriented fleshy lobes, or flukes, composed of connective tissue. The posterior half of the trunk provides most of the thrust for locomotion. The forelimbs are paddlelike, with the phalanges embedded in the distal end. Hind limbs and girdles are either mere vestiges embedded in the body wall or, as in some species, they are entirely missing. Early Cenozoic ancestors had small, but complete, hind limbs. A dorsal fin, also of fibrous tissue, is present in some species.

The frontal and nasal bones are short, with the result that the nostrils are on top of the head, sometimes united to form a single large blowhole. The nostrils have valves that close underwater. The thoracic diaphragm is exceptionally muscular, and the whale's waterspout, composed mostly of water vapor, may last from three to five minutes. The vapor forms in cold air when warm respiratory air is expelled at the water's surface. A heavy layer of subcutaneous fat (blubber) conserves body heat in the cold ocean.

Bottle-nose dolphin

Blue whale

Harbor porpoise

FIGURE 4.49
Representative cetaceans.

Most cetaceans have teeth, but whalebone whales have, instead, frayed horny sheets of whalebone (baleen) hanging from the palate (see fig. 6.27). These strain several tons of small fish, crustaceans, or plankton, depending on the species, out of the sea each day. A few hairs occur on the muzzle of cetaceans, but elsewhere the skin is naked. Babies are born in the sea. The mother or an assistant bites the umbilical cord to separate the neonate from the afterbirth, and the mother nudges the baby to the surface for the first inhalation of air. Thereafter, the baby hangs onto one of the four inguinal teats as the mother swims about.

Cetaceans have good eyesight, but they also scan their environment by echolocation, which can tell them even the texture of another object. They have an excellent sense of taste but no sense of smell, since their olfactory nerves do not develop. They would be useless, anyway. If the animal were to inhale under water, it would drown. Cetaceans communicate with one another by distinct whistles (at least 16 in bottle-nosed dolphins), snapping the jaws, and slapping the water with the flukes.

VARIATION AMONG INDIVIDUALS

In this chapter we have had a glimpse of the diversity of vertebrate life. However, variation is not restricted to differences between species. Variation occurs within populations and even among littermates. Indeed, no two products of sexual reproduction are identical. The number of kinds of gametes or of homozygous genotypes possible within n pairs of genes is 2^n. The number of different kinds of genotypes possible is 3^n. Even if there were only 100 pairs of genes, 48 digits would be needed for writing the number of possible genotypes. And there are thousands and thousands of pairs of genes. In addition, there may be as many as 400 mutational sites on a single gene. Therefore, the number of possible combinations of hereditary traits staggers the imagination. For example, the gastrohepatic artery in a population of domestic cats is sometimes 20 mm long, sometimes 1 mm long, or the gastric and hepatic arteries may arise independently, hence no gastrohepatic artery. The number of variations that actually occur in arterial channels alone in any population is such that one could identify a single shark, or a single human being, from the arterial channels alone. In practice, each of us does better than that! We identify a single human being simply by looking at a face or at a print made by a fingertip. We could, of course, recognize an individual organism by looking at a map of the sequence of nucleotides in its DNA. Ongoing, random genetic mutations and recombinations of genes resulting from sexual reproduction maintain variation in populations. These processes then provide the raw material for evolution. Genetic variation, coupled with natural selection, is the matrix out of which new species evolve.

SUMMARY

1. Agnathans (jawless vertebrates) are the extinct ostracoderms and the living cyclostomes (hagfishes and lampreys). Ostracoderms are the oldest known vertebrates. They were covered with dermal armor and lacked paired appendages.
2. Acanthodians, named for their spiny fins, and placoderms, named for the bony plaques in the skin, were the earliest jawed fishes. They flourished in the Devonian.
3. Chondrichthyes (cartilaginous fishes) include elasmobranchs and holocephalans. The former have a spiracle, naked gill slits, and placoid scales that become teeth on the jaws. Holocephalans have nearly naked skin, a fleshy operculum, and bony plates on the jaws instead of teeth.
4. Bony fishes are in two classes, lobe-finned fishes (Sarcopterygii) and ray-finned fishes (Actinopterygii). They have ganoid, ctenoid, or cycloid scales, and a bony operculum. Most lobe-fins have internal nares.
5. Sarcopterygians include crossopterygians and dipnoans. The former gave rise to amphibians. The latter include the modern lungfishes.
6. Actinopterygians include chondrosteans (sturgeons, paddlefishes, and other living relics of the Paleozoic) and neopterygians (bowfins and gars, which are surviving Mesozoic fishes, and modern teleosts). In modern fishes, thin flexible scales have replaced the bony armor of early fishes.
7. Amphibians include three subclasses: labyrinthodonts (the earliest tetrapods), lepospondyls (upper Carboniferous), and lissamphibians (Triassic and modern amphibians).
8. Modern amphibians are anurans (frogs, toads), urodeles (tailed amphibians), and apodans (limbless species). Anurans and urodeles are scaleless, have a glandular skin, typically deposit their eggs in the water, and have an aquatic larval stage. A few are viviparous. Apodans have minute dermal scales.

9. Reptiles, birds, and mammals are amniotes. Embryos develop within an amniotic sac in association with extraembryonic membranes.
10. Reptiles are ectotherms with cornified epidermal scales, plaques, or scutes, and with claws. They typically lay large yolky eggs surrounded by a porous shell. Some are viviparous.
11. There are three subclasses of reptiles: anapsids, diapsids, and synapsids. Anapsids have no temporal fossa. Anapsids include cotylosaurs (stem reptiles) and turtles. Diapsids have two temporal fossae and include all living reptiles except turtles. Synapsids have one temporal fossa and were ancestral to mammals.
12. Archosaurs are a subtaxon of diapsids. They include winged reptiles (pterodactyls), dinosaurs, and crocodilians.
13. Lepidosaurs are another subtaxon of diapsids. It includes sphenodons and squamates (lizards, snakes, and amphisbaenians). Other subtaxa include large extinct aquatic reptiles (plesiosaurs and ichthyosaurs).
14. Birds (Aves) are endothermal vertebrates with feathers. The oldest known birds (Archaeornithes) appeared in the mid-Mesozoic. All others are Neornithes. Living birds are in two subtaxa, ratites (Palaeognathae) and carinates (Neognathae). The former are flightless. The latter have a large sternal carina.
15. Mammals are vertebrates with hair and mammary glands. They include one group of oviparous mammals (Monotremata: platypuses and echidnas), one group that employs the yolk sac as the fetal portion of a placenta (Marsupialia: opossums, kangaroos, others), and the remaining mammals, which have a chorioallantoic placenta ("true placentals"). The order Primates includes Old and New World monkeys, apes, and extinct and modern humans.

SELECTED READINGS

Ahlberg, P. E.: Origin and early diversification of tetrapods, Nature **368**:507, 1994.

Brodal, A., and Fange, R., editors: The biology of Myxine, Oslo, 1963, Universitetsforlaget (Norway).

Chen, J.-Y., Dzik, J., Edgecombe, G. D., Ramsköld, L., Zhou, G.-Q.: A possible Early Cambrian Chordate, Nature **377**:720.

Chiappe, L. M.: A diversity of early birds, Natural History, p. 52, June 1995.

Dineley, D. I., and Loeffler, E. J.: Ostracoderm faunas of the Delorme and associated Devonian-Silurian formations, North West Territories, Canada, Palaeontological Society, London, 1976.

Forey, P., and Janvier, P.: Agnathans and the origin of jawed vertebrates, Nature **361**:129, 1993.

Forey, P., and Janvier, P.: Evolution of the early vertebrates, American Scientist **82**:554, 1994.

Gingrich, P. D.: The whales of Tethys, Natural History, p. 86, April 1994.

Hickman, C., and Roberts, L.: Animal diversity, Dubuque, Wm. C. Brown Publishers, 1995.

Hou, L., Zhou, Z., Martin, L. D., and Feduccia, A.: A beaked bird from the Jurassic of China, Nature **377**:616, 1995.

Lee, M.: The turtle's long lost relatives, Natural History, p. 63, June 1994.

Lombard, R. E., and Sumida, S. S.: Recent progress in understanding early tetrapods, American Zoologist **32**:609, 1992.

Nelson, J. S.: Fishes of the world, ed. 3, New York, 1994, John Wiley and Sons.

Norell, M., Chiappe, L., and Clark, J.: New limb on the avian family, Natural History, p. 39, September 1993.

Stock, D. W., and Witt, G. S.: Evidence from 185 ribosomal RNA sequences that lampreys and hagfishes form a natural group, Science **257**:787, 1992.

Szalay, F. S.: Evolutionary history of the marsupials and an analysis of osteological characters, 1994, Cambridge University Press.

Szalay, F. S., et al., editors: Mammal phylogeny: Placentals, vol. 2, New York, 1993, Springer-Verlag.

Wilson, E. O. The diversity of life, Cambridge, The Belknap Press/Harvard University Press, 1992.

EARLY VERTEBRATE MORPHOGENESIS

In this chapter we will learn that all vertebrates pass through similar, though not identical, early formative stages. We will discover the steps whereby a hollow ball of totipotent embryonic cells is converted into an early embryo with three indispensable germ layers, and the contribution of each layer to the adult body. At the end, we will examine the extraembryonic membranes and their roles before hatching or birth.

OUTLINE

Specimens studied in vertebrate morphology laboratories are viewed in a single time frame, the immediate present. A bit of knowledge from two additional time frames will enrich our understanding of morphology: the specimen's remote past, which requires knowledge of its ancestors; and its recent past, which includes its early ontogeny. Chapter 4 introduced the ancestors. The present chapter recounts the highlights of early embryogenesis. The total picture—phylogeny, ontogeny, and morphology—is more meaningful than is morphology alone, if only because the first two account for the last.

VERTEBRATE EGGS

Egg Types

Eggs of vertebrates vary in the amount of yolk they contain and in the distribution of yolk within the egg. Eggs with very little yolk, such as those of the amphioxus and placental mammals, are **microlecithal**. Eggs with moderate amounts of yolk, such as those of freshwater lampreys, ganoid fishes, lungfishes, and amphibians, are **mesolecithal**. Eggs with massive amounts of yolk, such as those of marine lampreys, elasmobranchs, teleosts, reptiles, birds, and monotremes, are **macrolecithal**.

The yolk in microlecithal eggs is evenly distributed throughout the cytoplasm as fat droplets and small yolk globules. Eggs with an even distribution of yolk are said to be **isolecithal**. In mesolecithal and macrolecithal eggs, the large yolk mass tends to be concentrated at one end, the **vegetal pole** (fig. 5.1, *red*). The opposite pole, containing the nucleus and relatively yolk-free cytoplasm, is the **animal pole**. Eggs in which cytoplasm and yolk tend to accumulate at opposite poles are **telolecithal**.

Oviparity and Viviparity

Animals that spawn (lay their eggs) are said to be **oviparous**. Eggs of oviparous species contain sufficient nourishment in the form of yolk, and sometimes albumen, to support development into a free-living organism that is able to take food orally. If the yolk is massive, the young may hatch essentially fully formed, as in birds. If there is less yolk, the young hatch in a larval state, as in frogs. When there is very little yolk, as in the egg of the amphioxus, the free-living, self-nourishing state must be achieved very quickly after the egg is deposited. Accordingly, the amphioxus hatches into an externally ciliated, free-swimming embryo 8 to 15 hours after fertilization, at which time the notochord is a mere ridge in the roof of the primitive gut and there are no gill slits.

Animals that give birth to offspring are said to be **viviparous**. The relationship between mother and embryos varies from one in which the mother provides only protection and oxygen, nourishment having been

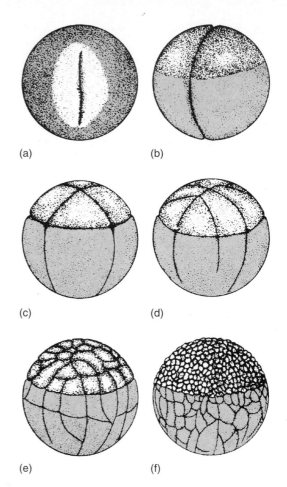

FIGURE 5.1

Segmentation and blastula formation in a mesolecithal egg (amphibian). (*a*) Initial cleavage of egg, seen from above. (*b–f*) Successive cleavages resulting in a blastula. *Gray* indicates animal pole, the cells of which contain little yolk; *red* is vegetal pole. In (*f*), gastrulation by epiboly is about to commence.

stored in the egg, as in viviparous sharks, to one in which the embryos are dependent on maternal tissues for all nourishment, for oxygen, and for carrying away waste products of metabolism, as in placental mammals. The term **ovoviviparity** has been coined to designate the former viviparous condition. **Euviviparity** designates the viviparous condition in which the embryo cannot develop without nourishment being constantly supplied by the mother from maternal tissues. There are many intermediate conditions. Reptilian embryos that are initially ovoviviparous may require nourishment from the mother before pregnancy has terminated. A reptilian egg in a viviparous species has a shell, but it is an uncalcified semipermeable membrane.

Viviparity in one degree or another has evolved in every class of vertebrates except agnathans and birds. It has developed independently more than 12 times in teleost fishes, at least 100 times in lizards, and at least 6 times in snakes. In fact, 20 percent of all squamate species are viviparous. There are no viviparous crocodilians or turtles. Viviparity occurs in 40 families of sharks and rays, and there is fossil evidence for viviparity in extinct holocephalans and chondrosteans. Unborn fossilized young have been found in the uteri of the mesozoic marine reptile *Ichthyosaurus.* There are viviparous urodeles and anurans that inhabit a terrestrial environment rather than a pond and give birth to fully metamorphosed young, and there are viviparous aquatic apodans.

The dogfish shark, *Squalus acanthias,* is an example of an ovoviviparous organism. The unborn pups are nourished entirely by the yolk from their own yolk sac (fig. 5.12), but they receive oxygen from greatly enlarged and highly tortuous blood vessels in the uterine lining of the mother. The pups may be removed from the mother two to three months before birth (the gestation period is 20 to 22 months) and will complete their development, utilizing the nourishment from their yolk sac, provided they are supplied with oxygenated seawater.

In viviparous teleosts the egg may be fertilized and the young may develop in the ovarian follicle, there having been no ovulation. The young of *Gambusia* develop in this way, or the embryos may develop in the ovarian cavity (fig. 15.33). Among adaptations of the embryo for development within the follicle is temporary enlargement of the *embryo's* pericardial sac, gut, or some other organ, which absorbs oxygen and nutrient-rich juices from ovarian tissues. Some species that develop in the ovarian cavity sprout long villuslike projections **(trophotaeniae)** from the lining of their embryonic gut; these protrude through the vent into the surrounding nutrient-rich medium within the cavity. In still other teleosts the young develop for a time in the follicle and exhibit an enlarged pericardial sac, then pass into the ovarian cavity and develop absorptive villi. In some species, developing eggs or larvae, which are present in enormous numbers in teleosts, are ingested by other larvae, thus guaranteeing a source of nourishment for a portion of the larval population.

Maternal tissues, under the influence of hormones, exhibit diverse adaptations for viviparity. The wall of the ovarian follicle occupied by teleost embryos may develop vascular folds or villi that secrete nutrients, and these may even project into the mouth or opercular cavity of the embryos. In *Dasyatis americana,* a cuviviparous sting ray, the gravid uterus is lined with villi 2 to 3 cm long, which produce a copious secretion that nourishes the embryo. Secretions of uterine glands provide nourishment for temporarily unimplanted blastocysts of mammals and perhaps for implanted blastocysts throughout pregnancy in perissodactyls and artiodactyls. **Histotrophic (embryotrophic) nutrition** is the term applied to nutrition by *glandular secretions* from maternal tissues, as contrasted with nutrition supplied via a placenta.

Internal and External Fertilization

In viviparous vertebrates, fertilization takes place within the body of the female. Fertilization is also internal whenever eggs are covered by an impenetrable shell before being extruded, as in reptiles and birds.

In oviparous fishes, frogs, and toads, external fertilization is the rule despite an aquatic environment that dilutes the sperm. This is possible only because millions of sperm are shed over the eggs as the latter are being extruded. In apodans and urodeles, however, fertilization is usually internal even in oviparous species. Some male urodeles deposit a gelatinous packet of sperm **(spermatophore)** in the immediate vicinity of congregated females during the mating ritual. The spermatophore is generally picked up by the lips of a female's cloaca, although in some species it is placed in the cloaca by the male. The sperm eventually escape from the jelly and fertilize eggs, which are passing down the oviduct. In some species the sperm are stored in crypts **(spermathecae)** in the female tract and fertilize eggs that are ovulated at a later date when the environment is more suitable for embryonic development. The adaptation has a genetic basis. Packaging sperm in spermatophores makes it possible to convey sperm to females despite a lack of copulatory organs.

EARLY DEVELOPMENT OF REPRESENTATIVE CHORDATES
Cleavage and the Blastula

Early cell divisions of the zygote are initiated by fertilization of the egg, and are referred to as **segmentation,** or **cleavage.** As a result, the zygote becomes subdivided into smaller and smaller cells that form a usually hollow sphere, or **blastula.** Each cell of the blastula is a **blastomere,** and the cavity is the **blastocoel** (fig. 5.2).

When there is very little yolk in an egg, the blastomeres are approximately of equal size (fig. 5.2, blastula). When cell division is impeded by a moderate amount of yolk, as in amphibian eggs, the yolk-laden cells at the vegetal pole divide more slowly and are larger (fig. 5.1). When there is a massive yolk, as in the eggs of reptiles and birds, segmentation is confined to the animal pole, and results in a **blastoderm** perched like a skullcap on the massive yolk (fig. 5.6). The embryo develops from this blastoderm.

The eggs of placental mammals exhibit an animal-vegetal polarity, just as do the eggs of reptiles and birds, even though they have almost no yolk. The first cleavage division divides the egg into two blastomeres, one representing the animal pole, the other the vegetal pole (fig. 5.10, cleavage). Further cell division results in a blastula. The descendants of the vegetal pole rapidly produce a nutritional membrane, the **trophoblast,** which absorbs nourishment from the uterine fluids. The

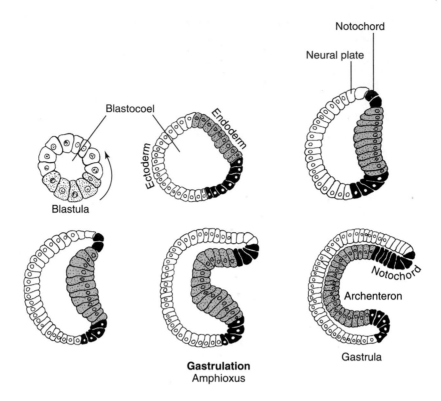

FIGURE 5.2

Blastula and gastrulation by involution in amphioxus, sagittal section. Cells in *black* surround the blastopore (entrance to archenteron) and are the most active, mitotically. *Red* indicates presumptive endoderm in the blastula, definitive endoderm in the gastrula. The notochord establishes the dorsum and the long axis of the body.

animal pole cells become the blastoderm, better known in mammals as the **inner cell mass.** Because of the cyst-like nature of the mammalian blastula, it is called a **blastocyst.**

As a result of cleavage, vertebrates typically pass through a hollow, spherical stage of development known as a blastula. During this stage, areas of distinctive developmental potential—the rudiments of future ectoderm, endoderm, mesoderm, and notochord—are established.

Gastrulation: Formation of the Germ Layers and Coelom

Gastrulation in chordates is a dynamic process of cellular movements whereby presumptive endoderm, mesoderm, and notochord cells of the blastula flow to the interior, thereby generating the three germ layers from which all future tissues and organs form. In the process, bilateral symmetry is established. These cellular migrations are referred to as **formative,** or **morphogenetic, movements.** Rapid cell proliferation maintains a continuing supply of additional cells for these formative movements. The developmental processes that constitute gastrulation are induced and regulated by inductor substances (morphogens) that will be discussed shortly. Because gastrulation in an amphioxus is unencumbered by yolk, we will examine the process in this chordate first.

Amphioxus

In an amphioxus, presumptive endoderm from the surface of the blastula invaginates into the blastocoel (fig. 5.2, *red*). The process, **involution,** produces the earliest gut, or **archenteron,** and results in a gastrula (fig. 5.2). The entrance to the archenteron is the site of involution, the **blastopore.** Soon, presumptive notochord cells from the rim of the blastopore flow inward to become the beginning of the notochord, which is located temporarily in the roof of the archenteron (fig. 5.2, gastrula). Then, presumptive mesoderm cells form bands of undifferentiated mesoderm lateral to the notochord (fig. 5.3*a*). At this stage, the amphioxus embryo consists chiefly of an outer layer of ectoderm and an inner layer of endoderm that surrounds the archenteron. The roof of the archenteron is the elongating notochord and the parallel bands of mesoderm. (The notochord is generally regarded as mesoderm.)

Once mesoderm has been established in the dorsolateral roof of the archenteron, the actively mitotic bands fold upward, separate from the endoderm, and form a series of hollow **coelomic pouches (mesodermal pouches)** commencing at the *anteriormost* end of the band (fig. 5.3*a* and *b*). About the time two pouches have formed, the embryo hatches into a free-swimming larva. The larva continues to elongate partly as a result

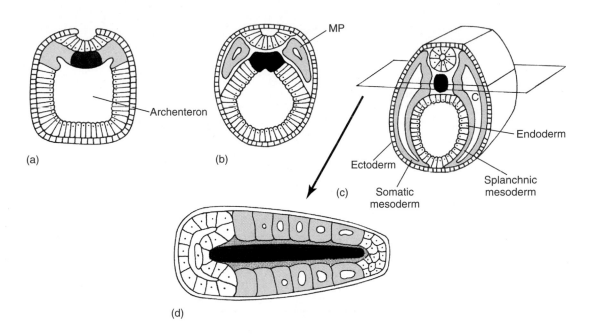

FIGURE 5.3

Mesoderm, *light red*, and coelom formation in an amphioxus. (*a*) Longitudinal bands of mesoderm lie in the dorsolateral wall of the archenteron lateral to the notochord (*dark red*). (*b*) Mesodermal pouches (**MP**) have formed. (*c*) Pouches have grown ventrad between ectoderm and endoderm to form a coelom, **C**. (*d*) Early larva in frontal section showing segmented coelom.

of proliferation of additional cells from the rim (**dorsal** and **ventral lips**) of the blastopore. The mesodermal bands elongate and additional coelomic pouches form.

After a coelomic pouch has formed, it grows ventrad, pushing between the ectoderm and endoderm (fig. 5.3*c*). The mesoderm of the two pouches, left and right, eventually meet underneath the gut. The outer wall of each pouch lies against the ectoderm and is **somatic mesoderm.** Together with the ectoderm it constitutes the **somatopleure,** or **body wall.** The inner wall of each pouch lies against the endoderm and is **splanchnic mesoderm.** Together with the endoderm it constitutes the **splanchnopleure,** which gives rise to the digestive tract. The cavity between somatic and splanchnic mesoderm is the **coelom** (fig. 5.3*c*).

The coelom of a larval amphioxus is segmented for a while because of its origin from a series of pouches (fig. 5.3*d*). Later, the cavities within successive pouches become confluent, *starting at the anterior end,* establishing a single coelomic cavity on each side. Still later, the right and left coelomic cavities become confluent beneath the gut. The germ layers are now positioned to participate in the formation of organs, a process called **organogenesis.**

Frog

In vertebrates other than placental mammals, gastrulation is complicated by the presence of moderate to large quantities of yolk in the cells of the vegetal pole. This unwieldy yolk impedes simple involution such as seen in an amphioxus. Still, a gastrula forms by alternative cell movement. In frogs, which have mesolecithal eggs, a process known as **epiboly** occurs in which the small cells of the animal pole grow downward over the larger cells of the vegetal pole, thus effectively tucking the yolk inside and forming a blastopore (fig. 5.4*a–c*). Yolk cells then constitute the archenteric floor (figs. 5.4*d–f* and 5.5).

Notochord and Dorsal Mesoderm As involution gets under way, a stream of undifferentiated cells flows over the dorsal lip of the blastopore to establish in the roof of the archenteron a narrow band of **chordomesoderm** (fig. 5.4*f*). A notochord organizes from the midline portion of the chordomesoderm and remains, temporarily, in the roof of the archenteron. The chordomesoderm immediately lateral to the notochord gives rise to a pair of longitudinal bands of mesoderm that parallel the notochord. These bands become the **dorsal mesoderm.** They subsequently segment and reorganize to form hollow **mesodermal somites** (fig. 5.9*b* and *c*). These are the source of the body wall musculature. The formation of the notochord and dorsal mesoderm in frogs is essentially the same as in an amphioxus. Both arise from streams of undifferentiated cells that flow over the dorsal lip of the blastopore, and both become established in the roof of the archenteron. The rest of the mesoderm forms in a slightly modified manner possibly dictated by the presence of the yolk.

Endoderm and Ectoderm At the same time that the notochord and dorsal mesoderm are differentiating, presumptive endoderm streams to the interior, fans out, and

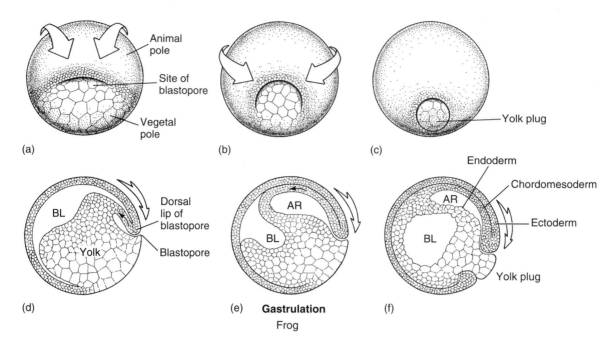

FIGURE 5.4

Epiboly, involution, and the blastopore of a frog. (*a*)–(*c*) Epiboly to yolk stage. (*a*) is an advanced stage of fig. 5.1. (*d*)–(*f*) Hemisections of (*a*)–(*c*) showing involution at blastopore. (*f*) Compare fig. 5.5. **AR**, archenteron. **BL**, blastocoel. Arrows indicate growth of cells.

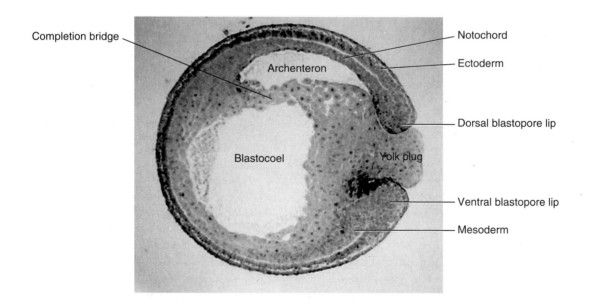

FIGURE 5.5

Early frog gastrula. Notochord is differentiating from chordomesoderm. Redistribution of weight has rotated embryo 90 degrees counterclockwise from orientation in fig. 5.4*c*.

From Phillips, JB: *Development of Vertebrate Anatomy* 1975 (Fig. 7.5c page 137).

provides, with the yolk, a lining for the archenteron. The cells on the surface of the gastrula, along with additional ones that are being rapidly proliferated, become the ectoderm.

Intermediate and Lateral Mesoderm   The cells that give rise to mesoderm other than that of the somites reach the interior by flowing over the lateral and ventral rims of the blastopore. Once inside, they stream forward as an advancing sheet of rapidly expanding **lateral-plate mesoderm,** pushing its way cephalad between ectoderm and endoderm. Unlike dorsal mesoderm, the lateral-plate mesoderm does not segment and, except for a narrow strand paralleling the dorsal

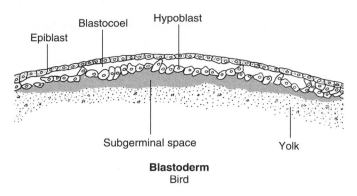

FIGURE 5.6

Blastoderm at animal pole of avian egg, sagittal section. Future head end is to left. Only a fraction of the massive uncleaved yolk is shown.

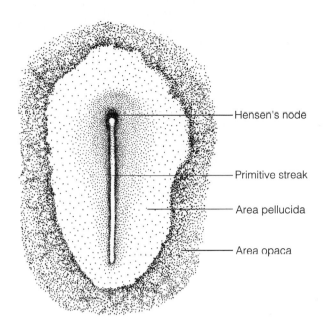

FIGURE 5.7

Primitive streak stage of chick embryo at 16 hours of incubation. The tissue seen is the animal pole of the egg. It was removed from the yolk with scissors, fixed, stained, and rendered translucent. The area opaca is the site in the yolk sac where blood islands are forming. It is therefore opaque when viewed with transmitted light. The area pellucida was not attached to the yolk; hence it transmits light readily.

mesoderm, it soon splits into two lamina—an outer sheet, the **somatic mesoderm,** and an inner sheet, the **splanchnic mesoderm** (fig. 5.9b). The cavity between the two sheets is the **coelom.** As in an amphioxus, somatic mesoderm plus ectoderm constitute the **somatopleure (body wall);** splanchnic mesoderm plus endoderm constitute the **splanchnopleure.** The strand of lateral-plate mesoderm immediately lateral to the mesodermal somites (that which doesn't participate in the splitting process) is **intermediate (nephrogenic) mesoderm** (fig. 5.9b). It gives rise to the kidney tubules and to the ducts of the urogenital system.

Chick

We will use the chick to illustrate gastrulation in amniotes with macrolecithal eggs. But first, we must review the term blastoderm.

Blastoderm: Epiblast and Hypoblast   A macrolecithal egg has a massive yolk. Epiboly would take a long time to tuck it into the interior. As an adaptation, nature has created an alternate means of achieving a **triploblastic** (three germ-layered) **organism.** The adaptation is a **blastoderm,** a result of segmentation of the cytoplasm at the animal pole only. At first, the blastoderm is simply a thin multi-layered mound of cells perched on top of the massive yolk like a skullcap (fig. 5.6). This blastoderm then organizes into an upper sheet of cells, the **epiblast,** and a lower sheet, the **hypoblast,** as seen in illustration. Cells from the periphery of the hypoblast slowly grow outward over the surface of the yolk, then downward around it to become the **endodermal lining** of a yolk sac. The epiblast is then chiefly presumptive ectoderm and the hypoblast is presumptive endoderm. This process of endoderm formation is known as **delamination.**

Notochord and Mesoderm Formation   Before looking at initial notochord and mesoderm formation, it is necessary to examine two additional structures that

develop within the blastoderm: (1) a multilayered longitudinal **primitive streak** (fig. 5.7); and (2) **Hensen's node,** a thickened nodule of closely packed blastoderm cells that defines the caudal end of the future embryo (figs. 5.7 and 5.8). The primitive streak and Hensen's node are the functional equivalent of the blastopore of eggs with less yolk.

From Hensen's node a **notochordal process** pushes forward beneath the epiblast (fig. 5.8), while other cells stream forward alongside the notochord to establish the dorsal mesoderm (fig. 5.8, dark red). Concurrently, lateral-plate mesoderm streams from both sides of the primitive streak, between the ectoderm and the endoderm (epiblast and hypoblast), and onto the surface of the yolk in all directions except the region of the future head (fig. 5.8, arrows). While the dorsal mesoderm is segmenting to form somites (fig. 5.9c), the lateral mesoderm is splitting into somatic and splanchnic sheets (fig. 5.9d). The resulting cavity between the two is the coelom.

By the end of the second day of incubation, the splanchnopleure, which has spread outward onto the surface of the yolk to form a yolk sac, has produced the first blood cells and a network of delicate **vitelline (omphalomesenteric) vessels** in the **area opaca** (fig. 5.9c). These vessels collect yolk globules and transport them to the simple, twitching, S-shaped heart

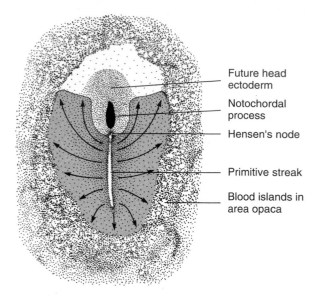

FIGURE 5.8

Chick embryo at 18 hours of incubation. Arrows indicate morphogenic streaming of lateral-plate mesoderm (*light red*) from the primitive streak. *Dark red* indicates dorsal mesoderm from Hensen's node. Fig. 5.9*c* shows primitive streak in an embryo about 8 hours older.

Future head ectoderm
Notochordal process
Hensen's node
Primitive streak
Blood islands in area opaca

(fig. 16.6). The heart then propels this early stream of cells and nutrients to the rapidly growing tissues within the embryo.

Overview As a result of complex morphogenetic movements and accompanying cell proliferation, the chick establishes ectoderm and endoderm; dorsal, intermediate, and lateral mesoderm; a coelom; and the beginning of an extraembryonic yolk sac. For a look at a later stage showing the relationship between the splanchnopleure, yolk sac, and trunk, you may wish to look at figure 5.13*a*.

The processes of gastrulation described above for vertebrates with mesolecithal and macrolecithal eggs differ in details among the vertebrate groups that exhibit them. Variations are most numerous among bony fishes, since there are so many divergent species.

Placental Mammals

At the time gastrulation commences in mammals, the blastocyst exhibits an **inner cell mass** derived from the animal pole of the blastula (fig. 5.10, blastocysts). The wall of the blastocyst of mammals does not contribute to formation of the embryo; the embryo develops from the inner cell mass. The blastocyst wall, called a **trophoblast,** establishes contact with, or implants itself into, the lining of the maternal uterus (in humans, at about 14 days). It is the first membrane through which the **conceptus** (embryo plus extraembryonic membranes) receives nourishment and oxygen directly from

maternal tissues. Any earlier nutrition is histotrophic (see Oviparity and Viviparity, above).

The method of formation of the first endoderm is not precisely the same in all placentals. It is, however, a product of the inner cell mass, whether by cell migration or delamination. In either case, the inner cell mass eventually is converted into epiblast and hypoblast. It is then called an **embryonic disk** or **blastodisk.** Thereafter, there is much streaming of ectoderm, presumptive mesoderm, and endoderm from the embryonic disk, resulting in establishment of the germ layers and, simultaneously, of the extraembryonic membranes. Among the latter is a yolk sac (fig. 5.10, gastrula) for yolk that doesn't exist. The trait bespeaks the reptilian ancestry of mammals.

Before concluding the discussion on vertebrate gastrulation, a final word must be added. Because of the diversity of the specific processes whereby various vertebrates achieve a bilaterally symmetrical, three germ-layered stage with notochord and neural tube rudiments, it is not possible to specify a common end point for the process of gastrulation that would be applicable to all species. Gastrulation places the germ layers in their respective positions. When this has been accomplished, gastrulation is complete and organogenesis can proceed.

Neurulation

Presumptive neural ectoderm occupies a wide band on the dorsal surface of the elongating gastrula, overlying the notochord. As elongation of the gastrula continues, the band thickens to become a **neural plate (floor plate,** in the terminology of the developmental biologist). The lateral borders of the neural plate become elevated to form a pair of neural folds bounding a neural groove. The embryo is now a **neurula** (fig. 5.9*a*). The anterior end of the neural groove is widest, being the future brain. The rest is future spinal cord. The floor plate sinks into the dorsum of the embryo, and the neural folds grow toward one another above the groove, meet in the middorsal line, then fuse, converting the neural groove into a **neural tube.** The process of closure commences near the caudal end of the brain and sweeps cephalad and caudad. Finally, only a small opening remains at each end of the tube **(anterior** and **posterior neuropores).** These later close. As a result of formation of a neural tube by closure of the neural groove, the central nervous system is ectodermal, dorsally located, and hollow. The cavity is the **neurocoel.** It remains as the ventricles of the brain and central canal of the spinal cord. The process that establishes the neural tube is **neurulation.**

Neurulation differs among the vertebrate classes in minor details. The major variation is seen in cyclostomes and ray-finned fishes, in which neural folds do not form, and the floor plate does not roll into a tube. Instead, it becomes a wedge-shaped **neural keel** that invades the

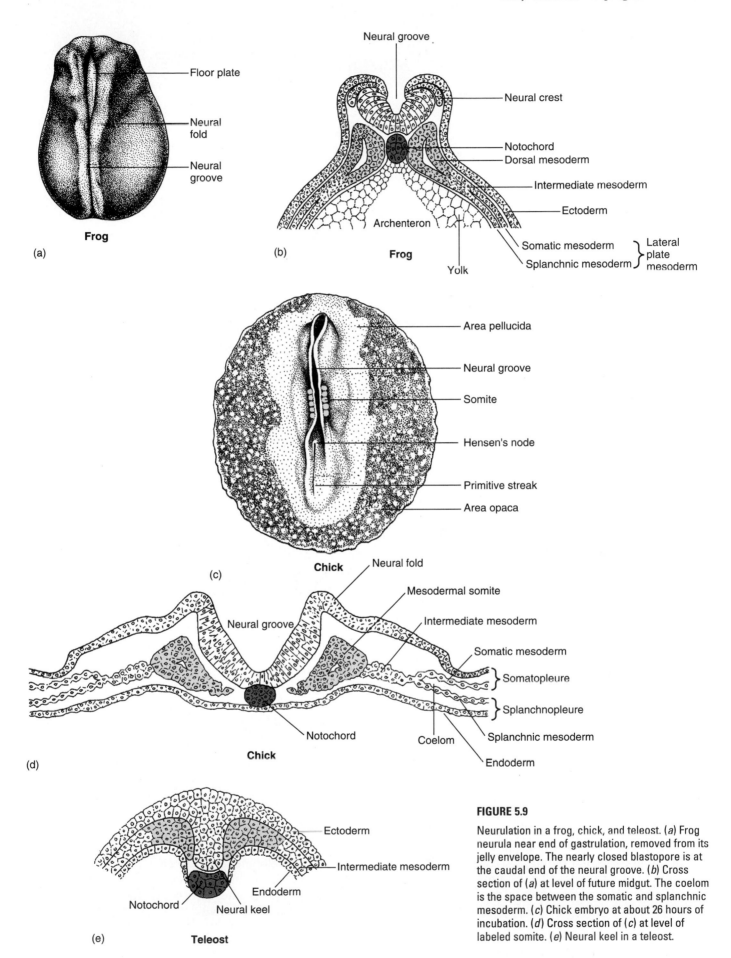

Floor plate

Neural fold

Neural groove

Frog

(a)

Neural groove

Neural crest

Notochord

Dorsal mesoderm

Intermediate mesoderm

Ectoderm

Archenteron

Somatic mesoderm } Lateral plate mesoderm
Splanchnic mesoderm }

Yolk

(b)   **Frog**

Area pellucida

Neural groove

Somite

Hensen's node

Primitive streak

Area opaca

(c)   **Chick**

Neural fold

Mesodermal somite

Intermediate mesoderm

Somatic mesoderm

Somatopleure

Splanchnopleure

Splanchnic mesoderm

Endoderm

Neural groove

Notochord

Coelom

Chick

(d)

Ectoderm

Intermediate mesoderm

Endoderm

Notochord   Neural keel

(e)   **Teleost**

FIGURE 5.9

Neurulation in a frog, chick, and teleost. (*a*) Frog neurula near end of gastrulation, removed from its jelly envelope. The nearly closed blastopore is at the caudal end of the neural groove. (*b*) Cross section of (*a*) at level of future midgut. The coelom is the space between the somatic and splanchnic mesoderm. (*c*) Chick embryo at about 26 hours of incubation. (*d*) Cross section of (*c*) at level of labeled somite. (*e*) Neural keel in a teleost.

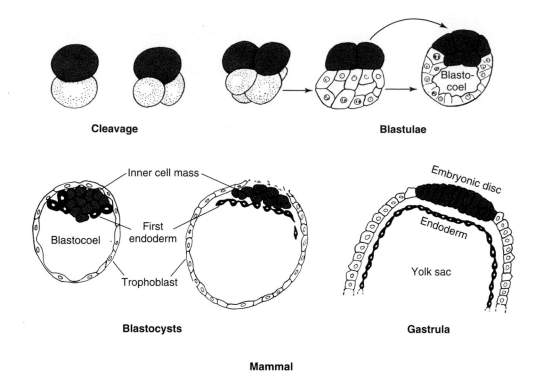

FIGURE 5.10

Cleavage, blastula, and gastrulalike stage of a placental mammal. At cleavage (upper left) the red cell at the animal pole gives rise to the embryo. The other cell gives rise to the trophoblast. The first endoderm comes from the inner cell mass. The yolk sac contains no yolk.

dorsal body wall above the notochord (fig. 5.9e). Subsequently, the wedge becomes detached from the overlying ectoderm and organizes a neurocoel by rearrangement of the interior cells. The result is a typical dorsal hollow central nervous system.

Induction of Differentiation: Morphogens

Early in the twentieth century, Hans Spemann, using transplantation of tissues in amphibian gastrulas to induce differentiation of an organ in an ectopic (abnormal) location, demonstrated that the dorsal lip of the blastopore is an **organizer area** that establishes the long axis of the vertebrate body. This was the first demonstration that preexisting cells govern the fate of contiguous ones. Experimental biologists using similar techniques later demonstrated the organizer effects of one embryonic tissue after another in species from fruit flies to fishes to chicks to mice. Chapter 17 cites induction of the lens of the eye from embryonic ectoderm in the presence of a retina. What could not be answered with techniques then available was how induction brought about its effect. Obviously, it had a genetic basis; but how, specifically, were genes expressed? Answers are now coming from molecular biology.

Genes similar to those that are important in fruit fly development are the key molecular choreographers of the major events of gastrulation, neurulation, and cer-

tain later embryonic developments in vertebrates. In fruit fly embryos, these genes cause the ventral cuticle region to take on a bristly appearance like that of a hedgehog. These genes in vertebrates have therefore been dubbed **hedgehog** (also **homeobox,** or "similar signaling") **genes.** Hedgehog genes have been shown, thus far, to participate in differentiation of the notochord, the early neural tube, and fins or limbs in vertebrates. The molecular basis for induction of differentiation is a protein encoded by the genes. The protein is called a **morphogen.** Without morphogen there is no induction. There is evidence that in at least some inductive processes, more than one morphogen is involved. Whether or not these together participate in producing a common final inductor, or whether they regulate parallel processes, is not yet known.

Chordin is a secreted morphogen encoded by genes in the dorsal lip of the blastopore during early proliferation of presumptive notochord cells. It therefore has potent body axis-forming activity in the early stages of gastrulation. The transcription of chordin is activated by growth factors that also induce formation of dorsal mesoderm in the *neurula stage*. At the *tailbud stage* (two days of development in *Xenopus,* the experimental amphibian), chordin is expressed only in the tailbud region, which is the last population of cells with chordin-organizing activity. The fact that chordin is a protein consisting of 941 amino acids, and that it is only one of several morphogens regulating differentiation in the dorsal axial region of the early gastrula, attests to the complexity of the problem of exploring induction processes.

Formation of the first fundament of the central nervous system, the floor plate (fig. 5.9), requires inductive signals from the notochord. The morphogen in this instance is not a diffusible secretion; induction requires direct contact of the notochord with the overlying ectoderm. Differentiation of the components of the neural tube, which occurs later, is accomplished by diffusible morphogens. The floor plate of the gastrula participates with the notochord in inducing differentiation of motor neurons in the ventral half of the neural tube. (This half, illustrated in fig. 16.7, becomes the **basal plate of the spinal cord,** and is so labeled. It is not to be confused with the floor plate of the gastrula.) It is this area of the neural tube that is in contact with the notochord. Ventral portions of the developing brain are also responsive. Induction of components of the dorsal half of the tube appears not to be the province of the notochord.

The same genes that act early in gastrulation initiate, later, the development of pectoral fins, chick wings, and the anterior limbs of mice. (Limb and fin buds are illustrated in figs. 11.17 and 11.18.) Participating with the hedgehog genes is a growth factor produced in the thickened ectodermal ridge overlying the mesoderm of the limb bud. Removal of the ridge halts limb bud development; but the (fibroblast) growth factor, when implanted, restores the function of the epidermal ridge, inducing the continued proliferation of mesenchyme that keeps the limb bud growing outward from the body. Substituting anterior limb bud mesoderm for the blastema of the posterior bud in chick embryos has resulted in reversal of the position of wings and hind limbs. Research on development of the hand of a chick indicates the probability that successive morphogens are responsible for the development of successive digits.

If a homeobox gene is defective, the result may be a severe developmental anomaly. For instance, failure of dorsal closure of a section of the neural tube, a condition known as **spina bifida** in human infants, is a result of failure of a normal inductive process early in ontogeny. Anomalies of the arm, hands, or digits have a similar etiology.

Mesenchyme

Mesenchyme is an unorganized accumulation of dendritic (branching), undifferentiated, embryonic cells that comprise much of an early embryo. During organogenesis, under the influence of morphogens, mesenchyme becomes organized in aggregates known as **blastemas.** These gradually assume the general outlines of a future organ, and the cells therein commence to differentiate to form appropriate tissues. As organogenesis proceeds, mitosis of the differentiating cells participates in furthering the growth of the organ.

Although mesenchyme is primarily embryonic, some mesenchyme formation continues into adulthood, albeit at an increasingly slower rate as aging progresses. These are reserve cells that can replace some varieties of worn or damaged tissues, especially connective tissues. Many specialized tissues—epithelial, for instance—are replaced solely by mitosis of existing cells in that tissue. Epidermal cells are replaced only by mitosis of the already differentiated cells of the basal layer of the epidermis. Most mesenchyme arises from mesoderm. Mesenchyme of ectodermal origin, a minor source, is known as **mesectoderm.** If the source is neural ectoderm (neural crests, for example), it is referred to as **neurectoderm.** A few of the many varieties of connective tissue that arise from mesenchyme are listed in figure 7.1.

Blastema cells are totipotent—they can be induced to become any specialized cell that the inductor substance specifies. This accounts for ectopic organs induced in transplantation experiments on embryos.

FATE OF THE ECTODERM

Embryonic ectoderm gives rise to (1) the entire nervous system and some of its membranes, (2) the lens and retina of the lateral eyes and the sensory epithelium of the other special sense organs that monitor the external environment, (3) the epidermis and its derivatives, including glands that lie within the dermis, (4) the lining of the stomodeum and proctodeum and their derivatives, and (5) a broad spectrum of other tissues, particularly in the head and pharyngeal regions, that result from migration of mesenchyme from neural crests and ectodermal placodes. These other tissues include cartilage, bone, and other connective tissues, which elsewhere arise from mesoderm.

Stomodeum and Proctodeum

Formation of the oral cavity commences with a **stomodeum,** a midventral invagination of the ectoderm of the embryonic head that develops at the anterior end of the foregut (figs. 1.1 and 17.6a). The two cavities are temporarily separated by a thin partition, the **oral plate,** the rupture of which provides an opening to the digestive tract from the exterior. Stomodeal ectoderm lines the anterior portion of the oral cavity and covers at least the anterior part of the tongue. There is, however, some overriding of stomodeal ectoderm by endoderm, so we cannot say with certainty to what extent either ectoderm or endoderm contributes to some of the oral glands that evaginate near the junction of the two.

Stomodeal ectoderm provides the cells that secrete the enamel of mammalian teeth; however, the enameloids of some lower vertebrates are secreted by cells that are thought to be of neural crest origin. An evagination (sometimes a club-shaped growth) known as **Rathke's pouch** evaginates from the roof of the stomodeum before the oral plate ruptures, and the pouch gives rise to the adenohypophysis (fig. 18.6*a*).

An ectodermal invagination, the **proctodeum,** similar to the stomodeum, develops in relation to the embryonic hindgut. The proctodeum and hindgut are separated temporarily by a **cloacal plate** (fig. 1.1).

Neural Crests

As the neural groove sinks into the dorsal body wall and the neural folds meet above the groove to form a tube, some of the ectoderm at the site of union of the folds separates from the surface ectoderm and from the neural tube to form a pair of mesenchymal ribbons paralleling the neural tube on each side. Shortly thereafter, the ribbons become segmented, forming a cluster of cells in each metamere and several clusters lateral to the midbrain and hindbrain. These neurectodermal clusters are **neural crests** (fig. 5.9*b*). They are induced by protein growth factors evidently synthesized in the dorsal walls of the neural groove.

Neural crest cells proliferate rapidly, giving rise to a large number of descendants, many of which migrate to areas remote from the crests. Some of the cells in the head stream ventrad into developing pharyngeal arches where they form blastemas for the skeleton of the arches, including that of the jaws. Others contribute to the anteroventral portion of the neurocranium and, in chicks at least, to some of the membrane bones that ensheathe the neurocranium. Still others invade the anlagen of the thyroid, parathyroid, thymus, and ultimobranchial bodies that are developing from the epithelium of the pharyngeal pouches or pharyngeal floor. The lingual cartilages and branchial basket of agnathans are products of neural crest mesenchyme. Induction of many of these cranial derivatives appears to be initiated by morphogens from head ectoderm or endoderm. Some of the derivatives of cranial neural crests, such as cartilage and bone, are of mesodermal origin elsewhere in the body.

Neural crest mesenchyme in the head or trunk streams ventrad around the notochord to form autonomic ganglia and the specific cells of the adrenal glands that produce adrenaline (fig. 16.35). Other emigrants aggregate around the developing neural tube to invest it with meninges other than the dura mater. Still others become pigment cells not only in the skin, but in muscles and in many internal organs other than the retina, brain, and spinal cord.

One might suppose, with all this emigration of cells, that there would be few remaining in the neural crests. This is not so. Mitosis continually replenishes the supply until organogenesis is well underway. A very large number in the crests remain there to become the cell bodies of sensory neurons in discrete groupings (ganglia) on the

pathway of spinal and cranial nerves (fig. 16.5*a* and *b*). Others migrate distally and centrally along the course of the sensory nerve fibers, contributing to the neurilemma, a living membrane that ensheathes all nerve fibers outside the central nervous system.

Ectodermal Placodes

Ectodermal placodes are paired localized thickenings of the embryonic ectoderm that sink beneath the skin to one degree or another and give rise to neuroblasts and to the sensory epithelia of certain sense organs. Any other ectodermal thickenings that are characterized as placodes—those, for example, that become the lens of the lateral eyes (lens placode, fig. 17.6*b*)—do not give rise to neurectoderm. Presumptive neurectodermal placodes can be grouped as follows:

1. A pair of **nasal (olfactory) placodes** forms just above the stomodeum at the anterior end of the brain. These placodes sink into the head and become part of the lining of a pair of nasal pits that open to the exterior via nostrils (fig. 13.7). Some of the cells of these placodes become neurosensory cells in the olfactory epithelium. Their processes grow into the olfactory bulb of the brain and constitute, collectively, the olfactory nerves.
2. A pair of **otic placodes** in the ectoderm lateral to the hindbrain sink deep into the head to become otocysts (fig. 17.6*a* and *b*). These are precursors of the sensory epithelia of the membranous labyrinth (inner ears).
3. A series of **epibranchial placodes** on the side of the head (named for their location in fishes) sink into the head to a location alongside the midbrain and hindbrain, where they might be mistaken for neural crests. These placodes, like neural crests, contribute neuroblasts that become sensory cell bodies in one pair or more of the ganglia on cranial nerves V, VII, IX, and X. The number varies with the species, augmenting the cell bodies coming from neural crest cells.
4. A **linear series of placodes** extending the length of the trunk and tail of fishes and amphibians, and continuing onto the head in a variable pattern, sink into the dermis to become the neuromast organs of the cephalic and lateral-line canal systems. Other placodes on the head of fishes, not part of the linear series, become thermoreceptive and electroreceptive epithelia.

From the discussions on neurulation, neural crests, and ectodermal placodes, it may be evident that the components of the entire nervous system, central and peripheral, many of their supportive and nutritive cells, and the sensory epithelia of all the special sense organs other than taste buds are of either neural plate, neural crest, or ectodermal placode origin. It is also evident that the embryonic ectoderm exerts a much greater influence on differentiation and organogenesis than might be supposed from its superficial position in early embryos.

FATE OF THE ENDODERM

Endoderm gives rise to the epithelium of the entire alimentary canal and its evaginations between the stomodeum and proctodeum. The epithelioid components of the parathyroid glands, ultimobranchial glands, thymus, and the lining of the auditory tube and middle ear cavity are endodermal because of their origin from pharyngeal pouches. Midventral evaginations of the pharynx, including thyroid, lungs, and swim bladders and their ducts, if any, are lined with endoderm or have endodermal components. Caudal to the pharynx the endoderm evaginates to form crop sacs, liver, gallbladder, pancreas, and various gastric and intestinal ceca. Most urinary bladders and the urogenital sinus of mammals are lined with endoderm, because these are derived from the cloaca. Any other structure that arises as an evagination of the embryonic foregut, midgut, or hindgut has endodermal components.

FATE OF THE MESODERM

Dorsal Mesoderm (Epimere)

The somites collectively constitute the dorsal mesoderm. They are aligned beside the notochord and neural tube the entire length of the embryonic trunk and tail, and develop in the head in varying numbers (figs. 5.9c and d, and 16.6). A somite exhibits three regions, myotome, sclerotome, and dermatome (fig. 5.11). **Myotomes** contribute mesenchyme that migrates into the body wall of the trunk and tail to form skeletal muscles. **Sclerotomes** give rise to the vertebral column and proximal portion of ribs; those of head somites give rise to parts of the neurocranium, and to one or more sense capsules, depending on the species. **Dermatomes** contribute mesenchyme that forms the dermis of the skin of the dorsum. (Most of the dermis arises from lateral-plate mesoderm or, in the head, from neural crests.)

Lateral-Plate Mesoderm (Hypomere)

Lateral-plate mesoderm is confined to the trunk and consists of somatic and splanchnic mesoderm (figs. 5.9b and d and 5.11). **Somatic mesoderm** gives rise chiefly to the connective tissues and blood vessels of the body wall and to the skeleton of the body wall, girdles, and limbs. The dermis of the body wall is its outermost product, and the parietal peritoneum (fig. 1.2, 5) is the innermost. **Splanchnic mesoderm** gives rise to smooth muscles and connective tissue of the digestive tract and of its outpocketings, to the heart, and to the blood vessels of the viscera. The visceral peritoneum (fig. 1.2, 6) is its outermost

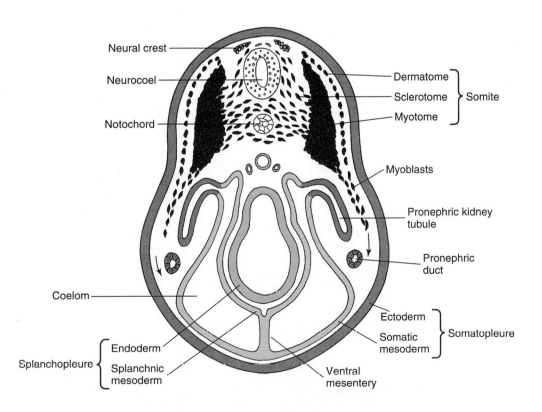

FIGURE 5.11

Vertebrate embryo in cross section showing fate of the mesoderm. *Dark red*, mesodermal somite (dorsal mesoderm); *medium red*, derivatives of intermediate mesoderm; *light red*, lateral mesoderm. Sclerotome cells are streaming toward notochord and neural tube to form a vertebra; myotomal cells (myoblasts) are streaming into lateral body wall (*arrows*) to form myotomal muscle; dermatome will form dermis of skin of dorsum.

derivative. Lateral-plate mesenchyme streams into the developing head in chicks and at least some other vertebrates to contribute to the developing skull and pharyngeal arches.

Intermediate Mesoderm (Mesomere)

Intermediate (nephrogenic) mesoderm consists of a pair of longitudinal ribbons of unsegmented mesoderm extending the length of the trunk just lateral to the somites (fig. 5.9b and d). It gives rise to kidney tubules and the longitudinal ducts of the urogenital system.

SIGNIFICANCE OF THE GERM LAYERS

There appears to be nothing intrinsic to the earliest cells of the germ layers that determines the direction of differentiation of their descendants. The early generations are totipotent. The factors that drive differentiation are the morphogens, but we have no idea what induces Spemann's organizer area to start putting them out. That puzzle may not be solved for a long time. The germ layers, positioned by an inherent genetic code, constitute the initial cellular organization that enables organogenesis to proceed.

EXTRAEMBRYONIC MEMBRANES

Most embryonic vertebrates are provided with special membranes that extend beyond the body. These extraembryonic membranes arise early in ontogeny and perform important services for the embryo until hatching or birth. The chief extraembryonic membranes are the **yolk sac, amnion, chorion,** and **allantois.** The last three are found only in reptiles, birds, and mammals. The yolk sac is the most primitive.

Yolk Sac

The yolk sac surrounds the yolk (figs. 5.12 and 5.13). It empties into the midgut and is usually lined by endo-

derm, although not in bony fishes. The sac is highly vascular, and its vessels **(vitelline arteries** and **veins)** are confluent with circulatory channels within the embryo proper. Yolk particles in the sac are usually digested by enzymes secreted by the lining of the sac and then transported to the embryo by vitelline veins. (In sharks, yolk also enters the intestine directly from the sac, propelled by rapidly beating cilia that line the sac and stalk.) The yolk sac becomes smaller as the embryo grows larger. As it shrivels, it slowly disappears into the ventral body wall. The intracoelomic remnant of the sac is finally incorporated into the wall of the midgut, or it may remain as a small diverticulum.

In embryonic sharks, a large diverticulum of the yolk sac develops within the coelom close to where the sac empties into the duodenum. In pups ready to be born and with the yolk sac almost completely retracted, the diverticulum is easily demonstrated. It is distended with yolk, which has also spilled over into the spiral intestine. This remaining yolk nourishes the newborn pup for several days until it is able to obtain food from the environment.

Despite the fact that no yolk is present in placental mammals, the embryos develop a yolk sac—a reminder of their genetic relationship with egg-laying reptiles. In humans, a vestige of the yolk sac (Meckel's diverticulum) remains in about 2 percent of the adult population. Its average position is on the small intestine, 30 cm above the ileocolic valve. Its average length is about 5 cm.

Because the yolk sac in viviparous fishes and amphibians is highly vascularized and lies close to maternal tissues, it often serves as a membrane for absorbing oxygen from the parent. After the yolk in the sac is depleted, or if there is none, the sac may absorb nourishment from maternal tissues. When functioning in either capacity, it constitutes a **simple yolk sac placenta.**

Amnion and Chorion

The embryos of reptiles, birds, and mammals develop within two membranous sacs, an amniotic sac composed of an **amnion,** and a chorionic sac composed of a **chorion.**

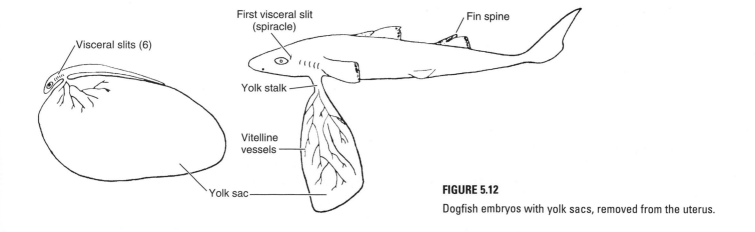

FIGURE 5.12

Dogfish embryos with yolk sacs, removed from the uterus.

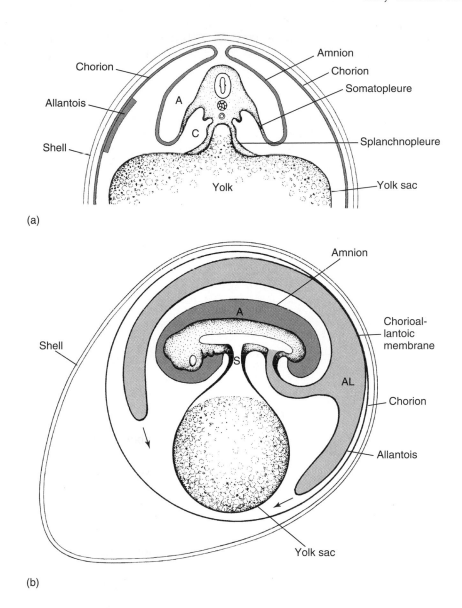

(a)

(b)

FIGURE 5.13

Extraembryonic membranes of an egg-laying amniote. (*a*) Cross section of embryo showing amnion and chorion arising from amniotic folds of somatopleure. Position of the allantois is indicated on left of drawing. (*b*) Relationships of the allantoic sac (**AL**), which is growing in direction of arrows. **A,** amniotic cavity; **AL,** allantoic cavity; **C,** intraembryonic coelom; **S,** yolk stalk, an evagination of the midgut.

These two membranes are formed when upfoldings of the embryonic somatopleure meet above the embryo and fuse, forming the amniotic sac around the embryo. The chorion forms with continued growth of somatopleure and surrounds both the amniotic sac and the yolk sac (fig. 5.13*a*). Figure 15.14 shows fetuses within these sacs in the uterus of a pregnant dog. The amnion and chorion, along with the semiporous eggshell, enabled ancestral reptiles and their descendants to lay their eggs on land, which had become their primary environment.

Surrounding the embryo within the amniotic sac is a slightly salty **amniotic fluid** consisting chiefly of metabolic water from embryonic tissues. Turtle eggs that are laid in moist locations absorb considerable water from the environment through the shell, but this is not a common source of amniotic fluid. When the embryonic kidneys begin to function, nitrogenous wastes are added to the fluid. Amniotic fluid buffers the fetus against mechanical injury and, in oviparous species, helps prevent dessication of the embryo.

The chorion lies in intimate relationship with either the eggshell or the lining of the mother's uterus, depending on the species. Thus, the amnion with its amniotic fluid protects the fetus, and the chorion keeps it in communication with its source of oxygen and, in viviparity, its source of nutrients.

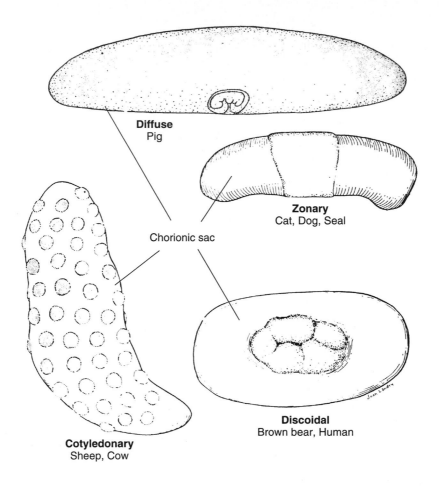

Diffuse
Pig

Zonary
Cat, Dog, Seal

Chorionic sac

Cotyledonary
Sheep, Cow

Discoidal
Brown bear, Human

FIGURE 5.14

Chorionic sacs of mammals showing placental areas. Pig embryo surrounded by the amnion can be seen through the exceptionally thin chorionic sac.

Allantois

The **allantois** is an extraembryonic membranous sac that arises typically as a midventral evagination of the embryonic cloaca (fig. 5.13*b*). It ordinarily grows until it comes in contact with the inner surface of the chorion to form a **chorioallantoic membrane** (fig. 5.13*b*). The chorioallantoic membrane in reptiles, birds, and monotremes is in contact with the eggshell where, except in those viviparous squamates that depend on the mother late in pregnancy for at least some nourishment, it serves chiefly as a *respiratory organ*. In mammals above marsupials, the chorioallantoic membrane is in direct contact with the lining of the maternal uterus, where it constitutes the fetal part of a **chorioallantoic placenta.** Here it performs two additional functions, being also the site of *transfer of nutrients* from maternal tissues to the embryo and of *transfer of metabolic wastes* from embryo to mother.

The base of the allantois—the part closest to the cloaca—becomes the **urinary bladder** of amniotes (fig. 14.38). In mammals, the part extending between the tip of the bladder and the umbilicus may remain after birth as a **middle umbilical ligament,** or **urachus** (fig. 15.49*f*); the portion within the umbilical cord is discarded at birth. It is likely that the allantois is the old amphibian urinary bladder with added functions.

Placentas

The term "placenta" in its broadest sense refers to any region in a viviparous organism where maternal and embryonic tissues of any kind are closely apposed, and which serves as a site for exchanges between parent and embryo. In this sense, viviparous fishes that develop in an ovarian follicle sometimes exhibit a placenta. In a more restricted sense, a placenta is an organ composed of (1) the highly vascular region of an extraembryonic membrane (yolk sac, choriovitelline membrane, chorioallantoic membrane, or chorion alone), and (2) the associated highly vascular lining of the maternal uterus.

The yolk sac frequently serves as part of a placenta in viviparous amphibians and fishes such as the dogfish shark. These vertebrates have no amnion, chorion, or allantois, and a **simple yolk sac placenta** lies in immediate contact with the maternal uterus. In reptiles that depend on viviparity near the end of fetal life, a soft, thin eggshell lies between the fetal membranes and the uterine lining, but it does not impede the essential functions of a placenta. In most marsupials, the yolk sac, rather than the allantois, lies against the chorion as part of a **choriovitelline placenta.** As stated earlier, mammals above marsupials have a **chorioallantoic placenta.** These mammals have an umbilical cord connecting the fetus with the placenta (fig. 14.38). In some of them (cats, rabbits, humans, others), the allantois extends only part way out the umbilical cord before dwindling and terminating as a blind sac, but its vessels, the allantoic (umbilical) arteries and veins, continue toward, and vascularize, the chorion.

The intimacy of the relationship between fetal and maternal tissues in mammals varies greatly. In marsupials and most ungulates, the extraembryonic membranes lie in simple contact with the uterine lining (**endometrium**), and at birth the fetal membranes simply peel away without any shedding of the lining. This is a **contact** or **nondeciduous placenta.** In more intimate relationships, **chorionic villi,** which are fingerlike outgrowths of the chorionic sac, become rooted to one degree or another into the endometrium, or even dangle into uterine blood sinuses, as in humans. When the fetal part of such a placenta disengages at parturition, the *invaded portion* of the uterine lining, the **decidua,** is shed, and some bleeding occurs. This is a **deciduous placenta.** In any case, the extraembryonic membranes and any sloughed uterine tissues are delivered as the afterbirth.

Chorionic villi are distributed on the surface of the mammalian chorionic sac (fig. 5.14) in isolated patches (**cotyledonary placenta**), in a band encircling the sac (**zonary placenta**), in a single large discoidal area (**discoidal placenta**), or diffusely over the entire surface of the chorion (**diffuse placenta**). The mammalian placenta is one source of hormones essential for the maintenance of pregnancy.

SUMMARY

Definitions may paraphrase those in the text.

1. Eggs are micro-, meso-, or macrolecithal, depending on the quantity of yolk. They are isolecithal or telolecithal, depending on the distribution of cytoplasm and yolk within the egg.

2. Oviparity is a condition in which unfertilized or fertilized eggs are shed to the exterior for development and hatching.

3. Viviparity is a condition in which offspring are born alive. The developing organisms may or may not be dependent on the parent for nourishment before birth.

4. Viviparity is found in some members of every vertebrate class except agnathans and birds.

5. Euviviparity is a viviparous condition in which offspring are dependent on the parent for nourishment throughout pregnancy.

6. Ovoviviparity is a viviparous condition in which offspring are not dependent on the parent for nourishment throughout pregnancy.

7. Fertilization is internal in viviparous species, in urodeles and apodans, and in species that cover the egg with a shell.

8. Cleavage follows fertilization. It produces a blastula with a blastocoel.

9. Gastrulation is characterized by notogenesis, neurulation, establishment of bilateral symmetry, and formation of the germ layers. The basis of the process is morphogenetic streaming of undifferentiated cells.

10. Gastrulation is initiated by involution in cephalochordates, epiboly in species with mesolecithal eggs, and delamination of a blastoderm in species with macrolecithal eggs.

11. In placental mammals an inner cell mass gives rise to the embryo. Gastrulation is by delamination or a modification thereof.

12. A trophoblast membrane develops from the blastula wall, providing initial embryo and uterine contact.

13. Neurulation results in formation of a brain and spinal cord.

14. Morphogens are inductor (signaling) proteins that induce organization of the germ layers and participate in differentiation of specialized tissues from embryonic mesenchyme. Morphogens are the expression of specific homeobox (hedgehog) genes.

15. Mesenchyme is an embryonic tissue consisting of undifferentiated, totipotent, embryonic cells that require inductor factors for differentiation.

16. Ectoderm contributes epidermis and its derivatives, the entire nervous system, epithelia of special sense organs such as lateral eyes and inner ears, the lens of paired eyes, the lining of the stomodeum and proctodeum, and a broad spectrum of miscellaneous tissues that differentiate from neural crests.

17. Neural crests give rise to the sensory ganglia on the roots of cranial and spinal nerves, to autonomic ganglia, to many of the skeletal components of the skull and pharyngeal arches, to most pigment cells, and to numerous miscellaneous tissues.

18. Ectodermal placodes give rise to neurosensory cells of the olfactory epithelium, to sensory epithelia of the inner ear, to a variety of mechano-, thermo-, and electroreceptive organs of fishes, and to some neuroblasts in the ganglia of cranial nerves V, VII, IX, and X.

19. Endoderm gives rise to the epithelium of the digestive tract and to the epithelioid components of the organs that evaginate from it.

20. Dorsal mesoderm forms mesodermal somites consisting of sclerotome, dermatome, and myotome. These give rise, respectively, to vertebrae and the proximal parts of ribs, the dermis of the dorsum, and skeletal muscles other than those of the pharyngeal arches.

21. Lateral-plate mesoderm gives rise to intermediate and lateral mesoderm. The latter splits into somatic and splanchnic layers.

22. The somatopleure consists of somatic mesoderm plus ectoderm. The splanchnopleure consists of splanchnic mesoderm plus endoderm.

23. The coelom is the cavity between the somatic and splanchnic mesoderm. It is initially segmented in cephalochordates.

24. Intermediate mesoderm gives rise to kidney tubules and to the ducts of the urogenital system.

25. In viviparity, highly vascularized embryonic tissues (pericardial sac, gills, villi, opercular lining, extraembryonic membranes) are sites for absorption of essential substances.

26. The chief extraembryonic membranes are yolk sac, amnion, chorion, and allantois.

27. The yolk sac serves as a simple yolk sac placenta in viviparous anamniotes and as part of a choriovitelline placenta in marsupials.

28. Amniotic fluid bathes the amniote embryo, preventing desiccation and helping to prevent mechanical trauma.

29. The chorion is applied to the eggshell or to the lining of the maternal uterus. In that latter case it is part of the placenta.

30. The allantois is a midventral evagination of the cloaca of amniotes. It participates in formation of chorioallantoic placentas. The part proximal to the cloaca becomes the urinary bladder. The distal part becomes the urachus in some mammals.

31. Chorioallantoic placentas are found in eutherian mammals. They may be deciduous or nondeciduous. Villi may be arranged in zonary, cotyledonary, discoidal, or diffuse patterns.

SELECTED READINGS

Bolker, J. A.: Comparison of gastrulation in frogs and fish, American Zoologist **34**:313, 1994.

Graveson, A. C.: Neural crest: contributions to the development of the vertebrate head, American Zoologist **33**:424, 1993.

Horner, J. R., and Weishampel, D. B.: Dinosaur eggs: the inside story, Natural History, p. 61, December 1989.

Maderson, P. F., editor: Developmental and evolutionary aspects of the neural crest, New York, 1987, John Wiley & Sons.

Meier, S.: Development of the chick embryo mesoblast: morphogenesis of the prechordal plate and cranial segments, Developmental Biology **83**:49, 1981.

Moore, J. A.: Science as a way of knowing, Developmental Biology, American Zoologist **27**:415, 1987.

Müller, G. B., and Wagner, G. P.: Homology, *Hox* genes, and developmental integration, American Zoologist **36**:4, 1996.

Nelsen, O. E.: Comparative embryology of the vertebrates, New York, 1953, The Blakiston Co., Inc. This comprehensive work has not been equaled.

Patten, B. M.: Early embryology of the chick, ed. 5, New York, 1971, McGraw-Hill Book Co.

Patten, B. M.: Early embryology of the pig, ed. 3, New York, 1948, McGraw-Hill Book Co.

Patten, B. M., and Carlson, B. M.: Foundations of embryology, ed. 5, New York, 1988, McGraw-Hill.

Roelink, H., et al.: Floor plate and motor neuron induction by vhh-1, a vertebrate homolog of hedgehog expressed by the notochord, Cell **76**:761, 1994.

Sasai, Y., et al.: *Xenopus* chordin: a novel dorsalizing factor activated by organizer-specific homeobox genes, Cell **79**:779, 1994.

Shine, R.: Young lizards can be bearable, Natural History, p. 34, January 1994. Patterns of viviparity in lizards.

Wourms, J. P.: Viviparity: the maternal-fetal relationship in fishes, American Zoologist **21**:473, 1981.

CHAPTER SIX

INTEGUMENT

In this chapter we will study the basic structure of the integument. We will examine the epidermis of aquatic vertebrates, then see how it has been modified for life in air. We will also note some of the remarkable structures that form from epidermis. After that, we will examine the dermis and learn that bone is present in much modern skin and that its absence is a specialization. As a recapitulation, we will review the integument class by class. Finally, we will be reminded of some of the roles the integument plays in survival.

OUTLINE

t would be difficult to imagine an organism the surface of which is not structured specifically for its role as an interface between organism and environment. Even amoebae have plasma membranes that wall off the protoplasm from the surrounding water. The integument of multicellular animals subserves many functions that contribute to the survival of the organism. It covers all external surfaces, including the exposed portion of the eyeball, where it is usually transparent and is called the **conjunctiva;** it covers the eardrum; and it is directly continuous with the mucous membranes that line all passageways opening onto the surface.

The integument of vertebrates from fishes to mammals conforms to a basic morphologic pattern, consisting of a multilayered **epidermis** derived from ectoderm and a **dermis** derived chiefly from mesoderm. Variations in the morphologic features of the epidermis and dermis of the several vertebrate classes involve (1) the relative number and complexity of skin glands, (2) the extent of differentiation and specialization of the most superficial layer of the epidermis, and (3) the extent to which bone develops in the dermis. The skin of an amphioxus exhibits epidermis and dermis, but the epidermis is only one cell thick (fig. 6.1).

FIGURE 6.1

Skin of a young amphioxus.

PREVIEW: SKIN OF THE EFT

As an introduction to the integument we will take a quick look at a skin that is unencumbered by scales, feathers, or hair—the skin of the red eft, the juvenile land stage of the aquatic urodele *Notophthalmus* (fig. 6.2). It does not reflect an ancient feature of vertebrate skin, the presence of bony plates or scales in the dermis, but it illustrates the contrasting adaptations of vertebrate skin for life on land as opposed to life in water.

The epidermis of the eft is a stratified (multilayered) epithelium. The columnar cells in the basal, or **germinal,** layer are constantly undergoing mitosis, replacing those lost from the surface. Proliferation from the basal layer causes older cells to be pushed outward. As they approach the surface they synthesize **keratin,** a scleroprotein that is insoluble in water, and they become flattened **(squamous).** They are then said to be **keratinized,** or **cornified.** Keratinization causes the cells to die. In efts, the cornified layer is thin in contrast to the thicker cornified stratum of terrestrial vertebrates.

The glands of the skin develop from the epidermis and, in efts, are simple multicellular sacs that bulge into the dermis, where they are in the immediate vicinity of capillaries that supply nutrients and oxygen, and carry away metabolic wastes. They are able to synthesize mucus, but in the eft stage they are quiescent, although not totally inactive.

The dermis of efts consists chiefly of connective tissue that supports the bases of the glands, blood vessels, lymphatics, small nerves, and pigment cells. It adheres closely to the underlying body wall muscle.

The life cycle of *Notophthalmus,* described in chapter 4, dramatically illustrates the direct correlation between a terrestrial habitat, lack of surface mucus, and the phenomenon of keratinization. The larvae of this species live in water, have many active mucous glands, and no cornified epidermis. When they metamorphose and assume

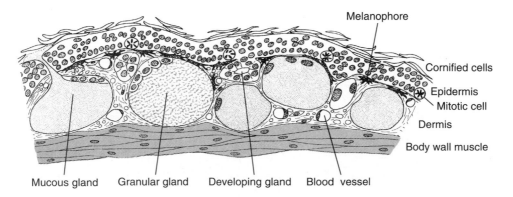

FIGURE 6.2

Skin of the red eft *Notophthalmus.*

life on land, most of the skin glands become quiescent, keratinization of the epidermis commences, and it persists as long as the eft lives on land. At the onset of sexual maturity the eft migrates back to water, the mucous glands again become active, and the cornified cells are shed and do not reappear. The skin of larvae and aquatic adults resembles that of fishes, whereas the skin of efts is like that of terrestrial amphibians.

THE EPIDERMIS

Having completed our preview, we are better able to study the integument in detail. We will study the epidermis and the dermis separately because the two have different roles. The avascular epidermis is the interface between organism and environment, which is reflected in its variable structure from fishes to mammals. The highly vascular dermis provides physiologic support for the interfacing epidermis.

Two kinds of nonliving coverings overlie the living epidermis of vertebrates. In fishes, especially teleosts, and in aquatic amphibians, it is a thin coat of **mucus** that is continually being replenished. The role of this mucus cuticle is speculative. In terrestrial vertebrates, the covering is a layer of dead, water-impervious cornified cells, the **stratum corneum** (fig. 6.3). Its survival value is clear. It minimizes water loss through skin that is exposed to air.

The epidermis is glandular to some degree in all vertebrates. The epidermis of most fishes and amphibians is exceptionally so. The most common epidermal glands of fishes and larval amphibians are unicellular, whereas those of metamorphosed amphibians and amniotes are predominantly multicellular. Although of epidermal origin, multicellular skin glands from fishes to humans invade the dermis, where they are in the immediate vicinity of blood capillaries that provide for their metabolic needs.

Epidermis of Fishes and Aquatic Amphibians

The predominant feature of the epidermis of most fishes and aquatic or semiaquatic amphibians is the abundance of epidermal glands. Scales are not in the epidermis; they are beneath it.

Epidermal Glands

Fishes Most of the integumentary glands in fishes are single cells, such as **goblet cells** (named for their shape) that secrete only mucus, and a variety of more specialized single cells, some with a granular cytoplasm (figs. 6.4 and 6.5). The **granular cells** of fishes secrete mucus and additional ingredients of mostly unknown functions. Multicellular glands (fig. 6.6) are not abundant in fishes,

FIGURE 6.3

Keratinized (cornified) cells of the stratum corneum of human skin.

Courtesy Johnson & Johnson Research, New Brunswick, NJ.

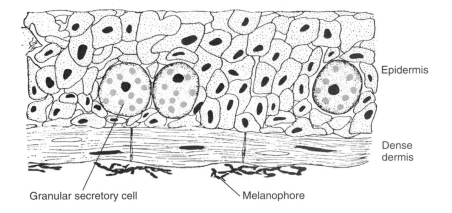

FIGURE 6.4

Skin of a larval lamprey.

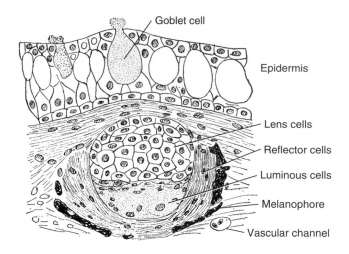

FIGURE 6.5

Light organ (photophore) of a luminous fish.

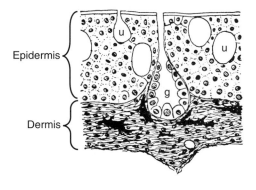

FIGURE 6.6

Glandular skin of the dipnoan *Protopterus*. Multicellular, **g**, and unicellular, **u**, glands in the epidermis. Melanophores are seen in the dermis.

but those that are present contribute mucus and other ingredients including, in one species or another, thick slime or alkaloids that repel predators. Alkaloids, which are products of some of the granular cells, are not common in aquatic vertebrates. They lose effectiveness when diluted with environmental water. Slimy mucus is secreted in quantity in response to stressful external stimuli, particularly when life is endangered by a predator. This is especially applicable to hagfishes, aptly called slime eels, whose slime glands are surrounded by striated muscle fibers. The lungfish *Protopterus* surrounds itself with a slimy dessication-retarding cocoon before aestivating in a burrow during the dry season. Small amounts of nonslimy mucus are secreted more or less constantly, especially in teleosts, forming the mucus cuticle. Embedded in the cuticle may be some of the alkaloid or other ingredients from granular cells. Some female teleosts secrete a nutritious mucus that is eaten by the hatchlings. The mucus secreted by cells at the base of the stinger in sting rays contains an irritating toxin. The role of the mucus secreted by the lining of the ampullary neuromast organs of fishes (fig. 17.2*a*) is unknown.

Aquatic Amphibians The epidermal glands of aquatic amphibians are mostly multicellular mucous or granular glands (figs. 6.7 and 6.8). Tailed amphibians that are semiaquatic, that is, those whose primary habitat is a pond or stream but who may emerge on occasion, have the largest number of integumentary glands. Their mucous secretions keep the skin moist in air, enabling it to function as a respiratory membrane, as it does in water. Mucous glands on the digits of some frogs and tree toads serve as holdfasts, and the swollen mucous glands on the thumb pads of male anurans during the reproductive season assist in restraining the female during amplexus.

FIGURE 6.7

Frog integument. The multicellular glands are in various stages of development. A thin stratum corneum covers the living epidermis.

FIGURE 6.8

Undersurface of skin reflected from the underlying muscle in tail of a necturus. Mucous glands (*arrow*) are seen projecting into subcutaneous tissue just external to the tail musculature.

Photophores

Some of the multicellular glands of deep-sea teleosts have become light-emitting organs, or **photophores.** Like other multicellular skin glands, photophores arise in the epidermis and invade the dermis. In one variety (fig. 6.5), the upper part of the gland consists of modified mucous cells that serve as a magnifying lens. Beneath the lens, other cells are the source of the light. Surrounding the base of the photophore, in the dermis, are a blood sinus and a concentration of melanophores (melanin pigment cells). The blood sinus is the source of raw materials needed for bioluminescence. The emitted light most often is produced in the manner of fireflies: the enzymatic action of luciferase on the phosphorus-containing compound luciferin. The "cold" light emitted is not intense and is of many hues. It serves for species and sex recognition, sometimes as a lure, a warning, or an aid to concealment by countershading. Photophores are not limited to teleosts, but they are rare in other fishes. The beautiful color patterns of the skin of fishes seen in aquaria are the result of pigment cells in the dermis.

Keratin

Most fishes synthesize little or no keratin in their epidermis, but some aquatic urodeles, especially those that venture onto land, have a thin desiccation-impeding stratum of cornified cells. In cyclostomes, conical cornified epidermal spines and teeth develop in the buccal funnel and on the rasping lingual cartilage (fig. 6.9). Anuran tadpoles have rows of horny tooth-like structures surrounding the entrance to the mouth that enable them to feed by scraping vegetation during their herbivorous larval stage. Tadpoles shed these at metamorphosis. Calluslike caps develop on the toes of aquatic urodeles subjected to buffeting in mountain streams. In general, however, keratin is a feature of the skin of terrestrial vertebrates, not of fishes and aquatic amphibians.

Epidermis of Terrestrial Vertebrates

The stratified epithelium that constitutes the epidermis of terrestrial vertebrates reflects a long history of gradual adaptation from life in water to life in air. According to Peterson's (Unnatural) Law, "Anything that can happen will happen." And it practically did! Could one have predicted feathers? hair? mammary glands? sweat? It was epidermal adaptations such as these, and less complicated ones, that participated in fostering the diversity of vertebrate life on land. Let us look, therefore, at the two primary epidermal features of terrestrial vertebrates—integumentary glands and the stratum corneum.

Epidermal Glands

The epidermal glands of well-adapted terrestrial vertebrates are either **saccular** or **tubular.** The sacs are known as **alveoli.** A variety of simple and compound saccular (**alveolar**) and tubular glands are illustrated in figure 6.10. Saccular glands are variously complex expansions of the simple saccular glands of dipnoans and amphibians. Tubular glands are uncommon in the skin below mammals but abundant in mammals. However, they are ubiquitous in the digestive tract from fishes to human beings.

Glands vary in the way their cells liberate the substances they synthesize. There are three basic categories. The cells of **merocrine glands,** the most common variety,

FIGURE 6.9

Lamprey teeth and buccal funnel. (*a*) Cornified tooth (*black*). The stellate tissue may be the beginning of a replacement tooth. (*b*) Buccal funnel. The horny teeth are shown as white against a dark background. The tip of the retracted tooth-bearing lingual cartilage (tongue) occupies the rear of the cavity.

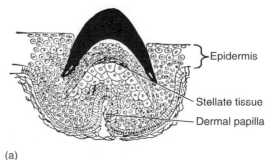

Epidermis

Stellate tissue

Dermal papilla

(a)

(b)

FIGURE 6.10

Major morphologic varieties of multicellular glands. (*a*) Simple tubular. (*b*) Coiled tubular. (*c*) Simple branched tubular. (*d*) Compound tubular. (*e*) Alveolus of a simple saccular gland. (*f*) Simple branched saccular. (*g*) Compound saccular (compound alveolar). *Red* shows secretory epithelia. Not to scale.

secrete their products via the cell membrane, and the cell remains intact. Most sweat glands of humans are merocrine. **Holocrine glands** are those in which the cells themselves constitute the secretion. Oil glands of birds and the sebaceous glands of mammals are holocrine. **Apocrine glands** exhibit an intermediate condition. The secretion accumulates in the apical portion of the cell, which is then pinched off along with some cytoplasm. The cell then repairs itself. Mammary glands are apocrine. **Goblet cells** do not fit precisely in any of these three categories. They rupture at the apex and the mucus oozes out. The cell eventually repairs itself numerous times before being discarded.

Mucous Glands   Mucus-secreting epidermal glands have practically disappeared except in mammals, in which they are confined to a few sites where lubrication of a surface is essential. Unicellular glands probably lost survival value when, as in efts, they became covered by a continuous sheet of cornified cells that rendered them functionless. Also, the synthesis of mucus other than in limited quantities would dehydrate a vertebrate lacking constant access to drinking water or to moisture that could be absorbed through the skin. Among early verte-

brates that ventured onto the land, those that had fewer mucous glands and a thicker stratum corneum surely had an advantage. Natural selection would have favored such a mutant.

Granular Glands   Granular glands are present with little variety in toads, which are terrestrial amphibians, and in a larger variety but not abundantly in reptiles. They are absent in birds and mammals. **Granular glands** secrete irritating or toxic alkaloids that are defensive, and ward off predators. Granular glands are the source also of many, but not all, pheromones. **Pheromones** are substances that, when released into the environment, affect the behavior or physiology of other organisms of the same or different species. They may signal the sex of an organism, identify members of the same population, or leave trails on a substrate for the information of other members of the species. Some reptilian pheromones attract insects on which the reptile then feeds, and some sound alarms.

Granular glands are generally restricted to a localized area of the body. In toads they are associated with warty skin, chiefly on the legs and back. The secretion can be irritating to human skin and highly distasteful to

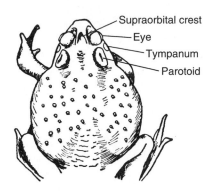

FIGURE 6.11

Warty skin of toad (*Bufo*). A granular gland, the parotoid, is seen behind the eye.

predators. Many toads also have a prominent granular **parotoid gland** behind the eye (fig. 6.11). Granular glands surround the vent of lizards and snakes, and most of these secrete pheromones. The only other integumentary glands in lizards are **femoral glands** on the medial aspect of the hind limbs of males. These secrete a substance that hardens to form temporary spines that restrain the female during copulation. Musk turtles exude a yellowish fluid from two glands on each side of the trunk at the lower border of the carapace. In crocodilians, a row of glands of unknown function, probably pheromonal, extends along the back.

Avian Oil Glands   Very few integumentary glands are found in birds, and the secretions of all have an oily component. The **uropygial gland** is a prominent swelling at the rump immediately behind the pygostyle (fig. 8.20). It is largest in aquatic birds and prominent also in domestic fowl. The secretion, which is water-repellent, is transferred to feathers during preening. Smaller oil glands line the outer ear canal and encircle the vent in some species.

Sebaceous Glands   Integumentary glands reach a peak in complexity and specialization in mammals. All appear to be variants of two functional types, sebaceous and sudoriferous. **Sebaceous glands** are alveolar glands with an oily exudate. They are present wherever there are hairs, and the secretion, **sebum,** is usually exuded into hair follicles (fig. 6.12). Fur, including human hair, glistens after brushing because of the oil. In the outer ear canal, **ceruminous glands** secrete **cerumen,** which in association with hairs traps insects that otherwise might go deeper into the canal and touch the painfully sensitive eardrum. **Meibomian glands** assist in moistening the conjunctiva of the eye. They are embedded in a plate of dense connective tissue (the palpebral tarsus) in each eyelid and their long ducts open via a row of minute

foramina along the border of the lid just internal to the eyelashes. If a duct (there are up to 30) occasionally becomes occluded, it causes a **chalazion,** an inflamed swelling on the conjunctival surface of the lid. Sebaceous glands open independent of hairs on the lips, glans penis, labia minora, and skin surrounding the nipples.

Sudoriferous Glands   Sudoriferous (sweat) glands are coiled tubular glands that extend deep into the dermis of mammals (figs. 6.10*b* and 6.12). Their secretions ooze onto the surface via tortuous canals that open as pores. The cooling effect of the evaporation of sweat, their predominant secretion, is thermoregulatory. Sweat glands in furry mammals are confined to the least furry regions, for example, the feet of cats and mice, lips of rabbits, the side of the head of bats. In hippopotami, they are only on the ears, which are above the water line when a hippo is in its preferred habitat. Some mammals lack sweat glands. Among these are pangolins, whose epidermis is scaly (fig. 4.36); sirenians and cetaceans, which are marine mammals; and echidnas, which are monotremes. Ciliary glands that open into the follicles of eyelashes are sudoriferous glands. Humans, with less hair than most mammals, have the largest number of sweat glands per centimeter of surface area.

Scent Glands   Sebaceous and sudoriferous glands both produce a variety of scents, all of which may be pheromones. Glands on the feet of goats leave scent trails that other members of the species are able to recognize; anal glands of skunks drive away enemies; and anal glands of male musk deer signal the sex. Kangaroo rats have sebaceous glands on the back, where they are exposed when the back is arched in defense. Male elephants have a temporal gland that swells during breeding seasons. Natives say that elephants are dangerous to humans at that time. A gland above the eye of a peccary looks like a navel. Some male lemurs have a hardened patch of skin on the forearm, under which is a gland the size of an almond. Most of the odors at a well-maintained mammalian zoo are caused by scent glands, not by unhygienic conditions in the pens. One species, *Homo sapiens,* takes the pheromone from the anal gland of a musk deer, adds other odorants, and dabs it behind the ears. It is called perfume, but it performs the role of a pheromone. Not all pheromones are products of integumentary glands. Mammalian urine, for example, contains substances that also serve as pheromones.

Mammary Glands   Mammary glands are compound alveolar glands that develop in both sexes from **milk lines,** a pair of elevated ribbons of ectoderm that extend along the ventrolateral body wall of the fetus from axilla to groin (fig. 6.13*a*). Patches of future mammary tissue develop at one or more sites along each milk line,

FIGURE 6.12

(*a*) Mammalian skin. Supporting the constituents of the dermis are bundles of collagenous connective tissue. Epidermal derivatives are *red*. Sweat glands are tortuous, hence are seen only in cross section. They are longer than the dermis is deep. The touch corpuscle occupies a dermal papilla. The pacinian corpuscle is in the subcutaneous connective tissue. (*b*) Enlarged section of hair. A cortex of cornified cells surrounds a medulla core (*red*). The cuticle is a thin membrane of overlapping squamous cornified cells.

FIGURE 6.13

(*a*) Milk line and nipples in a 20 mm pig embryo.
(*b*) Sites of supernumerary nipples in a human.

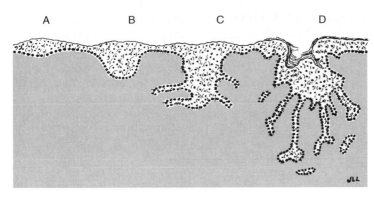

Mammary gland

Morphogenesis

FIGURE 6.14

Successive stages of mammary gland development.
A, Equivalent to a human embryo at 6 weeks; **B,** equivalent to a human embryo at 9 weeks and to a mouse embryo at 16 days; **C,** intermediate stage; **D,** at birth. *Gray* area represents dermis.

depending on the animal, and invade the dermis (fig. 6.14). They later spread *beneath* the dermis, and a nipple forms above each patch. As females approach sexual maturity, rising titers of female steroid hormones cause the juvenile duct system to spread and branch. Later, during pregnancy, a battery of hormones induces the formation of enormous numbers of alveoli at the ends of the branching duct system (figs. 6.10g and 6.15b and c). Like sebaceous glands, from which mammary glands of viviparous mammals appear to have been derived, mammary glands produce a secretion that includes lipids.

The number and location of mammary glands, and therefore of nipples, depend on the number of young that is typical of the species and on the survival value of one location over another. Cats, dogs, pigs, rodents, eden-

tates, and many other mammals have axillary, thoracic, abdominal, and inguinal nipples that develop along most of the milk line. During pregnancy in these species, adjacent glands may expand toward each other until they have formed two long continuous masses of considerable weight. Species with smaller litters have fewer nipples. Insectivores and some lemurs have one thoracic and one inguinal pair. Flying lemurs and marmosets have a single pair, in the armpits. Monkeys, apes, and humans have one pair located where the nursing baby can be protected in the mother's arms while she monitors the environment for enemies. Cetaceans have a single pair near the groin, where the baby porpoise or whale can hold on and nurse while the mother feeds, surfaces, and dives. Nutrias, which are at home in water, have four nipples on the back, and the babies ride along above the water line while nursing. The nipples originate from a ventrolateral milk line but are displaced dorsad by differential growth rates on the two sides of the milk lines during ontogenesis. Supernumerary nipples may develop in any mammal (fig. 6.13b). Although only three ducts are shown in the nipple in figure 6.15b, there are actually as many terminal ducts in a human nipple as there are lobes in the gland, which number from 15 to 25. Each lobe consists of many compound lobules. In ungulates, ducts from all lobes empty into a single terminal duct. At the base of the nipple, terminal ducts have a **cystern** in which milk accumulates after having been "let down" from the lobes. Suckling drains the cysterns. The hormone **oxytocin** from the pituitary (chapter 18) is responsible for the smooth muscle contractions that cause milk letdown.

Monotremes do not develop mammary glands like those of viviparous vertebrates, and they do not have nipples. Instead, in both sexes, what appear to be modified sweat glands produce a nutritious secretion that is lapped off an overlying tuft of hairs by the young (fig. 6.15a).

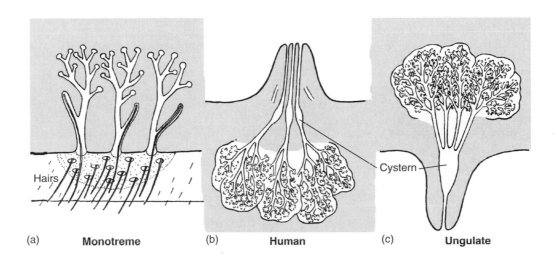

Hairs

(a) **Monotreme**   (b) **Human**   Cystern   (c) **Ungulate**

FIGURE 6.15

Mammary glands, ducts, and nipples. The monotreme lacks nipples, and the glands resemble modified sweat glands.

Stratum Corneum

Keratinization provides protection against desiccation on land. With the acquisition of an amnion, which completed the liberation of vertebrates from the water, the stratum corneum became increasingly specialized in various regions of the body for protection against abrasion, for offense and defense, and finally, as an adjunct to thermoregulation. Early specializations—scales, claws, and horny protuberances—were followed by hair and feathers, the latter being the most remarkable of all. A mammalian stratum corneum is seen in figure 6.12.

Epidermal Scales Epidermal scales are repetitious thickenings of the stratum corneum that are found only in amniotes. In squamates—lizards and snakes—the stratum corneum is disposed on overlapping folds of the epidermis (figs. 6.16 and 6.17). The cells on the exposed surface of each fold are especially heavily cornified, forming overlapping scales. Thinning of the stratum corneum at scale joints permits mobility of the skin. The scales of lizards sometimes assume bizarre shapes (figs. 6.18 and 17.14).

(a)

(b)

4
3
2

1

(c)

FIGURE 6.16

Squamate scales. (*a*) Collared lizard. The entrance to the outer ear canal is seen as a dark crescent behind the corner of the mouth. (*b*) Diagrammatic section of skin of a snake or lizard. **1**, Dermis; **2**, actively mitotic layer of epidermis; **3**, newly cornified layer of epidermis; **4**, older cornified layer to be shed at next molt. (*c*) Scales from the banded water snake, *Nerodia fasciata*.

(*a*) © John Cancalosi/Peter Arnold, Inc.

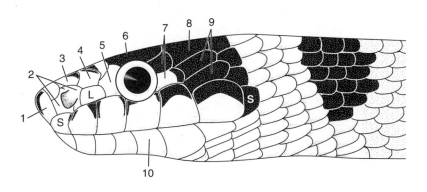

FIGURE 6.17

Epidermal scales on head of milk snake *Lampropeltis triangulum*. **1**, rostral; **2**, nasal; **3**, internasal; **4**, prefrontal; **5**, preocular; **6**, supraocular; **7**, postoculars; **8**, parietal; **9**, anterior temporals; **10**, fourth of seven infralabials; **L**, loreal; **S**, first and last supralabials.

FIGURE 6.18

Warty and spiny skin of the Coast horned lizard, *Phrynosoma*. Skin color and pattern resemble its substrate habitat.

© H. Armstrong Roberts.

The continuity of the stratum corneum is readily demonstrable in crocodilians, which have small, heavily cornified, nonoverlapping scales on the limbs, and larger, more or less regular thickenings on the rest of the body (fig. 6.35*a*).

Large, thin, quadrilateral or polygonal scales are called **scutes.** Snakes have them on the belly, where they are used for locomotion. Turtles have thick scutes on the plastron, which slides along the ground, thinner ones on the carapace (fig. 6.19), and small scales elsewhere. Turtle scutes and scales do not overlap.

Epidermal scales in birds develop where there are no feathers: the facial area, legs, and feet. Armadillos have hair and scales interspersed over the entire body (figs. 4.35 and 6.20), but the scales of most mammals are confined to the legs and tail, as in rats and beavers. The scales of pangolins appear to be agglutinated hairs, and are of recent origin rather than inherited from reptilian ancestors (fig. 4.36).

Lizards and snakes have two distinct layers of stratum corneum, an inner layer in the process of being deposited and an outer layer that will be shed at the next molt (fig. 6.16*b*). In lizards the outer layer flakes off in large patches, but in snakes the outer layer of the entire body, including the spectacle, a thick lenslike conjunctiva, is shed in one piece. Crocodilians and turtles do not molt; their stratum corneum wears off gradually, like that of humans.

Claws, Hoofs, and Nails   Claws, hoofs, and nails are modifications of the stratum corneum at the ends of digits. Claws first appeared in reptiles and have persisted in birds and most mammals. (The "claws" of the African clawed toad, an amphibian, and reptilian claws are not

(a)

(b)

FIGURE 6.19

Epidermal scales of the turtle *Chrysemys.* (*a*) Carapace. (*b*) Plastron. On the carapace the scutes shown are **c,** costals; **m,** marginals all around the periphery, including the nuchal (**nu**), and pygals (**p**); **n,** neurals. On the plastron, the scutes shown are, **g,** gulars; **h,** humerals; **p,** pectorals; **a,** abdominals; **f,** femorals; **an,** anals.

homologous.) Claws evolved into nails in primates, and into hoofs in ungulates. Claws, hoofs, and nails have the same basic structure. They consist of two curved parts, a horny dorsal plate, the **unguis,** and a softer ventral plate, the **subunguis** (fig. 6.21). The two plates wrap partially around the terminal phalanx, which is usually pointed when associated with a claw, blunt when associated with a hoof or nail. A still softer calluslike, cornified pad, the **cuneus** (called the "frog" by horsemen), is frequently present in ungulates, partially surrounded by the subunguis. The thick, horny unguis of a hoof is U- or

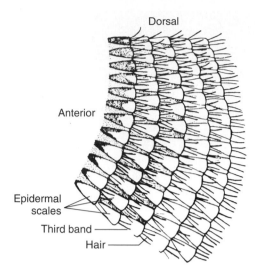

FIGURE 6.20

Epidermal scales and interspersed hairs of armadillo skin. See also fig. 4.35.

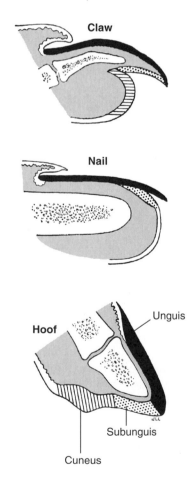

FIGURE 6.21

Claw, nail, and hoof with terminal phalanx, sagittal section.

V-shaped, and, because it consists of dead cells, a shoe can be nailed into it. The unguis of a nail has become flattened, and the subunguis is much reduced. As a result, nails cover only the dorsal surfaces of digits; but, if permitted to grow, nails become clawlike.

Although claws in birds are often thought of as associated only with the feet, sharp claws are frequently borne on one or more digits of the wings (ostriches, geese, some swifts, and others). Young hoatzins use claws on the wings for climbing about on the bark of trees, but the claws cease growing and disappear at maturity. *Archaeopteryx* had three claws on each wing. Only squamate claws are shed. Claws, hoofs, and nails of other animals are worn down by abrasion.

Feathers   Feathers are remarkably complicated cornified epidermal appendages of the integument. There are three morphologic varieties: contour feathers, down feathers (plumules), and hairlike feathers (filoplumes). The roles of feathers are described in chapter 4.

Morphologic Varieties of Feathers   Contour feathers are the conspicuous feathers that give a bird its contour, or general shape (fig. 6.22). A typical contour feather consists of a horny **shaft** and two flattened **vanes.** The base of the shaft is the **calamus, (quill).** The vane-bearing segment is the **rachis.** Each vane consists of parallel rows of **barbs** that have **barbules** and **flanges** (fig. 6.22*b*). The barbules have **hooklets** that interlock with the flanges on the next barb, stiffening the vane (fig. 6.22*c*). When a contour feather is ruffled, the barbules have become unhooked. Preening rehooks the hooklets. On the smaller contour feathers of the wing (covert feathers), hooklets are missing from the lower barbules, so this region of the feather is fluffy. In ostriches and a few other birds, all feathers are fluffy.

Arising from the **superior umbilicus,** a notch at the base of the rachis of a contour feather, is an **afterfeather** (fig. 6.22*a*). It is usually shorter than the main feather, but in emus and cassowaries the two are the same length, resulting in "double feathers."

Although contour feathers cover most of the body, the follicles from which they grow are usually disposed in feather tracts, or **pterylae** (fig. 6.23). A few birds, including ostriches and penguins, lack these.

Inserting on the walls of the feather follicles in the dermis are smooth erector muscles **(arrectores plumarum),** which, along with extrinsic integumentary muscles, enable a bird to fluff its feathers.

Down feathers are small, fluffy feathers lying underneath and between contour feathers. Very young birds lack contour feathers and are covered with down. Down feathers have a short calamus (fig. 6.22*d*) with a crown

FIGURE 6.22

(*a*) Contour feather from a grouse. (*b*) Two successive barbs showing two barbules interlocked by hooklets with flanges. (*c*) Cross section of a flange showing interlock. (*d*) A down feather. (*e*) A filoplume.

FIGURE 6.23

Feather tracts.

of barbs that lack hooklets. Eiderdown, used in pillows, is the down feathers of the eider duck.

Filoplumes are hairlike feathers consisting of a shaft and a few barbs and associated barbules at the tip (fig. 6.22*e*). They are scattered across the skin among contour feathers. Their follicles are richly supplied with tactile nerve endings. The long colorful feathers of a peacock are conspicuous examples of filoplumes. **Bristles** resemble filoplumes but lack any terminal barbs. A few barbs do occur at the base of the shaft. Bristles occur on the head and neck where they screen eyes and ear and nasal openings of foreign matter. Those around the mouth also are tactile receptors, conveying information such as a possible insect meal for a flycatcher.

Development of a Feather   Feather development is initiated by formation of a **dermal papilla,** which is a mound of mesodermal cells in the dermis that indents the undersurface of the epidermis and induces mitotic activity in its basal layer (fig. 6.24*a*). Growth of the papilla,

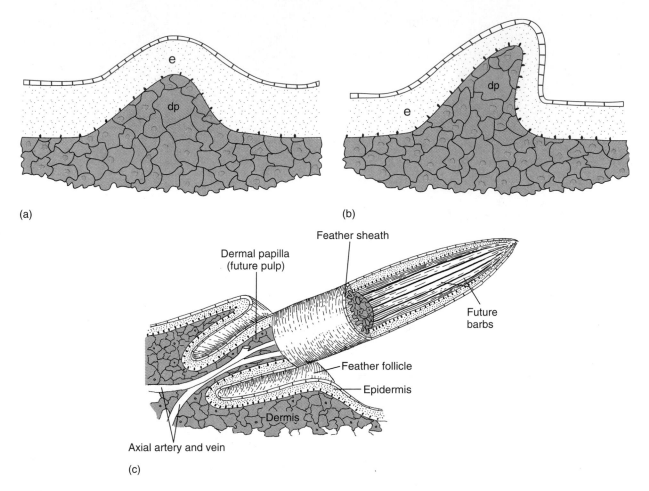

FIGURE 6.24

Growth of a contour feather. (*a*) Germinal dermal papilla **dp.** (*b*) Feather primordium. (*c*) Pin feather. **e** is epidermis.

and its inductor effect on the overlying epidermis, results in a pimplelike elevation, the **feather primordium,** on the surface of the skin (fig. 6.24*b*). This is the first indication that a feather is organizing. The dermal papilla becomes vascularized, and thereafter it is the source of nutrients and oxygen for the developing feather. It is also a collecting site for metabolic wastes that are then carried away in the blood stream. As the feather primordium elongates, a pit lined with epidermis, the **feather follicle,** develops around its base (fig. 6.24*c*).

At the base of the feather follicle a mitotically active growth zone proliferates tall columns of epidermal cells that push toward the distal tip of the growing feather between the dermal papilla and the epidermis, now a **feather sheath.** These epidermal columns separate from one another, cornify, and develop into barbs. A growing feather still surrounded by its sheath is a **pinfeather.** When the feather sheath splits open, the fluffy barbs stretch out of their cramped quarters and the shaft elongates.

When the feather is full grown, the dermal papilla in the shaft dies and becomes pulp. The living basal portion of the papilla withdraws from the base of the shaft, leav-

ing an opening, the **inferior umbilicus** (fig. 6.22). New feathers develop from reactivated dermal papillae that have already given rise to feathers. In many birds—perhaps in all—old feathers are passively pushed out of the follicles by incoming feathers.

Origin of Feathers   One hypothesis about how feathers originated is that they arose from reptilian scales, because early developmental stages are similar. Development of both is initiated by formation of a vascularized dermal papilla, after which the overlying epidermis thickens and then forms a scale or feather. However, subsequent development of the two is dissimilar, and the hypothesis is viewed with skepticism. The only other hypothesis that is justified at present is that, like hairs, they may be new evolutionary structures lacking a reptilian precursor.

Hair   Hairs, like feathers, are keratinized appendages of the skin. They may form a dense, furry covering over the entire body, or there may be only one or two bristles on the upper lip, as in some whales. Where fur is dense, there are usually short, fine hairs (underfur) as well as long, coarse ones.

Hairs have an insulating effect when dense enough, and are also sensitive tactile organs. The root of each is surrounded by a basketlike network of sensory nerve endings, and displacement of the root initiates a train of sensory impulses to the brain. Disturb a single hair on the back of your hand and note the sensation evoked. Vibrissae (stiff long whiskers on the face of many mammals) perform this role exclusively.

Morphology of a Hair   Hairs grow from hair follicles and elongate as a result of continual mitosis in the **bulb** (fig. 6.12), within which a highly vascular dermal papilla provides nutrients and oxygen required for growth. The **root** is the part within the follicle where the hair cells are cornifying and dying but where the hair has not yet separated from the follicular wall. The remainder of the hair is the **shaft,** which commences just below the openings from the sebaceous glands. Within the follicle the shaft is surrounded by sebum from the sebaceous gland.

A single hair consists of dense keratin from disintegrated or partially disintegrated cornified cells, trapped air vacuoles, and varying quantities of melanin granules that were released when the hair cells disintegrated. (The source of the granules and an explanation of hair color are discussed later under "Dermal pigments.") Each hair is covered by a membranous **cuticle** composed of thin, transparent, cornified squamous cells (scales) arranged like shingles with free edges directed away from the root. Coarse or spiny hairs contain a **medulla** composed of

irregular, shrunken cornified cells separated by large amounts of air and connected by intercellular bridges of keratin (fig. 6.12*b*).

Inserted on the wall of each hair follicle is a tiny smooth muscle, the **arrector pili** (fig. 6.12). When the arrectores pilorum contract, the hairs are elevated and the skin is pulled into tiny mounds resembling "gooseflesh." Elevation of the hairs gives carnivores a ferocious appearance. It is also a device for thermoregulation, increasing the insulating effect of the fur.

The scales of pangolins, and perhaps the horns of rhinoceros, may be evolutionary products of agglutinated hairs. Other modifications of hair include bristles, the spines of spiny anteaters, and porcupine quills.

Development of a Hair   The first manifestation of a developing hair is a cylindrical ingrowth of epidermis into the dermis (fig. 6.25*a*). Soon thereafter a dermal papilla, histologically identical with that of a developing feather, organizes at the base of the ingrowth, becomes vascularized, and the two become intimately associated. Thereafter, the papilla plays the same role as in feather development. Utilizing oxygen and nourishment provided by the vessels of the dermal papilla, the hair primordium grows deeper and deeper into the dermis. The base of the primordium eventually becomes bulbous, cornified cells begin to form at that site, and a hair shaft begins to rise out of a hair follicle (figs. 6.12 and 6.25*b*).

Hairs are not permanent. After an interval of time— one or many months, and often seasonally, depending on the species—proliferation of hair cells in the follicle

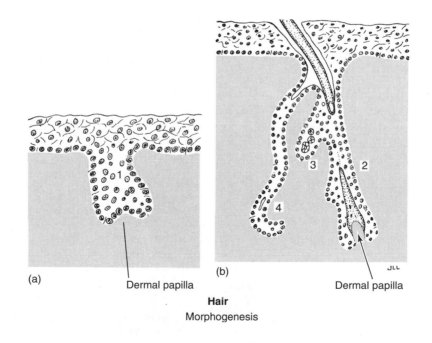

(a)   Dermal papilla

(b)   Dermal papilla

Hair
Morphogenesis

FIGURE 6.25

Successive stages in the development of hair and associated glands. **1,** initial epidermal ingrowth into dermis, dermal papilla not yet organized; **2,** hair follicle surrounding growing hair shaft; **3,** developing sebaceous gland; **4,** developing sweat gland. *Gray area,* dermis.

ceases temporarily, and the hair becomes loose in the follicle. Subsequently, when formation of hair in the follicle resumes, the old hair is shed if it hasn't already fallen out. When all follicles in an area permanently cease activity, the area becomes bald.

Origin of Hair   The phylogenetic origin of hair is conjectural. Hairs grow in isolated groups of two dozen or more where there are no scales, and grow between scales in areas such as the tail and legs of rodents. The interscalar distribution of hair in armadillos is seen in figure 4.35. Among anthropoids, some monkeys have hair in groups of three, apes have groups of five, and humans have groups of three to five, arranged linearly. Examining the back of a man's hand, near the base of the thumb, will reveal the linear arrangement of the groups in that location. Where scales are present, the interscale distribution of hairs makes possible the receipt of tactile stimuli despite a scaly skin. This distribution of tactile hairs is reminiscent of the interscale distribution of sensory pits in lizards, where each pit houses a tactile bristle (**protothrix**). The interscale distribution of hairs may reflect the reptilian origin of mammals, but tactile hairs and tactile protothrixes seem to be products of convergent evolution. Feathers and hairs may be **neomorphs** (structures that have no ancestral precursor).

Feathers, Hair, and Dermal Papillae   We have seen that dermal papillae are essential to the early formation of feathers and hair. In the morphogenesis of a feather, a dermal papilla appears first, and this is followed by activity in the ectoderm. In the morphogenesis of a hair, an ectodermal invagination appears first, and this is followed by formation of a dermal papilla. Feathers have been induced on the feet of chicks where scales ordinarily develop if dermis from a potential feathery area is transplanted to a site under the epidermis of an embryonic foot at a critical time during organogenesis. Even the extraembryonic chorionic epithelium of a chick embryo has been induced to form feather papillae or embryonic scales by carefully timed transplants of appropriate dermis. Whether or not the dermis induces the initial ectodermal ingrowth that signals development of a hair is not known.

The development of placoid scales and vertebrate teeth is also dependent on dermal papillae (fig. 12.12). The dermal papillae at the interface of the epidermis and dermis of fully differentiated mammalian skin (fig. 6.12) should not be confused with the embryonic dermal papillae we have been discussing here.

Horns and Antlers   Many ungulates are endowed with organs of offense, defense, and display, the horns and antlers. The term "horn" means that the surface is composed of keratin. Three varieties of mammalian horns meet this criterion: **bovine horns** and **hair horns** (fig. 6.26), and the **horns of pronghorn antelopes.** Antlers and giraffe horns are not horns.

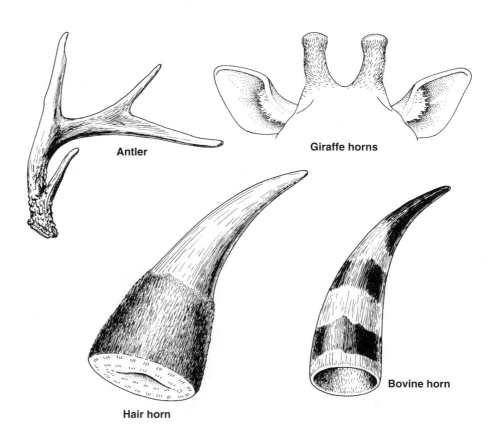

FIGURE 6.26
Mammalian horns and antlers.

Antler

Giraffe horns

Hair horn

Bovine horn

Bovine Horns and Pronghorns Artiodactyls of the family Bovidae (oxen, cows, sheep, goats, true antelopes, others) and pronghorn antelopes have true horns. They consist of a *core of dermal bone covered by a sheath of horn.* Bovine horns are usually curved or recurved, and are never shed. When the sheath is removed from the underlying bone by surgery the sheath is hollow (fig. 6.26). Bovine horns are usually found in both sexes, but there are exceptions. Polled cattle have lost their horns by selective breeding. The chief difference between bovine horns and pronghorns is that pronghorns are branched, and the horny covering, but not the bony core, is shed annually.

Hair Horns Rhinoceroses have hair horns (fig. 6.26). These differ from other horns in being composed of agglutinated keratinized hairlike epidermal fibers that form a solid horn perched on a roughened area of the nasal bone. Both sexes have them, and they are not shed. Some African rhinos have two horns, one behind the other.

Antlers and Giraffe Horns Antlers are characteristic of the deer family. They are not cornified structures but dermal bone attached to the frontal bone. New growing antlers are said to be "in velvet" because they are covered with a soft vascular skin and velvety hair. Antlers develop only in males, with the exception of caribou and reindeer. Typically at the approach of autumn, when antlers are full-grown, the blood supply to the velvet is occluded at the base of the antlers, and the skin, unable to be maintained, rubs off, leaving bare bone. Later, at the end of the rutting season, when male territory no longer must be defended and testosterone concentration declines in the circulation, the antlers are shed with some help from the deer, who evidently find them annoying. They are replaced annually.

The "horns" of giraffes resemble stunted antlers. They are short bony projections of the frontal bones that remain in velvet throughout life.

Baleen and Other Cornified Structures Toothless whales have from 100 to 400 broad, thin, horny sheets of oral epithelium called baleen, or whalebone (fig. 6.27), that hang into the oral cavity from the palate along its length. Each sheet is fringed along the edge, and the fringes act like combs or sieves that strain food out of the water passing between them. The sheets of the huge right whale exceed 3 m in height. Differences in the arrangement and configuration of the sheets are correlated with feeding habits. A blue whale swims up to a dense swarm of small fish or crustaceans, opens its mouth, and takes in water and food until a huge pouch under the tongue, pharynx, and chest wall is filled with about 70 tons of food and water. The pouch is then emptied by forcing the water through the sieve and back into the sea. The collected food is swallowed leisurely through a very small throat. Other baleen whales skim the surface for plankton with the mouth open or feed near the bottom. Baleen is continually being worn away and replaced, like hair or fingernails.

(a)

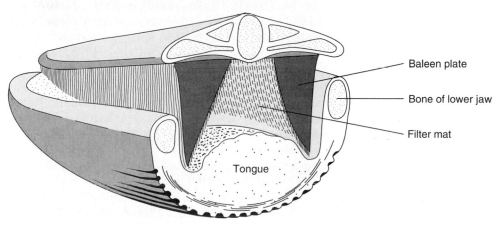
(b)

Baleen plate

Bone of lower jaw

Filter mat

Tongue

FIGURE 6.27

(*a*) Humpback whale snout showing outside edges of approximately 125 baleen plates. (*b*) Diagrammatic section through the head of a rorqual whale. The filter mat lines the oral cavity with the frayed edges of numerous baleen plates.

(*a*) © Thomas Kitchin/Tom Stack & Associates. (*b*) From Nigel Bonner, *Whales of the World,* 1989. Copyright © 1989 Cassell PLC, London. Reprinted by permission.

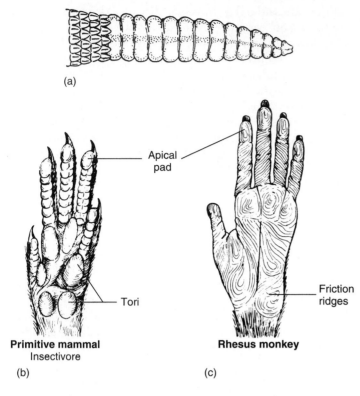

FIGURE 6.28

Products of the stratum corneum. (*a*) Rattlesnake rattle. (*b*) Tori. (*c*) Friction ridges.

Rattlesnake **rattles** are rings of horny stratum corneum that remain attached to the tail after each molt (fig. 6.28*a*). **Beaks** are covered with a horny sheath, and roosters' **combs** are covered with a thick, warty stratum corneum. Monkeys and apes sit on thick **ischial callosities,** and camels kneel on **knee pads. Tori** are epidermal pads that most mammals other than ungulates walk on (fig. 6.28*b* and *c*). Cats "pussyfoot" by retracting their claws and walking stealthily on tori. At the ends of digits, tori are called **apical pads. Corns** and **calluses** are temporary thickenings of the stratum corneum that develop where the skin has been subjected to unusual friction.

The structure of the mammalian epidermis is described later in the chapter under "The Integument from Fishes to Mammals."

THE DERMIS

The basic component of dermis, whether of fishes or human beings, is collagenous connective tissue that holds the other components in place and provides tensile strength (fig. 7.2). Other components include, universally, blood vessels, small nerves, and pigment cells; and in one species or another, lymphatics, naked and encapsulated exteroceptors, the bases of multicellular glands, and in endotherms, the bases of hairs or feathers and

their erector muscles. In addition, *the dermis has an ancient and persistent potential to form bone,* and bone is a constituent of the dermis in some members of every vertebrate class except birds. Early fishes had so much bone in the skin that they are called armored fishes. When bone is lacking in the dermis, it is because bone salts were not deposited on the collagenous matrix. Cowhide and pigskin, from which leather is made, the dermis of leatherback turtles, the skin of cyclostomes—all are exceptionally tough because the collagen bundles are densely packed as though prepared for a shower of bone salts that, for lack of an enzyme or some other reason, never materialized.

The Bony Dermis of Fishes

The bony armor of early fishes was disposed as broad bony plates or smaller bony scales over most of the body, immediately beneath the epidermis (figs. 4.4 and 4.9). The structure of this **dermal bone** varied in histologic details, but a generalized pattern consisted of lamellar bone, spongy bone, dentin, and a surface layer of an enamellike substance of one physical nature or another (fig. 6.29*a*). Frequently, the surface of the armor was sculptured by knobby or spiny elevations of the dentinal layer known as **denticles** (3 plus 4 in fig. 6.29*a*). Lamellar bone is compactly structured bone that has been deposited in successive layers, or lamellae, a common method of bone deposition. Spongy bone, on the other hand, is penetrated by blood channels of macroscopic size, which give the bone a spongy appearance. Dentin is yet another variety of bone. These varieties of bone are discussed in more detail in the next chapter.

Dermal armor was protective, but it may also have served as a storage site for calcium and phosphates that could be retrieved by the blood stream when needed. In time, the large plates on the trunk and tail gave way to the smaller and thinner scales of later fishes, while those on the head became part of the skull and those immediately behind the last pharyngeal arches became a part of the skeleton of the shoulder girdle. Dermal plates and scales are classified as **cosmoid, ganoid, placoid,** and **modern.** The latter have evolved two forms, **ctenoid** and **cycloid,** named for their appearance.

Cosmoid plates and scales were present on lower and middle Paleozoic fishes, including early lobe-fins and resembled the dermal armor seen in figure 6.29*a*. The dentin in cosmoid scales is sometimes called cosmine, which accounts for their name. No living fishes have cosmoid scales.

Ganoid plates and scales covered Paleozoic fishes when they were at their zenith. The earliest ganoid scales (paleoniscoid, fig. 6.29*c*) are found today only on the chondrosteans *Polypterus* and *Calamoichthyes* and the crossopterygian *Latimeria,* although in the latter the dentin layer has been reduced. The Mesozoic brought

FIGURE 6.29

Dermal bony scales through the ages. In (*a*): **1,** lamellar bone; **2,** spongy bone; **3,** dentin; **4,** enameloid. **3** plus **4** constitute a denticle. In (*e*): **1,** cellular lamellar bone; **5,** fibrous plate characteristic of modern fishes. The constituents of placoid and chondrostean ganoid scales (*b* and *c*) are similar. Figures are arranged to denote relative age.

(e)   **Cycloid/Ctenoid**
(Teleostean)   **Cenozoic**

(d)   **Ganoid scale**
(Holostean)   **Mesozoic**

(c)   **Ganoid scale**
(Chondrostean)   **Upper Paleozoic**

(b)   **Placoid scale**
(Chondrichthyan)

(a)   **Ancient armor**   **Lower and Middle Paleozoic**

modifications of these scales, which are mirrored in the two genera of living holosteans. *Amia* and gars have lost the spongy bone and dentin (fig. 6.29*d*); but these very hard ganoin-covered scales still constitute a continuous bony armor over the entire trunk and tail (fig. 6.30). On the head they form bony plates and smaller scalelike bones (fig. 9.11*a*). The scalelike bony plates of *Amia* lack ganoin and are confined to the head. Modern flexible scales cover the rest of the body.

Placoid scales were present on Paleozoic sharks, and are still present on today's sharks, skates, and rays. They have the same structure as paleoniscoid scales, consisting of lamellar bone, dentin, and an enameloid covering (fig. 6.29*b*). But arising from the flat bony **basal plate (root)** in the dermis is a **spine** of enamel-covered dentin that erupts through the epidermis (fig. 6.31). The spine is a specialized denticle. Placoid scales become teeth at the edges of the jaws.

(a)

(b)

FIGURE 6.30

Ganoid scales. (*a*) Scales on trunk of a garfish. The tail is to the right. (*b*) Canaliculi and lacunae within a single scale (microscopic).

Cycloid and ctenoid scales are found on teleosts and modern lobe-finned fishes. There is little difference between them except that the ctenoid scales have a comblike (ctenoid) free border (fig. 6.32). These modern scales consist of a very thin layer of acellular lamellar bone underlain by a plate of dense collagen (fig. 6.29e). As a result they are flexible and translucent. The overlying epidermis is very thin. Their arrangement in the dermis is seen in figure 6.34.

The ability to form scales has been lost by cyclostomes, catfish, bony eels, and some other fishes; nevertheless, scale anlagen develop transitorily in the embryos. In contrast to scaleless teleosts, sea horses and pipefish have dense bony plates in their dermis over most of the body surface.

Dermal Ossification in Tetrapods

The outstanding difference between the dermis of one vertebrate class and another is the extent to which each contains bone. When the first tetrapods lumbered onto land they brought with them versions of the cosmoid scales of their lobe-finned ancestors. Some labyrinthodonts, and even some cotylosaurs, had large bony plates in the dermis; others had minute bony scales. The latter are called **osteoderms** in tetrapods.

Among amphibians, caecilians and some tropical toads still have osteoderms. Those between furrows in caecilian skin are microscopic, are perpendicular to the surface, and lie in circumferential bands separated by glandular skin. They are just large enough to be seen with the naked eye when loosened with a scalpel (fig. 6.33).

(a)

(b)

FIGURE 6.31

(*a*) Unerupted placoid scale in skin of embryonic shark. Dermis consists of compact layer, **dc,** and loose layer, **dl; dp,** dermal papilla (pulp) within root and spine; **e,** enamel; **gl,** unicellular epidermal gland; **m,** melanophore; **s,** spine, composed of dentin, enamel, and a core of pulp. (*b*) Erupted placoid scale, a spiny denticle.

Cycloid

Ctenoid

FIGURE 6.32

Modern flexible fish scales. The upper edges are caudal free borders.

FIGURE 6.33

A single osteoderm from caecilian skin.

Crocodilians have oval osteoderms, especially along the back, where they are often associated with cornified crests (figs. 6.34, alligator, and 6.35). The young of some species of lizards have osteoderms under the epidermal scales of the head until some time after hatching or birth, after which they usually fuse with the underlying membrane bones of the skull. Turtles other than leatherbacks are truly armored, being boxed in by large bony plates that meet in immovable sutures (fig. 6.36). The carapace and plastron are united by bony lateral bridges that must be sawed through to expose the viscera.

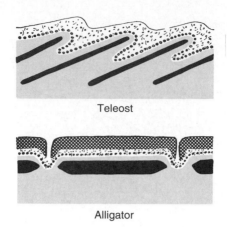

FIGURE 6.34

Site of dermal scales (*black*) in a teleost and of osteoderms in a crocodilian.

FIGURE 6.35

Alligator skin. (*a*) From dorsum of neck, showing epidermal scales. (*b*) Same section turned upside down to show osteoderms (dermal scales) embedded in skin beneath the tall crests. The middorsal line may be used as a reference point.

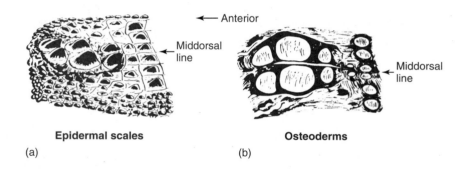

FIGURE 6.36

Dermal plates of the carapace (*a*) and plastron (*b*) in the turtle *Chrysemys,* from an internal view. On the carapace the plates shown are **c,** costals, united with ribs; **n,** nuchal; **p1** and **p2,** precaudals; **v,** vertebrals (six of the eight are labeled). Marginals encircle the periphery and include the pygal, **py.** On the plastron, the plates shown are **e,** epiplastrons; **en,** entroplastrons; **h,** hyoplastrons; **hp,** hypoplastrons; **xi,** xiphiplastrons.

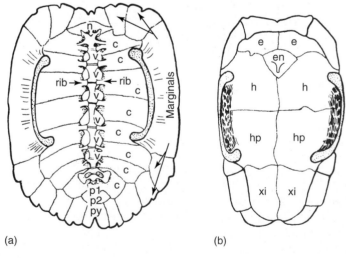

Among mammals, armadillos alone have dermal armor. It consists of identical small polygonal bones immovably united and extending almost to the midventral line. The bone is covered by epidermal scales (fig. 6.20). Bone forms in the skin of other mammals, including human beings, as a pathologic condition. An earlier observation is worth repeating: *the dermis of vertebrates has an ancient and persistent potential to form bone.*

Dermal Pigments

Chromatophores are cells that contain pigment granules. They are found in many locations of the body, but those in the skin are in the dermis and are responsible for skin coloration. All have processes that are permanent extensions from the cell body, and in some the pigment granules can be dispersed into the processes or aggregated in the vicinity of the nucleus by neutrotransmitters and by hormones such as intermedin and melatonin. Chromatophores with dispersible granules are responsible for **physiologic color changes,** which are rapid changes such as are seen in chameleons.

Chromatophores are named for the color of their granules. **Melanophores** contain melanin granules, which are varying shades of brown. These are located in the cytoplasm in organelles known as **melanosomes.** **Xanthophores** contain yellow granules. **Erythrophores** contain red granules. Xanthophores and erythrophores are sometimes called lipophores because the granules are soluble in lipid solvents. **Iridophores** contain a prismatic substance, guanine, which reflects and disperses light, producing silvery or iridescent skin.

The pigment granules in hair and feathers, and any in the epidermis, are not in the cells that produced them for the following reason: deep within the dermis the branched processes of dermal melanophores ramify among the epidermal cells of the growth zones of feather and hair follicles and actively *inject* pigment granules into cells that are being added to the growing hair or feather. Similarly, melanin granules in the mammalian epidermis have been injected into cells of the germinal layer.

Hairs receive only melanin granules, and hair color (black, brown, red, blonde) is attributable to the specific distribution and density of the granules and to the number of air vacuoles in the medulla of the hairs. Gray or white hair is the result of large numbers of air vacuoles and few melanin granules.

Feathers receive brown, yellow, and red pigments. Despite what appear to be blue feathers, vertebrates have no blue granules. When viewed under a light microscope by *transmitted* light, the "blue" feather is brown, the color of the melanin granules beneath the prismatic layer. The blue color observed in *reflected* light is a dispersion phenomenon, like the blue of the sky. A red feather is red even when viewed in transmitted light because of red pigment granules. The iridescence of feathers is also a dispersion phenomenon.

Physiologic color changes occur only in ectotherms, chiefly in teleosts, anurans, and some lizards. Dispersal of granules creates a pigmentary blanket that masks any underlying pigments; aggregation exposes the latter. Inasmuch as not all varieties of chromatophores respond similarly to the same stimuli, various color combinations result. A functional association of dermal chromatophores is illustrated in figure 6.37.

Most vertebrates cannot change color reflexly because the granules in their chromatophores cannot aggregate and disperse. These animals change color only as pigment granules are synthesized in response to long-term environmental or hormonal stimuli such as exposure to sunlight ("getting a tan," for example) or when hairs or feathers are shed and replaced with others. These are **morphologic color changes.** They are generated most commonly by seasonal biological rhythms that regulate hormone synthesis (fig. 4.41*a*).

FIGURE 6.37

Dermal chromatophore unit of the frog *Hyla cinerea* in transverse section. There are three types of pigment cells: xanthophores, **X**; iridophores, **I**; melanophore, **M**. Arrows show melanosomes **(MS)**, dark granules, moving into four of the six-finger **(F)** processes alongside the iridophores. Dispersal of melanosomes into fingers will temporarily darken the skin in a darkened environmental background.

Pigment cells are not confined to the skin. They are found in many deep locations, including the meninges and striated muscles. Pigment cells in these locations contain only melanin granules and are usually referred to as **melanocytes**. The granules are permanently dispersed. All pigment cells arise from neural crests.

Not all skin color is due to pigment. The deep scarlet color of the face, forehead, and scalp of red ouakari monkeys from Amazon River valley (fig. 4.39a) is due to an exceptionally rich network of blood capillaries lying immediately under the epidermis. When circulation ceases, the color disappears.

THE INTEGUMENT FROM FISHES TO MAMMALS: A CLASS-BY-CLASS OVERVIEW

Agnathans

Epidermis   Living agnathans lack scales, but their multilayered epidermis is characterized by an abundance of unicellular mucous glands, which earned hagfishes the descriptive name "slime eels" (fig. 6.4). All layers are mitotic, including the surface cells. The only cornified structures are horny denticles in the buccal funnel (fig. 6.9). These are periodically shed and replaced.

Dermis   The dermis is thinner than the epidermis, but it is exceptionally tough because of very compactly interwoven bundles of collagenous connective tissue. It contains many melanophores, and adheres tightly to the underlying body wall musculature, especially at the myosepta.

Cartilaginous Fishes

Epidermis   The epidermis of cartilaginous fishes (fig. 6.31) consists of more layers of cells than does that of cyclostomes, the cells are compactly disposed, and unicellular glands are less abundant, except in chimaeras. An aggregate of goblet cells at the base of the stinger in sting rays have been modified to secrete toxins. Multicellular glands are few in number and in restricted locations, such as at the base of the claspers of males. Light-emitting organs, or photophores, in the dermis in a few species are modified multicellular epidermal glands that have invaded the dermis, then lost their connection with the epidermis.

Dermis   The dermis is considerably thicker than the epidermis and is composed of two more or less well-defined layers. The epidermis of elasmobranchs is unique in the presence of placoid scales, inherited unchanged from paleozoic sharks. Placoid scales have a bony basal plate in the dermis and a spine that penetrates the epidermis from below and projects above the surface of the skin, giving it a sandpaper texture. Just under the epidermis is a continuous sheet of melanophores, which is more dense on the back than on the belly, except in deep-sea species. This distribution makes a shark in partially illuminated waters less visible from both above and below. Other types of pigment cells may also be present.

Chimaeras have lost the scales over most of the surface of the body; and, since they have many more mucous glands than elasmobranchs, their skin is slippery, rather than having a sandpaper texture.

Bony Fishes

The skin of most extant bony fishes, especially dipnoans and teleosts, have a relatively large number of mucous glands, and they have dermal scales of one kind or another, except placoid.

Epidermis   **Epidermal glands** are chiefly unicellular mucous glands that maintain a constant thin mucous coating on the surface of the skin (fig. 6.5). Multicellular glands (fig. 6.6) are relatively sparse. Some of them secrete the mucoid cocoon that envelopes aestivating lungfishes in dry seasons, impeding desiccation. A few are granular glands that secrete an irritating or, to some species, poisonous alkaloid. Many teleosts that inhabit the deep, dark recesses of the seas have variously disposed and sometimes bizarre photophores that facilitate species recognition or serve as lures or warnings (fig. 6.5).

Dermis   The dermis is characterized by the presence of ancient ganoid scales, or modern cycloid or ctenoid scales (figs. 6.29, 6.30, and 6.32). Ganoid or modified ganoid scales are found on several relict chondrosteans and holosteans such as *Polypterus* and garfishes, often in the form of plates on the head and scales on the body and tail. A few close relatives, however, have modern fish scales on some parts of the body. Teleosts, except for a few naked ones, have ctenoid or cycloid scales (sometimes both) on the same fish; and some scaleless teleosts develop abortive scales during embryonic life.

Amphibians

The skin of amphibians differs from that of fishes in three major respects: (1) scales are absent except in a few species; (2) epidermal glands are generally abundant and predominantly multicellular; and (3) except in toads, the epidermis exhibits only an incipient stratum corneum.

Epidermis   The epidermis is glandular, and highly so in aquatic or semiaquatic species (figs. 6.2 and 6.7). Unicellular cells secrete mucigens, but most mucous glands are multicellular. Urodeles and anurans differ in the proportion of time they spend on land and in the water. In aquatic amphibians that are infrequently exposed to air, many of which breathe partly with gills, integumentary mucus keeps the skin moist while the animal is out of the water. Produced in abundance when the animal is being attacked, mucus makes the body slimy, facilitating escape from prey. Granular glands are common but not abundant in toads (fig. 6.11, paratoid). Their irritating or toxic secretions aid in the survival of these land-adapted anurans.

The epidermis of species that spend a considerable time on land exhibits a layer of cornified cells of varying thickness that impede desiccation. Toads have the thickest layer; permanently aquatic species have little or none. Efts have a thin covering until they return to water as mature adults, at which time the cornified cells are shed, epidermal glands hypertrophy, and the skin becomes smooth and slippery.

Cornified appendages are rare. Calluslike caps form on the ends of the digits of urodeles subjected to buffeting in mountain brooks. Tadpoles have horny teeth that rasp aquatic mosses and algae, which is their larval diet. At metamorphosis, the horny teeth are shed and the adult becomes insectivorous.

Dermis   The dermis is firmly attached to the underlying musculature over much of the body in apodans and urodeles, but the relationship of skin and muscle in anurans is unique. Separating the skin from the underlying muscles in anurans are voluminous broad subcutaneous lymph sinuses, the survival value of which is conjectural (fig. 14.42). The skin of anurans is therefore less intimately attached to the underlying tissues than in other vertebrates. Dermal chromatophores in some species are capable of eliciting physiologic color changes (fig. 6.37).

Tiny bony scales are embedded in the dermis of some caecilians (fig. 6.33), often in bands alternating with bands of glandular epidermis. A few tropical toads have bony scales in the dermis of the back.

Reptiles

The skin of reptiles has a thick stratum corneum with a variety of specialized cornified features. Integumentary glands are relatively sparse, and the dermis of many reptiles has bony deposits varying from small dermal scales (**osteoderms**) to large bony plates.

Epidermis   The epidermis is unique in having a thick, water-impervious stratum corneum with many localized modifications, including epidermal scales, scutes, beaks, rattles, claws, plaques, and spiny crests (figs. 6.16 through 6.18, 6.19b, and 6.28a). These features represent the ultimate adaptation of vertebrate skin to life in an arid and often hostile environment. All these are modifications of a continuous sheet of stratum corneum.

The scales of squamates overlap because they form on the surface of overlapping folds of epidermis (fig. 6.16). Periodically, snakes shed the entire outer layer of stratum corneum in one piece, or "cast," in a process called **ecdysis.** Lizards slough it off in patches. Other reptiles wear it off by abrasion. The scales of amphisbaenians are disposed in concentric rings that encircle the body,

imparting an annulate appearance to the body similar to that of caecilians. Scutes on the ventral surface of snakes and on the plastron of turtles are smooth, quadrilateral, or polyhedral sheets of stratum corneum that protect the ventral surface from abrasion. (The scutes on the plastron and carapace of turtles are identified by name in fig. 6.19.) The stratum corneum of crocodilians is warty, crested, or composed of nonoverlapping scales or thick plaques, depending on the region of the body. In all species, small nonoverlapping epidermal scales cover the limbs, tail, neck, and other surfaces that are not otherwise protected.

Epidermal glands are almost entirely granular and are restricted to localized regions of the body where their protective and pheromonal secretions will be most effective. Inasmuch as granular glands secrete only in response to specific discontinuous stimuli, the skin of reptiles is exceptionally dry.

Dermis   The dermis is notable for the presence of dermal bone. Most conspicuous is the shell of turtles, consisting of carapace, plastron, and lateral bridges that connect the two. Carapace and plastron are composed of large dermal plates arranged in a pattern, but one that does not coincide with that of the overlying scutes (fig. 6.36). Soft-shelled and leatherback turtles have a flexible leathery skin because of the absence of ossification in the dermis. Crocodilians have osteoderms in localized regions of the body (fig. 6.35). Along the dorsum they are overlaid by prominent cornified crests that give the back an awesome prehistoric appearance. A few lizards have small osteoderms underlying the epidermal scales. Osteoderms are absent in snakes.

Birds

The term "thin skinned," sometimes applied figuratively to humans, is literal when applied to birds. The epidermis and dermis constitute a delicate membrane that is loosely applied to the underlying muscle, and is therefore quite mobile. Only on the feet and head is the skin relatively thick and intimately attached. In these regions are the horny beak, epidermal scales, spurs, and claws. Elsewhere the body is clothed in feathers, the most complex of all the derivatives of the vertebrate stratum corneum (figs. 6.22 to 6.24).

Epidermis   Epidermal scales, similar to the small scales of turtles, are limited to the feet and the base of the beak. The beak is covered with a horny sheath similar to that of turtles. It may consist of several segments, and it may extend dorsally over the nasal region to form a frontal shield. The beak sometimes bears horny toothlike protuberances that are correlated with specific diets, and which take the place of teeth, which are missing in today's birds. Sharp claws are present on the toes and one or two are found on digits of the wings of species whose habits include climbing over bark, as do young hoatzins, or hanging from crags, as do chimney swifts.

Integumental glands are lacking except for two varieties, a prominent uropygial gland at the base of the foreshortened tail, best developed in water birds, and small oil glands that line the outer ear canal of some domestic fowl and sometimes encircle the vent. The oily secretion of the uropygial gland is applied by the beak to the feathers during preening.

Dermis   The dermis supports the feather follicles and erector muscles. Its exceptional thinness, and the mobility of the skin as a whole, may be correlated with the thermoregulatory role of feathers, which can be fluffed to provide insulation for the body. Birds have no osteoderms. Some, however, such as male gamecocks, develop a spur of dermal bone on each ankle, fused to the tarsometatarsus. In some other species, both sexes have a spur on the carpometacarpus of the wrist.

Mammals

The notable features of mammalian skin are (1) hair, (2) a greater functional variety of epidermal glands than in any other vertebrate class, (3) a relatively thick cornified epidermis, and (4) a dermis that, unlike earlier vertebrates, is many times thicker than the epidermis.

Epidermis   The mammalian epidermis in general consists of three strata, an actively mitotic **stratum germinativum** (the basal layer), a relatively thin **stratum granulosum,** and a **stratum corneum** (fig. 6.12). The granulosum, when present, contains granules of keratohyaline, a precursor of the keratin found in the stratum corneum. The stratum corneum is thickest on the parts of the forelimbs and hindlimbs that come in contact with the substrate (hoofs, apical pads, knee pads, and tori, for example). In humans, it is thickest on the palms and on the soles. Although corns and calluses are a result of friction due to usage, the thick stratum corneum of the palms and soles is hereditary. It develops in the fetus. A fourth cellular layer, the **stratum lucidum,** is present in the epidermis of the palms and soles. It is a layer of cells rendered translucent by the dissolution of the keratohyaline granules of the granular layer in those locations.

Hairs are cornified epidermal appendages that arise from hair follicles that extend deep into the dermis (fig. 6.12). Hair color results from combinations of varying concentrations of melanin granules and air vacuoles within the shaft.

Epidermal glands are of two major types, sebaceous, which synthesize oily secretions, and sudoriferous, which are predominantly sweat glands. Mammary glands of viviparous mammals appear to be modified sebaceous glands. The milk glands of monotremes appear to be modified sudoriferous glands. The variety of glands found at junctions between cornified and noncornified integumentary epithelia—glands that lubricate the eyeball, for example—are mostly sebaceous.

The stratum corneum gives rise in various species to scales, claws, and horns. Epidermal scales are found on the tail and feet of opossums and many rodents; they also develop in fetal bears and European hedgehogs but are shed into the amniotic fluid before birth. Claws have become hoofs in ungulates and flat nails in primates and a few other orders (fig. 6.21). Horns are either composed entirely of keratin, as are the unique hair horns of rhinoceros, or they have a horny sheath overlying a bony core, as in bovines and pronghorn antelopes (fig. 6.27).

Scalelike skin unlike any other is seen in armadillos and pangolins. Armadillos have a continuous sheet of horny scales overlying a similarly sculptured sheet of dermal bone, making armadillos truly armored mammals (figs. 4.35 and 6.20). Pangolins have epidermal scales that give the impression of having originated as clumps of agglutinated hairs (fig. 4.39). The horny baleen of toothless whales is an adaptation for straining its food of small organisms out of seawater (fig. 6.27).

Dermis   The exceptional thickness of mammalian dermis is a result of the presence of many hair follicles, erector muscles of the hairs, sweat and sebaceous glands, the requisite supportive connective tissue, and its exceptional vascularity. The rich blood supply is essential to satisfy the metabolic demands of these structures. It also regulates heat loss from the body surface through vasoconstriction and vasodilation. Encapsulated sensory nerve endings are abundant in dermal papillae that indent the underside of the epidermis (fig. 6.12). The dermal bone of armadillos was referred to earlier. Antlers and giraffe horns are dermal derivatives.

The skin of most mammals is hidden by fur; dermal pigments other than those in hairs play little role in skin coloration. The vividly colored ischial callosities of baboons are among the exceptions. In human beings, melanophores protect against the rays of the sun in proportion to their density.

Mammalian skin is separated from the underlying muscle by a cushion of loose connective tissue, the **superficial fascia.** Contour-shaping adipose tissue (fat), of which whale blubber is an extreme example, accumulates in this subcutaneous tissue.

SOME ROLES OF THE INTEGUMENT

The vertebrate integument is an organ, just as are the liver and gill, and it plays a variety of roles in the survival of a species. Some of the roles are primary, others are supplementary. Among the roles in one species or another are protection, exteroception (sensitivity to environmental conditions), respiration, excretion, thermoregulation, participation in locomotion, maintenance of homeostasis (regulation of the ratio of water to salts in body fluids), nourishment of the young, sexual and species signaling, and additional miscellaneous functions. The paragraphs that follow bring together ideas that have been advanced in other contexts throughout the chapter.

The **protective role** of the integument is primary. Dermal armor protects the internal organs from mechanical injury, glands secrete slimy or noxious substances, pigments provide protective coloration and a barrier to the rays of the sun, the bristling fur of a threatened mammal and the ruffled plumage of a bird give these animals an ominous appearance, and claws, horns, spiny protuberances, needles—all confer advantages in the struggle for existence.

The **exteroceptive role** of the integument is protective in its most primitive state. Naked nerve endings are stimulated when a foreign object comes in contact with the skin. This specific information is a danger alert; and, in cyclostomes, is the only role of cutaneous receptors. In higher fishes and tetrapods, more complex nerve endings provide additional information; but even in humans, some of this is essential for survival. Cutaneous receptors are discussed in chapter 17.

Respiration via the skin supplements that via gills or lungs in many amphibians, particularly aquatic urodeles, which may acquire as much as three-fourths of their oxygen from the water in this manner. Plethodontid salamanders rely entirely on skin, having neither gills nor lungs. Scales and a thick stratum corneum are not conducive to cutaneous respiration.

Excretion of carbon dioxide in some aquatic amphibians is entirely via the skin, even in species in which oxygen is being acquired via gills. Sweat glands are a supplementary route for excretion of nitrogenous wastes in mammals that have them.

Thermoregulation is partly a function of the skin of endotherms. Fur and feathers insulate against cold, sweat cools by evaporation, and dilation of small blood vessels within the dermis increases heat loss by radiation. When heat conservation is necessary, these same vessels constrict.

Locomotion is subserved by adhesive pads, by claws that assist in climbing, scutes that assist in slithering, and feathers that provide an airfoil for flight. Integumentary webs stretched between the digits of many species that frequent ponds or the seashore facilitate locomotion in sand, muck, or water. Other integumentary webs permit flight in bats.

The **maintenance of homeostasis** is assisted in some fishes by dermal scales, which serve as reservoirs for calcium and phosphate molecules that can be drawn on as needed. Chloride-secreting glands dispose of excess chloride ions acquired from a marine environment, and the stratum corneum of terrestrial species conserves water. Under the influence of posterior pituitary hormones, the skin of aestivating lungfishes, and of toads and some other vertebrates holed up in moist locations during dry weather, absorbs water from moist surroundings, thereby preventing desiccation.

Nourishment is provided to some teleost hatchlings that feed on mucus secreted by the mother's skin, and mammary glands nourish newborn mammals.

Pheromones and sex-linked **skin coloration** signal the sex or serve for species identification by other members of a population. Some pheromones serve as alarms. **Vitamin D** is synthesized from ergosterol in the skin of some higher animals, including primates. One can no doubt list other roles performed by the integument. No wonder that an earlier biologist characterized the integument as "a jack of all trades."

SUMMARY

The section entitled "The integument from fishes to mammals" provides an additional summation.

1. The vertebrate integument consists of a nonvascular epidermis and a vascular dermis.
2. The epidermis of most fishes has a thin mucous cuticle. That of terrestrial vertebrates is covered by keratinized cells that retard desiccation.
3. Unicellular mucous glands abound in the skin of most fishes; multicellular ones appear in aquatic amphibians. Granular glands produce irritating alkaloids; they are characteristic but sparse in toads and reptiles. Oil glands are few in birds, abundant in mammals, where they are chiefly sebaceous.

Mammary glands of viviparous mammals are probably modified sebaceous glands. Those of monotremes resemble sudoriferous glands.

4. Photophores are multicellular light-emitting epidermal glands.
5. The stratum corneum of amniotes produces a variety of cornified appurtenances, including epidermal scales, scutes, claws, hoofs, nails, horns, baleen, rattles, horny teeth, and beaks, feathers, and hair.
6. Contour feathers typically consist of an afterfeather and a main feather, each with an axial shaft supporting two vanes. Vanes consist of barbs, barbules, flanges, and hooklets. Down feathers lack hooklets

and are therefore fluffy. Filoplumes are little more than a threadlike shaft. Feathers are erected by arrectores plumarum.

7. Hairs arise out of hair follicles, and consist of a bulb, root, and shaft. Arrectores pilorum insert on the walls of the follicles.
8. Hairs have become modified as spines, quills, bristles, and vibrissae, and components of the scales of scaly anteaters. Rhinoceros horns may be agglutinated hairs.
9. The basic constituent of dermis is collagenous connective tissue. It supports blood vessels, lymphatics, nerves, chromatophores, and, in one species or another, the bases of multicellular glands, encapsulated receptors, feather or hair follicles, and their erector muscles.

10. The dermis has an ancient potential to form bony plates or scales, which persists in one or more living members of every vertebrate class except birds.
11. The dermal scales of extant fishes are placoid (cartilaginous fishes), ganoid (some chondrosteans and holosteans), and flexible ctenoid or cycloid scales (modern fishes). Cosmoid scales were present in early armored fishes.
12. Chromatophores are permanently branched dermal pigment cells. Physiologic color changes are reflexive and result from movement of pigment granules into or out of the processes. Morphologic color changes are slow and depend chiefly on seasonal pigment synthesis.

SELECTED READINGS

Appleby, L. G.: Snakes shedding skin, Natural History **89**(2):64, 1980.

Birch, M. C., editor: Pheromones, New York, 1974, Elsevier/Excerpta Medica/North Holland.

Goss, R. J., illustrated by Wendy Andrews: Deer antlers, regeneration, function, and evolution, New York, 1983, Academic Press.

Nelson, D. O., Heath, J. E., and Prosser, C. L.: Evolution of temperature regulatory mechanisms, American Zoologist **24**:791, 1984.

Rao, K. R., Fingerman, M., and others: Chromatophores and color changes, American Zoologist **23**:461, 1983.

Regal, P. J.: The evolutionary origin of feathers, The Quarterly Review of Biology **50**:35, 1975.

Sengel, P.: Morphogenesis of skin, Cambridge, England, 1976, Cambridge University Press.

CHAPTER SEVEN

MINERALIZED TISSUES:
AN INTRODUCTION TO THE SKELETON

In this chapter we will be introduced to the various types of skeletal tissues, see how they are formed, and look briefly at their invertebrate origins. We will discover that mineralized tissues have a role in homeostasis, and we will learn why growing skeletons must undergo continual remodeling.

OUTLINE

he skeleton is composed of mineralized connective tissue and of ligaments, tendons, and bursae. The mineralized tissue for the most part is bone, but there is also dentin (often considered a variety of bone), cartilage, and enamel or enameloid substances. Osteoblasts produce bone, odontoblasts produce dentin, chondroblasts produce cartilage, and ameloblasts produce most enamels. These specialized cells arise from less-differentiated scleroblasts that arise from mesenchyme (fig. 7.1).

A preliminary step in formation of skeletal tissues is the synthesis of collagen by fibroblasts. Collagen is a proteinaceous **fibril** demonstrable only by high-power electron microscopy. Fibrils aggregate to form **collagen fibers** that are visible with light microscopy. The fibers form dense **collagen bundles** that are woven into a compact network of dense connective tissue like that found in dermis, tendons, and ligaments (fig. 7.2). It is on such a network that minerals are deposited to form cartilage and bone.

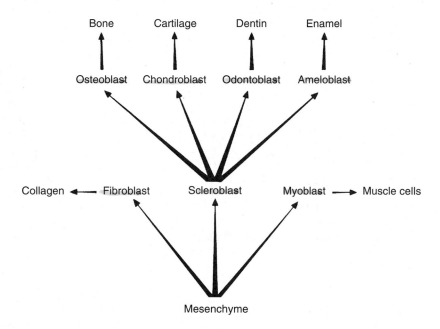

FIGURE 7.1

Some products of mesenchyme.

FIGURE 7.2

Bundles of collagen fibers in tendon. The individual fibers in each bundle appear as dots in cross section. The long nuclei are in fibrocytes (mature fibroblasts).

BONE

The gross structure of a long bone is shown in figure 7.3. Bone tissue is composed of a matrix of collagenous fibers, the spaces between which have been impregnated with **hydroxyapatite crystals** that are composed of calcium, phosphate, and hydroxyl ions [$3Ca_3 (PO_4)_2 \cdot Ca(OH)_2$]. The crystals are deposited under the influence of osteoblasts. A **cementing substance** composed of water and mucopolysaccharides binds the crystals to the collagenous matrix. In most bone, the osteoblasts ultimately become trapped by the bone they have laid down around them, and, as osteocytes, occupy tiny fluid-filled pools, or **lacunae,** of interstitial fluid (figs. 7.3a–c and 7.4). Interconnecting the lacunae are fluid-filled canals, **canaliculi,** that house protoplasmic processes extending from the osteocytes. The fluid in the lacunae and canaliculi contains calcium and phosphate ions, which are constantly being deposited on the collagenous matrix or withdrawn from it, depending on the level of serum calcium.

Compact Bone

Compact bone consists of layers **(lamellae)** of mineralized collagenous bundles arranged concentrically around a **haversian canal** (fig. 7.3a). Haversian canals contain an arteriole, a venule, a lymph vessel, and nerve fibers. The canal and its surrounding lamellae constitute an **osteon,** or **haversian system** (fig. 7.4). The blood vessels are responsible for the configuration of the haversian systems, and because the vessels branch and rebranch, haversian systems do likewise. The vessels in the canals are continuous with those in the bone marrow.

FIGURE 7.3

(a) Compact bone in cross section. *Red,* arterioles in haversian canals. (b) Osteon bone in longitudinal section. (c) Nonlamellar membrane bone.

Lacuna

Haversian
canal

Canaliculi

FIGURE 7.4

Cross section of one osteon (haversian system).

© R. Calentine/Visuals Unlimited.

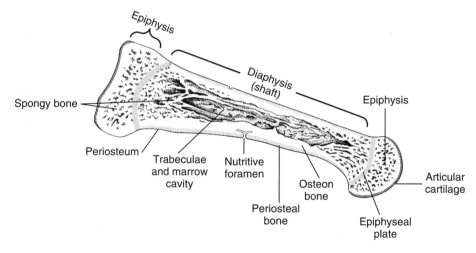

Epiphysis

Diaphysis
(shaft)

Epiphysis

Spongy bone

Periosteum

Trabeculae
and marrow
cavity

Nutritive
foramen

Osteon
bone

Articular
cartilage

Periosteal
bone

Epiphyseal
plate

FIGURE 7.5

A metatarsal bone in longitudinal section. Epiphyseal plates in *light red*.

In the shaft of long bones (fig. 7.5), the haversian systems more or less parallel the long axis of the bone (fig. 7.3b), an architectural feature that minimizes the likelihood of fracture under normal external stress. The lamellae of **periosteal bone** are formed by osteoblasts on the inner surface of the **periosteum,** the dense fibrous membrane that covers all bones except at their articular surfaces. Haversian systems are characteristic of amniotes. Most amphibians, a few reptiles, and some small insectivores and rodents lack them.

Spongy Bone

Spongy, or **cancellous, bone** consists of bony trabeculae and bone marrow (fig. 7.5). **Trabeculae** are an assemblage of beams, bars, and rods that, like architectural trusses, form a rigid framework that provides maximum strength at areas of stress. They are composed of irregularly arranged lamellae without haversian systems. **Marrow** occupies the cavities between trabeculae. It consists of a reticulum of connective tissue fibers that support blood vessels, nerve fibers, and adipose tissue (yellow marrow). The hemopoietic tissue (red marrow) in some bones produces red blood cells and some types of white blood cells. The marrow cavities are lined with a thin connective tissue membrane, the **endosteum.** The endosteum has many features of a periosteum, and has the capacity to deposit bone, or remodel it. A core of spongy bone and marrow sandwiched between two layers of compact surface bone characterizes **flat bones,** such as ribs, scapulae, and membrane bones of the skull.

Dentin

Dentin has the same constituents as compact and spongy bone, but the odontoblasts are not trapped in lacunae during osteogenesis. They retreat as they deposit dentin, hence are always at its inner border (fig. 7.6). However, they leave behind protoplasmic processes in canaliculi. In dentin the canaliculi are called **dentinal tubules.** The tubules extend all the way to the surface of the dentin. Dentin forms only in the outer layer of the dermis, just beneath the epidermis, and is frequently coated on its surface by enamel or enameloid (fig. 6.29a,b,c). The enameloid of placoid scales is produced by odontoblasts, hence is a very hard dentin. Although a constant feature of the dermis of the earliest vertebrates, dentin is found today only in the scales of ganoid and elasmobranch fishes and in teeth.

Acellular Bone

There is one type of bone in which the osteoblasts not only retreat as they deposit bone but in addition leave behind no processes or canaliculi. It is the thin layer of **acellular bone (aspidin)** that constitutes the fibrous plates of the flexible scales of modern fishes and the cementum of vertebrate teeth.

Membrane Bone and Replacement Bone

Before bone or cartilage can be deposited, a preskeletal blastema must develop. A **blastema** is any aggregation of mesenchyme that, given the appropriate stimulus, differentiates into some tissue, such as muscle, cartilage, or bone. Once a preskeletal blastema has formed, some of the mesenchyme cells become fibroblasts and secrete collagen. Others become either osteoblasts or chondroblasts and secrete the enzymes essential for formation of bone or cartilage.

Membrane Bone

Bone deposited directly within a membranous blastema without having been preceded by a cartilaginous model

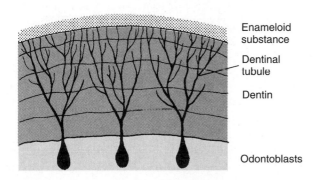

FIGURE 7.6

Dentin covered with enameloid. The dentinal tubules contain processes of the odontoblasts.

Enameloid substance

Dentinal tubule

Dentin

Odontoblasts

is **membrane bone.** Intramembranous ossification gives rise to certain bones of the lower jaw, skull, and pectoral girdles; to dentin and other bone that forms in the dermis of the skin; to vertebrae in teleosts, urodeles, apodans only; and to a few miscellaneous bones. Membrane bone may be compact or spongy, and lamellar or non-lamellar. Because of the arrangement of the blood vessels that participate in its deposition, it lacks haversian canals (fig. 7.3c). Periosteal bone is membrane bone.

Any bone derived ontogenetically or phylogenetically from the dermis of the skin may be called a **dermal bone.** The term specifies its history, not its histological features. It is membrane bone. Many, though not all, membrane bones are of dermal origin, as for example, the bones that constitute the dermatocranium, the bones that invest Meckel's cartilage of the lower jaw, and the membrane bones of the pectoral girdles. These bones are phylogenetic derivatives of the integument. Not much bone remains in the dermis of recent vertebrates, but that which does—the bony plates and scales of lower vertebrates—arises in situ.

Replacement Bone

Replacement bone is deposited where hyaline cartilage already exists (fig. 7.7). In the process the existing cartilage undergoes degenerative changes and disappears. The processes of endochondral and intramembranous ossification are the same in that both consist of impregnation of a collagenous matrix with hydroxyapatite crystals. The difference is, in endochondral ossification cartilage must be removed before bone can be deposited. In both cases the immediate result is formation of temporary spongy bone. The latter, in turn, is eroded and replaced by compact bone, spongy bone, or a marrow cavity, depending on its location.

The process of endochondral ossification of a typical long bone of a tetrapod limb commences midway in the shaft **(diaphysis)** of a miniature cartilaginous model of the bone-to-be, and procedes toward the two ends, or **epiphyses.** Soon after the diaphyseal center becomes active, one or more endochondral ossification centers appear in each epiphysis. Additional cartilage continues to be deposited in advance of all zones of ossification, and the rate of deposit of cartilage equals the rate of ossification. As a result, the shaft continues to grow in length while simultaneously becoming ossified. As the individual grows older, however, deposition of cartilage gradually slows, eventually leaving only a narrow cartilaginous **epiphyseal plate** between the diaphysis and each epiphysis (figs. 7.5 and 7.8). Chondrogenesis continues within this plate for a time, and the bone continues to elongate so long as the epiphyseal plates are elaborating cartilage. In ectotherms this is throughout life. In birds and mammals the epiphyseal centers cease elaborating cartilage shortly after the individual attains sexual maturity, the epiphyseal plates ossify, the epiphyses become sutured to

New matrix —

Old matrix —

Marrow —

Osteoblasts —

Lacuna —

Spongy bone

Dechondrification and
bone deposit

Hyaline cartilage

FIGURE 7.7

Site of endochondral ossification. *Red,* blood vessels of bone marrow invading hyaline cartilage (old matrix) and digesting it in preparation for deposit of new matrix (*light gray*) by osteoblasts.

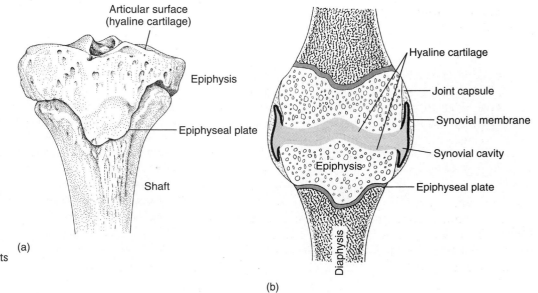

Articular surface
(hyaline cartilage)

Epiphysis

Epiphyseal plate

Shaft

Hyaline cartilage

Joint capsule

Synovial membrane

Synovial cavity

Epiphysis

Epiphyseal plate

Diaphysis

(a)

(b)

FIGURE 7.8

(*a*) Proximal end of tibia of youth aged 16 years. The epiphyseal plate is unossified, hence the bone is still elongating.
(*b*) Diagrammatic section of a simple diarthrodial joint, ligaments removed. *Red* is synovial fluid.

the shaft, and the bone stops growing. Bones of other shapes exhibit other combinations of ossification centers (fig. 9.5). Growth in diameter of the bones takes place when bone is deposited from the inner surface of the periosteum.

While endochondral ossification is going on within a bone, periosteal bone is being deposited at the surface. As a result of this activity combined with continual resorption and remodeling (to be discussed shortly), a bone reaches adult shape, size, and proportions. Details of the histogenesis of bone will be found in textbooks on histology.

Cartilaginous fishes either no longer have the genetic codes necessary for endochondral ossification or there is some factor that prevents gene expression for ossification.

CARTILAGE

Cartilage resembles bone in that it is formed within a matrix of collagenous connective tissue and the cells lie in lacunae. The intercellular matrix, however, contains a sulfated mucopolysaccharide instead of hydroxyapatite crystals. Unlike bone, cartilage has no canaliculi demonstrable by light microscopy and no blood vessels of its own. The lacunar cells **(chondrocytes)** are supplied with oxygen and nutrients by diffusion from the nearest capillaries adjacent to the cartilage.

Cartilage formation is initiated when chondroblasts deposit their mucopolysaccharide on a pre-existing collagenous matrix. The mesenchyme surrounding the area of chondrification organizes a bounding membrane, the **perichondrium.** Thereafter, additional cartilage is laid down within the blastema by mesenchyme from the perichondrium, and by fibroblasts and chondroblasts that have arisen by mitosis from parent cells within the developing mass. Chondroblasts become chondrocytes after being trapped in lacunae. The perichondrium is constantly restructured as the mass enlarges and changes shape.

Hyaline cartilage is the least differentiated variety and is the precursor of replacement bone. It is translucent because all components have the same refractive index to light. Little hyaline cartilage remains after the growth of replacement bones has ceased. Thereafter, it is found chiefly on the articular surfaces of bones within the joints of tetrapods (fig. 7.8*b*). Most any other hyaline cartilage that forms, whether in fishes or tetrapods, is likely to be transformed eventually into fibrocartilage, elastic cartilage, or calcified cartilage.

Cartilage with exceptionally thick, dense collagenous bundles in the interstitial matrix is **fibrocartilage.** The intervertebral disks of mammals are fibrocartilage. **Elastic cartilage** contains, in addition to collagenous fibers, a network of elastic fibers. In mammals it is found in the pinna of the ear, in the walls of the outer ear canal, in the epiglottis, and elsewhere. **Calcified cartilage** is

formed when calcium salts are deposited within the interstitial substance of hyaline cartilage or fibrocartilage. It is often mistaken for bone. The jaws of sharks, no matter how large, are calcified cartilage.

SKELETAL REMODELING

Bones not only provide mechanical support for the vertebrate body but, with scales and teeth, constitute important storage places for calcium and other mineral salts. They therefore participate in maintaining homeostasis. Calcium is constantly being deposited and withdrawn in response to dietary intake and cellular demands. When serum calcium levels are rising, excess calcium that is not excreted is deposited along with inorganic phosphate as hydroxyapatite crystals. When serum calcium is falling, calcium is withdrawn from the skeleton and other depots, thereby maintaining a normal serum calcium level. Withdrawal is regulated by parathyroid hormone and calcitonin; deposit may not be under endocrine control.

In addition to bone resorption in response to lowered serum calcium levels, cartilage and bone are constantly being resorbed and replaced in another process, skeletal remodeling. For example, a skull the size of that of a newborn child could not accommodate a 21-year-old brain without growing (fig. 7.9). But growth by accretion would not provide a larger brain cavity; it would only provide a thicker skull. The entire skull must be continuously remodeled by constant resorption of existing cartilage and bone and replacement of these in a new alignment to provide a wider, higher skull with a cavity large enough to accommodate the growing brain. This entails not only remodeling of the cranium but also of the facial bones and lower jaw. Cartilage and bones elsewhere in the body are also subject to remodeling. Although

FIGURE 7.9

Comparative size of skull of newborn and 21-year-old human being. Continued remodeling enlarges the brain cavity, hence the size and shape of the growing skull.

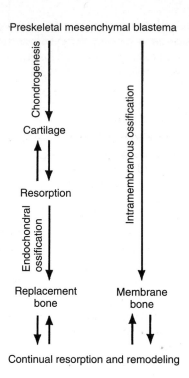

Preskeletal mesenchymal blastema

FIGURE 7.10
Steps in osteogenesis and remodeling.

remodeling is a more active process in growing skeletons, it continues to some degree throughout life, accelerating during periods of skeletal repair (fig. 7.10).

Remodeling of bones also occurs in response to mechanical stress resulting from continuing use of inserting muscles or from weight bearing. A bone becomes thicker at sites where stress is high, thin when stress is withheld. Roughened surface areas, bony ridges, and other prominences to which muscles attach enlarge with sustained muscle use. The arrangement of the latticelike components of spongy bone (fig. 7.5) bears a striking resemblance to lines of force generated by the bone's natural load.

Just as mechanical stress results in bone remodeling, affecting the shape and architecture of existing bones, a correlation exists between mechanical stress and ossification in certain tendons, ligaments, and connective tissue seams. Heterotopic bones, to be discussed shortly, are instances of stress-correlated ossifications.

TENDONS, LIGAMENTS, AND JOINTS

Tendons and ligaments consist chiefly of closely packed bundles of collagen (fig. 7.2). **Tendons** connect muscles with bones and have a shiny white appearance. The collagen bundles are so arranged as to offer maximal resistance to the tension created when a muscle contracts. Where the tendon is attached to muscle, the collagenous bundles of the tendon are directly continuous with those of the fibrous muscle sheath (**epimysium**); and where it attaches to cartilage or bone, they are continuous with the collagenous bundles of the perichondrium or periosteum. **Ligaments** connect bone to bone. The arrangement of the collagen bundles is less regular, but they are directly continuous with those of the periosteum. Tendons and ligaments that are flat and wide are **aponeuroses.** The mammalian scalp consists largely of an aponeurosis, the gala aponeurotica. The longest ligament in mammals is the nuchal ligament in the back of the neck of grazing herbivores. Extending between the occiput of the skull and the neural spines of some of the thoracic vertebrae, it provides a resilient mechanical support for the head. It is least developed in mammals with short necks.

The term "ligament" is sometimes applied to fibrous mesenteries or cords that support visceral organs or hold them in place, as for instance, the falciform ligament of the liver and the round ligament of the ovary. However, these are not skeletal structures.

In some species certain tendons and ligaments become mineralized. Turkeys, for example, have ossified tendons in their legs, and ornithischian dinosaurs had them millions of years ago. **Sesamoid cartilages** or **bones** are mineralized nodules in tendons or ligaments. Best known is the patella, or kneecap. It is endochondral bone in some species, membrane bone in others.

A joint, or **arthrosis**, is the site where two bones or cartilages meet. The joint is a **diarthrosis** (fig. 7.8b) if it is freely movable in one or more planes and the articular surfaces are covered with hyaline cartilage, and if the joint is enclosed in a fibrous capsule lined by a synovial membrane that secretes a lubricatory fluid. Ligaments hold the bones in their proper alignment in the joint. The articulations between the upper and lower jaws of early bony fishes were diarthroses. These same joints and the hinge joint at the elbow and knee of mammals are diarthroses.

An **amphiarthrosis** permits limited movement. Resilient fibrocartilage unites the components of the joint, and fibrous joint capsules keep the bones properly aligned. There may or may not be a cavity within the capsule, but there is no synovial membrane. The joints between the centra of mammalian vertebrate are amphiarthroses.

A sutured joint is a **synarthrosis.** A suture is an irregular jagged seam at the junction of two bones that renders the joint immovable. The joints in the roof of the mammalian skull are synarthroses (fig. 9.27). If the suture between two bones becomes obliterated during development, as in the skull of birds, the condition is an **ankylosis.** The premaxilla and maxilla of the human embryo ankylose and as a result cannot be distinguished as two bones in an adult. In cats they are sutured but not ankylosed. A **symphysis** is a joint in the midline of the body, in which bilateral bones are separated by a pad of fibrocartilage and movement is severely restricted, if not impossible. The pubic symphysis of pregnant female mammals becomes movable by hormone-initiated dissolution of the fibrocartilage shortly before the onset of labor.

The terms "diarthrosis," "amphiarthrosis," and "synarthrosis" were coined by human anatomists to classify mammalian joints. Subsequent studies of lower vertebrates revealed considerable diversity in the anatomy of joints that permit a degree of movement in one plane or another but lack synovial cavities. Examples include the hyostylic joints between the hyomandibular cartilages and the skull of modern sharks and the joints between segments of the entoglossal and paraglossal bones in the tongue of many birds. Special terms have been coined to designate these other varieties. The range of articulations in birds is much broader than in mammals.

MINERALIZED TISSUES AND THE INVERTEBRATES

Mineralized tissues are not limited to vertebrates. In fact, two-thirds of the living species of animals that contain mineralized tissues are invertebrates. The matrix is collagen, and it is as old as the sponges, but the inorganic crystals are more often calcium carbonate than calcium phosphate. Cartilage is found among invertebrates, including squids, some gastropods, and protochordates; and bone, dentin, cartilage, and enameloid were all present in Ordovician ostracoderms. For that reason it cannot be said that one of these is phylogenetically older than another.

REGIONAL COMPONENTS OF THE SKELETON

The skeleton may be divided regionally into the following components, which include associated ligaments and tendons. These are the subjects of the next three chapters.

Axial skeleton
    Notochord and vertebral column
    Ribs and sternum
    Skull and visceral skeleton
Appendicular skeleton
    Pectoral and pelvic girdles
    Skeleton of paired fins and limbs
    Skeleton of median fins of fishes

HETEROTOPIC BONES

In addition to the skeletal components outlined above, miscellaneous **heterotopic bones** develop by endochon-

dral or intramembranous ossification in areas subject to continual stress in amniotes. They are usually missing from routinely prepared skeletons. Among heterotopic bones are an **os cordis** in the interventricular septum of the heart of deer and bovines, a **baculum** (os penis) in the septum between the spongy bodies of the penis of dogs, lower primates, and many other mammals (fig. 7.11), and an **os clitoridis** in many female mammals. In walruses the baculum is nearly 60 cm long.

A heterotopic bone forms in the gizzard of some doves, in the tongue of at least one species of bats, in the gular pouch of a South American lizard, in the muscular diaphragm of camels, in the syrinx of some birds (pessulus, fig. 13.16), and in the upper eyelid of crocodilians (adlacrimal, or palpebral, bone). A similar plate of fibrous tissue, the tarsal plate of the eyelid, develops in human beings. A rostral bone develops in the snout of swine and is used in routing in soil, and a cloacal bone develops in the ventral wall of the cloaca of some lizards. All these and others, including sesamoid bones, are incidental bones located at foci of stress and lacking phylogenetic significance.

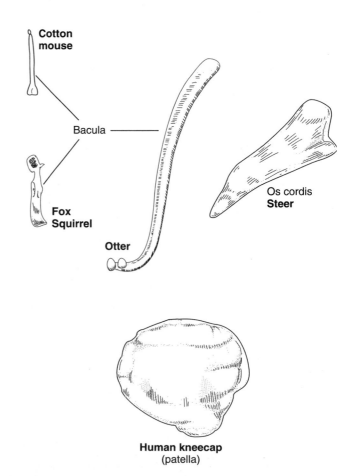

FIGURE 7.11

Some heterotopic bones.

SUMMARY

1. The chief mineralized tissues are bone (from osteoblasts), cartilage (from chondroblasts), and enameloids (from ameloblasts or odontoblasts).

2. Collagen is a proteinaceous fibril disposed in dense bundles that form a compact matrix upon which the mineralizing constituents of cartilage or bone are deposited.

3. Bone is a tissue consisting of collagen, hydroxyapatite crystals, cementing substance, and osteocytes in lacunae that are usually connected by canaliculi.

4. Bone tissue is compact or spongy (cancellous). The lamellae of deep compact bone (osteon bone) are concentric around haversian canals. The lamellae of periosteal bone are not part of a haversian system.

5. Spongy bone is characterized by the presence of trabeculae and marrow cavities lined by an endosteum. It constitutes the core of all bones.

6. Ossification may be intramembranous, producing membrane bone, or endochondral, replacing preexisting cartilage and yielding replacement bone.

7. Endochondral ossification of the long bones of tetrapod limbs proceeds from a single diaphyseal center in the shaft and from epiphyseal centers in each epiphysis. The final site of ossification, if any, is in the epiphyseal plates.

8. Bones are surrounded by a membranous periosteum.

9. Dentin is a form of bone with peripheral odontoblasts, dentinal tubules, and no lacunae. It is found in the skin and teeth.

10. Mature acellular bone lacks cells. It is found in modern fish scales.

11. Dermal bones are membrane bones originating ontogenetically or phylogenetically from the dermis of the skin.

12. Cartilage consists of a collagenous matrix, a mucopolysaccharide in the interstices of the matrix, chondrocytes in lacunae, and canaliculi. It may be hyaline, fibrocartilage, or elastic cartilage and is sometimes calcified. Cartilage is surrounded by a membranous perichondrium.

13. Bone and cartilage are constantly being deposited and resorbed, thereby facilitating orderly growth, homeostasis, and tissue repair.

14. Tendons attach muscles to bone. Ligaments connect bone to bone. Both may be calcified or contain sesamoid bones.

15. Joints are diarthroses, amphiarthroses, synarthroses, or symphyses. Ankyloses are junctions between bones in which the sutures have been obliterated.

16. Cartilage, bone, dentin, and enameloid were present in ostracoderms, and it cannot be said which of these may be phylogenetically older.

17. Heterotopic bones form in miscellaneous locations by endochondral or intramembranous ossification.

SELECTED READINGS

Burt, W. H.: Bacula of North American mammals, University of Michigan Museum of Zoology Miscellaneous Publications 113:1–76, 1970.

Chinsamy, A., and Dodson, P.: Inside a dinosaur bone, American Scientist 83:174, 1995.

Currey, J.: Comparative mechanical properties and histology of bone, American Zoologist 24:5, 1984.

Kemp, N. E.: Organic matrices and mineral crystallites in vertebrate scales, teeth, and skeleton, American Zoologist 24:965, 1984.

Moss, M. L.: Skeletal tissues in sharks, American Zoologist 17:335, 1977.

Patterson, C.: Cartilage bones, dermal bones and membrane bones, or the exoskeleton versus the endoskeleton. In Andrews, S. M., and others, editors: Problems in vertebrate evolution, New York, 1977, Academic Press.

CHAPTER EIGHT

VERTEBRAE, RIBS, AND STERNA

In this chapter we will look at some typical and some not-so-typical vertebral columns of fishes and tetrapods, discuss their ontogeny and evolutionary history, and note the advantages conferred by regional specialization of the column in tetrapods. Next, we'll study the ossified myosepta commonly known as ribs. Last, we will examine the diversity and roles of amphibian and amniote sterna and speculate concerning their phylogenetic origins.

OUTLINE

The vertebral column, skull, ribs, and sterna, along with their ligaments, constitute the major components of the axial skeleton. The vertebral column forms around, and sometimes invades, the embryonic notochord during ontogenesis. Ribs and the tetrapod sternum form in the lateral and ventral body walls, respectively.

VERTEBRAL COLUMN

The vertebral column is the keystone of the vertebrate skeleton. It is a segmented, more or less flexible, arched rod flanked by axial musculature, to which the head is attached, from which the rest of the body is suspended in fishes, and from which the trunk is suspended between anterior and posterior limbs in most tetrapods. It provides a protective bony tunnel for the spinal cord, and it is an essential architectural component of the axial locomotor apparatus in vertebrates other than agnathans. In no vertebrate is its locomotor role more readily demonstrable than in fishes and limbless tetrapods.

Fishes are surrounded by an environment that offers considerable resistance to forward locomotion and a buoying effect on the body. During locomotion, resistance is almost universally overcome in fishes by pushing laterally against the water with rhythmical, undulating, side-to-side movements of the trunk and tail (fig. 11.8). These movements are brought about by segmental muscles attached to the vertebrae, or to the ribs, and myosepta. The articulations between vertebrae in most fishes permit only side-to-side flexibility of the column. When vertebrates ventured onto land they brought with them the fishlike method of locomotion, but it was an awkward way to move about. Unlike water, land is beneath the animal rather than surrounding it, it is not level, and it is littered with obstacles that must be avoided or clambered over. Eventually, selective forces altered the articulations in the column in such fashion as to provide dorsoventral flexibility, which is more suited to locomotion on land, while also providing a strong skeletal bowlike arch for suspending the trunk above the ground between forelimbs and hind limbs. These changes were achieved at the expense of some side-to-side flexibility and they were correlated with regional specialization of the vertebral column. These and other adaptations are discussed in the sections that follow.

Centra, Arches, and Processes

Modern vertebrae typically consist of a **centrum, neural arch,** and one or more processes, or **apophyses,** that project from the arch or centrum (fig. 8.1a–g). Centra occupy the position occupied early in ontogeny by the notochord. Neural arches are perched on the centra, and the successive arches and their interconnecting liga-

ments enclose a long **vertebral canal (neural canal)** occupied by the spinal cord. **Hemal arches,** known also as chevron bones in amniotes, are inverted beneath the centra of the tail and house the caudal artery and vein (fig. 8.1b, c, and e).

Transverse processes (diapophyses) are the most common. They articulate with ribs that project laterad into the horizontal skeletogenous septum separating the epaxial and hypaxial muscles of the body wall (figs. 1.2 and 8.5). Transverse processes serve as attachments for some of the muscles that extend or flex the vertebral column. **Zygapophyses** are paired processes at the cephalic end of trunk vertebrae **(prezygapophyses),** and at their caudal end **(postzygapophyses),** chiefly in tetrapods (fig. 8.1d–f). The articular facets of prezygapophyses face dorsad and articulate with the ventrad-facing facets of the postzygapophyses just ahead of them. This interlocking arrangement limits dorsoventral flexion of the column in the region of the trunk where they are present. Tetrapod tails are highly flexible because zygopophyses are absent. **Parapophyses** are lateral projections from the centra of a few tetrapods. When present, they serve as the articulation site for the capitulum of a bicipital rib, as illustrated in figure 8.23b. (This is not the usual site, however, as will be seen when we study ribs.) **Hypapophyses** are prominent midventral projections from the centra of snakes and a few other amniotes (fig. 8.1d). They are sites of attachment for certain muscles or tendons. Still other processes develop in occasional species.

Morphogenesis of Vertebrae

A typical vertebra arises from mesenchyme cells that stream out of the sclerotomes of mesodermal somites, surround the notochord and neural tube, and produce the blastema for a future vertebra (fig. 5.11). Chondroblasts that differentiate within the blastema subsequently deposit a cartilaginous centrum and neural arch, and, in the tail, a hemal arch. The result is a cartilaginous vertebra with the notochord constricted within each centrum. Later, except in cartilaginous fishes, the cartilage is generally removed and bone is deposited. Remnants of the notochord may remain in the centrum to varying degrees, depending on the species. In Chondrichthyes the cartilaginous vertebrae are never replaced by bone. In teleosts, urodeles, and apodans, membrane bone rather than cartilage is deposited in the perichordal blastema (fig. 8.3), but the arches arise from cartilage.

In many fishes, and in amphibians except anurans, not only is cartilage or membrane bone deposited *around* the notochord **(perichordal cartilage or bone),** but chondroblasts invade the notochord sheath and deposit cartilage in the sheath and sometimes in the notochord itself. Consequently, the centra of these vertebrates contain **chordal cartilage** that may calcify or ossify (figs. 8.2, 8.3, and 8.8). The result of these variations in

FIGURE 8.1

Selected modern vertebrae. (*a*) and (*b*) Transverse sections from the center of a trunk vertebra and at the cephalic end of a tail vertebra of *Squalus*. (*c*) Amphicelous vertebra of a teleost. (*d*) and (*e*) Procelous vertebrae from a python, cephalic and lateral views. (*f*) Heterocelous vertebra from a bird, head end to the left. (*g*) Acelous trunk vertebra from a porpoise. Note persistent notochord in (*a*)–(*c*). **ca, cv,** canals for caudal artery and vein, respectively.

the process of centrum formation can be illustrated by contrasting centra of urodeles with those of anurans. The centra of urodeles (and also of apodans) consist of perichordal membrane bone, chordal cartilage, and, within the core, unaltered notochordal tissue (fig. 8.3). The centra of anurans, on the other hand, consist entirely of replacement bone deposited in a perichordal blastema. The notochord disappears. The difference has been interpreted by some paleontologists as supporting evidence for a diphyletic origin of amphibians (chapter 4).

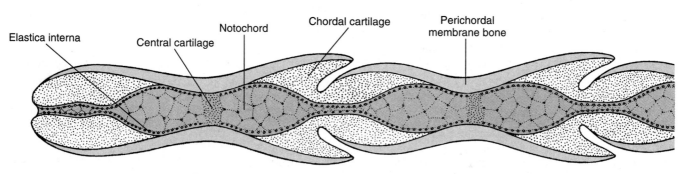

FIGURE 8.2

Caudal vertebral column of *Squalus,* sagittal section. Calcified chordal cartilage has been deposited in the notochord sheath (*dark red*). Intercalary plates are seen from lateral view in fig. 8.5.

FIGURE 8.3

Two trunk vertebrae from the plethodontid salamander *Gyrinophilus,* frontal section, head to the left. The chordal cartilage is within the notochord sheath. The elastica interna is the innermost derivative of the sheath.

Source: Data from J.S. Kingsley, *Outline of Comparative Anatomy of Vertebrates,* 1920, The Blakiston Company, Philadelphia, after Widersheim.

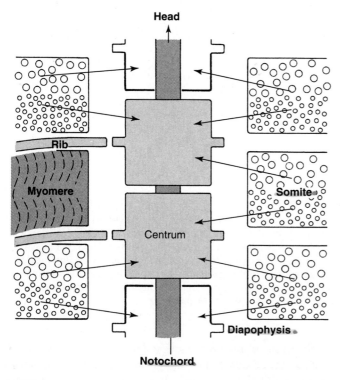

FIGURE 8.4

Intersegmental (intersomitic) position of centra and myosepta or ribs. Each centrum is a contribution from two successive somites (*arrows*).

In forming centra, scleroblasts from the caudal half of one somite and the cephalic half of the next somite stream to an intersegmental location around the notochord to establish the perichordal blastema. Consequently, centra are typically intersegmental, and myomeres are segmental (fig. 8.4). However, chiefly in the tail of some fishes and in the tail of a few primitive tetrapods, there are two centra per segment, a condition known as **diplospondyly.** The condition may provide the tail with increased flexibility for locomotion. Generally, in fishes, flexibility of the vertebral column is facilitated by elasticity of the intervertebral ligaments rather than by intervertebral joints, as in tetrapods.

The Vertebral Columns of Fishes

Fish vertebrae are highly diverse, as might be expected in view of the enormous number of species that evolved—greater than in any other class of vertebrates—and the opportunity for subsequent genetic mutations during the lengthy Paleozoic Era. We will look at just a few interesting morphological variants. They are attributable to varying developmental factors among which are the extent to which scleroblasts invade and thicken the notochord sheath and the notochord proper, the extent to which the notochord persists within the adult column, the number of chondrification and ossification centers that develop in the perichordal blastemas, and the extent to which separate centers remain independent or coalesce during ontogeny. As for a single species, however, the vertebral column exhibits only two major morphologic regions of specialization, trunk **(dorsals)** and tail **(caudals).** The vertebrae in each region generally show gradual modifications between the head and vent, where there may be a fairly abrupt change, and between the vent and the end of the tail.

Agnathans have a strange vertebral column if, indeed, they may be said to have one. The only skeletal elements are **lateral neural cartilages** (fig. 1.5), one or two pairs per body segment, depending on the species. Caudally, the lateral neural cartilages fuse to form a single longitudinal dorsolateral cartilaginous plate perforated by foramina for spinal nerves. In hagfishes, lateral neural cartilages are limited to the tail. Lateral neural cartilages may be vestigial vertebrae or primitive vertebrae, or they may bear no phylogenetic relationship to vertebrae. Whether you call them vertebrae depends on how you wish to define "vertebra." How would *you* define the term?

In sharks the notochord is present throughout the length of the adult vertebral column and is constricted within each centrum (fig. 8.2). The centra, composed of chordal and perichordal cartilage, are concave at each end. The centrum that is concave at each end is said to be **amphicelous** (fig. 8.1*a,b*). The vertebral canal (fig. 8.5) consists of paired **dorsal plates** that constitute a neural arch, paired **dorsal intercalary plates** between the arches, and, in some species, paired wedge-shaped **supradorsal cartilages** (fig. 8.5). *Squalus acanthias* lacks supradorsals. The neural arches and intercalary plates of *Squalus* are perforated by foramina for the dorsal and ventral roots of spinal nerves and for blood vessels. In *Scyllium* they emerge between plates. Hemal arches consist of paired **ventral plates. Ventral intercalary plates** are common between hemal arches. A fibroelastic ligament, present in all fishes, overlies and connects the neural spines along the entire length of the vertebral column.

A few fishes have exceptional columns. Dipnoans, chondrosteans other than *Polypterus,* and the coelacanth *Latimeria* develop no centra. The notochord is present and unconstricted, and its thick fibrous sheath contains little cartilage or bone (fig. 8.6). Associated with the notochord in each body segment are paired **basidorsal, basiventral, interdorsal,** and **interventral cartilages** that are products of separate chondrification centers in

Figure labels: Dorsal intercalary plate, Dorsal plate (neural arch), Supradorsal cartilage, Ventral root of spinal nerve, Dorsal root ganglion, Elastic ligament, Vertebral canal, Notochord, Postcardinal vein, Dorsal aorta, Dorsal mesentery, Dorsal rib, Horizontal skeletogenous septum, Centrum, Spinal nerve

FIGURE 8.5

Vertebral column and associated structures of trunk of *Scyllium,* a shark. Each rib articulates with a transverse process of a centrum.

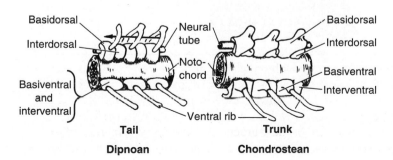

Figure labels: Basidorsal, Interdorsal, Basiventral and interventral, Neural tube, Notochord, Ventral rib, Tail, Dipnoan, Basidorsal, Interdorsal, Basiventral, Interventral, Trunk, Chondrostean

FIGURE 8.6

Vertebral components in adult dipnoan (*Neoceratodus*) and chondrostean (sturgeon). Arrow in the dipnoan indicates canal occupied by a supportive longitudinal ligament.

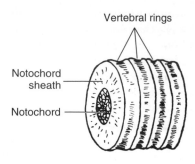

FIGURE 8.7

Vertebral rings from an extinct holocephalan. Neural arches have been removed.

the embryonic perichordal blastema. An extinct holocephalan's "vertebral column" consisted of thick rings of calcified cartilage that had been deposited in the notochord sheath (fig. 8.7). The rings were more numerous than the body segments.

Teleosts have well-ossified amphicelous vertebrae (fig. 8.1c). The core of each centrum is a dumbbell-shaped vacuole where the notochord previously existed, and the space between successive vertebrae is occupied by a porous cartilagelike material that probably includes notochordal remnants. The centra and neural arches of successive vertebrae are interconnected by a complex system of collagenous and elastic ligaments that facilitate lateral undulation of the body for locomotion. In the posterior trunk and tail these ligaments are frequently ossified to form long, delicate, bony rods. The neural spines are often very tall, and they are sometimes surmounted by **supraneural bones** (fig. 8.24, *Perca*). A variety of processes, mostly unique to fishes, protrude from the arches and centra.

Elasmobranchs and numerous bony fishes have two centra and two sets of neural and hemal arches in each metamere of the tail. A few have duplication of these in the trunk also. Consequently, the number of centra in those regions is double the number of myomeres and spinal nerves, a diplospondylous condition. *Amia* has what appear to be two centra per body segment (fig. 8.8), but only one bears arches. The other is termed an intercentrum, although it may not be homologous with intercentra of other fishes. The embryonic basidorsal and basiventral cartilages are incorporated in the centra, and the interdorsal and interventrals are incorporated in the intercentra.

No movement is possible in fishes between the first vertebra and the skull. They are united by cartilage or nonelastic connective tissue. The caudal-most vertebrae are part of the skeleton of the caudal fin, as described in chapter 10.

Evolution of Tetrapod Vertebrae

Unlike the vertebral column of modern tetrapods, which in the trunk consists of a centrum and a neural arch in each body segment, the columns of some crossopterygians and early labyrinthodonts consisted of several bones per segment. Generally, these were **a hypocentrum,** a large, median, U-shaped anterior bone that cradled the notochord (fig. 8.9); a pair of **pleurocentra,** small wedges of bone overlying the notochord dorsolaterally; and independent left and right laminae of bone lateral to the spinal cord that, collectively, provided a **neural arch.** Each lamina rested in a notch between the pleurocentrum and the hypocentrum. The notochord was continuous throughout the length of the body but constricted at the level of each hypocentrum. A "vertebra" of this nature is said to be **rachitomous** (consists of several pieces). Later tetrapod vertebrae with the exception of those of urodeles and apodans appear to be modifications of the rachitomous condition. The change appears to have been characterized by increased prominence of pleurocentra and concomitant reduction of hypocentra (fig. 8.10).

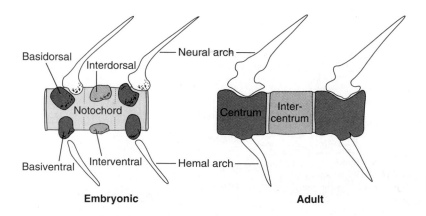

FIGURE 8.8

Tail vertebrae of *Amia*. A centrum and intercentrum develop in each body segment.

Prezygapophysis

Postzygapophysis

Neural arch

(a)

(b)

Hypocentrum

Pleurocentrum

(c)

Hemal arch

(d)

(e)

FIGURE 8.9

Rachitomous vertebrae from a primitive labyrinthodont, left lateral view. (*a*) and (*b*) Trunk. (*c*)–(*e*) Caudal.

Source: Data from E.S. Goodrich, *Studies on the Structure and Development of Vertebrates,* 1930, Macmillan and Company, Ltd., London.

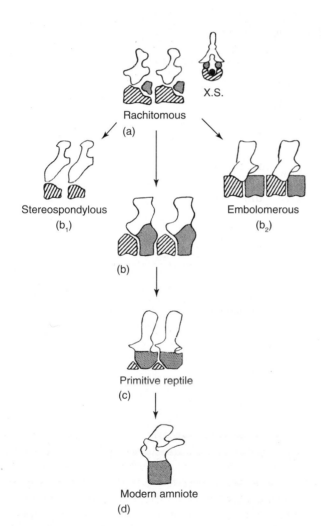

X.S.

Rachitomous
(a)

Stereospondylous
(b₁)

Embolomerous
(b₂)

(b)

Primitive reptile
(c)

Modern amniote
(d)

FIGURE 8.10

Postulated modifications leading from labyrinthodont to modern amniote vertebrae. (*a*), (*b*), (*b₁*), and (*b₂*), four varieties of labyrinthodont vertebrae. **X.S.,** Cross section of (*a*), notochord in *black*. The rachitomous type was present in rhipidistian crossopterygians and in the earliest tetrapods. (*b*) is from an extinct amphibian thought to be in the reptile line. For phylogeny of (*b₁*) and (*b₂*), see fig. 4.15. *Diagonal lines,* hypocentrum; *red,* pleurocentrum.

Today, as in the distant past, each centrum begins ossification at several paired and median loci in the perichordal blastema. As separate ossification centers expand, they coalesce, unlike in Devonian times. The findings of paleontology, comparative anatomy, and developmental biology all point to the same conclusion: a modern adult tetrapod vertebra is a composite structure.

The vertebrae of urodeles and apodans lack evidence of pleurocentra of hypocentra, and for this and numerous other reasons, the phylogenetic history of these amphibians is in doubt. (For a further discussion, see chapter 4 under "Lepospondyls.") Whether or not anuran centra are pleurocentra or hypocentra has not been ascertained.

The vertebrae of generalized urodeles, apodans, *Sphenodon,* and primitive lizards such as geckos are amphicelous, as are the vertebrae of most fishes. The centra are concave at each end, and vestiges of the notochord persist in the intervertebral joints. In most other tetrapods, the concavity has disappeared at one or both ends of the centra, resulting in **opisthocelous, procelous,** or **acelous** vertebrae (fig. 8.11).

The vertebrae of anurans and modern reptiles are procelous. This condition is thought to have evolved from a buildup of an intercentrum of chordal cartilage between two centra and the eventual coalescence of the intercentrum with the centrum ahead of it. This would have resulted in procelous vertebrae. Coalescence of the intercentrum with the centrum immediately behind it would have resulted in the opisthocelous centra seen in salamanders (fig. 8.3). The vertebrae of mammals are acelous, having a concavity at neither end. Instead, between successive vertebrae is fibrocartilaginous **intervertebral disc,** and in its core is a remnant of the embryonic notochord known as a **pulpy nucleus.** Humans are aware of the disc only when it becomes displaced and presses on a spinal nerve. These modified intervertebral joints of tetrapods permit more flexibility to the column

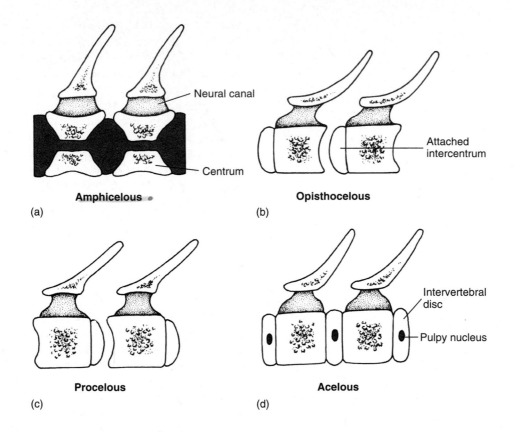

FIGURE 8.11

Vertebral types based on articular surfaces of centra. Midsagittal section, head to the left. The complex heterocelous vertebra of birds is seen in fig. 8.1. The *red* areas are occupied by the notochord in (*a*) and by notochordal vestiges in (*d*).

FIGURE 8.12

Ichthyostega, an early labyrinthodont 3 feet long, reconstructed from recovered fossil skeletal parts.

Source: Data from Jarvic, *Scientific Monthly,* Vol. 80, p. 152, March 1955.

than do the amphicelous vertebrae of fishes. An intervertebral joint permitting still more flexibility will be described when we study the highly specialized heterocelous neck vertebrae of birds.

Regional Specialization in Tetrapod Columns

With the advent of life on land, the vertebral column underwent regional specialization. Tetrapod limbs push against the earth. Initially, they did so purely for locomotion; later they were also supporting the body above the ground. The opposing physical force exerted by the earth in response to the push of a tetrapod posterior appendage is transmitted via the pelvic girdle to one or more of the

hindmost trunk vertebrae (fig. 8.12). These vertebrae became appropriately modified and are now called **sacral** (figs. 8.13, 8.14, 8.17, 8.21, and 10.11). The pelvic fins of fishes have no such relationship to the vertebral column.

Survival on land with its irregular contours and scattered obstructions to vision, such as rocks or vegetation, was facilitated in tetrapods by increased mobility of the head with its special sense organs, including eyes that can scan the horizon. This was achieved to a small degree in amphibians by development of a somewhat more mobile joint between the first vertebra and the skull. The innovation was improved in many reptiles by shortening or eliminating ribs on some of the more anterior vertebrae,

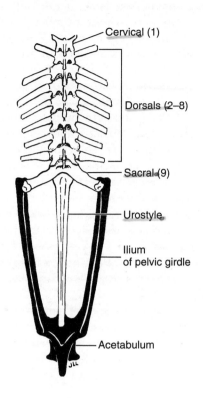

FIGURE 8.13

Vertebral column and pelvic girdle of an anuran. The long ilium of the girdle is braced against the sacral vertebra.

and by increasing the mobility of the intervertebral joints in that region, which became the neck. These are **cervical vertebrae** (fig. 8.14). In amniotes other than turtles and snakes, long ribs have become restricted to vertebrae in the anterior portion of the trunk immediately behind the neck, thereby preserving a bony cage that houses the thoracic viscera and participates in external respiration. These are **thoracic vertebrae.** The remaining vertebrae anterior to the sacral are **lumbar** (fig. 8.14). Vertebral columns that have become modified by life in water are shown in figure 10.32.

Among tetrapods, snakes have the longest columns with as many as 400 or more vertebrae and almost no regional specialization. Such a column provides snakes with an elongated, multijointed body that can be coiled or drawn into sinuous curves for locomotion without limbs. Apodans, also limbless, have as many as 250 or more vertebrae, and some urodeles have 100. The awkward leaping anurans have the shortest columns, and limited flexibility. In turtles and birds only the cervical and caudal segments of the column are flexible; most of the trunk vertebrae are rigidly fused to the synsacrum in birds (figs. 8.19*b*, 8.20), and to the carapace in turtles (fig. 6.36). The role of the vertebral column and its associated musculature in locomotion is discussed in chapters 10 and 11.

The Craniovertebral Junction and Neck Vertebrae

The first and only cervical vertebra of modern amphibians lacks processes, and its cranial end bears two smooth concave facets that articulate with two knoblike occipital condyles. The condyles of a primitive urodele are illustrated in figure 9.12*b*. The absence of processes and the nature of the amphibian craniovertebral joint permits limited dorsoventral rocking of the skull, a function that is not possible in fishes. The two-facet articulation is a change from the earliest amphibians, who had only one occipital condyle.

Amniotes have a larger number of cervical vertebrae than amphibians, which provides them with a long, flexible neck. The first two vertebrae are modified in such a fashion as to permit considerably more independent

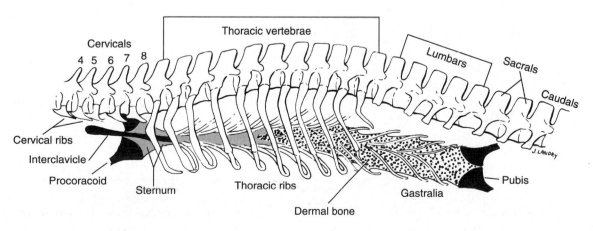

FIGURE 8.14

Vertebrae and ribs of an alligator.

movement of the head. The first vertebra, or **atlas,** is ringlike because its centrum is missing (fig. 8.27). On its cephalic end are one or two deep concavities (**superior articular facets**) for articulation with the single occipital condyle of reptiles and birds, or with the two of mammals. These are condyloid joints in which the skull rocks as in nodding "yes." Except in snakes, the centrum of the atlas has become the **odontoid process** of the second vertebra, or **axis.** The process projects forward to rest on the floor of the atlas where in mammals, it is held in place by a **transverse atlantal ligament** (fig. 8.15a). The skull and atlas rock as a unit on the odontoid process, which is the axis of rotation. Rocking is facilitated by the reduction or absence of zygapophyses on the atlas. The odontoid process of *Sphenodon* chondrifies as an independent element and unites with the axis only after hatching.

A bony **proatlas,** resembling a neural arch and derived from a pair of bilateral cartilaginous blastemas, is interposed between the skull and atlas in crocodilians, *Sphenodon,* and the primitive insectivorous spiny hedgehogs (figs. 8.15b and 8.16). This element appeared first in cotylosaurs, and was present in the synapsid reptilian ancestors of mammals. Some modern lizards have a fibrous membrane instead of a bone at that location.

A flexible neck is a valuable asset in gathering food and avoiding enemies on land. It not only extends the horizon, but also increases the number of degrees in its arc. A human, for example, moving nothing but the head, can sweep the horizon over an arc of more than 180 degrees.

Flexibility of the neck is exceptional in birds because of the nature of the articulations of their vertebrae. The caudal ends of the centra are saddle-shaped, having a convexity in the right-left axis and a concavity in the dorsoventral axis (fig. 8.1f). The cephalic end of the next more posterior centrum is shaped to accommodate this configuration. Such a vertebra is said to be **heterocelous.** These vertebrae enable both lateral and dorsoventral flexion of the neck. Heterocelous centra, combined with the atlas-axis complex and the largest number of cervical vertebrae in any tetrapod—12 commonly, 25 in swans—enable some birds to turn their heads 180 degrees to the rear, enlarging the arc of their horizon to that of a complete circle.

Turtles also have a uniquely flexible neck. Ball-and-socket joints between successive procelous centra enable the entire head and neck to be completely retracted into the shell by flexing the neck into several folds dorsoventrally or, in "side-necked" turtles, sidewise, during

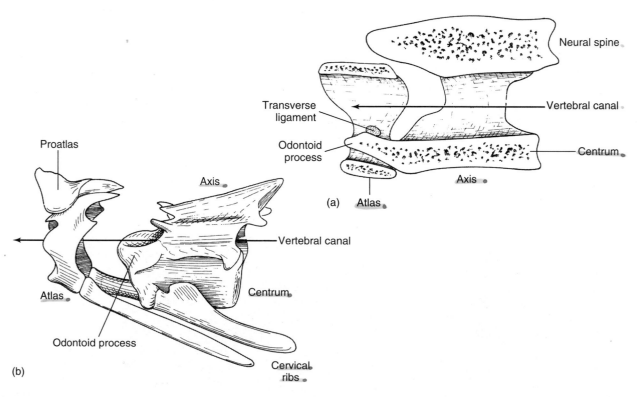

FIGURE 8.15

Atlas, axis, and odontoid process of a cat, (*a*) and a crocodilian, (*b*) head to the left. (*a*) is a sagittal section. The transverse atlantal ligament has been severed where it overlies the odontoid process. In (*b*), the vertebrae have been slightly separated for clarity. Note that the second cervical rib is bicipital.

FIGURE 8.16

The eight cervical vertebrae, eight cervical ribs (*red*), and proatlas, **P**, of an alligator, left lateral view. **1,** atlas and attached first rib. Immediately behind the first rib is the rib of the axis. The ribs are fused to transverse processes.

retraction. However, like other living reptiles, turtles have only eight cervical vertebrae. In contrast to the high flexibility of the neck, the remainder of the column of birds and turtles, except a short section of caudal vertebrae, is totally rigid.

Mammals almost always have seven cervical vertebrae. This is as true of the stubby, rigid neck of whales as it is of the neck of the tallest giraffe. The sole exceptions are three edentates (two sloths and the Great Anteater) with six, eight, or nine, and manatees with six. The cervical vertebrae of marsupial moles are shortened and more or less fused together (an adaptation to life in a burrow?). The same is true of armadillos and most cetaceans. It is the length of the centra, not their number, that determines neck length in mammals.

Stabilizing the Hind Limbs: Sacrum and Synsacrum

Sacral vertebrae bear stout transverse processes that are strong enough to bear the thrust of the pelvic girdle as the hind limbs push against the earth in locomotion (fig. 8.13). Ankylosed to the distal ends of the transverse processes may be equally stout ribs (fig. 10.11). Amphibians have 1 sacral vertebra; living reptiles, most birds, and opossums have 2; and most mammals have 3 to 5. However, there are perissodactyls with up to 8, and edentates with 13. The sacral vertebrae of mammals usually ankylose to form a single bony complex, the **sacrum** (fig. 8.17). Sacral vertebrae do not differentiate in tetrapods that lack hind limbs.

In modern birds the last thoracic vertebra, all lumbars, the sacrals, the first few caudals, and their ribs ankylose to form one adult complex, the **synsacrum.** The latter fuses with the pelvic girdle. Thus, there is formed a compact pelvis (figs. 8.18, 8.19, and 8.20) that provides a rigid brace for the teeter-totterlike stance of birds. The thoracic vertebrae anterior to the synsacrum also unite more or less completely, so there is little flexibility in the avian backbone caudal to the neck. This axial rigidity minimizes the number of muscles needed to keep the body streamlined during flight, thereby

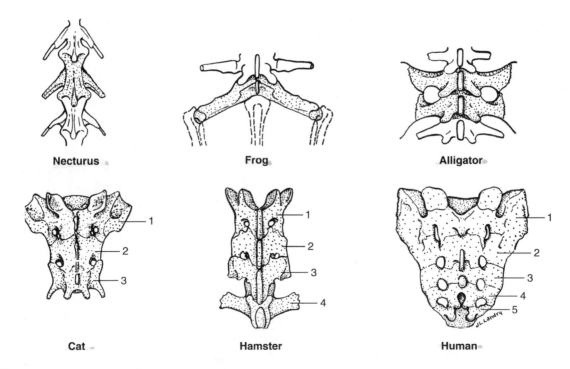

FIGURE 8.17

Sacral vertebrae (*stippled*) of selected vertebrates, dorsal views. They have ankylosed to form a sacrum in mammals. **1–5,** Sacral vertebrae of mammals.

FIGURE 8.18

Diagram of vertebral column of pigeon. **T,** Thoracic; **L,** lumbar; **C,** caudal; **P,** pygostyle, composed of four fused vertebrae.

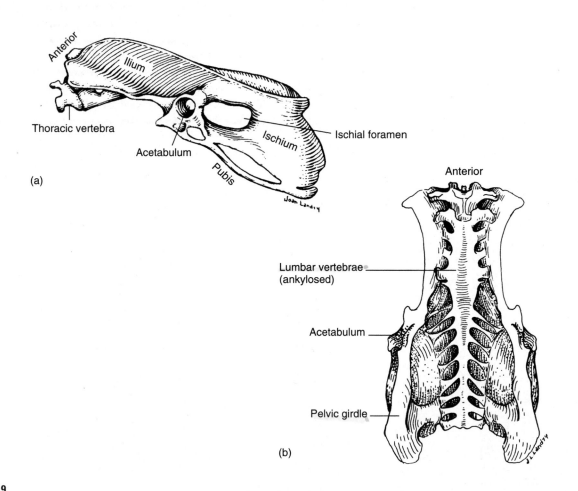

FIGURE 8.19

Synsacrum and pelvic girdle of the guinea hen. (*a*) Left lateral view. (*b*) Ventral view. The acetabulum accommodates the head of the femur.

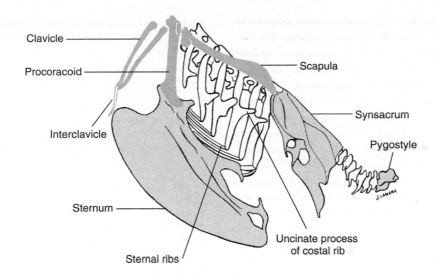

FIGURE 8.20

Skeleton of trunk, tail, and pectoral girdle of pigeon, head to the left. Bones of the pectoral girdle are *red*. The ventral keel of the sternum is the carina. The clavicles and interclavicle comprise the furcula, or wishbone. The first rib illustrated is the second and last of the two cervical ribs.

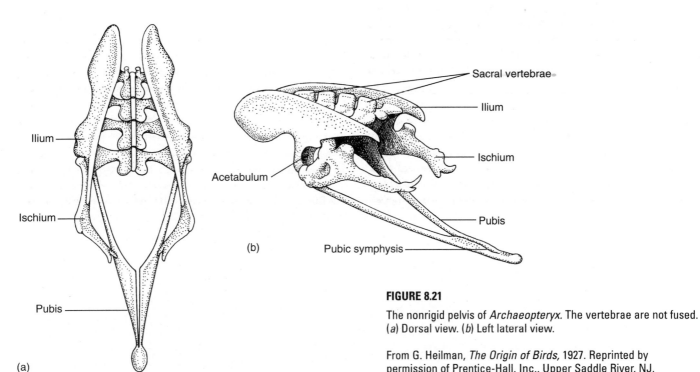

FIGURE 8.21

The nonrigid pelvis of *Archaeopteryx*. The vertebrae are not fused. (*a*) Dorsal view. (*b*) Left lateral view.

From G. Heilman, *The Origin of Birds,* 1927. Reprinted by permission of Prentice-Hall, Inc., Upper Saddle River, NJ.

reducing the cost in terms of energy expended. However, the historical selective forces responsible for it may have been quite different from what appear today to be its advantages. *Archaeopteryx* provides us a view of a rudimentary synsacrum (fig. 8.21). The trunk vertebrae were not fused, and the pelvic girdle (ilium, ischium, and pubis) was proportionately smaller than in today's birds. The weight of the body when standing was counterbal-

anced to a considerable degree by the long tail shown in figure 4.29.

Among mammals, armadillos have a synsacrum. It consists of up to 13 fused sacral and caudal vertebrae.

Tail Vertebrae: Urostyle, Pygostyle, and Coccyx
Caudal vertebrae in early tetrapods probably numbered 50 or more. In modern ones, the number is much

reduced and highly variable, dictating how effective the tail may be for one use or another. Toward the end of the tail in all tetrapods, the arches and processes become progressively shorter and rudimentary until finally the vertebrae consist solely of small cylindrical centra (fig. 8.22).

Anurans have a unique **urostyle** at the end of the vertebral column (fig. 8.13). It develops from a continuous elongated perichordal cartilage at the base of the larval tail, and it grows and ossifies after the tail is lost at metamorphosis. That it is composed of postsacral vertebrae is evident from the one or more coalesced centra, vestigial arches, transverse processes, and nerve foramina that are part of the cranial end of the urostyle in some species.

Among reptiles the caudal vertebrae of lizards deserve mention. Many lizards, when captured by the tail, break off to the end distal to the site of capture and scurry away. The tail then regenerates. This **autotomy** is implemented by a zone of soft tissue that divides each tail vertebra into cephalic and caudal sections, the location being at the level of a myoseptum. Sudden opposing reflex muscle contractions on each side of the septum cause the break.

Archaeopteryx had a long tail, and modern birds still have a tail, although it is not prominent. Pigeons have 15 caudal vertebrae. Five are ankylosed with the synsacrum, 6 are independent, and the last 4 are fused to form a **pygostyle,** which is the skeleton of the visible part of the tail (fig. 8.20). The pygostyle develops as 4 separate cartilaginous centra.

Among mammals, there are as few as 3 caudal vertebrae and as many as 50. The tail of sperm whales has 24. Apes and humans have 4 or 5 vestigial caudal vertebrae comparable to the pygostyle of birds. These caudal vertebrae lack arches, but most of them have rudimentary transverse processes. The last 3 or 4 diminish progres-

sively in size and, in humans, they fuse with one another to form a rigid **coccyx** at about 25 years of age. The centra of the coccyx are still identifiable, but the last one is a mere nodule of bone. People whose "tailbone" is recurrently sore have fractured the coccyx at some earlier date in a fall that caused them to land on the end of their spine. In contrast to hominoids, adult rhesus monkeys have a prehensile tail the length of which averages nearly 50 percent of the monkey's sitting height.

RIBS

Ribs articulate with vertebrae and extend into the body wall. The part adjacent to a vertebra is formed intersegmentally by scleroblasts from two successive mesodermal somites in the same manner as centra (fig. 8.4). The remainder of the rib arises from lateral-plate mesoderm. Most bony ribs are of endochondral origin. The so-called abdominal ribs, or **gastralia,** of some reptiles (fig. 8.14) are not ribs, but riblike membrane bones that may be remnants of an ancestral bony dermal exoskeleton.

Fishes

The chondrostean *Polypterus* and some teleosts (perch, for example) have two sets of ribs, dorsal and ventral, associated with each trunk vertebra (fig. 8.23). **Dorsal ribs** pass laterad into the horizontal skeletogenous septum that separates the epaxial and hypaxial muscles (figs. 1.2 and 8.15). **Ventral ribs** develop in myosepta and arch ventrad in the lateral body wall just external to the parietal peritoneum as in the mammalian thorax. In fishes, they stop short of the midventral line. Commencing at the vent, beyond which there is no coelom, each pair of ventral ribs, or their basal stumps,

FIGURE 8.22

Complete set of tail vertebrae from a hamster, left lateral view.

FIGURE 8.23

Dorsal and ventral ribs of *Polypterus* and bicipital rib of a tetrapod. A more usual articulation site of the capitulum of tetrapods is seen in fig. 8.25. The basal stump in (*a*) and the parapophysis in (*b*) are not homologues.

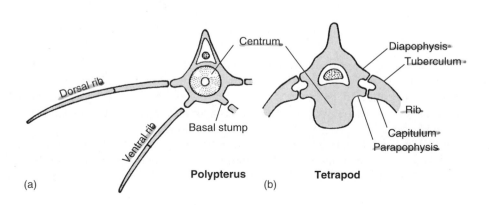

(a)   Polypterus

(b)   Tetrapod

approach one another beneath the centrum and unite to form the hemal arches of the tail (fig. 8.1c).

Most fishes have only ventral ribs (fig. 8.24, except perch). Sharks and a few others have only dorsal ribs. Skates, chimaeras, and a few unusual teleosts such as sea horses have no ribs. Neither have cyclostomes, but this is probably correlated with the absence of centra. Fish ribs strengthen the myosepta to which the longitudinally disposed locomotor muscles of the myomeres (muscle segments) are attached (fig. 11.5c).

Tetrapods

In early tetrapods, ribs were associated with vertebrae from the atlas to nearly the base of the tail (fig. 8.12). They later became restricted in length and number, the short ones often ankylosing with transverse processes (fig. 8.16). The long ribs of the anterior thoracic region acquired a ventral anchorage, the sternum. Typical long tetrapod ribs occupy the same position as the ventral ribs of fishes, but the two have not been shown by appropriate criteria to be homologous.

Most tetrapod ribs are **bicipital,** having a dorsal head, or **tuberculum,** and a ventral head, the **capitulum** (fig. 8.25). This is solely a tetrapod feature. The tuberculum in early tetrapods articulated with the extremity of a transverse process and the capitulum articulated with the hypocentrum. With reduction in size of the hypocentrum that took place in amniotes over time, the articulation site of the capitulum became altered. Thereafter, the capitulum articulated in one of several fashions: (1) on two adjacent demifacets (half-facets), one on the caudal end of a centrum and the other on the cephalic end of

FIGURE 8.24

The skeleton in actinopterygian fishes. *Glaucolepis* was a Triassic chondrostean; *Caturus* was a Mesozoic holostean; *Clupea* is a generalized teleost, and *Perca* (perch) is a specialized teleost.

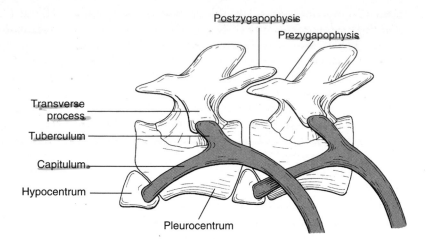

FIGURE 8.25

Two thoracic ribs (*red*) from a primitive tetrapod showing articulation with vertebral column.

Source: Data from E.S. Goodrich, *Studies on the Structure and Development of Vertebrates,* 1930, Macmillan and Company, Ltd., London.

the next more posterior centrum; (2) with a parapophysis, when the latter was present (fig. 8.23*b*); or (3) in a facet on a single centrum. The site often changes along the length of the trunk. In some mammals the tuberculum becomes reduced to a vestige in one region of the column or another. In some other tetrapods, crocodilians, for example, there was a trend toward fusion of the two heads until only the tuberculum remained. In figure 8.16 the tuberculums of a crocodilian are fused to the tips of transverse processes.

Typical thoracic ribs of amniotes consist of two parts, a **costal rib** adjacent to the vertebrae, and a distal, more ventral, **sternal rib** (fig. 8.21). At least some of the ventral ribs generally articulate with the sternum. Sternal ribs may remain cartilaginous, as in man, in which case they are also referred to as costal cartilages. A skeletal enclosure for the thoracic viscera of amniotes is provided by the vertebral column, costal ribs, sternal ribs, and the sternum. No such skeletal cage exists in amphibians.

Amphibians

All the ribs of anurans and urodeles have become very short and, in anurans, ankylosed to transverse processes. By contrast, the ribs of the limbless, burrow-dwelling apodans are quite long. They are present in association with all except the first vertebra and a few of the more posterior ones and, as in snakes, they play a vital role in locomotion. (Recall that apodans have no postanal tail.) Among amphibians, only the ribs of anurans (fig. 8.13) are not bicipital.

Reptiles

Lizards and crocodilians have long ribs on many of the trunk vertebrae and short ribs in most of the neck. Crocodilian ribs are illustrated in figure 8.14. *Sphenodon,* a primitive lizardlike reptile, has long ribs in most of the trunk and ribs continue into the tail. Geckos, which are relatively unspecialized modern lizards, have ribs associated with every cervical vertebra; more specialized

lizards usually lack them on the atlas and axis. A half dozen or more of the posterior ribs of the trunk of the gliding lizard *Draco* are greatly elongated. They can be rotated outward to elevate a broad fold of body wall skin, the **patagium,** which then becomes a winglike membrane on which the lizard soars. When not in use, the patagium is folded against the side of the body.

Turtles have no cervical ribs, and the ribs of the trunk are fused with the costal plates of the carapace (fig. 6.36). The two sacral ribs are not fused with the carapace. They are short, and their expanded distal ends are ankylosed to the ilia of the pelvic girdle (fig. 10.11). They brace the girdle against the vertebral column. Slender, riblike processes may be associated with some of the caudal vertebrae.

Snakes have long, curved ribs beginning at the second vertebra and continuing far into the tail. There is no sternum for ribs to attach to; however, the ventral ends of the ribs have ligamentous connections with integumentary scutes. These ribs participate in locomotion. Long, curved ribs in the neck of cobras can be rotated outward like those of the trunk of *Draco,* causing a fold of loose skin in the neck to "spread" (fig. 8.26). The fold is then inflated with air from the lungs.

Birds

The first two pair of ribs in many carinates articulate with the last two cervical vertebrae at the base of the neck. They are generally movable, and lack a sternal segment. The next five pairs are thoracic ribs with bony sternal segments. Thoracic ribs constitute the major skeleton of the thoracic basket (chest cavity). These ribs are thin, flat, and broad, and most of them bear broad **uncinate processes** (fig. 8.20). Thin, broad ribs and uncinate processes provide birds with a lightweight but sturdy thoracic body-wall skeleton for the attachment of the powerful muscles required for flight. Ribs caudal to the thoracic basket are ankylosed to the underside of the synsacrum, as are all trunk ribs in turtles. (There is an

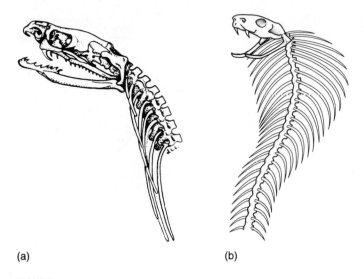

(a)   (b)

FIGURE 8.26

Cervical ribs of a cobra, a "spreading adder." (*a*) Lateral view.
(*b*) Frontal view with ribs spread.

old saying that birds are turtles with feathers and wings!
Sounds good, but it is phylogenetically untenable. Birds
evolved from archosaurs.)

Although ribs are confined to the trunk in adults,
developing embryos exhibit transitory rudimentary ribs
deep into the tail. Uncinate processes are found also in
some lizards, and were present in some early labyrinthodonts (fig. 8.12).

Mammals

Recognizable ribs in mammals are generally confined to
the thorax. The number ranges from 9 pairs in some
whales to 24 in some sloths; 12 pairs are common.
Where the number is larger than 10, the rest are "floating," that is, the (**costal cartilages**) sternal ribs fail to
reach the sternum. Humans and apes other than orangutans have 12 pairs of thoracic ribs, and apes have an
additional 2 pairs of lumbars. Humans frequently have

an additional rib, cervical or lumbar, as an anomaly, but
it usually remains undiscovered unless revealed by an
x-ray examination.

Studies of mammalian embryos reveal that two heads
of a vestigial bicipital rib develop in association with
each cervical vertebra. (A bicipital rib is illustrated in
fig. 8.25.) One of the heads is attached to a transverse
process; the other to a centrum. The two heads create
between them a **transverse foramen** on each cervical
vertebra (fig. 8.27). The consecutive foramina provide a
bony tunnel, the **arteriovertebral canal,** which transmits a vertebral artery from its source at the base of the
neck to the brain. Transverse foramina develop in birds,
but the number of vertebrae involved depends on the
species. The vertebral artery exists in reptiles but is not
enclosed in a canal.

TETRAPOD STERNUM

The sternum is a tetrapod structure and predominantly
amniote (fig. 8.28). It is of endochondral origin. Its presence, size, and anatomical relationships are correlated
with the extent to which the forelimbs are employed in
locomotion. It serves chiefly as a base against which the
pectoral girdles and, in amniotes, the ribs are braced, and
as the anatomical origin of the ventral muscles of the
forelimbs. The sternum is largest in pterosaurs, carinate
birds, and bats, which require strong muscles for flight.
It is much reduced or absent in tetrapods lacking forelimbs. It has acquired additional roles in some tetrapods,
such as assisting in breathing movements in birds.

The sternum of amphibians is well differentiated only
in anurans (fig. 8.28, frog). Whether or not *Necturus* has
a sternum depends on one's definition. It consists of
more or less isolated nodules of cartilage associated with
the linea alba immediately caudal to the coracoid cartilages of the pectoral girdle and extending into the proximal ends of adjacent myosepta. A small but discrete
sternal element is present in salamanders at the same
location. Apodans have no trace of a sternum.

Atlas   **Axis**   **Cervical**
(caudal view)   (left side)   (caudal view)

FIGURE 8.27

Mammalian cervical vertebrae. **1,** Centrum; **2,** pedicle; **3,** transverse process; **4,** neural spine; **5,** spinal canal; **6,** transverse foramen;
7, odontoid process; **8,** vestige of a cervical rib with two heads enclosing a transverse process.

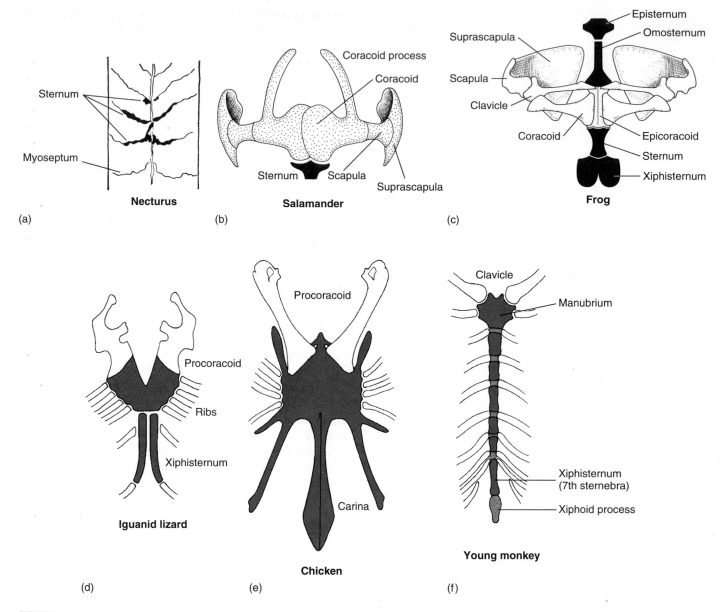

FIGURE 8.28

Tetrapod sterna (*red*). Stipple indicates cartilage.

Lizards are agile climbers and runners. As a correlate, their strong pectoral girdles are firmly braced against a substantial shield-shaped sternum of cartilage or replacement bone (fig. 8.28, lizard). The sternum of the lumbering crocodilians is a simple cartilaginous plate attached to the procoracoid of the pectoral girdle (fig. 8.14). Caudally, it is continuous with an expanse of a mineralized fibrous membrane incorporating gastralia. Turtles have no sternum. The two halves of the pectoral girdle are united ventrally by fibrous ligaments or by a mere wedge of cartilage.

The sterna of modern birds that are capable of flight have developed an enormous keel, or **carina,** to which the massive pectoral flight muscles are attached (fig. 8.20). The development of a carina in pterosaurs and birds is an instance of convergent evolution. The sternum of *Archaeopteryx* was small and lacked a keel, an indication that these first birds were not strong fliers.

The sternum of terrestrial mammals, including humans, is composed of a series of bony segments, or **sternebrae** (fig. 8.28, monkey) and is comparatively narrow in comparison to that of reptiles and birds. The last sternebra, **xiphisternum,** bears a cartilaginous or bony **xiphoid process.** Marine mammals tend to have a relatively short sternum, only a small number of ribs reach the sternum, and individual sternebrae tend to be

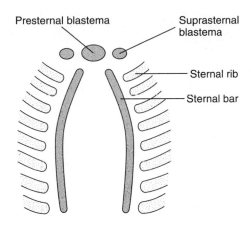

FIGURE 8.29

Mesenchymal blastemas that contribute to the amniote sternum (*red*). Presternal and suprasternal blastemas develop in mammals only. The ventral ends of developing ribs are also shown.

fused. This condition in cetaceans may be correlated with the galloping mode of swimming (chapter 10) necessitated by the lack of functionally competent anterior flippers.

Efforts to determine the nature of the earliest sterna have been frustrated by the fact that there is little evidence of a sternum in labyrinthodont amphibians. Either it was not preserved (perhaps because it was cartilage?), or it had not yet come into being. The tendency of collagenous tissues—which includes myosepta and the linea alba—to undergo ossification when subjected to continued mechanical stress was pointed out in chapter 7. It may be that ossification in the cephalic portion of the linea alba that ultimately produced sterna was a response to sustained mechanical stresses resulting from locomotion on land. A sternum may have arisen several times in the evolution of tetrapods, perhaps independently in urodeles, anurans, and in an amphibian line leading to amniotes.

The amniote sternum arises as paired mesenchymal bars that later unite and undergo chondrogenesis. In many mammals **presternal** and **suprasternal** blastemas also develop (fig. 8.29). The presternal blastema contributes to the manubrium, and the suprasternal centers sometimes do. The latter may give rise to independent **suprasternal ossicles**. Some humans have suprasternal ossicles, but they are not found unless there is an occasion to x-ray the sternoclavicular joint.

SUMMARY

1. Typical vertebrae consist of centra, neural arches, one or more processes, and, in many tails, hemal arches.
2. Transverse processes (diapophyses) are present in most vertebrates. Tetrapods of one species or another have basapophyses, hypapophyses, parapophyses, and zygapophyses.
3. Fishes have only trunk and tail vertebrae. The trunk vertebrae of tetrapods are subdivided into cervical, dorsal, and sacral vertebrae. Dorsals are further subdivided into thoracic and lumbar when long curved ribs are limited to the anterior trunk.
4. In fishes the notochord persists within the adult vertebral column from skull to tip of tail. It is usually constricted within centra, expanded between centra, producing an hourglass-shaped amphicelous vertebra.
5. Vertebra are constructed around, and sometimes within, the notochord by mesenchyme cells from the sclerotomes of mesodermal somites. The notochord sheath may be infiltrated with chordal cartilage and surrounded by perichordal cartilage that usually ossifies to become replacement bone. Membrane bone is deposited in teleosts, urodeles, and apodans. No replacement of cartilage occurs in Chondrichthyes.

6. Notochordal remnants often remain within the vertebral column in all vertebrate classes. In mammals a vestige persists in the intervertebral discs as a pulpy nucleus.
7. A few fishes have aberrant columns, lacking unified centra (dipnoans, some chondrosteans, *Latimeria*) or have calcified cartilaginous rings instead of typical centra (holocephalans)
8. Agnathans have no centra or arches. Lateral neural cartilages are perched above the notochord.
9. Diplospondyly is common in many fishes.
10. Early tetrapod vertebrae were rachitomous, consisting of independent arches, a hypocentrum, and two pleurocentra. The notochord was present and uninterrupted throughout the column.
11. Amphicelous centra are concave at both ends because the persistent notochord is constricted within centra and expanded between them. They are less specialized than other types and are found in most fishes and in urodeles, apodans, and some primitive lizards.
12. The centra of the remaining tetrapods are opisthocelous (salamanders), procelous (anurans and modern reptiles), heterocelous (in the neck of birds), and acelous (mammals). These increase the flexibility of the spine.

13. Amphibians have one cervical vertebra. The number is higher in reptiles and still higher in birds. Mammals typically have seven. The first two vertebrae in amniotes are atlas and axis. A proatlas is found in a few reptiles and mammals.

14. Sacral vertebrae bear stout transverse processes that brace the pelvic girdles against the vertebral column. Amphibians have one, reptiles and birds two, and mammals three to five sacral vertebrae. The number generally reflects the stress transmitted to the pelvic girdle from the hind limbs.

15. Sacral vertebrae in amniotes usually unite to form a sacrum. The sacrum of birds unites with adjacent lumbar and caudal vertebrae to form a synsacrum.

16. Caudal vertebrae often bear hemal arches or chevron bones. Neural and hemal arches and chevron bones become vestigial or disappear near the end of the tail.

17. Caudal vertebrae form a urostyle in anurans, a pygostyle in birds, and a coccyx in apes and humans.

18. Ribs, like myosepta, develop intersegmentally from cartilaginous blastemas. Most fishes have only ventral ribs, sharks and a few others have only dorsal ribs. A few fishes, such as *Polypterus,* have both. Agnathans have none.

19. Tetrapod ribs are generally bicipital with a tuberculum and capitulum. Most thoracic ribs have dorsal (costal) and ventral (sternal) segments.

20. Snakes have long movable ribs beginning at the axis and continuing far into the tail. They are essential for locomotion.

21. Short ribs are sometimes attached to the distal ends of transverse processes, as in anurans.

22. A vestigial head of a missing rib in the neck of mammals becomes the roof of a transverse foramen that, with others, creates a vertebroarterial canal housing a vertebral artery.

23. Sterna are tetrapod structures that facilitate life on land. They are reduced or missing in species lacking forelimbs, and in turtles. The sternum of birds capable of flight has a carina. Mammalian sterna exhibit sternebrae.

24. Amniote sterna arise as paired sternal bars that generally unite. An unpaired embryonic presternal blastema may contribute the manubrium sterni. Paired blastemas contribute independent suprasternal ossicles in some mammals. These occur as anomalies in humans.

SELECTED READINGS

Readings relating to the role of the vertebral column in locomotion are at the end of chapter 10.

Goodrich, E. S.: Studies on the structure and development of vertebrates, London, 1930, The Macmillan Co., Ltd. (Reprinted by The University of Chicago Press, Chicago, 1986.) Although the theoretical discussions are out of date, the work is rich in skeletal morphology and abundantly illustrated.

Heilmann, G.: The origin of birds, New York, 1927, Appleton-Century-Crofts, Inc.

Hoffstetter, R., and Gasc, J. P.: Vertebrae and ribs of modern reptiles. In Gans, C., Bellairs, A. d'A., and Parsons, T. S., editors: Biology of the reptilia, vol. 1, New York, 1969, Academic Press.

Jarvik, E.: Basic structure and evolution of vertebrates, vol. 1, New York, 1980, Academic Press. An in-depth study of *Amia* and the early fossil vertebrate *Eusthenopteron.*

Panchen, A. L.: The origin and evolution of early tetrapod vertebrae. In Andrews, S. M., Miles, R. S., and Walker, A. D., editors: Problems in vertebrate evolution, New York, 1977, Academic Press.

CHAPTER NINE

SKULL AND VISCERAL SKELETON

In this chapter we will find that vertebrates from fishes to human beings form their head skeleton out of three components: an embryonic cartilaginous braincase, ancient dermal armor, and contributions from the branchial skeleton. We will learn how these components are assembled, and note the new functions achieved by the branchial skeleton when vertebrates took up life on land.

he word "skull" is seldom misunderstood by the layperson. To him or her, it is the bony structure that Hamlet held in his hand and gazed at dolefully as he spoke his now famous words, "Alas, poor Yorick." To the morphologist, however, the term poses a problem because of the intimate relationship in fishes between the skeleton that protects the brain and special sense organs of the head and the skeleton of the jaws and branchial arches. The latter has been inherited in modified form by higher vertebrates including, alas, poor Yorick. For this reason, morphologists may avoid the term "skull" and refer instead to (1) the **neurocranium**, or **primary braincase**; (2) the **dermatocranium**; and (3) the **visceral skeleton**, or **splanchnocranium**. In this chapter, for practical purposes, "skull" will mean what the layperson probably thinks it means, but without the lower jaw. We can then classify the parts of the cranial skeleton of vertebrates as follows.

Skull (Cranial skeleton)
    Neurocranium
    Dermatocranium
  Visceral skeleton
    Embryonic upper jaw cartilage (palatoquadrate)
      and its replacement bones
    Embryonic lower jaw cartilage (Meckel's) and its
      replacement and investing bones
    Skeleton of the branchial arches

The upper jaw is visceral skeleton being part of or derived from the first visceral arch (fig. 9.1). In bony vertebrates it becomes incorporated into the skull as development progresses.

NEUROCRANIUM

The neurocranium (sometimes called endocranium, chondrocranium, or primary braincase), is the part of the skull that (1) protects the brain and certain special sense organs, (2) arises as cartilage, and (3) is subsequently partly or wholly replaced by bone except in cartilaginous fishes. The neurocrania of all jawed vertebrates develop in a similar manner, as described below. It commences as several independent cartilages that later expand and unite to form a cartilaginous braincase.

Cartilaginous Stage

Parachordal and Prechordal Cartilages and Notochord
The neurocranium commences as a pair of **parachordal** and **prechordal** cartilages (fig. 9.2a) underneath the brain. Parachordal cartilages parallel the anterior end of the notochord beneath the midbrain and hindbrain. Prechordal cartilages (**trabeculae cranii**) develop anterior to the notochord underneath the forebrain. The parachordal cartilages expand across the midline toward each other and unite. In the process, the notochord and parachordal cartilages are incorporated into a single, broad, cartilaginous **basal plate**. The prechordal cartilages likewise expand and unite across the midline at their anterior ends to form an **ethmoid plate**.

Sense Capsules
While parachordal and prechordal cartilages are forming, cartilage also appears in two other locations: (1) as an **olfactory (nasal) capsule** partially surrounding the

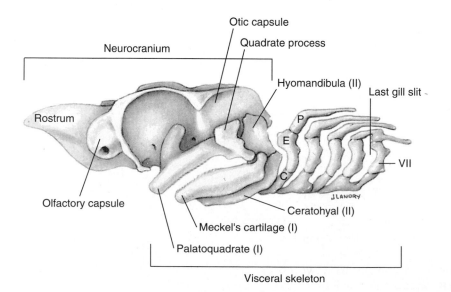

FIGURE 9.1

Skull and visceral skeleton of the shark, *Squalus acanthias*. **I**, **II**, and **VII**, skeleton of first, second, and seventh pharyngeal arches. **C**, ceratobranchial; **E**, epibranchial; **P**, pharyngobranchial cartilages of the second gill arch. Labial cartilages, gill rakers, and gill rays are omitted. The hyomandibula suspends the entire visceral skeleton from the skull at the otic capsule.

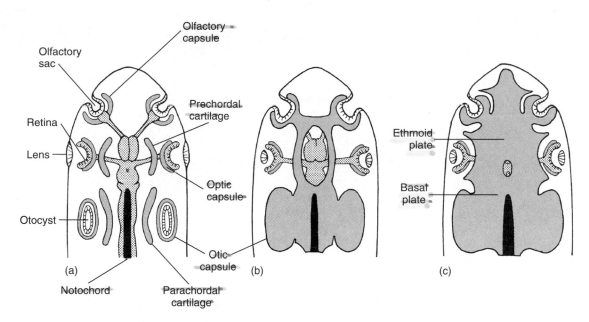

FIGURE 9.2

Early stages in development of a cartilaginous neurocranium, as seen from a ventral view. In (*a*) the notochord is seen underlying the midbrain and hindbrain. In (*b*) the notochord has been incorporated into the caudal floor of the neurocranium (basal plate). In (*c*) a cartilaginous floor has been completed beneath the entire brain, and a hypophyseal fenestra remains between the ethmoid and basal plates. The optic capsule becomes the sclerotic coat of the eyeball.

olfactory epithelium and (2) as an **otic capsule** completely surrounding the otocyst, which is the developing inner ear (fig. 9.2*a* and *b*). The olfactory capsules are incomplete anteriorly, since water (in fishes) or air (in tetrapods) must have access to the olfactory epithelium. The walls of the olfactory and otic capsules are perforated by foramina that transmit nerves and vascular channels.

An **optic capsule** forms around the retina but it is not the orbit, or skeletal socket, in which the eyeball lies. It is the **sclerotic coat** of the eyeball. Although this capsule is fibrous in mammals, cartilaginous or bony plates often form within the sclerotic coat (fig. 17.17). This is an ancient condition, having been present in crossopterygians. Because the optic capsule does not fuse with the rest of the neurocranium, the eyeball is free to move independent of the skull. Therefore the sclerotic coat is not conventionally considered part of the neurocranium.

Completion of the Floor, Walls, and Roof

The expanding ethmoid plate unites anteriorly with the olfactory capsules, and the expanding basal plate unites with the otic capsules, which are lateral to the hindbrain. The ethmoid and basal plates also expand toward one another until they meet to form a floor upon which the brain thereafter rests (fig. 9.2*c*). In the latter process there remains in the midline between the two plates a **hypophyseal fenestra** accommodating the hypophysis and internal carotid arteries, the latter being en route to the brain. Later the fenestra is usually reduced to a pair of foramina transmitting the arteries. Further development

of the neurocranium primarily involves construction of cartilaginous walls around the brain and, in lower vertebrates, a cartilaginous roof (**tectum**) above the brain with one or two prominent fenestrae (fig. 9.3*c*). The cranial nerves and blood vessels are already present by this time, and cartilage is deposited in such a manner as to leave foramina for these structures. The largest is the foramen magnum in the rear wall of the neurocranium. As the brain, blood vessels, nerves, and sense organs grow, remodeling of the neurocranial cartilage takes place.

The preceding basic pattern of development recurs in all vertebrate classes, producing a cartilaginous neurocranium that houses the brain, olfactory epithelia, and inner ear (fig. 9.3). The mesenchyme that gives rise to the neurocranium comes from at least two sources. The prechordal cartilages are formed from neural crest ectoderm that streams ventrad in front of the developing eyestalks (optic stalks) that connect the brain with the developing eyeball. The mesenchyme that forms the parachordal cartilages and encapsulates the inner ear comes from lateral-plate mesoderm. The source of the mesenchyme that forms the olfactory and optic capsules has not yet been definitely established.

Cartilaginous Neurocrania of Adult Vertebrates

Cyclostomes

The several cartilaginous components of the embryonic neurocranium remain more or less independent

FIGURE 9.3

Cartilaginous neurocrania of early ammocoete larva (*a*), young *Amphiuma* (*b*), recently metamorphosed frog (*c*), and young *Sphenodon* (*d*), (*a*) and (*b*) dorsal views; (*c*) and (*d*), left lateral views with associated visceral arch components. *Dark red,* ethmoid plate and its derivatives; *light red,* palatoquadrate and Meckel's cartilages; **Hy,** hyoid arch component; **M,** Meckel's cartilage; **Na,** nasal capsule; **Ot,** otic capsule; **P,** palatoquadrate cartilage; **Q,** quadrate region of palatoquadrate cartilage.

throughout life (fig. 9.35). An olfactory capsule, otic capsules, a basal plate, and a notochord (not fused with the basal plate) are identifiable, and several other cartilages of unknown homology are also present. The roof above the brain remains unchondrified, hence fibrous.

Cartilaginous Fishes

The neurocranium of *Squalus acanthias* is typical of cartilaginous fishes (fig. 9.1). It constitutes a high water mark in the development of neurocrania because the embryonic components unite to form a boxlike adult cartilaginous braincase or **chondrocranium** in which the basic components have almost lost their identity. Walls are fully developed, and the brain is completely roofed by cartilage. The last portion of the roof to chondrify is an area just behind the rostrum, and in young skulls this may still be an unossified fenestra. (Some elasmobranch species have a fenestra at this location throughout life.) The otic capsules are inextricably fused into the posterolateral walls of the braincase, and the olfactory capsules are firmly united to it anteriorly. The notochord is visible on the ventral aspect as a ridge extending cephalad from the base of the foramen magnum. The hypophysis is cradled in a cartilaginous

pocket, the sella turcica, beneath the brain. The neurocranium projects forward beyond the olfactory capsules as the rostrum.

On each side of the foramen magnum, an occipital condyle is the site of the immovable articulation between the occipital region of the neurocranium and the first vertebra. On the posterodorsal aspect of the neurocranium, a depression, the **endolymphatic fossa**, exhibits two pairs of foramina that house **endolymphatic** and **perilymphatic ducts**, respectively. The endolymphatic ducts open onto the surface of the head. A well-developed, totally cartilaginous adult skull, devoid of dermal bones, is found only in Chondrichthyes.

Lower Bony Fishes

Although a cartilaginous neurocranium is an embryonic structure in teleosts and tetrapods, it persists throughout life as cartilage in most chondrosteans, semionotiformes other than gars, and dipnoans. In order to see it one must strip away the dermal bones that overlie it (fig. 9.7). In teleosts and tetrapods, however, the embryonic cartilaginous neurocranium is replaced partially or wholly by endochondral bone as development progresses. The ossification centers where this takes place will be discussed next.

Neurocranial Ossification Centers

The process of endochondral ossification within the neurocranium occurs more or less simultaneously at numerous separate ossification centers. The specific number of centers varies among species, but four regional groups are universal. These groups—occipital, sphenoid, ethmoid, and otic—will be discussed next. They are illustrated in a fetal pig in figure 9.4. The composite bones that develop from these centers in humans are shown in figure 9.6.

Occipital Centers

The cartilage surrounding the foramen magnum may be replaced by as many as four bones. One or more endochondral ossification centers ventral to the foramen magnum produce a **basioccipital bone** underlying the hindbrain (figs. 9.4 and 9.5a). Centers in the lateral walls of the foramen magnum produce two **exoccipital bones.** Above the foramen a **supraoccipital bone** may develop. In mammals all four occipital elements usually fuse to form a single **occipital bone.** In modern amphibians one or more of these may remain cartilaginous, although they were bony in stem amphibians.

The neurocranium of tetrapods articulates with the first vertebra via one or two **occipital condyles.** Stem amphibians had a single condyle borne chiefly on the basioccipital bone. Living reptiles and birds still have a single condyle. Modern amphibians and mammals diverged from the early tetrapod condition, gradually shifting the condyles from the median basioccipital to the two exoccipitals. What the survival value may have been is purely speculative.

Sphenoid Centers

The embryonic cartilaginous neurocranium underlying the midbrain and pituitary gland ossifies to form a **basisphenoid bone** (anterior to the basioccipital) and a

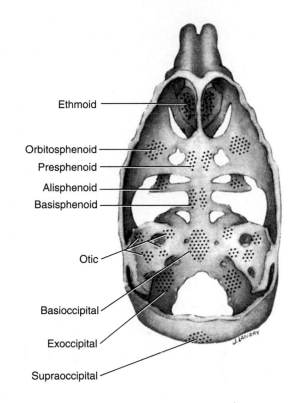

FIGURE 9.4

Cartilaginous neurocranium of a fetal pig with chief endochondral ossification centers of mammals superimposed (*dots*). The neurocranium is complete as shown, there being no cartilage above the brain. The ethmoid center becomes the cribriform plate perforated by olfactory foramina. The otic centers are in the otic capsule. The alisphenoid center may be derived from palatoquadrate cartilage in some mammals.

presphenoid (fig. 9.4). Thus a bony platform consisting of the basioccipital and the sphenoid bones underlies the brain. The side walls above the basisphenoid and presphenoid ossification centers form lateral sphenoid elements (**orbitosphenoid, pleurosphenoid,** and

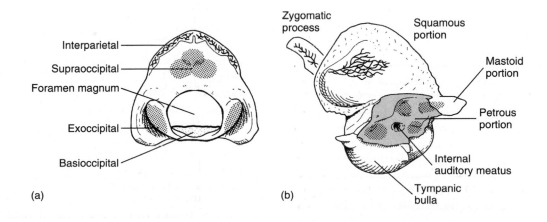

FIGURE 9.5

Typical mammalian endochondral ossification centers (*dots*) and intramembranous centers (black reticulum) in two cat skull bones. (*a*) Occipital bone (caudal view). (*b*) Right temporal (medial view). The petrous portion (*red*) houses the inner ear.

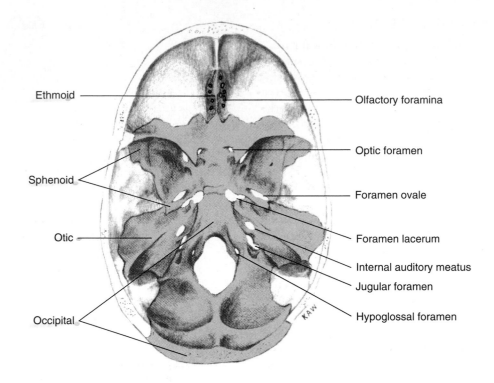

FIGURE 9.6

Bony neurocranium (*red*) of human skull. The calvarium (roof) of the skull has been sawed off, and view is looking down into skull from above. Major endochondral ossification centers are labeled at left. The sphenoid bone exhibits an anterior (lesser) wing enclosing the optic foramina and a posterior (greater) wing between the two label lines for the sphenoid bone. The olfactory foramina are in the cribriform plate of the ethmoid.

others—the alisphenoid of at least some mammals forms in the palatoquadrate cartilage rather than in the neurocranium). These may remain separate or unite with the basisphenoids and presphenoids to form a single adult **sphenoid bone** with "wings" (fig. 9.6). The pituitary rests in the sella turcica of the basisphenoid region. No replacement bones develop above the brain.

Ethmoid Centers

The ethmoid region lies immediately anterior to the sphenoid and includes the ethmoid plate and olfactory (nasal) capsules (fig. 9.3). Of the four major potential ossification centers in the cartilaginous neurocranium (occipital, sphenoid, ethmoid, and otic), the ethmoid tends to remain cartilaginous in tetrapods from amphibians to mammals. Most lower tetrapods develop no ethmoidal ossification centers whatsoever. Ossification centers in amniotes are chiefly **mesethmoid** bones, which contribute to the otherwise cartilaginous nasal septum of birds and mammals, and in birds, to the anterior portion of the interorbital septum; one or more scroll-like **turbinal** bones **(conchae)** in the walls of the nasal passageways of most reptiles, birds, and mammals; and in mammals, a **cribriform plate** perforated by **olfactory foramina** that transmit bundles of olfactory nerve fibers from the olfactory epithelium to the brain (fig. 9.6). In anurans a **sphenethmoid** is the sole bone arising in the

ethmoid and sphenoid regions. An **ectethmoid** develops in the lateral walls of the nasal passageway of *Sphenodon* (fig. 9.3d). It should be noted here, however, that not all cartilages in the nasal passages of amniotes are derived from ethmoid ossification centers. Among these that are not are the winglike alar and small sesamoid cartilages that stiffen the lateral walls of the human nose, thereby keeping the walls from collapsing against the nasal septum during inhalation.

Otic Centers

The cartilaginous otic capsule surrounding the membranous labyrinth is replaced by several bones with such names as **prootic, opisthotic,** and **epiotic.** One or more of these may unite with adjacent replacement or membrane bones. For example, in frogs and most reptiles the opisthotics fuse with the exoccipitals and in birds and mammals the prootic, opisthotic, and epiotics all unite to form a single **periotic** or **petrosal bone.** The petrosal, in turn, may unite with the squamosal, a membrane bone, to form a **temporal bone** (fig. 9.5b). Six ossification centers have been described in the otic capsule of a human fetus.

GENERALIZED DERMATOCRANIUM

The membrane bones of the skull constitute collectively the dermatocranium. After considering how the dermatocranium probably originated, we will examine its basic architecture in generalized vertebrates. The insight thus gained will enable us to relate the dermatocranium to the rest of the skull in modern tetrapods. Because the skulls of modern fishes are highly specialized and vary widely, we will mention them only occasionally.

How It May Have Begun

Much of the body of the earliest vertebrates was encased in a bony dermal armor. As far back as ostracoderms this armor varied widely in the extent to which it covered the body—head and anterior trunk only, head and entire trunk, or head, trunk, and tail; in the relative size of the bones or scales making up the armor—large shields, smaller plates, minute scales; and in the relative size of the plates or scales on the head contrasted with those on the rest of the body. There is also evidence that some ostracoderms—cephalaspids, at least—passed through one or more cycles of expansion and reduction of their dermal armor, going from small scales to increasingly larger plates, then more or less reversing the trend; and that the reduction was less complete on the head. The large bony plates in the skin of early *jawed* fishes also gave way to smaller scales, first on the trunk and tail and eventually on the head. However, bony plates are still present in the skin of the head in the relict *Amia* and gars. These plates are an integral part of the skull (fig. 9.7). Note also

Dermatocranium   **Cartilaginous neurocranium**

Amia

FIGURE 9.7

Skull of *Amia calva,* dorsal view. Dermal bones have been removed on the right side to reveal underlying cartilaginous neurocranium. **F,** frontal bones; **N,** nasal; **OP,** operculum; **P,** parietal; **PO,** postorbital; **PoT,** posttemporal; **PT,** pterotic; **T,** tabular. The bones anterior to the nasals are ethmoids. Premaxillae are not visible in this view.

the cheek plates in the skull of a gar (fig. 9.11). They are either skull bones or integumentary scales, whichever you prefer to call them. But whatever one calls them does not alter the fact that these bones in the skin are also part of the skull. The neurocranium is the endoskeleton; the dermal bones constitute an exoskeleton. It is the latter that constitutes the dermatocranium in these "living fossils."

In modern vertebrates from fish to humans, these membrane bones of the head no longer ossify from mesenchyme in the *dermis,* but from *subdermal* mesenchyme. The inductors that convert this mesenchyme to fibroblasts and scleroblasts may be the same ones that induced ossification in the dermis of earlier generations. *It is these membrane bones that constitute the modern dermatocranium from fish to humans.* Ongoing research on mesenchymal cell migration and inductor substances reinforces the concept that the dermatocranium is derived from the dermal armor of the earliest vertebrates.

Its Basic Structure

For convenience of discussion the generalized dermatocranium can be divided into (1) bones that form above and alongside the brain and neurocranium (roofing bones), (2) dermal bones of the upper jaw (marginal bones), (3) dermal bones of the primary plate, and (4) opercular bones.

Roofing Bones

An early pattern of roofing bones is seen in crossopterygian fishes, and this pattern was inherited almost intact by labyrinthodonts (fig. 9.8). Roofing bones in these vertebrates provided a protective shield over the brain and special sense organs with openings only for the nares, paired eyes, and a median (parietal) eye. In crossopterygians, a series of paired and unpaired scalelike bones extended along the middorsal line from the nares to the occiput, overlying the brain, olfactory capsules, and any other components of the neurocranium that developed in that area. In labyrinthodonts, the unpaired bones were missing and a series of paired **nasals, frontals, parietals,** and **postparietals (dermoccipitals)** took their place. The postparietals have disappeared as independent bones in modern amphibians, but they may have coalesced with endochondral supraoccipitals that appeared first in early reptiles. (The bone caudal to the parietal was named dermoccipital in recognition of this possibility.) A parietal foramen housing the median eye is still present in many fishes, amphibians, and lizards.

Forming a ring around the orbit in the generalized skull were a **lacrimal, prefrontal, postfrontal,** and **infraorbital (jugal).** The lacrimal bone derives its name from its relationship to the nasolacrimal duct of amniotes, which drains excess fluid, including tears (lachrymae), from the nasal corner of the eye to the nasal canal. At the posterior angle of the skull were **intertemporal, supratemporal, tabular,** and lower down,

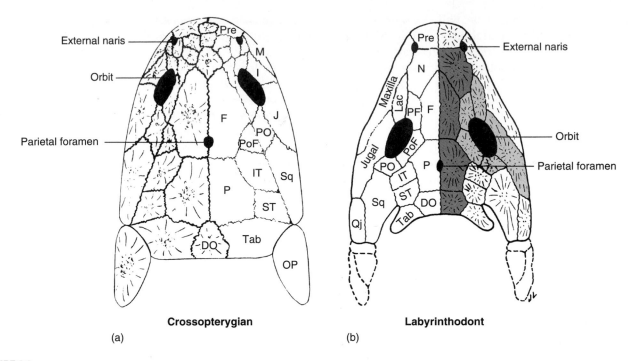

FIGURE 9.8

Early dermal bone patterns from which tetrapod dermatocrania probably evolved. (*a*) Skull of the crossopterygian fish *Eusthenopteron*. Note midline bones and small, scalelike bones in the rostral region. The location of the foramina for the median eye in the two skulls is correct. The parietal foramen houses a median eye. (*b*) Skull of a Carboniferous labyrinthodont. Broken lines indicate missing opercular bones. **DO,** postparietal (dermoccipital); **F,** frontal; **I,** infraorbital; **IT,** intertemporal; **J,** jugal; **Lac,** lacrimal; **M,** maxilla; **N,** nasal; **OP,** opercular; **P,** parietal; **PF,** prefrontal; **PO,** postorbital; **PoF,** postfrontal; **Pre,** premaxilla; **Sq,** squamosal; **ST,** supratemporal; **Qj,** quadratojugal; **Tab,** tabular. In (*b*), midline roofing bones are *dark red,* periorbital bones are *light red.*

squamosal and **quadratojugal** bones. Labyrinthodonts developed a longer facial area, or snout, than is seen in crossopterygians by elongating the bones between the external nares and orbits. This could have been correlated with altered methods of gathering or capturing food or manipulating it in the mouth.

Dermal Bones of the Upper Jaw

The palatoquadrate cartilages, the only upper jaws that cartilaginous fishes develop (fig. 9.1), are the embryonic precursors of the upper jaw of bony vertebrates. In the latter, these cartilages become overlaid, or ensheathed, by tooth-bearing dermal bones, the **premaxillae** and **maxillae,** which then become part of the margins of the dermatocranium (figs. 9.8*b* and 9.10). Thus the completed upper jaw complex of bony vertebrates becomes part of the skull. The lower jaw will be discussed with the other components of the visceral skeleton.

Primary Palatal Bones

The primary palate is the roof of the oropharyngeal cavity of fishes and of the oral cavity of lower tetrapods. In sharks it is cartilaginous, being the floor of the neurocranium on which the brain rests. In bony vertebrates, membrane bones are applied to the underside of the neurocranium and to any part of the palatoquadrate cartilages (upper jaw cartilages) occupying the area. These membrane bones then become the major components of the primary palate.

In crossopterygians and early tetrapods, these membrane bones of the palate were an unpaired **parasphenoid** beneath the sphenoid region of the neurocranium, paired **vomers** beneath the ethmoid region, and paired **palatines, ectopterygoids,** and **pterygoids** laterally (fig. 9.9). The last three invested the palatoquadrate cartilage to one degree or another. Primitively, teeth formed on all these bones, and several of them still bear teeth in lower vertebrates. Internal nares pierced the palate anterolaterally. A primary palate is still present with modifications in all tetrapods (fig. 9.13), but in those that also develop a secondary palate the primary palate remains in the roof of the nasal passageway (fig. 9.18).

Opercular Bones

The operculum is a flap of tissue that arises as an outgrowth of the hyoid arch and extends caudad over the gill slits. It is membranous in holocephalans and absent in elasmobranchs. In bony fishes, it is stiffened by squamous plates of dermal bone. Most constant are a large **opercular,** smaller **preoperculars** overlying the site of

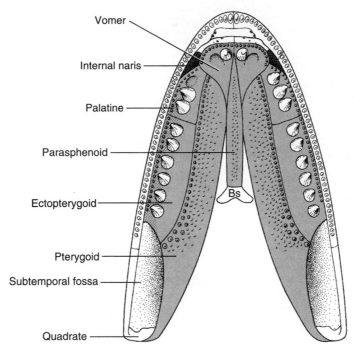

Vomer

Internal naris

Palatine

Parasphenoid

Ectopterygoid

Pterygoid

Subtemporal fossa

Quadrate

Bs

Crossopterygian

(a)

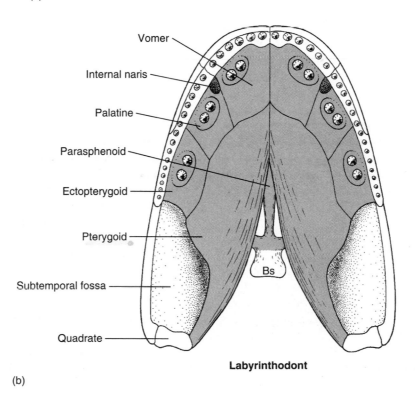

Vomer

Internal naris

Palatine

Parasphenoid

Ectopterygoid

Pterygoid

Subtemporal fossa

Quadrate

Bs

Labyrinthodont

(b)

FIGURE 9.9

Primary palate (*red*) of a rhipidistian crossopterygian and a late Paleozoic labyrinthodont. Small denticles cover the parasphenoid and pterygoid bones in (*a*).

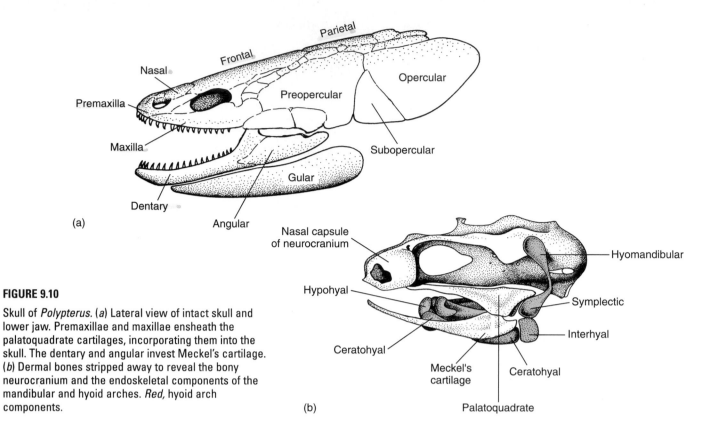

FIGURE 9.10

Skull of *Polypterus*. (*a*) Lateral view of intact skull and lower jaw. Premaxillae and maxillae ensheath the palatoquadrate cartilages, incorporating them into the skull. The dentary and angular invest Meckel's cartilage. (*b*) Dermal bones stripped away to reveal the bony neurocranium and the endoskeletal components of the mandibular and hyoid arches. *Red,* hyoid arch components.

articulation of the upper and lower jaws, **suboperculars**, and **interoperculars** (fig. 9.10). In some of the more primitive bony fishes, one or more **gular bones** (fig. 9.10) lie in the opercular membrane in the floor of the opercular chamber. They are represented in more specialized ray-finned fishes and dipnoans by **branchiostegal rays** located in caudally directed flaps, the **branchiostegal membranes** of the operculum. No opercular or gular plates or vestiges of them remain in tetrapods.

NEUROCRANIAL-DERMATOCRANIAL COMPLEX OF BONY FISHES

We will now examine the neurocranium and dermatocranium as they combine architecturally to contribute to the skulls of bony fishes. Later we will study the skeleton of the visceral arches to complete the picture.

Chondrosteans

Sturgeons and spoonbills (paddlefishes) have a skull that justifies the name Chondrostei. The neurocranium remains almost completely cartilaginous throughout life. The only traces of endochondral ossification in the neurocranium of sturgeons are isolated sites in the otic capsules and in the sphenoid bone where it contributes to the walls of the orbit. Elsewhere, the neurocranium remains cartilaginous. The spoonbill is not much different. Its unusual bill is an extension of the cartilaginous

rostrum. Numerous overlying dermal bones hide the cartilaginous neurocranium and the cartilaginous components of the upper jaws and hyoid arch.

Polypterus retains more of the primitive features of Paleozoic fishes than do sturgeons and spoonbills. The neurocranium is well ossified and is overlaid in the mid-dorsal line by paired nasals, frontals, and parietals of dermal origin (fig. 9.10). Lateral to these is a linear series of small, mostly unnamed bony dermal plates. Premaxillae and maxillae overlie the endochondral palatoquadrate bones of the first visceral arch, and a preopercular bone overlies the site of articulation of the upper and lower jaws. Other opercular bones complete the dermatocranium laterally, and a pair of primitive gular bones of dermal origin lie in the opercular membranes.

Semionotiformes (Holosteans)

The skulls of *Amia* and gars, like those of chondrosteans, bear little resemblance to the skulls of modern fishes (figs. 9.7 and 9.11*a*). The dermal bones are grooved and pitted, exhibiting the sculpturing effects of the overlying skin. They can be characterized equally well as skull bones or dermal scales. The cheek plates of gars are especially compelling evidence that the dermatocranium may have been derived from earlier dermal scales. The neurocranium of *Amia* remains highly cartilaginous; in gars it is bony. A detailed account of the anatomy of the skull of *Amia* has been provided by E. Jarvik (1980).

(a)   **Gar**

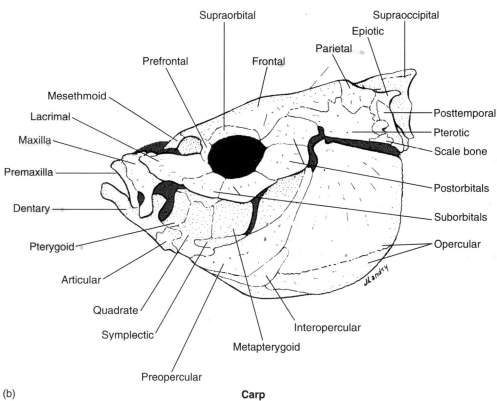

(b)   **Carp**

FIGURE 9.11

Skull of a holostean (gar) and of a modern fish (carp). Note the scalelike nature of the cheek plates in the gar. In the carp, *red* represents unossified cartilage.

Teleosts

The skulls of most teleosts are highly specialized and architecturally diverse. This is correlated with the diverse feeding habits of the group, which includes species that harvest plant tissues or capture animal life in every conceivable niche of the streams, lakes, and oceans of the world. A combination of highly maneuverable jaws and palates, coupled with the largest number of bones in any modern vertebrate skull, accounts for the anatomic diversity.

The skull of a carp is illustrated in figure 9.11. Like many teleost skulls, it is compressed laterally and vaulted dorsally. The names of most of the bones will be recognized from earlier discussions. The maxillae, premaxillae, dentary, articular, quadrate, and symplectic are associated with the jaws and hyoid arch and will be discussed with the visceral skeleton. The posttemporal is the dorsalmost segment of the pectoral girdle. It has been secondarily incorporated into the skull. The mesethmoid, epiotic, and pterotic often incorporate some membrane bone along with the usual replacement bone. Branchiostegal rays in paired branchiostegal membranes beneath the gill chamber take the place of the more primitive gular bones of early ray-finned fishes.

The neurocranium in most teleosts is fully ossified except for the olfactory capsules. However, in cyprinids, which include the carp, neurocranial ossification is less than complete, and the islands of cartilage seen on the surface of the carp skull are unossified regions of the neurocranium or, at some sites, new cartilage.

Despite the kinetic nature of many teleost palates and the architectural disruptions that this has entailed, the dermal bone complement of the palate reflects that of ancestral bony fishes. The vomer, parasphenoid, and pterygoid bones are common components.

As mentioned earlier, names given to bones of the skull many years ago have been assigned on the basis of location (palatine), shape (pterygoid), or some other characteristic (squamosal) rather than on the basis of whether they are endochondral or intramembranous in origin. A bone of the same name in two species may be of endochondral origin in one species and of dermal origin in another. In such instances, the bones are not homologues. On the other hand, homologues sometimes have different names in different species, which poses problems of synonymy. Some of the bones labeled on the carp skull in figure 9.11 are of doubtful homology.

Dipnoans

Following their divergence from ray-finned fishes in the early Devonian, dipnoans did not experience the burst of speciation subsequently undergone by teleosts. Consequently, changes in the architecture of dipnoan skulls have been more conservative. With reference to the dermatocranium, time has taken lungfishes in a direction quite different from that of teleosts, whose dermal bones increased in number and decreased in size. The dermatocranium of lungfishes has evolved from a large number of scalelike dermal bones in fossils to a relatively few broad bony plates in today's species. The neurocranium is largely cartilaginous like that of some chondrosteans and holosteans, rather than bony, as in teleosts. The palate accommodates openings of the nasal canal into the oral cavity just behind the mouth, as in crossopterygians. This trait may be as ancient as pharyngeal gills (see fig. 4.2).

NEUROCRANIAL-DERMATOCRANIAL COMPLEX OF MODERN TETRAPODS

We can now look at the neurocranial-dermatocranial complex as it has evolved in modern tetrapods. A disproportionate amount of time will be spent on reptiles because they have diversified more than other tetrapods and because during the Mesozoic the skulls of some of them underwent architectural changes that were incorporated into the skulls of birds and mammals.

Amphibians

The skulls of modern amphibians are considerably modified from those of labyrinthodonts, although they are still flattened (**platybasic**) compared with the vaulted skulls of amniotes. The neurocranium is incomplete dorsally (fig. 9.3c), and much of it remains cartilaginous except in apodans. The rigidity of the apodan skull may be correlated with burrowing. The only replacement bones in anurans and urodeles (excepting the columella, when ossified) are a sphenethmoid, two prootics, and two exoccipitals, each of the latter bearing a condyle. In some perennibranchiate urodeles even a sphenethmoid fails to ossify.

Abutting against the otic capsule in amphibians (and all higher vertebrates) is a skeletal rod, the **columella,** a middle ear ossicle that conducts sound waves from an eardrum to the capsule. The phylogenetic origin of the columella will be discussed under "Visceral skeleton."

The dermatocranium is quite incomplete. The bones that surrounded the orbit of early amphibians have been lost except for the lacrimals and prefrontals of primitive urodeles (fig. 9.12a). Also missing are the primitive bones of the temporal region from the orbits caudad (intertemporals, supratemporals, tabulars, and postparietal). This loss of ensheathing bones leaves the otic capsule (prootic) exposed dorsally and laterally. Only the squamosal and sometimes a quadratojugal remain in this region. Premaxillae and maxillae are usually present to ensheath the upper jaw cartilages at the margin of the dermatocranium, but in perennibranchiates even maxillae may fail to develop.

The primary palate has been altered. In anurans and a few urodeles, large palatal vacuities have evolved beneath the orbit, reducing the palatines to transverse splinters that brace the upper jaw against the palate anteriorly and reducing the pterygoids to a pair of bipartite bones that brace the upper jaw against the braincase posteriorly (figs. 9.12b, and 9.13, *Rana*). Because of the palatal vacuities, the large eyeballs of anurans, which protrude just above the water line to monitor the environment in an otherwise submerged frog, can be retracted until the membranous floor of the orbit bulges into the oral cavity. In urodeles, the parasphenoid has become exceptionally broad (figs. 9.12b, and 9.13, *Necturus*), and palatal vacuities are lacking in some orders.

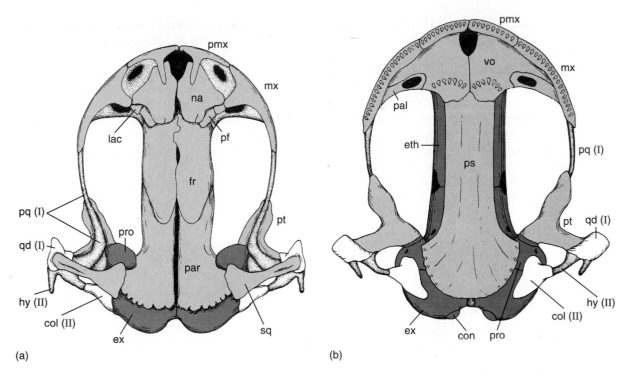

FIGURE 9.12

Skull of *Ranodon,* a primitive urodele in the family Hynobiidae. (*a*) Dorsal view. (*b*) Palatal view. *Light red,* dermal bones; *dark red,* neurocranial bones; *stipple,* cartilage. **col,** Columella; **con,** occipital condyle; **eth,** sphenethmoid; **ex,** exoccipital; **fr,** frontal; **hy,** a dorsal segment of the hyoid skeleton; **lac,** lacrimal; **mx,** maxilla; **na,** nasal; **pal,** palatine; **par,** parietal; **pf,** prefrontal; **pmx,** premaxilla; **pq,** palatoquadrate cartilage; **pro,** prootic; **ps,** parasphenoid; **pt,** pterygoid; **qd,** quadrate; **sq,** squamosal; **vo,** vomer; **I** and **II** indicate origin from the first or second visceral arch.

From I.I. Schmalhausen, *The Origin of Terrestrial Vertebrates.* Copyright © 1968. Academic Press, Orlando, FL. Reprinted by permission.

Seymouria

Amia

Rana

Primary Palates

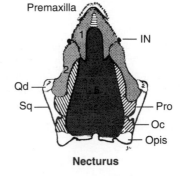

Necturus

FIGURE 9.13

Primary palates (in color) of a reptilelike amphibian (*Seymouria*), a lower bony fish (*Amia*), and two amphibians. Cartilage is indicated by diagonal lines; internal nares are black. **1,** vomer; **2,** palatine (in *Necturus,* palatopterygoid); **3,** ectopterygoid; **4,** pterygoid; **5,** parasphenoid; **6,** epipterygoid. **Oc,** Cartilaginous portion of otic capsule; **Opis,** opisthotic; **Pro,** prootic; **Qd,** quadrate; **Qj,** quadratojugal; **Sq,** squamosal; **IN,** internal naris.

Reptiles

The skulls of cotylosaurs (stem reptiles) were little changed from those of labyrinthodonts, and some cotylosaurian features are still present in living reptiles. Among these are a well-ossified neurocranium, a single occipital condyle, a large complement of membrane bones, and, in rhynchocephalians and many lizards, a parietal foramen housing a median eye (fig. 9.14c). Figure 17.19 is a photograph of the median eye of an iguana lizard. Major changes from stem reptile skulls included the appearance of temporal fossae and development of a partial or complete sec-

ondary palate. Temporal fossae and secondary palates were transmitted to birds and mammals.

The fully ossified neurocranium of an adult alligator consists of two exoccipital bones, a supraoccipital, a basioccipital that bears a single condyle for articulation with the atlas, pleurosphenoids, ethmoids, and several otics that arise from separate ossification centers in the otic capsules. None of the otic bones are on the surface, being overlaid by the squamosal, a dermatocranial bone. Some of the otic bones are coalesced with an adjacent occipital. The chiefly cartilaginous neurocranium of a juvenile *Sphenodon* is seen in situ in figure 9.14c. It will

(a)

(b)

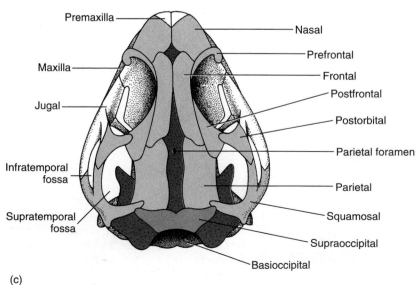

(c)

FIGURE 9.14

Roofing bones (*light red*) and marginal bones of the skulls of three reptiles. (*a*) Diapsid skull of alligator. (*b*) Derived anapsid skull of the snapping turtle, *Marcroclemys temminckii.* (*c*) Diapsid skull of juvenile *Sphenodon* (rhyncocephalia). *Dark red,* unossified regions of neurocranium. The supraoccipital and basioccipital are ossified parts of the neurocranium.

ossify further with age, but more cartilage remains in adult sphenodons and lizards than in other reptiles.

The roofing and marginal bones of the dermatocranium of three reptiles are seen in figure 9.14. Of living reptiles, crocodilians retain the largest number of membrane bones. Turtles have the most enigmatic skull, as will be explained shortly.

Temporal Fossae

A temporal fossa (temporal fenestra) is a cavernous opening in the temporal region of amniote skulls bounded by one or more bony arches (fig. 9.15). Stem reptiles had none, so their skulls are **anapsid,** that is, they lack temporal arches. Today, among living amniotes only turtles are anapsid (fig. 9.16). Synapsida, the extinct mammal-like reptiles, developed a single **lateral temporal fossa** surrounded by the postorbital, squamosal, and jugal bones, the last two forming an underlying **infratemporal (zygomatic)** arch. This synapsid skull was transmitted to mammals. In humans the cheekbones are the zygomatic arches (fig. 9.28).

Archosaurs and the predecessors of *Sphenodon,* lizards, and snakes had superior and inferior temporal fossae. When two fossae are present, the skull is said to be **diapsid.** (For derivation of the term see appendix 2.) The lower arch corresponds to the zygomatic arch of mammals. The upper arch, or **supratemporal**—the arch between the two fossae—consists of parts of the postorbital and squamosal bones. Crocodilians and *Sphenodon* still have diapsid skulls (figs. 9.14a and 9.17a), but modern lizards have lost part or all of the lower arch and snakes have lost both arches (fig. 9.17b and c). This loss has left a cavernous void in the posterolateral walls of squamate skulls. These are **modified diapsid skulls.** Reduction or loss of the arches along with the acquisition of intracranial joints facilitated **cranial kinesis,** the movement of one section of a skull independent of others.

Ichthyosaurs and plesiosaurs (fig. 4.24) had one temporal fossa, dorsally located, that resembled the dorsal fossa of diapsids in its anatomic relationships (fig. 9.15d). However, it is likely that this **euryapsid** condition is an instance of evolutionary convergence in lineage distinct from that leading to other reptiles. All euryapsid reptiles are extinct.

The temporal region of a turtle skull is an enigma. Absence of temporal fossae suggests a primitive condition. However, there has been extensive loss of dermal bones and excavation of the temporal region dorsal to the otic capsule (fig. 9.14b), a condition more pronounced in some families than in others. Supratemporal, tabular, and postparietal bones are missing, postorbitals and postfrontals have united, and parietals give the impression of having receded from the rear, leaving a wide vacuity that is best seen from a caudal or ventral view.

Temporal fossae provide space and surfaces in functionally advantageous positions for accommodating the powerful adductor muscles needed to operate the lower

Anapsid (stem reptile)

(a)

(b)   **Synapsid** (mammal stock)

Diapsid (bird stock)

(c)

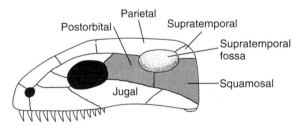

Euryapsid (ichthyosaurs and plesiosaurs)

(d)

FIGURE 9.15

Temporal fossae. *Light red,* infratemporal arch component; *dark red,* supratemporal arch component. Euryapsida is an extinct group.

FIGURE 9.16

Anapsid skull of small sea turtle, lateral view, lower jaw removed. **1,** premaxilla; **2,** maxilla; **3,** jugal; **4,** quadratojugal; **5,** quadrate; **6,** prefrontal; **7,** frontal; **8,** postfrontal; **9,** parietal; **10,** squamosal; **11,** supraoccipital; **12,** middle ear cavity.

(a)   **Sphenodon**

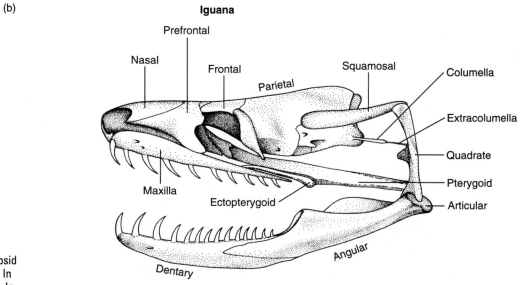

(b)   **Iguana**

FIGURE 9.17

Diapsid (*a*) and modified diapsid (*b, c*) skulls of three reptiles. In (*b*) the lower arch is missing. In (*c*) both arches are missing.

(c)   **Boa**

jaws of amniotes. In labyrinthodonts and cotylosaurs, the adductor mandibulae (chief levator muscle of the lower jaw) was confined to cramped quarters internal to the temporal region of the dermatocranium. (In cartilaginous fishes, as seen in figure 11.25a, it lies just under the skin of the head because there is no dermatocranium.) At best, this muscle enabled bony fishes and primitive tetrapods to seize food, bite off pieces, and close the mouth to prevent return of the food to the environment. The food was swallowed whole. Temporal fossae provided space for an expanding amniote adductor muscle to shorten and thicken during contraction; they provided access to the surface of the dermatocranium so that one portion of the adductor, the temporalis muscle, could acquire more leverage by spreading upward onto the temporal region of the skull; and they provided a zygomatic arch to which another portion of the adductor, the masseter, could acquire an anchorage. Figure 11.25c, shows the temporalis and masseter in relation to the zygomatic arch and temporal fossa of a primate. These powerful new adductors, assisted by other muscles of the mandibular and hyoid arches, make possible the complex side-to-side, forward-backward, and rotary chewing movements seen in herbivorous mammals that grind grasses or chew their cud and in carnivores that macerate flesh and crush bones.

Secondary Palates

A secondary palate is a horizontal partition that partially or completely divides the primitive oral cavity into separate oral and nasal passageways, thereby displacing the **internal nares (posterior choanae)** caudad. Embryonic development of a secondary palate in a mammal is illustrated in figure 9.18. In vertebrates with a secondary palate the primary palatal components remain in the roof of the nasal passageway, usually with a reduced complement in membrane bones. For example, no parasphenoid bone is present in mammals.

In crocodilians, medially directed, shelflike **palatal processes** of the premaxillae, maxillae, palatine, and pterygoid bones meet in the midline to form a long bony secondary palate with internal nares far to the rear (fig. 9.19). In other reptiles not all the palatal processes reach the midline, so the secondary palate is incomplete. Varying degrees of completion can be seen by examining several species of turtles (fig. 9.20). When the secondary palate is incomplete, the respiratory air stream is channeled in a fairly deep longitudinal groove, the **palatal fissure,** in the roof of the oral cavity (fig. 9.20b and c). The borders of the fissure are fleshy **palatal folds,** seen also in birds (fig. 12.4). In mammals, the secondary palate extends all the way to the pharynx but the caudal portion—the soft palate—lacks bone (fig. 12.3b).

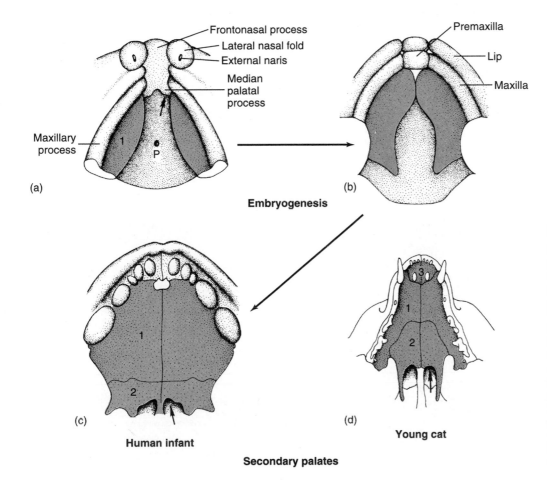

(a) Embryogenesis

(b)

(c) Human infant

(d) Young cat

Secondary palates

FIGURE 9.18

(a) to (c) Formation of secondary palate (red) in humans.
(d) Secondary palate of young cat for comparison. Arrows point to posterior choanae (internal nares).
1, palatal process of maxilla;
2, palatal process of palatine bone;
3, palatal process of premaxilla. In (a) (fetus approximately 18 weeks old) the palatal process of the maxillae are growing toward the midline, forming a secondary roof in the oral cavity. In (b) the processes have met anteriorly. In (c) the palate is complete.
P, evagination site of anterior pituitary.

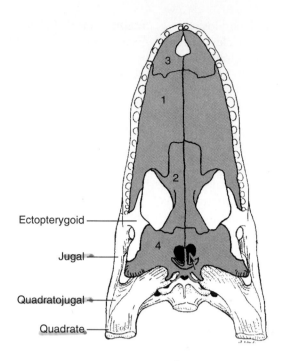

FIGURE 9.19

The long rigid secondary palate of an alligator (*red*). **1, 2,** and **3,** palatal processes of maxilla, palatine and premaxilla bones, respectively; **4,** pterygoid bone. Arrow indicates an internal naris.

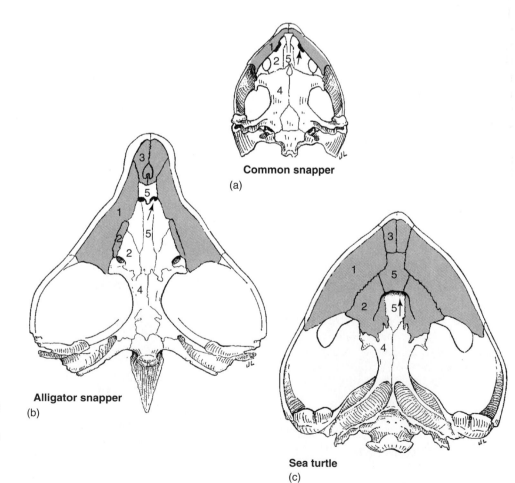

FIGURE 9.20

Species differences in the secondary palates (*red*) of turtles. (*a*) *Chelydra serpentina.* (*b*) *Macroclemys temminckii.* (*c*) *Lepidochelys olivacea,* temporal and supraoccipital regions omitted. **1,** maxilla; **2,** palatine; **3,** premaxilla; **4,** pterygoid; **5,** vomer. **1, 2, 3, 5** in *red* are secondary palatal processes only. Note that in (*a*), only the maxilla has palatal processes and these are rudimentary. Arrows point to an internal naris.

Common snapper
(a)

Alligator snapper
(b)

Sea turtle
(c)

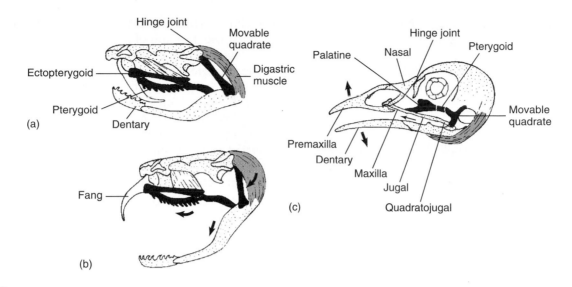

FIGURE 9.21

Cranial kinetism in the palate of snakes (*a* and *b*) and birds (*c*). Arrows indicate direction of movement of components. Digastric muscle (*red*) contraction lowers the mandible.

Cranial Kinesis

Cranial kinesis, or kinetism, is the movement of a functional component of a skull independent of another component. It is made possible by the presence of movable intracranial joints between the two components. Teleosts, lizards, snakes, and birds are especially well endowed with such joints. For instance, most of them raise the upper jaw and palate as a unit independent of the neurocranium when they open their mouth. In some teleosts, even the two sides of the upper jaw can be moved independently. These movements cannot be accomplished by crocodilians, for example, or by frogs, salamanders, or mammals. However, Paleozoic fishes, many labyrinthodonts, early reptiles, and the reptilian ancestors of mammals had kinetic skulls and large numbers of living vertebrates do. As a result of chance mutations, kinetism has probably evolved independently many times during vertebrate evolution.

Cranial kinesis (fig. 9.21) is correlated primarily with food-getting and manipulation of food within the oral cavity. It enables small vegetarian teleosts to scrape algae from underwater objects and it enables plankton-feeding and other small fish to obtain food by suction. (Details are described in chapter 12.) Cranial kinetism enables some squamates, especially vipers, to open their mouth wide enough to swallow prey larger than their own head. A movable quadrate, palate, upper jaw, and some of the bones in front of or above the orbit contribute to the process, along with a hinge joint between the kinetic unit and the rest of the braincase. The palate and temporal region of squamates and some species of birds have become so modified for feeding that the primitive dermatocranial-splanchnocranial complex has become greatly disrupted. The palatal complex illustrated in figure 9.22 exemplifies this disruption. The names applied to the

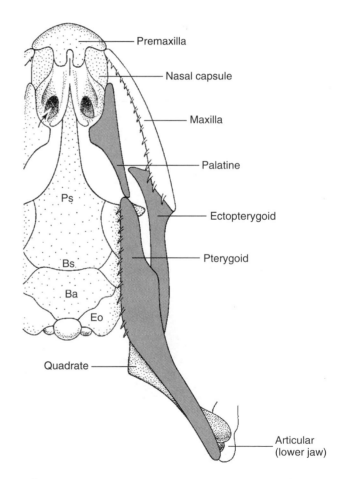

FIGURE 9.22

Kinetic palatal complex of the snake *Tropidonotus.* **Ba,** basioccipital; **Bs** and **Ps,** ankylosed basisphenoid and parasphenoids of primary palate; **Eo,** exoccipital. Vomer bones, not visible, lie anterior to the parasphenoid. The left and right palatine and pterygoid bound a median palatal fissure. Arrow indicates an internal naris.

bones in red are those assigned to a primary palate, but the parasphenoid of the primary palate is not kinetic and there are no other palatal bones.

Table 9.1 contrasts the skulls of early tetrapods with those of modern amphibians and modern reptiles with respect to the palates and other traits.

Birds

The avian skull is fundamentally reptilian but exhibits modifications correlated with flight, altered feeding habits, and a much larger brain. Some of the roofing bones have been lost and the remaining dermal bones are much thinner; in carinate skulls most of the sutures have been obliterated. Therefore, to identify individual bones, one must use the skull of a nestling or ratite.

One might think of the avian skull as comprising two functional regions (fig. 9.23*a* and *b*): at the rear, a solid bony box (neurocranium and dermatocranium) that houses the brain, olfactory organs, eyeball, and equilibratory and hearing complex, and therefore protects all structures needed for input and processing of information; and at the front, a food procuring and handling area, the elongated beak and palate.

The posterior component is highly vaulted, bulging outward and arching upward alongside the greatly enlarged brain. The orbits are situated where they

TABLE 9.1 Skulls of Early Tetrapods Contrasted with Those of Modern Amphibians and Reptiles with Reference to a Few Selected Characteristics

	Early Tetrapods	Modern Reptiles	Modern Amphibians
Neurocranium	Well ossified	Well ossified	Mostly cartilage
	One condyle	One condyle	Two condyles
Primary palate	Complete complement of dermal bones	Relatively complete	Fewer
	Parasphenoid small	Small	Large in urodeles
	Vacuity small	Small	Large in anurans
Secondary palate	None	Partial or complete	None
Temporal fossae	None	Present except turtles	None
Dermal roofing bones	Complete complement	Some reduction	Extensive reduction
Parietal foramen	Present	Present in some	Confined to larvae
Marginal bones	Complete complement	Usually complete	Fewer
Bones ensheathing Meckel's cartilage	Numerous	Numerous	Fewer

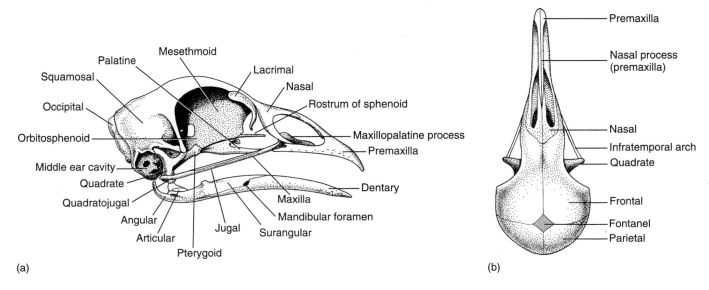

(a)   (b)

FIGURE 9.23

(*a*) Skull of adult chicken, lateral view. (*b*) Skull of young pigeon, dorsal view. The fontanel in (*b*) (*gray*) undergoes intramembranous ossification and sutures are obliterated as pigeon matures.

achieve maximal arcs of visibility, and they reflect the size of the massive eyeballs. The parietal foramen has closed. The neurocranium is incomplete dorsally and, like that of reptiles, well ossified, the only cartilage being in the olfactory capsule and in the mesethmoid component of the interorbital septum. As in reptiles, the neurocranium bears a single occipital condyle.

The skull is modified diapsid, the arch between the superior and inferior temporal fossae having been lost. The resulting fossa is broadly open to the rear and is confluent with the orbit in front. A preorbital fossa, separated from the orbit by the lacrimal bone, is present, as it was in dinosaurs. The infratemporal arch, consisting of a reptilian complement of jugal and quadratojugal bones, is intact but very slender (fig. 9.23).

The kinetic palate resembles that of squamates except that the ectopterygoids have been lost (fig. 9.24). When a bird with a kinetic palate opens its mouth by lowering the lower jaw, the quadrate bone is pushed forward and the motion is transmitted to the upper beak via either a movable palate, a movable zygomatic arch, or a combination of these. The parasphenoid is immobile, being fused to the basisphenoid.

There are a multitude of variations in the structure of beaks and palates and in their anatomic relationships to one another and to the braincase. For this reason, palatal structure has been used as one basis for establishing major avian taxa.

Not all bird skulls are equally kinetic. Birds such as woodpeckers that subject the beak to rough usage usually have minimal kinetism. The ethmoid region of the neurocranium and a narrow arch of roofing bones between the orbits bear the shocks generated when woodpeckers drill in solid wood. Some adaptations of the beak for procuring food are illustrated in figure 2.2.

Mammals

Major features that differentiate mammalian skulls from those of reptiles are emergence of the dentary as the sole bone of the lower jaw, modifications in the temporal region of the dermatocranium, an altered site of articulation of the lower jaw with the braincase, alterations in the secondary palate, and the presence of three bones (ear ossicles) in the middle ear cavity. The skull has become increasingly domed as the cerebral hemispheres have ballooned dorsally, laterally, and caudad.

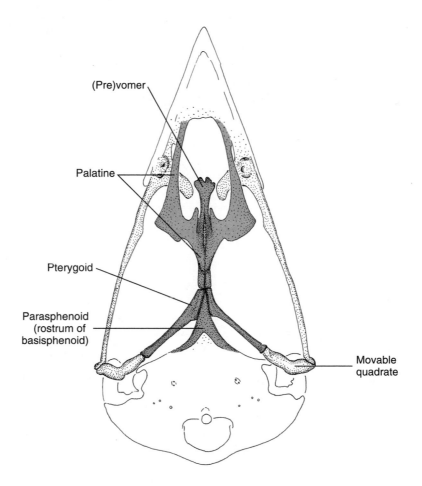

FIGURE 9.24
Palatal complex of a bird, the cotinga, ventral view. The parasphenoid is immovable. The homology of the prevomer is conjectural.

The neurocranium is incomplete dorsally (fig. 9.4), with the result that membranous soft spots, or **fontanels**, can be felt in the heads of newborn human infants until ossification of the roofing bones is complete (fig. 9.25). Fontanels enable the fetal skull to be molded as necessary during delivery through the narrow birth canal. One or more small **bregmatic bones** may ossify in the frontal fontanel in some species, and a single bregmatic bone may develop as an anomaly in human skulls. Paracelsus called this bone the "antiepileptic bone" because he believed it served as a pop-up valve for relieving pressure within the cranial cavity.

Ossification centers in the neurocranium of a fetal pig are shown in figure 9.4. The basioccipital, basisphenoid, and presphenoid bones form a floor on which the brain rests; exoccipitals, alisphenoids, and orbitosphenoids form partial side walls; and a supraoccipital completes, along with the basioccipital and exoccipitals, a bony ring that resembles a vertebra. Ethmoid cartilages and bones house the olfactory epithelium, underlie the olfactory bulbs of the brain, constitute much of the nasal septum (mesethmoid), and are perforated by foramina for olfactory nerve bundles (figs. 9.4, 9.6, and 9.30). Ossification centers in the otic capsules unite solidly to form petrosal (periotic) bones beneath the overgrown temporal lobes of the mammalian cerebral hemispheres. They are often incorporated into a temporal bone complex. Each exoccipital bone bears a condyle inherited from therapsid reptiles. The neurocranial derivatives of a generalized mammalian skull are diagrammed in figure 9.26.

The sequence in mammals of centrumlike basioccipital, basisphenoid, and presphenoid bones and their dorsal wings (the exoccipitals, alisphenoids, and orbitosphenoids) resembles a series of three successive vertebrae that are incomplete dorsally. It inspired Goethe's Vertebral Theory of Origin of the Skull. Goethe's observation was correct, but his conclusion lacked supporting data.

FIGURE 9.25

Two stages in the development of a human skull. (*a*) Intramembranous ossification is under way. The cartilaginous neurocranium (*black*) is incomplete lateral to and above the brain. (*b*) Intramembranous ossification has progressed, but "soft spots" (fontanels) remain where there is no membrane bone. Neurocranial bones are *red*.

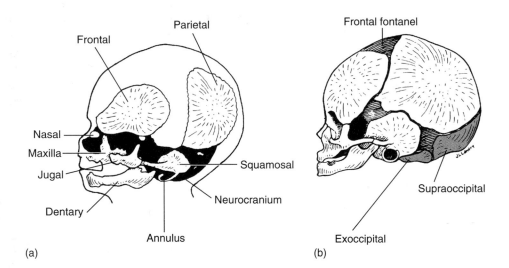

FIGURE 9.26

Chief endochondral bones (*red*) of a generalized mammalian skull. The vomer, palatine, and pterygoid are palatine, and pterygoid are parts of the primary palate. The premaxilla, maxilla, and palatine contribute horizontal processes to the secondary palate. **As,** Pleurosphenoid (alisphenoid); **Ba,** Basioccipital **Bs,** Basisphenoid **C,** Cribriform plate of ethmoid **E,** Ethmoid, perpendicular plate **Ex,** Exoccipital **F,** Frontal **I,** Interparietal **N,** Nasal **Os,** Orbitosphenoid **Ot,** Otic (petrous) **P,** Parietal **Pa,** Palatine **PM,** Premaxilla **Ps,** Presphenoid **Pt,** Pterygoid **So,** Supraoccipital **Sq,** Squamosal **Vo,** Vomer

The dermatocranium is represented in mammals by paired premaxillae, maxillae, jugals (malars), nasals, lacrimals, and squamosals; by paired or unpaired frontals and parietals; and by an unpaired interparietal, all of which are paired in embryos and, in some species, in neonates. A postparietal (dermoccipital) was present in *Homo erectus* (a likely immediate ancestor of humans) and is still present in some human populations, chiefly Mongolians. It is sometimes called an **Inca bone** because it was common among Inca Indians (fig. 9.27). Premaxillae are not identifiable in adult human skulls because they unite with the maxillae early in embryonic life, a discovery made by the poet-biologist Goethe, much to his "unspeakable joy." The zygomatic arch varies from massive to slender, depending on the magnitude of the force exerted on it by the masseter muscle. It has become extremely delicate, or even incomplete, in some insectivores.

The temporal complex of mammals consists of numerous components of intramembranous and endochondral origin (fig. 9.28). The **squamous portion** is the squamosal of lower tetrapods. The **tympanic portion,** new in mammals, surrounds the middle ear cavity with its contained middle ear ossicles and sometimes expands to form a **tympanic bulla** (fig. 9.29). The **petrous portion** is the ossified otic capsule (fig. 9.30). Associated with the tympanic portion and surrounding the tympanic membrane (eardrum) is a bony ring, the **annulus tympanicus.** It is derived, according to evidence from embryonic opossums, from the angular bone of reptiles. (The annulus of anurans is not a homologue. It is thought to be derived from the palatoquadrate cartilage.) A **mastoid portion** of endochondral origin is new in mammals (fig. 9.28). Although the tympanic and petrous portions are separate bones in some mammals, they frequently unite to form a **petrotympanic bone,** as in rabbits, and the petrotympanic may unite with the squamosal to form a **temporal bone.** A dorsal segment of the hyoid arch skeleton coalesces in some species with the temporal bone to become a **styloid process** (fig. 9.28). The composite nature of the temporal bone is diagrammed in figure 9.31.

The squamosal bone has become a new site of articulation of the lower jaw with the skull in mammals. The shift from the quadrate-articular joint of other vertebrates is correlated with expansion of the dentary bone during mammalian evolution (fig. 9.39).

Air-filled **cranial sinuses** are often found within the maxilla, frontal, sphenoid, and ethmoid bones (fig. 9.30**F** and **S**). The frontal sinuses of sheep and goats extend into the horns. When male goats butt heads at speeds of up to 35 miles (60 km) per hour as part of the mating ritual, the arched walls of the sinus act as a bony brace that shunts shock waves to the vertebral column and away from the brain via the bones of the skull. Inflammation of the sinuses (sinusitis) is a common ailment in human beings.

Of the primary palate inherited from reptiles, an unpaired vomer lies at the base of the nasal septum (fig. 9.30). The septum consists of a **mesethmoid** bone and more or less cartilage. A **nasal process of the palatine** develops in the lateral wall of the nasopharynx where it contributes to the wall of the orbit, and a **palatal process** of the palatine bone contributes to the secondary palate. The pterygoids are reduced to small, winglike, **pterygoid processes** of the sphenoid complex. They serve as the anatomic origin of the powerful pterygoid muscle, a derivative of the adductor mandibulae of

FIGURE 9.27

Postparietal (Inca) bone in a human skull from the Aleutian Islands. **O,** Occipital; **P,** parietal; **PP,** postparietal.

Courtesy William S. Laughlin.

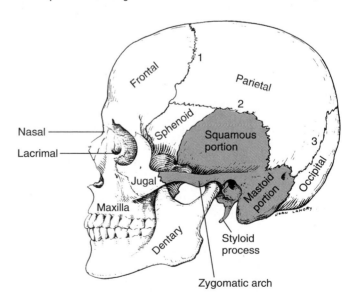

FIGURE 9.28

Temporal bone of human skull. **1, 2, 3:** coronal, squamosal, lambdoidal sutures, respectively.

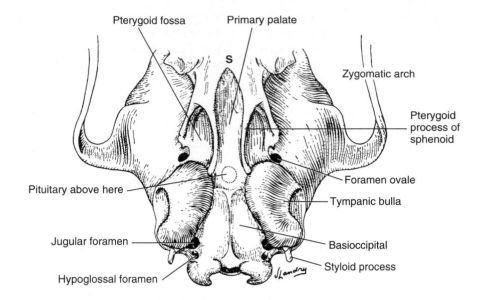

FIGURE 9.29

Hamster skull, caudal part, ventral view. **S,** secondary palate. The primary palate is the roof of the nasopharynx.

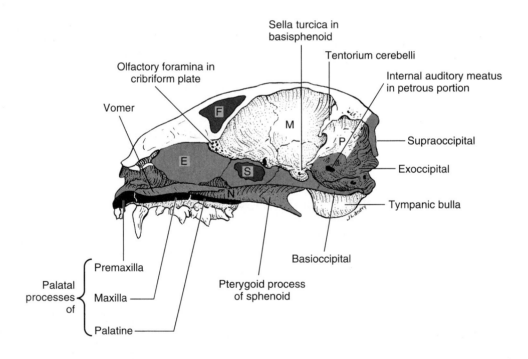

FIGURE 9.30

Sagittal section, cat skull, bony part of secondary palate in *black*. **E,** mesethmoid (perpendicular plate of ethmoid part of the nasal septum); **F,** frontal sinus in frontal bone; **M,** middle cranial fossa housing cerebral hemispheres; **N,** nasal passageway, *red;* **P,** posterior cranial fossa housing cerebellum; **S,** sphenoidal sinus in presphenoid. *Gray* designates ethmoid, sphenoid (pre- and basi-), otic (petrous), and occipital components of the neurocranium. Vomer and mesethmoid constitute the nasal septum.

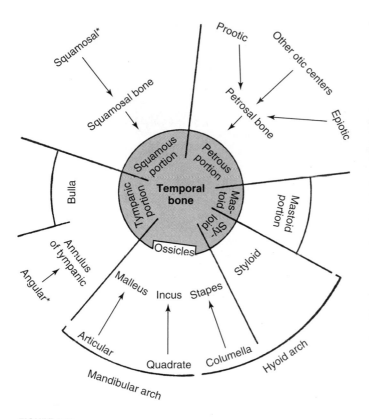

FIGURE 9.31

The multiple nature of the temporal bone of mammals. Note reduction in number of separate elements from the condition in reptiles (outer circle) to mammals (other circles). The two dermal elements have asterisks. The mastoid portion and tympanic bulla are mammalian innovations.

TABLE 9.2   Chief Foramina and Bony Canals in the Mammalian Skull

Foramen or Canal	Transmits
Incisive (anterior palatal)	Duct from oral cavity to vomeronasal organ
Infraorbital	Infraorbital branch of maxillary nerve and of internal maxillary artery
Jugular	Nerves IX, X, and XI; jugular vein
Lacerum	Internal carotid artery
Magnum	Spinal cord, vertebral arteries, and spinal roots of XI and XII
Mandibular	Inferior alveolar artery and nerve
Mental	Mental artery and nerve
Olfactory	Filia olfactoria
Optic	Optic nerve and ophthalmic artery from carotid plexus
Orbital fissure	Nerves III, IV, and VI and ophthalmic division of nerve V; in rabbits also transmits maxillary division of nerve V
Ovale	Mandibular division of nerve V
Rotundum	Maxillary division of nerve V; foramen is missing in rabbits
Stylomastoid	Exit of facial canal; transmits nerve VII
Facial canal	Nerve VII
Canal for auditory tube	Auditory tube enters tympanic bulla just lateral to foramen lacerum
Hypoglossal canal	Opens in caudal wall of jugular foramen; transmits nerve XII
Internal auditory meatus	Nerve VII to facial canal and nerve VIII to inner ear
Nasolacrimal canal	Nasolacrimal duct from orbit to nasal cavity

lower vertebrates. The parasphenoid and ectopterygoids have been lost. The mammalian secondary palate is illustrated in figure 9.18. The bony part (hard palate) consists of palatal processes of the premaxilla, maxilla, and palatine bones. Caudal to the bony part is a membranous soft palate (figs. 12.3 and 13.13*a*).

Mammals have three pairs of scroll-like **turbinal bones (nasal conchae)** in the medial walls of the nasal passageways (fig. 13.13*a*). The inferior (maxillary) concha is an independent scroll. The middle and superior conchae are appendages of the ethmoid bone. The two lower conchae are covered with **nasal epithelium,** the venous plexuses of which warm the air en route to the lungs. The superior concha is covered with **olfactory epithelium.** The scroll-like structure of the conchae increases the surface area of these epithelia. Most reptiles other than turtles have one pair of turbinals. Birds have two.

Unique in mammals is the presence of the ossified posterior tips of the palatoquadrate and Meckel's cartilages in the middle ear cavity, where they now serve as ear ossicles along with the columella (stapes). How they got there will be discussed shortly.

The chief foramina and bony canals in the mammalian skull are listed in table 9.2.

Reduction in Number of Cranial Bones during Phylogeny

The number of individual bones, especially membrane bones in the mammalian skull, has tended to be reduced during the evolution of tetrapods. With reference to figure 4.3, any tetrapod group at the end of an arrow has fewer bones in the skull than the group preceding it in the phylogenetic line. Labyrinthodonts had fewer than crossopterygians, cotylosaurs had fewer than labyrinthodonts, modern reptiles have fewer than cotylosaurs, and mammals have fewer than mammal-like reptiles. Modern amphibians have fewer than their ancestors, the labyrinthodonts. This generalization does not mean that *modern* reptiles have fewer bones than *modern* amphibians. In fact, reptiles have more. But then, modern reptiles were not derived from modern amphibians.

The reduction is a result of fusion of adjacent embryonic ossification centers, phylogenetic loss of ossification centers, and obliteration of sutures in young

TABLE 9.3   Reduction in Number of Dermal Bones Investing Meckel's Cartilage in Early Vertebrates Contrasted with Later Ones

Fishes			Tetrapods				
			Primitive	Modern			
Primitive	Crossopterygians	Modern	Labyrinthodonts	Reptiles and Birds	Amphibians	Mammals	
Dentary	Dentary	Dentary*	Dentary	Dentary	Dentary	Dentary
Angular	Angular	Angular†	Angular	Angular	Angular‡	
Surangular	Surangular		Surangular	Surangular		
Infradentary§	Splenial		Splenial	Splenial	Splenial‡	
Infradentary	Coronoid		Coronoid	Coronoid		
Infradentary	Prearticular	Derm-articular‖	Prearticular			
Infradentary			Intercoronoid			
Infradentary			Precoronoid			
Infradentary			Postsplenial			

Primitive forms had a larger number of bones than modern ones. Reptiles have retained more of the primitive elements than other modern tetrapods.

*Dentary incorporates mentomeckelian of endochondral origin in some teleosts.
†May be absent.
‡Sometimes incorporated in an angulosplenial.
§Variable number.
‖May include articular of cartilage origin.

animals. Reduction in the number of membrane bones in the mandible during phylogeny illustrates this trend (table 9.3).

Membrane bones often unite with adjacent replacement bones, giving rise to a single bone with a dual history. Postfrontals and supratemporals sometimes unite with replacement bones of the otic capsule to form sphenotic and pterotic bones; the squamosal unites with otic and other elements to contribute to a temporal bone. The mammalian interparietal, a membrane bone, may unite with the supraoccipital. Unions such as these have reduced the number of bones in mammalian skulls.

VISCERAL SKELETON

The visceral skeleton, or **splanchnocranium,** is the skeleton that develops within the pharyngeal arches. In fishes, therefore, it is the skeleton of the jaws and gill arches. In tetrapods, this skeleton has become modified to perform new functions on land.

The blastemas that give rise to the visceral skeleton come from neural crests, and they first produce cartilage. Later, the cartilage may be partly or wholly *replaced* by bone. Only in the first arch is it *ensheathed* by dermal bone. We will look first at a shark, in which no bone forms and in which the visceral skeleton is seen in its primitive role, that of supporting the jaws and gills.

Sharks

Squalus acanthias is a generalized vertebrate except that it lacks bones. The visceral skeleton consists of cartilages in each pharyngeal arch (fig. 9.1) and median **basihyal** and **basibranchial cartilages** in the pharyngeal floor (fig. 9.32). The skeleton in each arch conforms fairly closely to a basic pattern (fig. 9.33a), and all but the first and last in *Squalus* support gills. The first arch and, to a degree, the second are modified for procuring food.

The skeleton of the mandibular arch consists of two cartilages on each side, a **palatoquadrate cartilage** dorsally and **Meckel's cartilage** ventrally (fig. 9.33b). The left and right palatoquadrate cartilages meet in the midline dorsally to form the upper jaw, and Meckel's cartilages meet in the midline ventrally to form the lower jaw. Slender **labial cartilages** of unknown significance (not illustrated) extend from the angles of the mouth into a position where lips would be. The skeleton of the hyoid arch consists of paired **hyomandibular cartilages** dorsally and gill-bearing **ceratohyals** laterally (figs. 9.1 and 9.33c).

At the angle of the mouth, Meckel's cartilage and the palatoquadrate cartilage articulate with one another and with the hyomandibular cartilage in a movable joint united by ligaments (fig. 9.1). The dorsal end of the hyomandibula is bound by ligaments to the otic capsule and suspends the jaws and the entire branchial skeleton from the neurocranium. This is hyostylic jaw suspension.

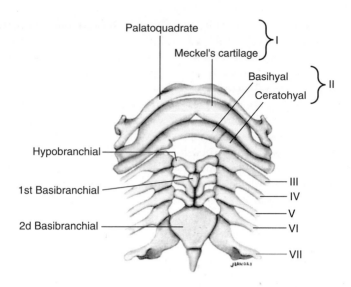

FIGURE 9.32

Visceral skeleton of *Squalus acanthias,* ventral view. **III** to **VII,** Ceratobranchial cartilages of the third to seventh pharyngeal arches. See also fig. 9.1.

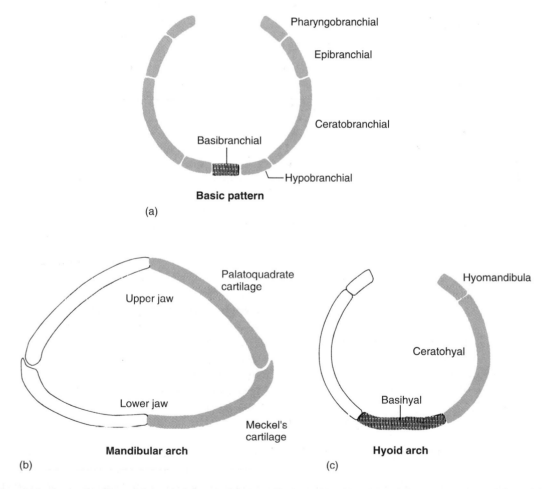

FIGURE 9.33

Skeletal components of a generalized branchial arch, (*a*), and of the mandibular and hyoid arches of *Squalus acanthias,* (*b*) and (*c*). The basihyal of sharks is paired in embryos.

Jaw Suspension in Fishes

The jaw-hyoid complex of fishes must necessarily be braced against some support. The nearest is the braincase. In most elasmobranchs (fig. 9.1) and most bony fishes, the hyomandibular cartilage is braced against the otic capsule, and the posterior end of the palatoquadrate cartilage is braced against the hyomandibular. This condition is known as **hyostylic jaw suspension,** or **hyostyly.** A more primitive condition is seen in some older sharks in which the hyomandibula and one or more processes of the palatoquadrate are braced independently against the braincase. This is known as **amphistyly.** A third variant, and perhaps the oldest, found in a very early shark and, with modifications, in chimaeras and lungfishes, is known as **autostyly.** Here the palatoquadrate is attached to the neurocranium, and the hyomandibula plays no role in jaw suspension. The structural details of jaw suspension among fishes, ancient and modern, vary greatly, and an expanded terminology has been devised for these variants. The anatomic relationship of the palatoquadrate cartilage to the hyomandibula and to the neurocranial-dermatocranial complex is correlated with the diet of the species.

Bony Fishes

The visceral skeleton of bony fishes resembles that of sharks in basic morphology (fig. 9.34). The major differences are that the embryonic palatoquadrate and Meckel's cartilages in bony fishes become invested by membrane bones during development, the hyoid skeleton may consist of a greater number of segments, and the embryonic cartilages of the gill arches are ultimately replaced by

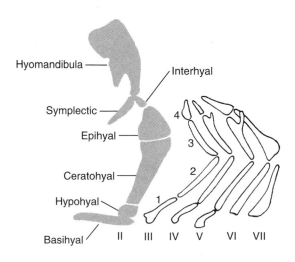

Branchial skeleton
Salmon

FIGURE 9.34

Visceral skeleton of a salmon, jaws removed. Hyoid cartilages are in *red.* The basihyal is unpaired. **1** to **4,** hypobranchial, ceratobranchial, epibranchial, and pharyngobranchial elements of the third arch. The hyomandibula articulates with the otic capsule.

bone. The site of articulation between the upper and lower jaws of bony fishes and of all other vertebrates except mammals is the same as in cartilaginous fishes; the posterior ends of the palatoquadrate cartilages, whether or not ultimately replaced by bone, articulate throughout life with the posterior ends of Meckel's cartilages, whether or not these ends are replaced by bone. To this day, there has been no change in this Paleozoic relationship in gnathostomes other than in mammals.

To gain a general impression of the fate of the embryonic **palatoquadrate cartilage in bony fishes,** the reader is urged to look again at figure 9.10. In most bony fishes, (1) this cartilage becomes ensheathed by two membrane bones, a **premaxilla** and **maxilla**; (2) the portion in the roof of the oropharyngeal cavity, the palatal portion, develops two or three sites of endochondral ossification including **palatine** and **ectopterygoid** bones (fig. 9.9*a*); (3) the posterior tip of the palatoquadrate cartilage ossifies to become a **quadrate** bone. The entire complex of ensheathing and replacement bones unites with the dermatocranium to one degree or another, depending on the species.

Meckel's cartilages ossify at their caudal ends to become **articular bones.** The remainder of the cartilage becomes invested by several membrane bones including a **dentary** and **angular.** The mandible of the more primitive *Amia* consists of **dentaryinfradentaries** (a single composite bone), angulars, surangulars, **prearticulars,** and four pairs of **coronoids.** All but the angular and surangular bear teeth.

The replacement bones of the **hyoid arch** of a relatively primitive teleost are shown in figure 9.34. The **symplectic** and **interhyal** are ossification centers in the hyomandibular cartilage, and the **epihyal** is an ossification center in the ceratohyal. The symplectic usually articulates with the quadrate bone or with the quadrate and lower jaw, where it participates in cranial kinetism. The hyomandibula has been lost in lungfishes.

The skeleton of a typical bony **gill arch,** like the gill arches of a shark, consists of four segments, **hypobranchial, ceratobranchial, epibranchial,** and **pharyngobranchial,** the latter in the pharyngeal roof. Only the middle two segments bear gills. The pharyngeal borders of the branchial arches are covered with bony dermal plates having flattened or pointed denticles. The size and arrangement of the plates and denticles are correlated with the nature of the diet and the method of food handling in the pharynx.

Feeding Mechanisms in Bony Fishes

Early voracious jawed fishes had wide mouths, the hinge of the jaws was far back under the skull, and the upper jaw was fused with the braincase and therefore incapable of independent movement. As a result, in feeding, the mandible was simply lowered and then snapped shut on prey in the manner of sharks. The portion of the prey that was inside the orobranchial chamber was swallowed.

With the appearance of the type of kinetism seen in teleosts, more complex mechanical means of procuring food became possible. As an accompaniment of these changes, the large adductor mandibulae muscle inserted farther and farther forward on the mandible, an additional array of jaw muscles developed, and the mouth became narrower and more oval. The effect of these changes was that the jaws of the most highly specialized teleosts can be thrust forward and employed in feeding by inertial suction. The mechanism works equally well in herbivorous and carnivorous fishes, although not in precisely the same manner.

Feeding by inertial suction in the herbivorous freshwater teleost *Petrotilapia* has been described by Liem. In its natural habitat this fish feeds on algae attached to submerged objects; in an aquarium it feeds equally well on commercial fish food. Abduction of the hyomandibula expands the orobranchial chamber, the lower jaw falls open, and the opercular cavity is then expanded, in that sequence, creating forward suction. As a result, zooplankton in the water immediately ahead are slowly drawn into the mouth. The lower jaw is then raised, the upper jaw remains briefly protruded, and the hyomandibula is adducted, which compresses the chamber. The two phases of the pharyngeal pump, expansion and compression, occupy together about 600 milliseconds. Activity of the muscles was recorded by eletromyography. By manipulating the hyoid skeleton and jaws appropriately, the protruded jaws can be directed upward for collecting food on the surface or downward for food on the substrate.

A similar mechanism operates in predatory fishes. When the hyomandibula is drawn forward, the symplectic forces the kinetic elements of the upper jaw and palate to slide forward, and the premaxillae and dentary bones close on the prey. A process of the maxilla usually becomes attached by a ligament to the dentary bone, so that movement of one participates in displacing the other. The details of the anatomic relationships and the mechanics of their operation are as numerous as the taxa that exhibit them. It should be pointed out, however, that snapping and biting has not gone out of style. One half or more of the teleosts have nonprotrusible jaws.

The theory has been advanced that vertebrate jaws are former gill arches that became modified, along with the hyoid arch, for predatorial feeding as an alternative to filter feeding, which was the ancestral method of obtaining nourishment. A student examining a shark and its visceral arches will see the basis for this idea. In fact, (1) the hyoid arch does contain a demibranch in *Squalus*, (2) the spiracle is a pharyngeal slit as are all gill slits, and (3) the nerves and arteries of the mandibular and hyoid arches are the most anterior of a series that is repeated in each gill arch, as will be seen in later chapters. The theory is an interesting speculation. It is unlikely that it will ever be validated or falsified.

Cyclostomes

The visceral skeleton of cyclostomes is totally unlike that of jawed fishes (fig. 9.35). There are no identifiable palatoquadrate cartilages, Meckel's cartilages, hyoid cartilages, or branchial arch cartilages. A V-shaped **lingual cartilage** (dental plate) bearing horny teeth and located in the floor of the buccal cavity moves in and out of the buccal funnel, serving as a rasping tonguelike organ. It is operated by protractor and retractor muscles that are attached to a narrow, segmented, immobile basal plate cartilage lying immediately beneath it (not illustrated).

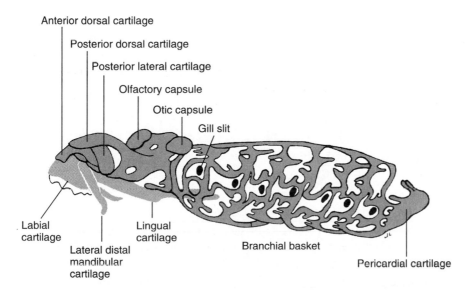

Neurocranium and visceral skeleton
Lamprey

FIGURE 9.35

Neurocranium, branchial basket, and feeding apparatus (*red*) of a lamprey. The olfactory capsule is a midline structure; otic capsules are paired. The lingual cartilage is also referred to as a tongue.

There is no evidence that any of the cartilages associated with the feeding apparatus are derived from ancestral visceral arches. The lingual cartilage, or "tongue," and associated cartilages are correlated with the unique method of feeding characteristic of cyclostomes. The rest of the pharyngeal skeleton consists of a branchial basket, less well developed in hagfishes, which is simply a fenestrated framework of cartilage immediately under the skin. The caudal end of it in lampreys is located at the level of the heart.

Tetrapods

With life on land, the visceral skeleton underwent profound changes from that of jawed fishes. Some previously functional parts were deleted, and those that persisted performed new and sometimes unexpected functions. It is these modifications, which took place over the course of millions of years, that we will examine in the pages that follow. But first, we will look at some of these changes in a frog tadpole, where they occur during the course of two or three days!

Larval frogs have six pairs of visceral cartilages, the last four of which bear gills (fig. 9.36a). The cartilages of the gill arches meet ventrally in a **hypobranchial plate.** During metamorphosis (fig. 9.36b) the third, fifth, and sixth visceral cartilages regress. The hypobranchial plate enlarges and, along with the first basibranchial, becomes incorporated into the body of the hyoid. The latter there-

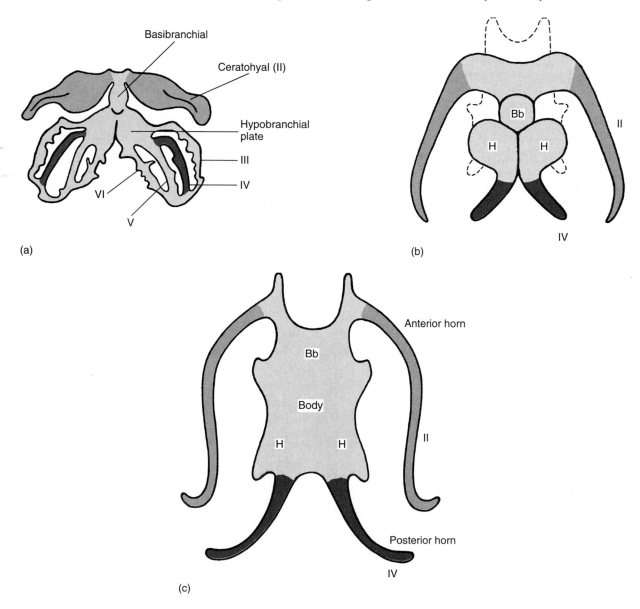

FIGURE 9.36

Changes in hyoid and branchial skeleton during metamorphosis in a frog. (*a*) Larval skeleton. **II–VI,** skeleton of second through sixth visceral arches. **III–VI,** bear gills. (*b*) Late metamorphosis. (*c*) Hyoid of a young frog. **Bb,** basibranchial contribution. **H,** hypobranchial contribution. Dashed line in (*b*) outlines new cartilage. *Red* shades designate homologous horns.

after is a broad cartilaginous and bony plate in the buc-copharyngeal floor. The ceratohyal cartilage of the second arch becomes a slender anterior horn, or cornu, of the hyoid apparatus, and the cartilage of the fourth arch becomes a posterior horn. As a result of these and other changes, a visceral skeleton initially adapted for branchial respiration becomes converted, in the span of a few days, to one adapted for life on land. Among its new roles is serving as an anchorage for the muscular tetrapod tongue.

The components of the branchial skeleton of a larval salamander are illustrated in figure 9.37. The number in urodeles is usually reduced at metamorphosis, as in frogs. The branchial skeleton of *Necturus,* a perennibranchiate urodele, is illustrated in figure 9.41*a*. In the pages that follow we will trace the adaptations that have taken place in the jaws and branchial arches of tetrapods and especially of amniotes, which abandoned a free-living, larval stage along with the gills that helped sustain it.

Fate of the Palatoquadrate and Meckel's Cartilages

We have seen in bony fishes that the embryonic palato-quadrate cartilages become ensheathed laterally by pre-maxillae and maxillae and ventrally by dermal bones of the primary palate; that the palatal portions may contribute in one fish species or another to the otherwise dermal primary palate; and that the posterior ends of the palatoquadrate cartilages usually ossify to become quadrate bones at the hinge of the jaws. Except for contributing to the primary palate, the story is the same for amphibians, reptiles, and birds. The quadrate remains as the site of articulation of the cranium with the lower jaw in all tetrapods below mammals. In that location in squamates and birds, it is part of the kinetic mechanism of the skull (fig. 9.21). The quadrate bone in mammals becomes the **incus,** a middle ear ossicle.

The embryonic Meckel's cartilages of amniotes may continue to grow and become a prominent core of carti-lage within the adult mandible, as in turtles and croco-dilians (fig. 9.38). More often, little remains in adults. However, the cartilages become ensheathed by dermal bones that included, primitively, on each side, a **dentary, angular, surangular, splenial,** one or more **coronoids,** and a **prearticular** (dermarticular). Modern amniotes have a smaller number, and mammals have only the dentary. As in bony fishes, the posterior end of each Meckel's cartilage ossifies to become an **articular bone** at the hinge of the jaws, except in mammals. In mammals it becomes the **malleus,** another middle ear ossicle.

Expansion of the Dentary and a New Jaw Joint in Mammals

As explained earlier, temporal fossae enabled the adductor mandibulae muscles of synapsid reptiles to enlarge, subdivide, and acquire anatomic origins on the temporal region of the skull and on the zygomatic arch. The increased mass of these new muscles was accompanied by expansion of the dentary bone and formation of a ramus on which the temporalis muscle inserted (figs. 9.39 and 11.25*c*). Eventually the other dermal bones of the mandible disappeared, the articular bone became an ear ossicle, and mammals were left with a mandible consisting solely of two dentary bones.

Expansion of the dentary brought it close to the **squamosal,** and there it eventually established a new

(a)

(b)

FIGURE 9.38

Meckel's cartilages (*red*) in lower jaw of embryonic lizard. (*a*), medial view; and within mandible of adult sea turtle, (*b*). The cartilage of the lizard is becoming ensheathed by dermal bones and will disappear with further development of the jaw. Its posterior end will become the articular bone. The dermarticular is partly endochondral in (*a*).

FIGURE 9.37

Hyoid and branchial skeleton of *Ambystoma* larva. **CB,** ceratobranchials; **CH,** ceratohyal; **HB,** hypobranchials, *black;* **HH,** hypohyal; basibranchials, **BB,** *white.*

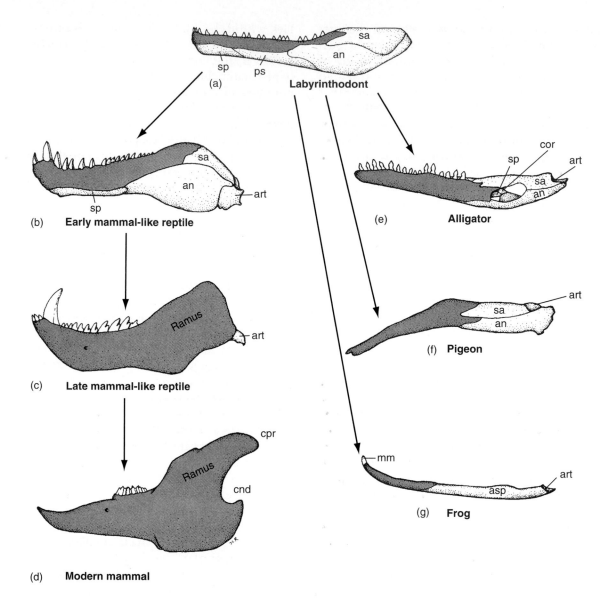

FIGURE 9.39

The mandibles of tetrapods, left lateral views, (a) to (d) Evolutionary stages leading to modern mammals. The dentary (*red*) became larger, other bones were reduced and finally lost. (e) to (g) Mandibles of three modern tetrapods. All are dermal bones except mentomeckelian and articular. **an,** angular; **art,** articular (cartilage in frog); **asp,** angulosplenial; **cnd,** condyloid process for articulation with squamosal; **cor,** coronoid; **cpr,** coronoid process; **mm,** mentomeckelian; **ps,** postsplenial; **sa,** surangular; **sp,** splenial; (b) Pelycosaur. (c) Late therapsid. (d) Rabbit.

site for articulating with the skull (fig. 9.28). For a while the lower jaw of early mammals articulated with the skull at two sites, the reptilian one (articular against the quadrate) and the new one (condyloid process of the dentary against the squamosal (fig. 9.39d). The two existed side by side in *Eozostrodon,* one of the oldest known mammals of the Triassic. Eventually the articular and quadrate were "captured" by the middle ear cavity and only the new articulation remained. The shape, slant, and relationships of the condyloid process of the mandible differ with the demands made on it by the feeding habits of the various mammalian orders.

Ear Ossicles from the Hyomandibula and Jaws

It will be recalled that the hyomandibular cartilage of sharks is interposed between the quadrate cartilage and the otic capsule and that the latter houses the inner ear (fig. 9.1). In an autostylic jaw, the hyomandibular bone can be dispensed with, and that is what happened to it in dipnoans. It persisted in tetrapods, the jaws of which are also autostylic, but became dissociated from the quadrate and attached, instead, to what was destined to become a tympanic membrane. It continued to abut against the otic capsule, which houses the inner ear, and it became surrounded by a portion of the first pharyngeal pouch, now known as the middle ear cavity. The hyomandibula thereby became the first ear ossicle, or **columella,** of tetrapods. Thereafter it conducted sound waves from the eardrum, across the middle ear cavity, to the inner ear. Figure 9.12*b* shows a segment of the hyoid cartilage (hy II) and columella (col II) still interposed between the quadrate (qd I) and the otic capsule (pro) in a primitive urodele whose middle ear cavity is vestigial and who lacks a tympanic membrane.

When the dentary bone of therapsid reptiles acquired an articulation with the squamosal, the articular and quadrate bones were freed to perform a new function or to disappear. There is no doubt that the caudal end of Meckel's cartilage becomes the **malleus** of mammals. The embryonic Meckel's cartilage in one mammalian species after another can be seen projecting into the area where middle ear cavitation is proceeding (fig. 9.40), and its cartilaginous posterior tip can be seen to separate, ossify, and become the malleus.

The evidence that the quadrate bone of synapsid reptiles became the **incus** of the middle ear of mammals is less direct, but the facts are these: (1) the articular and quadrate bones have been articulating in a diarthrosis since the time of the earliest jawed fishes, and were doing so in therapsid reptiles; (2) the articular has separated from the lower jaw and is now the malleus; (3) the quadrate has disappeared from the upper jaw; and (4) the articular still articulates in a diarthrosis with a bone (the incus), that, if it is not the quadrate, is of unknown homology (fig. 17.12). In the nineteenth century, C. Reichert proposed the theory that mammalian ear ossicles are derived from the jaws. Early in the twentieth cen-

tury, the theory was modified by E. Gaupp to include the origin of the stapes from the dorsal tip of the hyoid arch. Subsequent knowledge from paleontology and embryology has strengthened the Reichert-Gaupp theory, and it is now generally accepted (see E. Jarvik, 1980, 2:161).

Amniote Hyoid

The hyoid of amniotes consists of a **body** in the pharyngeal floor just anterior to the larynx and of two or three **horns (cornua)** in the pharyngeal walls (fig. 9.41*c–h*). We have seen that the hyoid of anurans is derived at metamorphosis from basibranchial and hypobranchial cartilages, from the skeleton of the hyoid arches, and from one of the larval gill-bearing arches (fig. 9.36). In amniotes it is derived from homologous anlagen. The cartilages of the second arch become anterior horns, and those of the third arches and sometimes the fourth become additional horns.

In lizards and birds (fig. 9.41*c* and *e*), an elongated bony process, the **entoglossus,** extends from the body forward into the long darting tongue. In some male lizards a similar process extends caudad into the dewlap, or gular pouch. Snakes have no hyoid. In fact, the entire branchial skeleton is vestigial.

The entoglossus and caudal horns of woodpeckers, in association with the tongue, is a remarkable tool for impaling grubs (figs. 9.42 and 12.6*a*). From the base of the entoglossus, two long, flexible caudal horns loop caudad then dorsad around the occipital region of the skull and forward under the scalp all the way to the **lore,** the space between the eyes and the bill, where both horns dip into the right nasal passageway. They then continue forward in the nasal canal a short distance farther and terminate. The termination of both horns in a single nasal passageway is surprising and lacks a functional explanation. When impaling grubs, the horns are straightened by an accelerator muscle that shoots the tongue into the prey. Elastic recoil of the horns immediately returns the tongue to the mouth with impaled food.

The hyoid of mammals (figs. 9.43 and 9.44) has cranial horns from the second arch and caudal horns from the third (fig. 9.41*g,h*). In dogs and cats, the cranial horns are the longer ones **(greater horns)** and are composed of four segments. The dorsalmost, or **tympanohyal** (fig. 9.43), ends in a notch in the tympanic bulla. (The stapes, which became detached from the tympanohyal during phylogeny, is inside the bulla.) In rabbits, the anterior horns are the short ones **(lesser horns),** and the tympanohyal is represented by a slender stylohyal bone embedded in the tendon of insertion of the posterior belly of the stylohyoid muscle. In humans, too, the anterior horns, equivalent to ceratohyals, are the lesser horns; an unossified stylohyoid ligament represents epihyal and stylohyal segments; and the tympanohyal is attached to the temporal bone where it is known to anatomists as the **styloid process** (figs. 9.28 and 9.45).

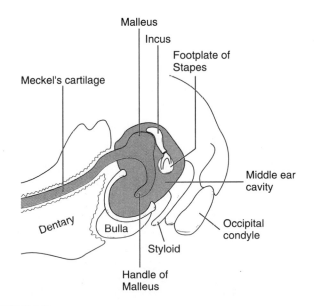

Malleus
Incus
Footplate of Stapes
Meckel's cartilage
Middle ear cavity
Dentary
Bulla
Occipital condyle
Styloid
Handle of Malleus

FIGURE 9.40

Caudal end of Meckel's cartilage surrounded by the developing middle ear cavity in a mammalian embryo.

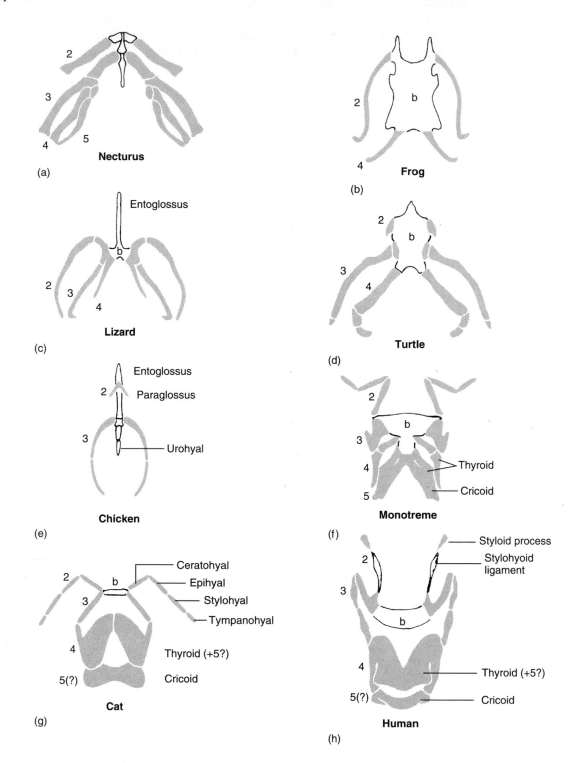

FIGURE 9.41

Skeletal derivatives of the second through fifth pharyngeal arches in selected tetrapods. **b,** Body of hyoid. **2** to **5,** Derivatives of arches **2** through **5** (*red*). The projections from the body in (*b*) through (*h*) are horns of the hyoid. In (*e*), the body of the hyoid extends forward into the tongue as an entoglossus to which are attached two paraglossals **(2).**

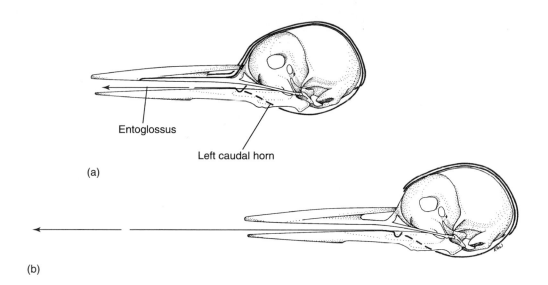

Entoglossus

Left caudal horn

(a)

(b)

FIGURE 9.42

Caudal horns and entoglossus of the hyoid of a hairy woodpecker. (*a*) Tongue retracted. (*b*) Tongue extended. The paired caudal horns lie close together and appear as one when extended.

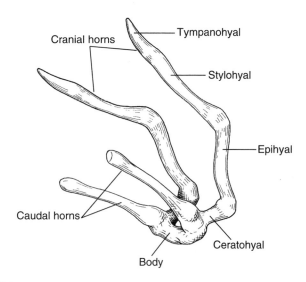

Cranial horns

Tympanohyal

Stylohyal

Epihyal

Caudal horns

Ceratohyal

Body

FIGURE 9.43

Hyoid of a dog. Caudal horns articulate with the cranial horns of the thyroid cartilage (fig. 13.12). Cranial horns terminate at the tympanic bullae.

The hyoid anchors the highly mobile tongue of tetrapods, is the skeleton for the buccopharyngeal pressure pump used in respiration in anurans, provides attachment for some of the extrinsic muscles of the larynx, has subtle effects on lower jaw movements, and is the site of attachment of muscles that participate in swallowing. The associated hypobranchial and branchiomeric muscles of amniotes approach the hyoid from many directions—lower jaw, larynx, sternum, clavicle, the temporal region of the skull, and elsewhere. These muscles stabilize the hyoid in a single position or move it forward, backward, up, or down.

Laryngeal Skeleton

Nearly all tetrapods have **cricoid** and **arytenoid cartilages** or replacing bones (fig. 13.11), and mammals have **thyroid cartilages** or **bones** in addition (figs. 9.44, 9.45, and 13.13*b*). Thyroid cartilages arise from mesenchyme of the fourth pharyngeal arch and, perhaps, the fifth. Cricoid and arytenoid cartilages are probably products of the fifth arch. Since the caudal end of the pharyngeal arch series has been subject to reduction during evolution, it is not surprising that problems are encountered in relating the caudalmost laryngeal cartilages to specific arches. The laryngeal skeleton is discussed in more detail in chapter 13.

PERSPECTIVE

It is evident that the visceral skeleton is a functional complex that was associated primitively with feeding and branchial respiration. In tetrapods it has been modified in part for transmission of airborne sounds, for attachment of tongue muscles, and for support of vocal cords. These adaptations illustrate how mutations alter ancient structures and perhaps produce new functions. Some of the skeletal derivatives of the visceral arches in representative living vertebrates are listed in table 9.4.

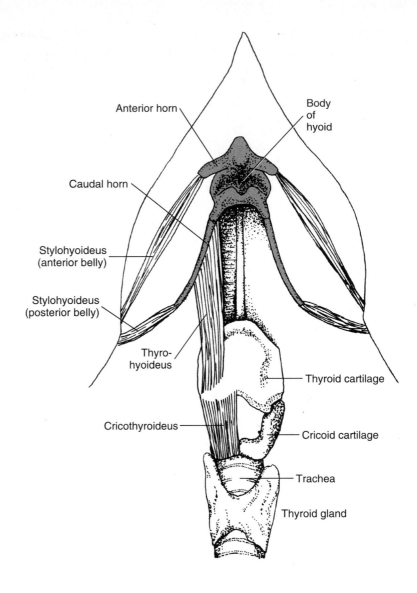

FIGURE 9.44

Hyoid (*red*), larynx, and associated structures of a rabbit, ventral view.

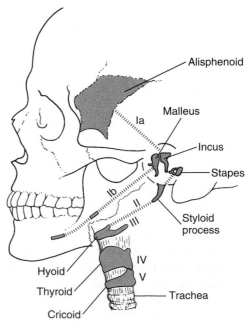

FIGURE 9.45

Derivatives of the human visceral skeleton (*red*). **Ia,** Broken line connects derivatives of the palatoquadrate cartilage. **Ib,** Broken line connects vestiges and derivatives of Meckel's cartilage. **II,** Broken line from stapes to styloid process to lesser horn of hyoid connects derivatives of hyoid arch. Section between lesser horn and styloid process is site of a stylohyoid ligament. **III–V,** Derivatives of the third, fourth, and fifth arches. **III** is at tip of the posterior (greater) horn.

Visceral skeleton

TABLE 9.4   Skeletal Derivatives of Pharyngeal Arches in Sharks and *Approximate* Homologues in Bony Vertebrates

Arch	Shark	Teleost	Necturus	Frog	Reptile and Bird	Mammal
I	Meckel's cartilage Pterygoquadrate	Articular* Quadrate Epipterygoid Metapterygoid	Articular Quadrate Palatal cartilage	Articular Mentomeckelian[†] Quadrate Annulus tympanicus (?)	Articular Quadrate Epipterygoid	Malleus Incus Alisphenoid
II	Hyomandibula Ceratohyal Basihyal	Hyomandibula Symplectic Interhyal Epihyal Ceratohyal Hypohyal Basihyal	Rudimentary Ceratohyal Hypohyals	Columella (stapes) Styloid process in mammals Anterior horn of hyoid Body of hyoid Entoglossus in reptiles and birds		
III	Pharyngobranchial Epibranchial Ceratobranchial Hypobranchial	Pharyngobranchial Epibranchial Ceratobranchial Hypobranchial	Epibranchial Ceratobranchial	2nd horn of hyoid Body of hyoid		
IV	Branchial skeleton	Branchial skeleton		Second horn of hyoid	Last horn of reptiles	Thyroid cartilages
V	Branchial skeleton			Cricoids and arytenoids (?) (precise homologies unknown)		
VI	Branchial skeleton	Not present				
VII	Branchial skeleton					

*Sometimes part of derm-articular.

†Of intramembranous origin in some species.

SUMMARY

1. During ontogeny the neurocranium is constructed of prechordal and parachordal cartilages, notochord, and cartilaginous olfactory and otic capsules knit together and completed by cartilaginous walls and, in lower vertebrates, a fenestrated roof above the brain. An optic capsule remains independent of the rest of the neurocranium and sometimes forms scleral rings.

2. In agnathans and cartilaginous fishes, the neurocranium remains cartilaginous throughout life. It ossifies at one or more sites in bony vertebrates.

3. The chief centers of ossification in the neurocranium are occipital, sphenoid, ethmoid, and otic. Basic replacement bones are exoccipital, basioccipital, supraoccipital; basisphenoid, presphenoid, orbitosphenoid, pleurosphenoid; mesethmoid, cribriform plate, ethmoturbinals; prootic, opisthotic, epiotic, or petrosal. These may be reduced in number by fusion.

4. The neurocrania of teleosts, apodans, and amniotes are well ossified. Those of dipnoans, lower ray-fins, and most amphibians retain considerable cartilage.

5. A single occipital condyle is found in ancient tetrapods and in modern reptiles and birds. Modern amphibians, synapsid reptiles, and mammals have two.

6. The membrane bones of the skull constitute the dermatocranium. They are vestiges of ancient dermal armor.

7. The chief dermal bones of primitive tetrapod skulls were (1) **roofing bones**—nasal, frontal, parietal, postparietal; intertemporal, supratemporal, tabular, squamosal, quadratojugal; lacrimal, prefrontal, postfrontal, postorbital, infraorbital (jugal); (2) **dermal bones of the upper jaw**—premaxilla and maxilla; (3) **primary palatal bones**—parasphenoid, vomer, palatine, endopterygoid, ectopterygoid; and (4) **opercular bones.** A partietal foramen was present.

8. Teleosts have the largest number of dermatocranial bones; dipnoans have large dermal bony plates. The dermatocrania of other bony fishes resemble those of their Devonian ancestors.

9. Modern amphibian skulls have lost many membrane bones. Anurans have large palatal vacuities. Apodan skulls have been modified the least.

10. Modern reptiles retain well-ossified neurocrania, a single occipital condyle, many membrane bones, and, in lizards, a parietal foramen. Among specializations are temporal fossae, a partial or complete secondary palate (turtles, some lizards, and crocodilians), and cranial kinesis.

11. Temporal fossae result in diapsid (crocodilians and *Sphenodon*), synapsid (therapsids), or euryapsid skulls (ichthyosaurs and plesiosaurs). Turtle skulls are anapsid; those of lizards, snakes, and birds are modified diapsid. Mammals have synapsid skulls.

12. Secondary palates arise as horizontal processes of premaxillae, maxillae, and palatine bones. They are complete to the midline and caudad only in crocodilians and mammals.

13. Cranial kinesis is found in skulls having intracranial joints. It is especially characteristic of teleosts, snakes, lizards, and birds. It is correlated with modes of procuring food.

14. Bird skulls are modified diapsid. Dermal bones are numerous; sutures have been obliterated except in ratites. The skulls are thin, domed, and have large orbits. Jaws are elongated to form a beak.

15. Mammalian skulls have a single temporal fossa. The dentary is the sole bone of the lower jaw, and it articulates with the squamosal. The quadrate and articular have become middle ear ossicles. The braincase is expanded, and the dermatocranial bones are reduced in number. There is a full complement of neurocranial bones with some fusions. A temporal complex combines numerous components that are separate in lower vertebrates.

16. The cyclostome visceral skeleton is unique. There are no palatoquadrate or Meckel's cartilages and no branchial arches. A cartilaginous branchial basket is located under the skin of the pharyngeal region.

17. The visceral skeleton of elasmobranchs consists of palatoquadrate and Meckel's cartilages, hyoid cartilages, branchial cartilages, and median ventral basihyal and basibranchial cartilages.

18. Jaw suspension in fishes is mostly hyostylic. Some older sharks were amphistylic. Chimaeras, lungfishes, and tetrapods are autostylic.

19. The hyoid skeleton of bony fishes consists of a larger number of components than in sharks, and the skeleton of the caudalmost gill arches may be reduced.

20. The embryonic palatoquadrate cartilages in bony vertebrates are invested by membrane bones. Their caudal ends become quadrate bones, except in mammals. Meckel's cartilages are invested, and their caudal ends become articular bones, except in mammals.

21. The articular and quadrate bones constitute the jaw joints in all vertebrates below mammals. In mammals, they become middle ear ossicles (malleus, incus).

22. In tetrapods below mammals, the hyomandibula becomes a columella (stapes). The remainder of the second arch skeleton, that of the third, and sometimes part of the fourth gives rise to horns of the hyoid. Part of the fourth arch and perhaps the fifth contribute to the skeleton of the larynx.

SELECTED READINGS

Readings on feeding mechanisms will be found also in Selected Readings, chapter 12.

Allin, E. F.: Evolution of the mammalian ear, Journal of Morphology **147**:403, 1975.

Carroll, R. L.: The hyomandibular as a supporting element in the skull of primitive tetrapods. In Panchen, A. L., editor: The terrestrial environment and the origin of land vertebrates, New York, 1980, Academic Press.

de Beer, G. R.: The development of the vertebrate skull, Oxford, 1937, The Clarendon Press. Reprinted with a foreword by Brian K. Hall and James Hanken by The University of Chicago Press, Chicago, 1985.

Drysdale, T. A., Elinson, R. P., Graveson, A. C., Hanken, J., Herring, S. W., Langille, R. M., Noden, D. M., Ramirez, F., Solursh, M., and Webb, J. F.: Development and evolution of the vertebrate head, a symposium, American Zoologist **33** (4):417, 1993.

Gans, C., and Parsons, T. S., editors: Biology of the reptilia, vol. 4, New York, 1973, Academic Press.

Goodrich, E. S.: Studies on the structure and development of vertebrates, London, 1930, The Macmillan Co., Ltd. Reprinted by The University of Chicago Press, Chicago, 1986.

Gorniak, G. C., Gans, C., Radinsky, L., Hylander, W. L., Herring, S. W., English, A. W., and Byrd, K. E.: Mammalian mastication: An overview, American Zoologist **25**:289, 1985.

Greaves, W. S.: The mammalian jaw mechanism—the high glenoid cavity, American Naturalist **116**:432, 1980.

Herring, S. W.: Formation of the vertebrate face: epigenetic and functional influences, American Zoologist **33**:472, 1993.

Jarvik, E.: Basic structure and evolution of vertebrates, 2 vols., New York, 1980, Academic Press.

Kuhn, H.-J., and Zeller, U., editors: Morphogenesis of the mammalian skull, New York, 1987, Paul Parey Scientific Publishers. A monograph in the series Mammalia depicta, a supplement to the International Journal of Mammalian Biology. Included are special problems of the skulls of monotremes, marsupials, and whales.

Langille, R. M.: Formation of vertebrate face: differentiation and development, American Zoologist **33**:462, 1993.

Liem, K. F.: Adaptive significance of intra- and interspecific differences in the feeding repertoires of cichlid fishes, American Zoologist **20**:295, 1980.

Lombard, R. E., and Bolt, J. R.: Evolution of the tetrapod ear: an analysis and reinterpretation, Biological Journal of the Linnaean Society **11**:19, 1979.

Radinsky, L. B.: Patterns in the evolution of ungulate jaw shape, American Zoologist **25**:303, 1985.

Reilly, S. M.: Ontogeny of cranial ossification in the eastern newt, *Notophthalamus viridescens* (Caudata: Salamandridae), and its relationship to metamorphosis and neoteny, Journal of Morphology **188**:315, 1986.

Reilly, S. M., and Lauder, G. V.: Atavisms and the homology of hyobranchial elements in lower vertebrates, Journal of Morphology **195**:237, 1988.

Rowe, T.: Co-evolution of the mammalian middle ear and neocortex, *Science* **273**:651, 1996.

Schmalhausen, I. I.: The origin of terrestrial vertebrates (Translated from the Russian by Leon Kelso), New York, 1968, Academic Press.

Smith, K. K.: Integration of craniofacial structures during development in mammals, American Zoologist **36**:70 (1996).

Stahl, B. J.: Vertebrate history: problems in evolution, New York, 1974, McGraw-Hill Book Company.

Trueb, L: A summary of osteocranial development in anurans with notes on the sequence of cranial ossification in *Rhinophrynus dorsalis* (Anura: Pipoidea: Rhinophrynidae), South African Journal of Science **81**:181, 1985.

CHAPTER TEN

GIRDLES, FINS, LIMBS, AND LOCOMOTION

In this chapter we'll focus on skeletal structures that, along with the axial skeleton, participate in locomotion. We'll see that the girdles of bony fishes were inherited almost without change by the first amphibians, and how the pelvic girdles of birds and mammals have been modified to facilitate egg laying or birthing. We'll find that the skeleton of all fins exhibit a common basic pattern that has diverged in several directions; and that the skeletons of all limbs consist of five jointed segments, the distal ones of which have been modified for life in the water, on land, or in the air. We'll see how fishes and fishlike mammals differ in their methods of swimming, and we'll examine locomotion in terrestrial vertebrates that have lost their limbs. Along the way we'll examine some hypotheses concerning the origin of fins and a plausible theory of the origin of limbs.

OUTLINE

he pectoral and pelvic girdles and the skeleton of fins and limbs constitute the **appendicular skeleton.** Girdles brace fins and limbs against the counterforces that appendages transmit from the water or from a substrate. (You may recall from a course in physics that every force evokes an equal counterforce.) The girdles, in turn, are braced against one or more components of the axial skeleton, thereby achieving stability. The forces transmitted to the girdles from the appendages are generally greatest in terrestrial amniotes because their limbs elevate the body above the ground.

Agnathans, moray eels, caecilians, snakes, some lizards, and all but one genus of amphisbaenians have no paired appendages. Having only an anterior pair are eels, numerous other teleosts, one family (Sirenidae) of urodeles, a lizard (*Bipes*), cetaceans, and sirenians. There are also lizards and birds with hind limbs only. Vertebrates that lack one or both pairs of appendages generally have an elongate body and are either aquatic or live in burrows. The evolution of efficient limbless locomotion has been accompanied by adaptations that will be described later.

Tetrapod limbs arise during embryogenesis as paired **limb buds** (fig. 1.10). Fins arise in the same manner and are called **fin folds.** It is not unusual for limb buds to appear transitorily during embryogenesis in limbless species or to develop incompletely and to persist throughout life as functionless vestiges. Inheritance of genes that have these effects bespeaks an ancestor with functional tetrapod limbs.

PECTORAL GIRDLES

A pectoral girdle is a skeletal complex in the body wall immediately behind the head that articulates with the anterior fins or limbs. The pectoral girdles of post-Devonian vertebrates—fishes and tetrapods—are variations of the girdles of Devonian fishes. In the latter, a pectoral girdle consisted of three pairs of replacement bones that were part of the endoskeleton, and at least four pairs of investing bones derived from dermal armor. These bones are named at the bottom of figure 10.1. The pectoral girdle of the chondrostean *Polypterus* exemplifies this basic pattern (fig. 10.2). The scapula and coracoid receive the forces transmitted to the trunk from the fins, the posttemporal braces the girdle against the caudal angles of the skull, and the clavicle is braced against the opposite clavicle in a midventral symphysis. Thus, the force generated by action of the fins is distributed to the body in such fashion as to diminish the impact.

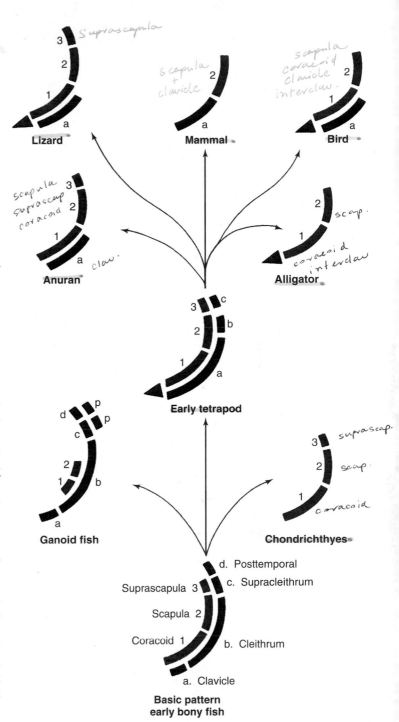

FIGURE 10.1

Pectoral girdle in selected phylogenetic lines. Dermal bones are *red;* cartilage and replacement bones are *black.* Triangles represent interclavicle. **P,** postcleithrum. Only one half of each girdle is illustrated, and relationships have been distorted when necessary to emphasize homologies. The basic pattern is based on *Eusthenopteron,* a crossopterygian. In lizard, alligator, and bird only, **1** is a procoracoid. The chondrichthyes girdle was derived from a bony ancestor not yet indisputably identified.

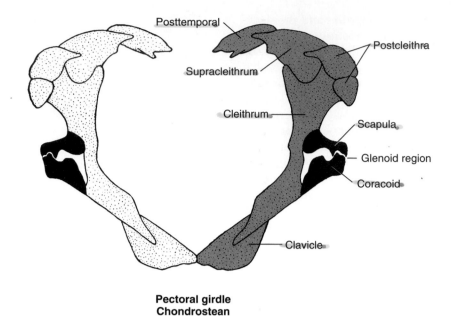

**Pectoral girdle
Chondrostean**

FIGURE 10.2

Pectoral girdle of *Polypterus.* Dermal bones are *red,* replacement bones are *black.* Compare fig. 10.1, ganoid fish.

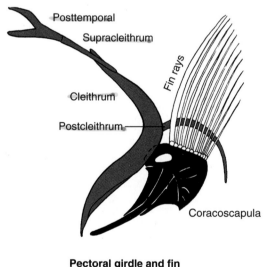

**Pectoral girdle and fin
Teleost**

FIGURE 10.3

Left half of pectoral girdle and the fin skeleton of a ribbonfish (teleost). Dermal bones are *red,* replacement bones are *black.*

In modern bony fishes—teleosts—the cleithrum has become the major bone of the girdle (fig. 10.3). They have lost the clavicle. Otherwise, modern bony fishes generally have a full complement of primitive dermal bones. The embryonic endoskeletal coracoid unites with the scapula to form a coracoscapula. Cartilaginous fishes, on the other hand, have only endoskeletal components and these do not ossify, although they become

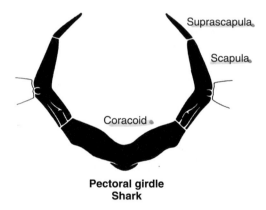

**Pectoral girdle
Shark**

FIGURE 10.4

Cartilaginous pectoral girdle of *Squalus*, anterior view.

hardened by calcification of the cartilaginous matrix (fig. 10.4).

The pectoral girdle of early tetrapods were hardly different from those of early bony fishes. If you compare early tetrapods with early bony fishes in figure 10.1, you will see that early tetrapods had acquired an additional membrane bone, the **interclavicle,** and that they had lost the posttemporal, which braced the girdle against the skull in fishes. The girdle of a stem amphibian is seen in figure 10.5. The supracleithrum is missing. The cleithrum, though present in this early amphibian, didn't last long once vertebrates became established on land.

The interclavicle of stem amphibians appears to have been persistent in lineages leading to amniotes. It is seen

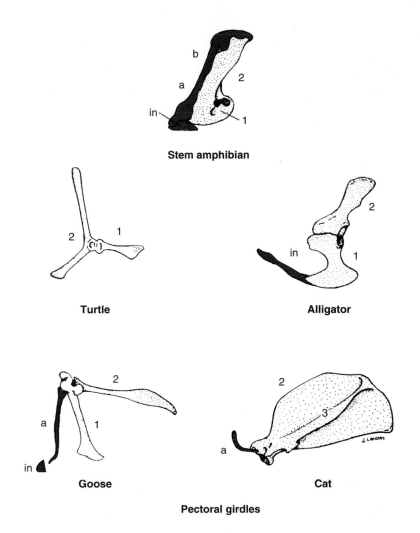

Stem amphibian

Turtle

Alligator

Goose

Cat

Pectoral girdles

FIGURE 10.5

Left half of the pectoral girdles of selected tetrapods, lateral views. Dermal bones are *red*. **1**, coracoid or procoracoid; **2**, scapula; **3**, scapular spine, **a**, clavicle; **b**, cleithrum; **in**, interclavicle. In turtles the clavicles and interclavicle are fused with the shell. The cat's clavicle is almost vestigial.

as an unpaired bone in an alligator between the base of the two procoracoids (fig. 8.14). In birds, it is the unpaired cartilage or bone at the tip of the "wishbone." Some paired bones associated ventrally with the clavicles or coracoids in reptiles, birds, and monotremes have also been named interclavicles by one authority or another. The interclavicles of a therapsid reptile and of a living monotreme are seen in figure 10.6. Whether or not all bones so labeled from amphibians to mammals are homologues is, at present, uncertain because of insufficient research data. We will now follow the fate of the clavicle, coracoids, and scapula as tetrapod diversity proceeds.

The fate of the **clavicles** is correlated with that of the coracoids. Either a clavicle or a coracoid are likely to brace the scapula against the sternum, and sometimes both do so (figs. 10.6*b* and 10.7). Clavicles are missing in urodeles and apodans and are uncommon in recent reptiles other than lizards (although vestiges develop

transitorily in crocodilian embryos). In birds they are the long bones of the **furculum** (wishbone). They are present in most orders of mammals.

The **coracoids** in tetrapods arise from an embryonic cartilaginous **coracoid plate** in the lateral body wall extending ventrad from the glenoid region of the scapula. Anterior ossification centers in this plate give rise to **procoracoids;** posterior centers give rise to coracoids (fig. 10.6). Neither procoracoids nor coracoids develop in Eutheria except for a vestige, the **coracoid process of the scapula** overhanging the glenoid fossa (fig. 11.22*b*). Coracoids and/or procoracoids assist the clavicles or replace them functionally in bracing the scapula against the sternum (fig. 8.28, iguanid and chicken).

The **scapula** is present in all tetrapods that retain any vestiges of anterior limbs, since it bears part or all of the glenoid fossa for articulation of the girdle with the head of the humerus. **Suprascapular** ossification centers

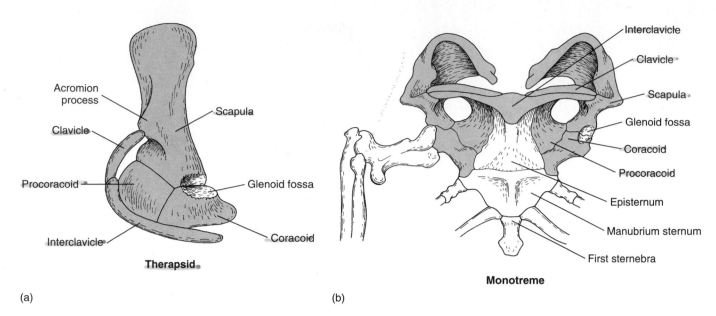

Therapsid

(a)

Monotreme

(b)

FIGURE 10.6

Pectoral girdles (*red*) of a mammal-like therapsid reptile and a primitive mammal, the latter from a ventral view. In (*b*), note the coracoid, procoracoid, clavicle, and interclavicle bracing the girdle against the sternum.

usually coalesce with scapular centers to become a single bone, but a suprascapula remains independent in urodeles and anurans (figs. 8.28, salamander, and 10.7).

The mammalian pectoral girdle is the result of evolutionary changes that gave rise to therapsids and their mammalian descendants. The therapsid girdle consisted of interclavicle, clavicle, procoracoid, coracoid, and scapula (fig. 10.6*a*). Monotremes have the same bones today (fig. 10.6*b*). In eutherian mammals the only bones remaining in the pectoral girdle are a scapula and, generally, a clavicle. As mentioned earlier, a vestige of the coracoid persists as a coracoid process of the scapula. In humans it remains unattached to the scapula until about 15 years of age.

The lateral aspect of the **mammalian scapula** is divided by a **scapular spine** into **supraspinous** and **infraspinous fossae** (fig. 10.5). These fossae are the anatomic origins of strong muscles that insert on the humerus. The scapular spine is also an insertion site for some of the appendicular muscles that arise on the vertebral column. An **acromion process** for muscular attachment forms near the glenoid fossa. The process in cats and humans is shown in figures 10.5 and 11.22, respectively.

The **mammalian clavicle** is large in the primitive monotremes and in insectivores and primates, the last two being generalized mammals. It also is large in mammals with strong forelimbs that are used in digging, climbing, or flying. In bats, strong clavicles brace the scapula against the sternum. On the other hand, there are mammals whose skeletomuscular system has undergone adaptations that have reduced the clavicle to a mere

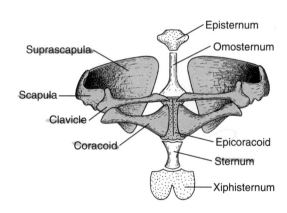

FIGURE 10.7

Pectoral girdle (*red*) and sternum of a frog, ventral view. A procoracoid (not illustrated) lies dorsal to the clavicle. *Stipple* is cartilage.

splinter, as in felines, or eliminated it altogether, as in cetaceans, ungulates, and some other mammals. In cats the clavicle is a vestigial splinter reaching neither the sternum nor the scapula (fig. 10.5). This enables cats to withstand the shock of landing upright on their forelimbs after a leap, since there are no rigid connections between the scapulae, which receive the initial impact, and any other part of the skeleton. The shock of the impact is dissipated in the musculature of the shoulders and in the muscles and joints of the forelimbs. Absence of a clavicle in ungulates facilitates grazing.

The phylogenetic history of the components of the pectoral girdle inspires this observation: *Dermal bones predominate in the pectoral girdle of bony fishes, whereas replacement bones predominate in tetrapods.* You may recall that dermal bones are derivatives of the bony armor of early fishes, in which they overlaid, like siding on a house, the replacement bones, the latter being the true (endo)skeleton of the girdle. Little wonder then that the dermal bones of the girdle became reduced or lost during the evolution of tetrapods. Dermal armor elsewhere in the body was undergoing the same fate. The phenomenon is discussed in chapter 6, under "The Bony Dermis of Fishes."

Two loose ends are worth attention. First, urodeles fail to develop any dermal bones in their girdles (or, in fact, elsewhere in their body except in the skull). The girdle of a salamander is seen in figure 8.28. Second, have you noticed that tetrapods almost never brace their pectoral girdles against the skull or vertebral column? It is said to have happened in a few pterosaurs. The story is quite different with respect to the pelvic girdle.

PELVIC GIRDLES

Pelvic girdles in most fishes consist of a pair of simple cartilaginous or bony **pelvic (ischiopubic) plates** that meet in a midventral **pelvic symphysis** and provide a brace for the pelvic fins (fig. 10.8, herring). In cartilaginous fishes and lungfishes the two embryonic cartilages unite to form one adult plate (fig. 10.8, shark). In teleosts that have a short trunk the pelvic plate lies immediately behind, or sometimes below, the pectoral girdle and is often attached to it. For this reason, pelvic

fins may be as far forward as the pectorals, or even appear to be anterior to them (fig. 10.17, perch). There are no dermal bones in the pelvic girdles of either fishes or tetrapods.

Tetrapod embryos also develop cartilaginous pelvic plates. Each plate ossifies at two centers to form a **pubis** (pubic bone) and a more posterior **ischium** (fig. 10.9). In urodeles, however, the pelvic plate may remain cartilaginous except for a small ischial ossification center (fig. 10.10, *Necturus*). Dorsal to the pelvic plate an additional blastema gives rise to an ilium (figs. 10.9 through 10.12). At the junction of the pubis, ischium, and ilium, a socket, the **acetabulum,** accommodates the head of the femur.

Dorsally, the tetrapod ilium is braced against the stout transverse processes of the sacral vertebrae, one in amphibians, two in reptiles, more in birds and mammals. Short sacral ribs intervene (fig. 10.11). Ribs often become ankylosed to the transverse processes, in which case the ribs may become unrecognizable except in embryos. Ventrally, except in birds, there is a symphysis between either the two pubic bones **(pubic symphysis),** the two ischia **(ischial symphysis),** or both (fig. 10.9). The symphysis is in the midventral body wall immediately anterior to the cloaca or its derivatives. The skeletal framework is such that the forces transmitted to the two acetabula as a result of gravity (weight bearing) or locomotion is distributed in two directions: to the vertebral column dorsally via the ilia, and to the symphysis ventrally via the ischial and pubic bones. The proportion distributed in each direction depends on the posture of the animal. The joint between the head of the femur and

FIGURE 10.8

Pelvic plates (*black*) and caudal fins of a bony fish and a male and female shark. The basal cartilages of the male fin have become claspers.

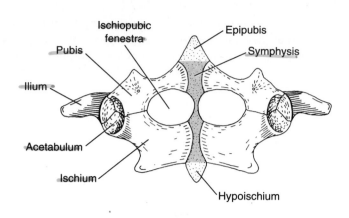

FIGURE 10.9

Pelvic girdle of *Sphenodon,* ventral view, showing ischiopubic symphysis.

FIGURE 10.10

Left halves of pelvic girdles of selected tetrapods, lateral views, except *Necturus,* which is a ventral view. **1,** ilium (*red*); **2,** ischium; **3,** pubis (unossified pelvic plate in *Necturus*). *Stipple* is acetabulum. In *Necturus* the ischium is an ossification center in a cartilaginous ischiopubic plate and a sacral rib is seen attached to the dorsal end of the ilium. There are no dermal bones in pelvic girdles.

Pelvic girdles

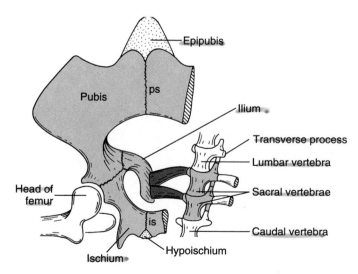

FIGURE 10.11

Pelvic girdle, the two sacral vertebrae, and the sacral ribs (*red*) of the freshwater turtle *Trionyx.* **is,** ischial symphysis; **ps,** pubic symphysis. The vertebrae and ribs are viewed obliquely upward through the pelvic cavity, from below.

Turtle

(a) **Primitive Tetrapod**

Ilium

Acetabulum

Pubis

Ischium

(b) **Alligator**

(c) **Saurischian dinosaur**

(d) **Monitor lizard**

(e) **Early synapsid**

(f) **Cat**

Acetabulum

Obturator foramen

(g) **Human**

Acetabulum

FIGURE 10.12

Left half of pelvic girdle of selected amniotes. (*a*)–(*f*) Lateral view, head to the left. (*g*), Anterior (ventral) view, body upright. The pubis and ischium are differentially colored *red*. *Stipple* in (*b*) is cartilage.

the girdle is stabilized by the joint capsule and by muscles that approach the femur from opposing directions. The sacrum and girdle in amniotes, often rigidly united, form a bony enclosure, the **pelvis**, that encircles the caudal end of the coelom. The resulting **pelvic cavity** contains the urogenital organs and the terminal portion of the large intestine.

Posture and mode of locomotion are correlated with the shape of the ilium, ischium, and pubis, the anatomic relationships of these bones, and their proportional size. A squatting, jumping frog has a different set of vectors affecting the pelvic girdle than does a bird, deer, kangaroo, or marine turtle. In frogs the ilia (fig. 8.13) are slender and greatly elongated, and they extend from the sacral vertebra to the end of the urostyle, where they meet the ischia and pubic bones and where the acetabulum is located. The joint between ilium and sacral vertebra is free to move when a frog pushes off at the start of a leap. As the frog lands, the joint, along with others in the leg, helps dissipate the force of the impact, acting as a shock absorber. In many tetrapods this **sacroiliac joint** is not freely movable.

Urodeles in general have weak posterior limbs that can scarcely lift the sagging belly off the substrate without the buoying effect of water. Consequently, the pelvic girdle is subject to relatively little stress as long as the animal is resting motionless on a substrate. The pelvic plate, mentioned earlier, differs little from that of fishes other than being braced against a single vertebra by a weak ilium. In urodeles that usually lose their gills and develop lungs, a slim median **prepubic (ypsiloid) cartilage** extends from the girdle forward in the linea alba for two or three somites. It furnishes attachment for certain muscles utilized in respiration.

The structure of the pelvic girdles of reptiles is correlated with their diverse body structure and with their mode of locomotion that varies from lumbering to highly agile. The ilium has become braced against an additional vertebra (fig. 10.11). In most reptiles, the pubis is directed away from the ischium, resulting in a triradiate girdle (fig. 10.12*b, c, d*). However, in ornithischian dinosaurs, the ischium and pubis are parallel and directed caudad (fig. 10.10). In some reptiles, a broad *ischiopubic fenestra* develops between the ischium and the pubis on each side (figs. 10.9 and 10.10, turtle). It should not be confused with the opening in the wall of the acetabulum of the alligator and saurischian dinosaur seen in figure 10.12. Those openings result from incompletion of the wall of the socket.

An **epipubic** and **hypoischial** bone frequently develops in association with the pelvic girdle of reptiles (fig. 10.9). One or both are present also in monotremes and marsupials. The epipubic bone is also called the **marsupial bone** in marsupials because it supports the marsupial pouch.

The ilia and ischia of modern birds are enormously expanded and united with the synsacrum (figs. 8.18 and 8.19). Such a girdle provides broad surfaces for attachment of hind limb muscles used in the bipedal stance and for the thrust required by some species for flight takeoff. The pubic bones are reduced to long splinters that are directed caudad parallel to the ischium, and they do not meet ventrally. Therefore, there is no ischial or pubic symphysis. Absence of symphyses provides a wide outlet from the pelvic cavity for laying the large-yoked avian eggs. The pelvic girdles of an ornithischian dinosaur and a goose are illustrated side by side in figure 10.10. Despite the similarity of pelvic girdles, birds are now thought to be descendants of some small saurischian dinosaur. (See chapter 4 for a phylogenetic discussion.) The pelvic girdle of the very early *Archaeopteryx* (fig. 8.21) was braced against four sacral vertebrae. There was no synsacrum.

In mammals, the ilium, ischium, and pubis ankylose early in postnatal life to form a left and right **innominate (coxal) bone** (fig. 10.12*f*). Dorsally, the ilium on each side ankyloses with the sacrum in an immobile **sacroiliac joint.** Ventroposteriorly, the pubis and, frequently, the ischia meet in a **pubic** or **ischiopubic symphysis** to complete the walls of a pelvic cavity. A pelvic outlet at the base of the tail provides access to the exterior.

Mammalian young are delivered through the pelvic outlet. In late pregnancy, the fibrocartilage separating the bones at the symphysis is softened by **relaxin,** an ovarian hormone, which then permits expansion of the pelvic outlet for delivery. In mice six days pregnant, the gap between the bones in the symphysis was shown on x-ray film to be only 0.25 mm. Thirteen days later, on the day of birth, the gap had widened to 5.6 mm.

FINS

Fins serve as steering devices for changing direction, as stabilizers that prevent the body from rolling or wobbling, and as devices that control the inclination (slope) of the body in swimming away from the horizontal. Paired fins also serve as brakes that slow or halt forward motion, but play little role in providing forward thrust. Forward thrust in most fishes is generated by lateral undulation of the posterior part of the trunk, tail, and caudal fin (fig. 11.8).

Fins, whether paired, median, or caudal, consist mostly of two surfaces of skin back to back, stiffened by flexible fin rays that radiate from a skeletal base (fig. 10.13). The rays are in the dermis and are of two varieties. The rays of bony fishes, **lepidotrichia,** consist of jointed *bony dermal scales* aligned end to end. **Ceratotrichia,** the rays of cartilaginous fishes, are long *horny rays* composed of a material similar to that comprising the dorsal spines of sharks.

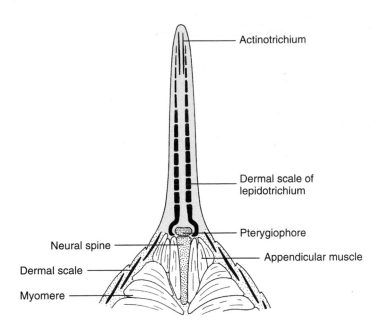

FIGURE 10.13

A pair of fin rays (lepidotrichia) in the dorsal fin of a teleost in diagrammatic cross section. A pterygiophore is a cartilage or bone at the base of rays.

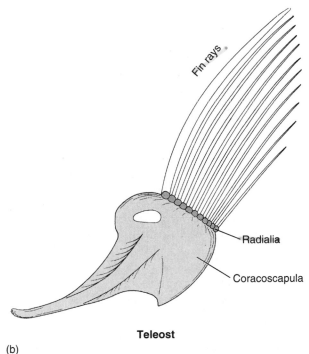

FIGURE 10.14

Pectoral fin skeleton of two ray-finned fishes. (*a*) *Polypterus*, **oc**, ossification center in central basal cartilage. The other basalia are fully ossified. (*b*) Specialized fin of a ribbonfish.

Short delicate **actinotrichia** develop distally in both groups. In elasmobranchs and a few other fishes, typical body scales continue onto the fin, growing smaller distally. These further stiffen the fin.

The skeletal base from which fin rays of generalized fishes radiate, whether of paired fins or median ones, consists of a row of cartilaginous or bony **basalia** and one or more rows of **radialia** (figs. 10.14–10.16). The specific anatomy of the base varies with the taxa. The chondrostean *Polypterus* is a generalized ray-finned fish whose pectoral fins (fig. 10.14*a*) have three large basals and two distal rows of radials. The paired fins of most specialized teleosts, on the other hand, have lost all basalia and retain mere vestiges of distal radials (fig. 10.14*b*). The basal and radial skeleton of median dorsal fins rests on the vertebral column (fig. 10.16). Dipnoans and crossopterygians exhibit a wide variety of bases in both paired and dorsal fins.

Striated muscle masses from several successive body segments extend into the base of a fin and insert on available skeletal elements. Intensive efforts in the past to identify an ancestral fin type, or **archipterygium,** that might have led to the fins of later cartilaginous and bony fishes have resulted only in speculation. There is too much diversity of morphology and too little evidence.

Paired Fins

Pectoral fins are braced against the glenoid fossa or glenoid region of the pectoral girdle (fig. 10.14*a*). Pelvic fins are braced against a prominence on the lateral or posterior aspect of the pelvic plate (fig. 10.8). Paired fins are grouped in three categories (fig. 10.17): (1) **lobed fins,** (2) **fin fold fins,** and (3) **ray fins.** These correspond to the three categories of jawed fishes that emerged from the Devonian and have persisted to the present: (1) sarcopterygians (dipnoans and crossopterygians), (2) chondrichthyans (elasmobranchs and chimaeras), and (3) actinopterygians (ray-finned fishes). (**Spiny fins,** characteristic of acanthodians, had disappeared by the end of the Paleozoic.)

Lobed fins consist of a fleshy proximal lobe containing the fin skeleton and its attached muscles, and a membranous distal portion that is stiffened by fin rays. The fin arises from a narrow base and resembles a paddle.

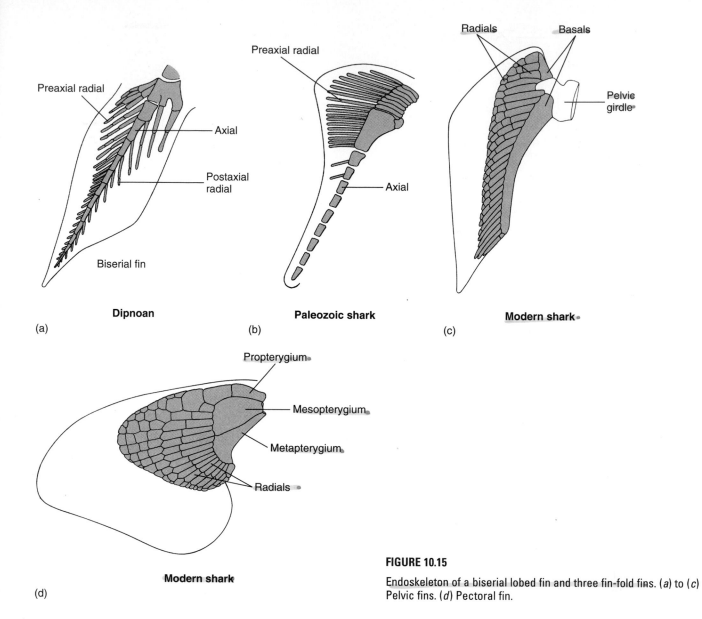

FIGURE 10.15

Endoskeleton of a biserial lobed fin and three fin-fold fins. (*a*) to (*c*) Pelvic fins. (*d*) Pectoral fin.

The paired fin skeleton of the dipnoan *Neoceratodus* (fig. 10.15*a*) consists of a jointed central axis, the bones of which function like basals, and a series of pre- and postaxial radials. Such a fin is said to be **biserial** because of the two series of radials. Crossopterygians have a wide variant in which the relationship of the axis and radials has been considerably disrupted (fig. 10.45).

Fin fold fins have a broad base. It is broad in modern sharks, and was broader in the Paleozoic. As in other fishes, the three basalia are termed **pro-, meso-,** and **metapterygia** (fig. 10.15*d*). In male Chondrichthyes, the basalia of the pelvic fin are modified to become an intromittent organ, the **clasper** (fig. 10.8, male shark).

The ray fins of teleosts have lost components of the basal skeleton in time (compare 10.14*a* and *b*), but the fins have become increasingly flexible. A number of ray-finned fishes have no pelvic fins. Tuna, which are among the largest fishes in the oceans, swim like whales, pro-

pelling themselves forward by dorsoventral undulation of the posterior trunk and tail rather than by lateral undulation as in most fishes.

It should be mentioned that a few fishes fly, although for short distances only. Characins, which are primitive voracious teleosts inhabiting fresh tropical waters, get an initial thrust out of the water with the caudal fin and, beating winglike pectorals, fly several yards using appendicular muscles that, for a fish, are exceptionally large. This behavior is said to occur only when the fish are alarmed. If so, it is a fishy way of performing a disappearing act.

Median Fins

Fishes have one, two, or a series of median **dorsal fins** and many have a midventral **anal fin** just behind the anus or vent. Dorsal and anal fins are seen in figure 10.17 and over a dozen paired and median fin patterns

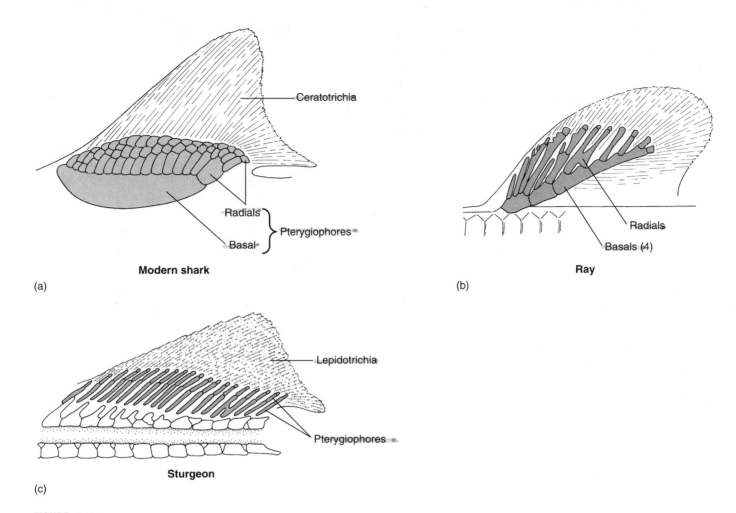

Ceratotrichia

Radials

Basal

}Pterygiophores

Modern shark

(a)

Radials

Basals (4)

Ray

(b)

Lepidotrichia

Pterygiophores

Sturgeon

(c)

FIGURE 10.16

Median dorsal fin skeletons. The pterygiophores (skeletal elements at the base of the rays) rest on the vertebral column.

Lobed fins (Latimeria)

Ray fins (Perch)

Fin fold fins (Cladoselache)

Spiny fins (Acanthodian)

FIGURE 10.17

The fins of fishes.

are illustrated in chapter 4. Median fins act as keels, keeping motionless fishes from rolling to the left and right. When the fish is swimming, median fins minimize deviation from a straight line that inevitably results from side-to-side thrusts of the locomotor tail. Lack of stabilizers of one kind or another would not only increase the cost of locomotion in terms of energy expended, but might place a species in jeopardy. Bottom-dwellers might be less disadvantaged. Because of the body shape of Rajiformes, one might expect less need for stabilizers; nevertheless, rays other than sting rays have two dorsal fins far back on the tail. However, there is no anal fin. Dorsal fins are long in lampreys and bony eels, compensating for the inadequacy or absence of paired fins.

The anal fin of the male of some viviparous teleost species has been elongated posteriorly and modified to transfer sperm to the female. This **gonopodium** is analogous to the clasper of cartilaginous fishes.

Caudal Fins

Caudal fins are classified on the basis of their shape and the direction taken by the terminal portion of the notochord and vertebral column, (fig. 10.18a to c). A tail that contains dorsal and ventral lobes and in which the notochord turns upward into a large dorsal lobe is said to be **heterocercal.** This was the predominant type in the Paleozoic, having been present in placoderms, Paleozoic sharks, and at least some acanthodians, and it is still seen in modern sharks

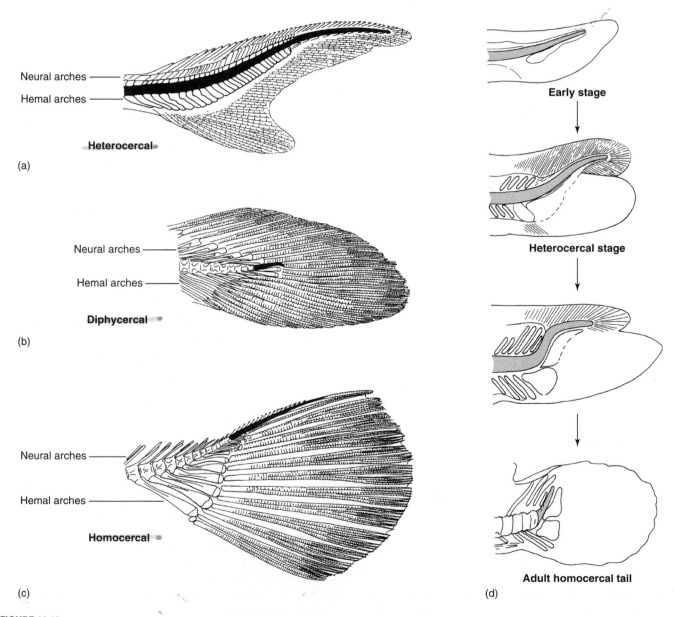

FIGURE 10.18

Caudal fins of fishes. (a) to (c) Major morphologic varieties, notochord in black. The heterocercal fin is probably the most primitive. (d) Successive developmental stages in the flounder *Pleuronectes,* notochord in *red,* showing transition from heterocercal to homocercal.

and two relict chondrosteans, namely, sturgeons and spoonbills (fig. 4.12). There is also a rare condition, **hypocercal,** in which the vertebral column turns downward. This was a prominent feature in ichthyosaurs—an example of evolutionary convergence.

Derived from the heterocercal condition were innumerable variants that make classification of caudal tails less than definitive. Two terms that have been used conventionally are diphycercal and homocercal. Both types are *symmetrical externally*, but the notochord and vertebral column terminate differently in the two. In **diphycercal** tails, the vertebral column ends with very little upbending, as in dipnoans and *Latimeria* (figs. 4.11 and 10.18*b*). In **homocercal tails,** the notochord, encased within a bony sheath, or **urostyle** (figs. 10.18*c* and 10.19) turns far dorsad. In the primitive teleost seen in figure 10.19, components (hypurals) of several terminal caudal vertebrae have been modified to become the basal skeleton of the fin. Homocercal tails are common in teleosts but not restricted to them. Specialized teleosts have only two hypurals and no epurals.

During embryonic development a homocercal tail is initially heterocercal (fig. 10.18*d*). Applying von Baer's theorem, we could conclude that modern caudal fins are modifications of a heterocercal condition. That conclusion is reinforced by the fact that heterocercal tails predominated during the Paleozoic.

There have been attempts to use the morphologic features of caudal fins in unraveling the phylogenetic relationships among early and subsequent fishes. The approach has been thwarted by enormous diversity, an incomplete fossil record, and the realization that similarities may be the result of convergent evolution.

Caudal fins have developed as adaptations to aquatic life in a few amniotes. The fishlike ichthyosaurs developed them, and cetaceans and sirenians acquired horizontal flukes rather than vertical fins. The vertebral column provided internal support for the caudal fins of ichthyosaurs, turning ventrad into the ventral lobe of their bilobed tails, but there were no fin rays. The flukes of marine mammals have neither fin rays nor endoskeletal support. The same is true of the caudal fins of larval amphibians.

Origin of Paired Fins

What was the nature of structures that might have given rise to paired fins? This is one of the more puzzling questions in the study of phylogeny, because to answer it requires more knowledge than we have of early vertebrates. This being so, the best we can do to satisfy our curiosity, which is the motivation for basic science, is to look for whatever clues we might find among ostracoderms and early jawed fishes, and then speculate.

The **fin fold hypothesis** states that paired fins are derived from a pair of continuous fleshy folds of lateral body wall analogous to the metapleural folds of an amphioxus. Some ostracoderms had such folds, although they were higher on the body wall. If such a structure were to become interrupted in the middle of the trunk and remaining sections were to be invaded by muscle buds at the base and endowed with an endoskeleton, the result would be paired fins like those of *Cladoselache* (fig. 10.17). But there is no evidence this happened, and the hypothesis is of historical interest only.

According to the **gill arch hypothesis** of C. Gegenbaur, pectoral and pelvic girdles are modified gill arches and the skeleton within the fin is an expansion of the gill rays. It is true that many students of comparative anatomy, seeing a shark skeleton suspended in a museum jar for the first time, think that the pectoral girdle is part of the pharyngeal skeleton because of its location immediately behind the last pharyngeal arch, and its U shape. It looks as though it might have been a gill arch, but this is quite unlikely.

The most promising hypothesis is the **fin spine hypothesis.** In early acanthodians (fig. 10.17, spiny fins), pectoral and pelvic appendages were the largest of a series of lateral hollow spiny appendages that extended the length of the trunk and supported a fleshy membrane. Evidently they were practically immobile, and in no way resembled the pectoral and pelvic fins of modern fishes. In later acanthodians, weak fin rays were present in the membranes of two pairs, those at the pectoral and pelvic levels only, and small radial elements supported the membranes at their base. In time, acanthodians

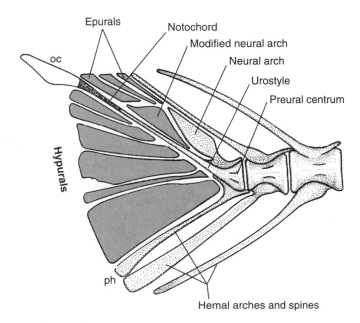

FIGURE 10.19

Skeleton of the homocercal caudal fin of the primitive teleost *Clupea.* The urostyle encloses all but the terminal segment of the upturned notochord. Fin rays extend from the hypurals and epurals. *Gray* indicates components of modified caudal vertebrae. **oc,** opisthotic cartilage; **ph,** parhyural.

tended to lose all spines except the two pairs containing fin rays. This may be a clue to the origin of paired fins, but it is not known that acanthodians were ancestral to later fishes. Among placoderms, arthrodires had a pair of fixed spines projecting behind the head, antiarchs had jointed, spiny, armored pectoral appendages (fig. 4.9), and some very early placoderms, probably bottom-dwellers, had pectoral fins resembling those of modern rays, an instance of evolutionary convergence. Although it is uncertain that placoderms are a natural group, they seem to have been "experimenting" with designs for later paired appendages.

No known protochordate has structures that conceivably could have given rise to paired vertebrate fins, even allowing the imagination free rein, and the earliest known vertebrates had none. Reliable clues to the origin of paired fins may be hidden forever in the obscurity of elapsed time.

TETRAPOD LIMBS

Although tetrapods typically have four limbs, some have lost one or both pairs and in others the forelimbs have been modified as wings or paddles. By employing limbs with appropriate modifications, tetrapods swim, crawl, walk, run, hop, jump, dig, climb, glide, or fly to avoid enemies, seek food and shelter, and find a mate. Each lifestyle is facilitated by some modification of the appendicular skeleton.

The limbs of early tetrapods were short, the first segment extended nearly horizontally from the trunk, and the second segment was perpendicular to the first, directed downward (fig. 10.46d). This posture persists to a considerable degree in urodeles, tortoises, and primitive lizards. In other reptiles and in birds and mammals, there has been a rotation of the entire appendage toward the body, so that the long axes of the humerus and femur more nearly parallel the vertebral column, the elbow being directed caudad and the knee cephalad (fig. 10.32a). Limbs so oriented are excellent shock absorbers. They also provide more support for a body suspended above the ground and greater leverage for locomotion on land.

Tetrapod limb skeletons consist of three segments: **propodium, epipodium,** and **autopodium.** In the forelimb, these correspond to the bones of the upper arm, forearm, and **manus,** or hand (fig. 10.20). Table 10.1 lists homologous components of the fore- and hind limbs. The skeleton within homologous segments in the various tetrapods is remarkably similar despite outward appearances; it is the orientation of the bones, the relative mobility of the joints, and the complexity of the appendicular muscles as much as the skeleton per se that makes possible the variety of locomotor activities of tetrapods. The most striking differences in skeletons are at distal ends of the appendages.

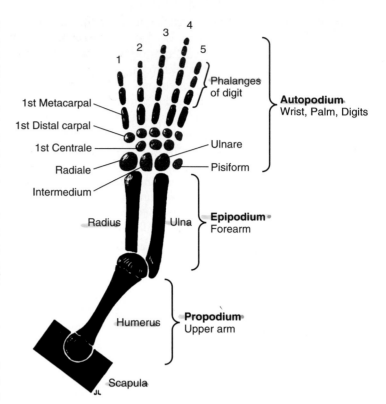

FIGURE 10.20

Skeletal organization of a generalized (primitive) right forelimb, palm down, wrist in *red.* **1–5,** digits.

Propodium and Epipodium

The humerus is the bone of the upper arm. The similarity of the humeri of all tetrapods is more striking than any differences (fig. 10.21). Variations in length, diameter, and shape are adaptive modifications. The odd humerus of the mole (fig. 10.22), for example, has expansions for insertion of massive shoulder muscles for digging. The humeri of carinate birds have a slender central cavity containing diverticula from the lungs.

The radius and ulna are bones of the forearm. The radius is a preaxial bone articulating proximally with the humerus and distally with wrist bones on the thumb side of the hand. It bears most of the force being transmitted from wrist to humerus. The ulna is a longer, postaxial bone articulating proximally with the humerus and radius and distally with wrist bones on the side opposite the thumb. The ulna sometimes fuses with the radius, or it may be vestigial, as in frogs and bats (fig. 10.21).

The femur is the bone of the thigh, and the tibia and fibula are bones of the lower leg (shank). The three bones differ relatively little from one tetrapod to another. A sesamoid bone, the **patella,** or kneecap, develops in birds and mammals. It ossifies in the tendon of insertion of the powerful extensor muscle of the thigh where the tendon passes over the complicated knee joint to insert on the tibia. The patella protects the joint from the abrasive

TABLE 10.1   Homologous Components of the Anterior and Posterior Limbs

	Name of Segment		Skeleton
Anterior Limb	1 Upper arm (brachium)		Humerus
	2 Forearm (antebrachium)		Radius and ulna
	3 Wrist (carpus)		Carpals
	4 Palm (metacarpus)	Manus	Metacarpals
	5 Digits		Phalanges
Posterior Limb	1 Thigh (femur)		Femur
	2 Shank (crus)		Tibia and fibula
	3 Ankle (tarsus)		Tarsals
	4 Instep (metatarsus)	Pes	Metatarsal
	5 Digits		Phalanges

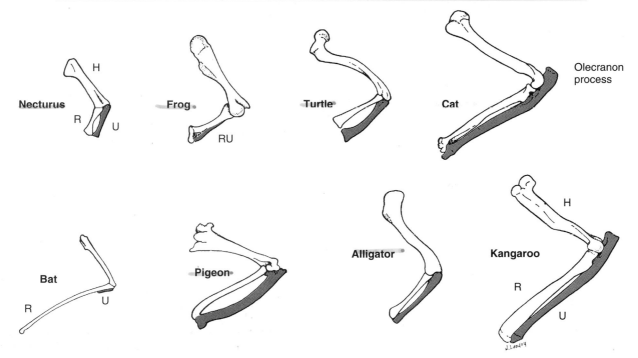

FIGURE 10.21

Humerus, radius, and ulna of the left forelimb (lateral views). **H,** humerus; **R,** radius; **U,** ulna (*red*). In the frog the radius and ulna have united to form a radioulna, **RU.** In bats the ulna is vestigial.

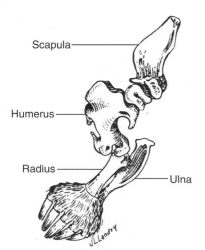

FIGURE 10.22

Right anterior limb of a mole, medial view. The palms turn outward as an adaptation for digging.

FIGURE 10.23

Left thigh and shank bones of representative tetrapods, lateral views, **1**, femur; **2**, tibia; **3**, fibula; **4**, patella; **5**, tibiotarsus; **6**, tibiofibula, fibular component in *red.*

action of the tendon. The fibula (fig. 10.23, *red*) may unite partially or completely with the tibia to form a **tibiofibula** as in frogs; it may be reduced to a splinter, as in birds; or it may be lost, as in deer and other ungulates. In birds the tibia fuses with the proximal row of tarsals to form a **tibiotarsus** (fig. 10.40).

Manus: The Hand

The wrist, palm, and digits constitute a functional unit— the hand **(manus).** Considering the multitude of modifications that have adapted hands (and feet, too) to the challenges of life on land, and eventually, in the air and sea, the skeletons of hands and feet are remarkably similar from frogs to eagles, and from moles to seals.

The forelimbs and hind limbs of the three oldest known tetrapods, recovered from the late Devonian, had from six to eight digits. The complete hind limb of *Ichthyostega,* one of the three, has seven digits, and a complete forelimb belonging to *Acanthostega* has eight (Coates and Clack, 1990). A third Devonian species, *Tulerpeton,* unearthed by a Soviet paleontologist, has six digits on its limbs. None of the three specimens has only five on either limb. The three digits in *Ichthyostega* that occupy the position of the big toe in later tetrapods are small and closely adherent, perhaps fulfilling the same locomotor role as the big toe of a **pentadactyl (five-digit) limb.** These facts suggest that sometime after the late Devonian, and before the amniote line became established, pentadactyl limbs, anterior and posterior, became dominant. They have persisted until this day in general-

ized amniotes. The primitive structure of the pentadactyl hand and foot and the adaptations that have modified these during geologic time are the subjects of this and succeeding sections.

The wrist **(carpus)** of a generalized pentadactyl hand consists of three more or less regular rows of carpal bones (fig. 10.20, autopodium). The proximal row has a radial carpal **(radiale)** at the distal end of the radius, an ulnar carpal **(ulnare)** at the end of the ulna, and an **intermedium** between the two. At the ulnar end of the proximal row in most reptiles and mammals is a sesamoid bone, the **pisiform.** The middle row of carpals consists of **centralia**—often three or four in early tetrapods, two in early reptiles—one of which is sometimes displaced to the proximal or distal row of carpals. The distal row is composed of five **distal carpals** numbered 1 to 5 commencing on the thumb side. Table 10.2 lists the names of the carpal bones.

Metacarpals are the skeleton of the palm. Primitively, the pentadactyl limb may have had as many distal carpals and metacarpals as there were digits.

Each digit consists of **phalanges.** The early phalangeal formula for a pentadactyl hand, commencing with the thumb, may have been something like 2-3-4-5-3. In late therapsids it had become 2-3-3-3-3, an almost universal formula for modern mammals with five fingers (fig. 10.25, human).

Modifications of the manus with few exceptions involve *reduction in the number of bones by evolutionary loss or fusion.* A less common modification is the *dispro-*

TABLE 10.2   Synonymy of Carpal Bones

Terms Preferred by Comparative Anatomists	Nomina Anatomica*	Anglicized Names and Synonyms
Radiale	Os scaphoideum	Scaphoid, navicular
Intermedium	Os lunatum	Lunate, lunar, semilunar
Ulnare	Os triquetrum	Triquetral, cuneiform
Pisiform	Os pisiforme	Pisiform, ulnar sesamoid
Centralia (0 to 4)	Os centrale	Central carpal(s)
Distal carpal 1	Os trapezium	Trapezium, greater multangular
Distal carpal 2	Os trapezoideum	Trapezoid, lesser multangular
Distal carpal 3	Os capitatum	Capitate, magnum
Distal carpal 4		
Distal carpal 5	Os hamatum	Hamate, unciform, uncinate

*Terms adopted by the Eighth International Congress of Anatomists at Wiesbaden.

1st Metacarpal

Radioulna

Frog

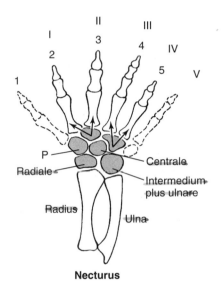

II   III

I

P

Radiale

Radius

Centrale

Intermedium plus ulnare

Ulna

Necturus

FIGURE 10.24

Hands of *Rana catesbeiana* and *Necturus,* dorsal views. Both have four digits. Which digit is missing in *Necturus?* Broken lines represent some of the missing primitive bones. Arabic numerals suggest that in a necturus the thumb, **1,** is missing and the little finger, **5,** is present. Roman numerals suggest that the thumb, **I,** is present and the little finger, **V,** and last distal carpal and metacarpal are missing (*broken lines*). Carpal bones are *red.* Arrows indicate existing muscle attachments, **P,** the prepollex or primitive distal carpal number **2,** depending on interpretation. In the frog illustrated, the history of the bone at the base of the first metacarpal is conjectural.

portionate lengthening or shortening of some of the bones. Least common is an *increase in the number of phalanges.* Centralia frequently unite with one of the proximal carpals or disappear. As a result, most reptiles and numerous mammals have a single centrale, and it is sometimes found among the proximal row of carpals. Fusion of distal carpals 4 and 5 is common and results in a **hamate bone.** Phalanges or entire digits may be lost. In the latter event the corresponding metacarpal becomes vestigial or lost.

Most amphibians have five digits on the hind limb and four on the forelimb, although a few have less. Members of the neotenic urodele genus *Amphiuma* have one to three digits. The metacarpals that correspond to any lost ancestral digits have become vestigial (fig. 10.24, frog), or they have been lost (fig. 10.24, *Necturus*). The number of wrist bones also is smaller in modern than in labyrinthodont amphibians. The

embryonic intermedium and ulnare often coalesce during ontogenesis, an embryonic proximal carpal frequently coalesces with an adjacent carpal, and fusion between embryonic centralia and proximal or distal carpals is common. In addition, some ancestral carpals must have been lost completely.

Since the line of descent from the earliest amphibians to urodeles is largely conjectural, there is no present way to ascertain homologies between specific bones of the urodele hand and foot and those of the earliest labyrinthodonts. In fact, we cannot be confident that *any* ancestor of urodeles passed through a pentadactyl limb stage. Indeed, there is a possibility that urodeles were derived independent of other tetrapods directly from a crossopterygian ancestor and that they underwent an evolution that was more or less parallel with that of labyrinthodonts and the latter's descendants (see Leptospondyly, chapter 4).

Assuming, for a moment, that the skeleton of the manus of *Necturus* is derived from a pentadactyl ancestor, the three distal carpals could be considered to be, commencing on the radial side (fig. 10.24, starting with P), distal carpal 2, distal carpal 3 (two arrows attached), and hamate (fused distal carpals 4 and 5, two arrows attached). This presupposes that the thumb (1, broken lines) is missing. An alternative would be that the fifth finger (V, broken lines), rather than the thumb, is missing. The three distal carpals, starting with P, would then be a **prepollex** (a bone of unknown homology that develops in some species near the thumb, or **pollex**), carpals 1 plus 2, and carpals 3 plus 4. Carpal 5 (broken line) would be missing. This interpretation takes into account that the bone labeled P has no muscle connecting it with a finger (no arrow), and that the muscle from the digits numbered 2 and 3 in figure 10.24 attaches to the second of the three distal carpals (arrows). This carpal also frequently has two separate ossification centers, suggesting that it is a composite. This is one approach used in attempting to determine homologies of the bones of the hand and foot. The technique used by Alberch and Gale (1985) is another.

The muscles that insert on the hands and feet of urodeles are neither strong nor well differentiated, and the joints between the epipodia and wrists or ankles and between the wrists or ankles and the metacarpals and metatarsals are capable of little mobility. Thus, neither the hands nor the feet of urodeles generate locomotor thrust. They are chiefly platforms, or podia, which, pressing on the substrate, provide friction while muscles higher on the limbs extend the legs. The same is true of the hands, but not the feet, of anurans.

The hands of living reptiles and primitive mammals such as insectivores and primates tend to remain pentadactyl and to have five metacarpals and a nearly full complement of carpals except central carpals, or centralia (fig. 10.25, turtle and human). In crocodilians, however, the wrist has been reduced to five adult bones (fig. 10.25, alligator), and in birds the entire manus has been reduced (fig. 10.26). When present in mammals, the centrale may lie in the distal row of carpals, as in rabbits (fig. 10.27), or it may unite with the radiale and intermedium to form a scapholunar bone of triple origin, as in cats. Human fetuses have a centrale until the third fetal month, when it fuses with the radiale. Among major modifications of the hand are those for flight, life in the ocean, swift-footedness, and grasping.

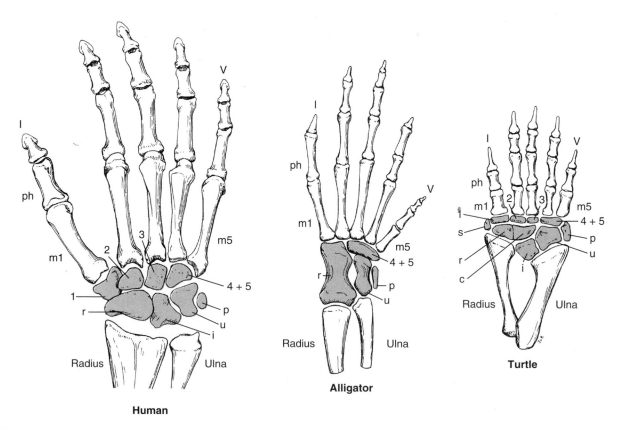

Human

Alligator

Turtle

FIGURE 10.25

Right manus of man, alligator, and turtle (dorsals views, carpal bones in *red*). **c,** centrale; **i,** intermedium; **m1** and **m5,** first and fifth metacarpals; **p,** pisiform; **ph,** proximal phalanx; **r,** radiale; **s,** radial sesamoid; **u,** ulnare; **1** to **5,** distal carpals; **l** and **V,** first and fifth digits. The alligator has an additional carpal that cannot be seen from this view.

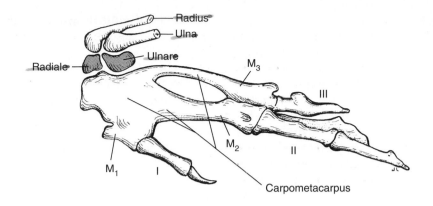

FIGURE 10.26

Left manus of a bird, **I** to **III,** Digits; **M₁** to **M₃,** metacarpals fused with three distal carpals to form a carpometacarpus.

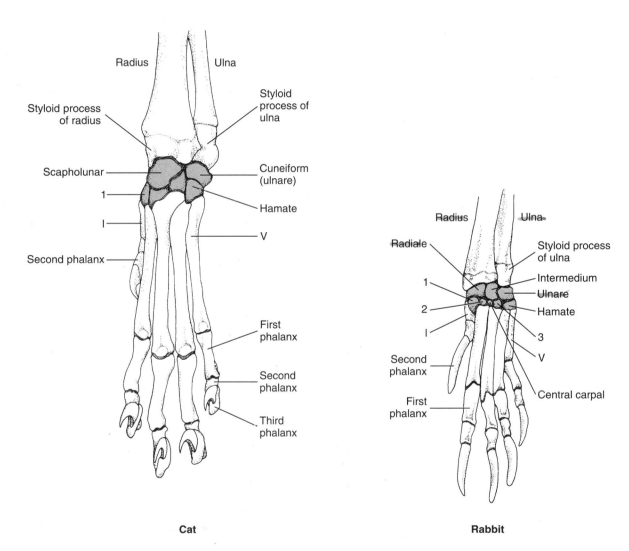

Cat   **Rabbit**

FIGURE 10.27

Left manus of cat and rabbit, anterior view, carpal bones in *red*. The pisiform bone is on the palmar side and cannot be seen. The scapholunar of cats combines radiale, intermedium, and central carpal, which are separate bones in kittens. The phalangeal formula of both animals is 2-3-3-3-3. **1** to **3,** first through third distal carpals. **I** and **V,** first and fifth metacarpals. Hamate incorporates fourth and fifth distal carpals.

Adaptations for Flight

In many birds, the hand has little independent role in propulsion in air, but being at the end of an airfoil, it has an aerodynamic effect. Loss and fusion of bones have reduced the hand to a rigid, tapering structure (fig. 10.26). Despite this, most of the basic components of tetrapod hands are identifiable in avian embryos. Two carpals (radiale and ulnare) form in the proximal row, and three in the distal row. As development progresses the three distal carpals unite with the three metacarpals to form a rigid **carpometacarpus.** Three fingers are usually present, and the number of phalanges has been reduced. (Terns often develop four embryonic digits, but only three persist.) The fingers often bear claws that are used in nonaerial locomotion, and like the rest of the hand, they are covered by feathers.

The first finger of birds that maneuver, alight, and take off in limited spaces is elongated, prominent, and independently movable and is called an **alula.** Songbirds have short, broad wings, and the feathered alula serves as an accessory airfoil at the leading edge of the wing as the bird flits among tree branches. Carnivorous birds have moderately long, broad wings adapted for slow-speed flight and quick landings in limited spaces. When these birds are braking, the alula is moved away from the rest of the wing, which creates a slot through which air rushes, making it possible to maintain stable low-speed headway until additional braking is applied with the main wings and tail feathers. There is a joint with considerable flexibility at the elbow, and another between the forearm and wrist, except in birds that hover. When flexed at the wrist, the hands exert a strong braking effect for landing, especially in birds with a large wingspread. Birds that hover have high-speed wings (up to 50 strokes per second in hummingbirds), the hand is as long as the arm or longer, and the entire wing is quite rigid.

Contrary to the condition in birds, the hand in pterosaurs and bats was, or is, the main part of the wing. Pterosaurs (fig. 10.28b) had four fingers, three of which were normal and bore claws. The fourth was embedded in the wing membrane (**patagium**) and consisted of four enormously elongated phalanges that made this finger as long as the entire body. The associated metacarpal was not elongated, but it was much enlarged. Bats have five fingers (fig. 10.28a). The thumb is normal and bears a claw. The other four fingers are elongated, and are associated with four greatly elongated metacarpals. The metacarpals and the phalanges of the four fingers constitute the skeleton of the patagium. The three proximal carpals are united in a single bone. Movement of the hand is responsible for takeoff and true flight in bats. No one has ever seen a pterosaur take off!

Gliding lemurs have a patagium, but it is less well developed than in bats and pterosaurs, and the fingers, although embedded in it, are not elongated. Gliding lemurs soar but are not capable of true flight. Patagiums in unrelated animals such as pterosaurs, bats, gliding lemurs, and gliding lizards are instances of evolutionary convergence.

Adaptations for Life in the Ocean

The hands have become paddlelike flippers in well-adapted marine amniotes. Among these are sea lions (fig. 4.44), cetaceans (fig. 4.49), sirenians (fig. 4.48), ichthyosaurs and plesiosaurs (fig. 10.29), seals (fig. 10.44), and penguins. Flippers are generally flattened and stout, and in several taxa the number of phalanges has greatly increased (fig. 10.30). In some ichthyosaurs there were as many as 26 phalanges per digit, and more than 100 in a single hand. Within the flippers of most other swimmers, however, the skeleton conforms closely to the generalized tetrapod pattern (fig. 10.31). Some aquatic mammals have lost all traces of hind limbs (fig. 10.32c).

Penguins obtain thrust for swimming solely from their flipperlike wings, their webbed feet serving as rudders. Sea turtles, fur seals, and sea lions swim by rowing action of their long front flippers. Some marine mammals, however, don't use their foreflippers to produce locomotor thrust when swimming. These include wriggling seals (true seals), walruses, cetaceans, and sirenians. How they swim, and how pinnipeds maneuver when on land will be discussed after we have studied the hind limbs.

Adaptations for Swift-Footedness

Mammals with pentadactyl hands and feet are usually **plantigrade,** which means that the wrists, ankles, and digits all rest on the ground. This is a primitive tetrapod stance not associated with fleetness. Generalized mammals—monotremes, marsupials, insectivores, primates—are plantigrade, as are some specialized mammals such as bears and the arboreal raccoons (fig. 10.33, monkey). Mammals in which only the first digit has been reduced or lost tend to be **digitigrade,** which means that they bear their weight on digital arches with wrist and ankle elevated (fig. 10.33, dog). Among digitigrade mammals are rabbits, rodents, and most carnivores. These mammals run faster, walk more silently, and are more agile than plantigrade species, and some, such as cats of various kinds, are the fastest runners of all. Cheetahs, for example, can sprint at an estimated 70 miles (112 km) per hour (Hildebrand, 1980). To a carnivore, speed makes eating more likely; to a herbivore such as a hare, it may mean reaching a burrow instead of being eaten.

The extreme modification of reducing the number of digits and walking on the tips of the remaining digits is seen in ungulates. Unguligrade mammals walk on four, three, two, or even one digit (fig. 10.33, deer), with wrists and ankles elevated well above the ground. The claws have become thick hoofs that bear the body weight and protect the living tissues of the toes from the abrasive action of the substrate (fig. 6.21, hoof). The metacarpals and metatarsals that correspond to the

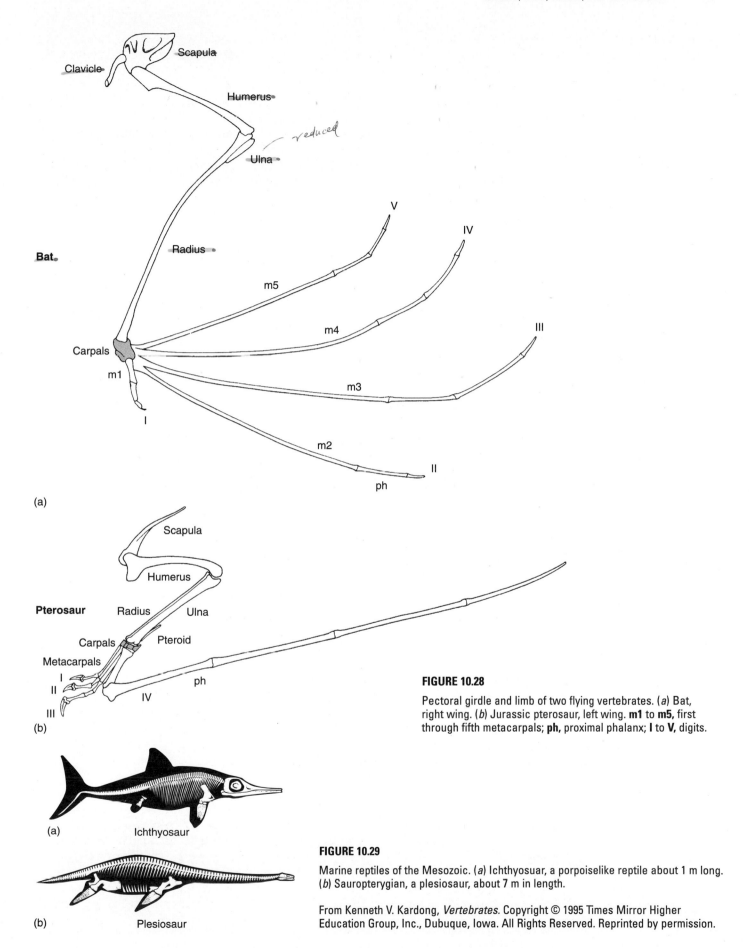

Bat.

reduced

Pterosaur

FIGURE 10.28

Pectoral girdle and limb of two flying vertebrates. (*a*) Bat, right wing. (*b*) Jurassic pterosaur, left wing. **m1** to **m5**, first through fifth metacarpals; **ph**, proximal phalanx; **I** to **V**, digits.

FIGURE 10.29

Marine reptiles of the Mesozoic. (*a*) Ichthyosuar, a porpoiselike reptile about 1 m long. (*b*) Sauropterygian, a plesiosaur, about 7 m in length.

From Kenneth V. Kardong, *Vertebrates*. Copyright © 1995 Times Mirror Higher Education Group, Inc., Dubuque, Iowa. All Rights Reserved. Reprinted by permission.

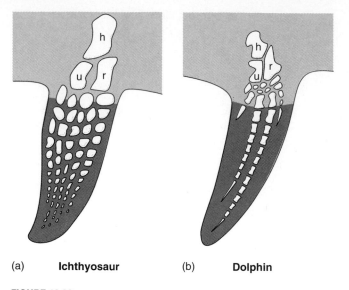

(a) **Ichthyosaur**   (b) **Dolphin**

FIGURE 10.30

Convergent evolution in anterior limbs. (*a*) Extinct, water-dwelling reptile. (*b*) Water-dwelling mammal. **h,** Humerus; **r,** radius; **u,** ulna.

FIGURE 10.31

Right scapula and forelimb of a beaked whale. A remarkable resemblance to basic tetrapod skeleton remains, although the limb has become paddlelike.

Negatives/transparencies #314469, Courtesy Department of Library Services, American Museum of Natural History.

FIGURE 10.32

(*a*) Skeleton of land-dwelling amniote. (*b*) and (*c*) Appendicular skeletal adaptations for life in the water. Appendicular skeleton is shown in *black,* (*b*) is a "wriggling seal" (*Phoca*). In the seal and porpoise, the hand is a paddle in which the phalanges are embedded.

FIGURE 10.33

Plantigrade, digitigrade, and unguligrade feet (from left to right). Ankle bones are *black;* metatarsals are *gray.*

Monkey   Dog   Deer

FIGURE 10.34

Right manus of several ungulates as seen from the front. **c2** to **c5,** distal carpals 2 to 5; **in,** intermedium; **m2** to **m5,** metacarpals 2 to 5; **pi,** pisiform; **ra,** radiale; **ul,** ulnare; **II** to **V,** digits. In horses, **m3** is the "cannon bone."

missing digits are vestigial or lost, and those that remain are much elongated and frequently united (fig. 10.34, horse, deer, camel). This has the effect of providing an additional functional segment to the limb. Although excelled in sprinting speed by some carnivorous digitigrade mammals, ungulates can sustain their speed for much longer. Not only are the feet of ungulates well suited for running, but they also function extremely well in rugged mountainous terrains. However, specialization has made their fingers and toes useless for anything else. This is the price of specialization.

Sequential evolutionary steps leading to the most specialized unguligrade stance may be illustrated by placing the fingers and palm flat on a tabletop, with the forearm perpendicular to the surface. This represents roughly the plantigrade position. Raising the palm off the table while keeping the fingers flat on the table illustrates roughly the digitigrade position. Unguligrade conditions may be illustrated by placing only the fingertips on the table and then raising the thumb, the little finger, the second finger, and finally the fourth finger, leaving only the third finger to bear the body weight, as in modern horses. Those fingers that fail to reach the table represent digits that have been successively reduced or lost in ungulates.

The horse underwent these successive changes commencing with the early *Eohippus*, which had four digits on the manus, and culminating in the modern *Equus*, which has a single digit. Despite extreme specialization of the manus of the modern horse, the proximal row of carpals (fig. 10.35) is intact, and the distal row lacks only the first carpal. With the loss of digits I, II, IV, and V, metacarpals 1 and 5 fail to develop and 2 and 4 have been reduced to splinters. Metacarpal 3, associated with digit III, has elongated.

Evolution among the ungulates progressed along two independent lines. In the line leading to **artiodactyls,** the weight of the body tended to be distributed equally between digits III and IV (fig. 10.35, camel). Thus arose the "cloven" hoof (fig. 10.36). Such a foot is said to be

FIGURE 10.36

Hoof of fetal pig. There is a cleft between the weight-bearing digits. Thus the hoof is "cloven." (*a*) As seen from in front. (*b*) As seen from behind.

paraxonic because the body weight is borne on two parallel axes. Artiodactyls of today have an even number of digits. In the evolutionary line leading to **perissodactyls,** the body weight is increasingly tended to be borne on digit III, the middle digit. This is a **mesaxonic** foot (figs. 10.34 and 10.35, horse). Most perissodactyls have an odd number of digits; tapirs, however, have a small fourth digit on the front feet, three on the rear. It is the mesaxonic foot and not the number of digits that defines the perissodactyls.

Adaptations for Grasping

Many mammals are able to flex their hand at the joint between palm and fingers. Rodents, for example, sit on their haunches and nibble on food held between *two hands,* which are flexed in this manner and which *face one another.* A further specialization is the ability to wrap the fingers *around* an object so that it is held securely in *one hand.* This is accomplished by flexing the fingers at each interphalangeal joint. Primates are among the few mammals that can do this.

Another step in the evolution of the mammalian hand was development of an opposable thumb—one that can be made to touch the tips of each of the other digits. This was accomplished by formation of a saddle joint at the base of the thumb where it meets the palm, by setting the thumb at increasingly wider and wider angles to the index finger, and by the evolution of strong adductor pollicis (thumb) muscles. True opposability is found in Old World monkeys, but even there the hand does not have the full range of functional capability that has evolved in human beings. Neither New World monkeys nor anthropoid apes have a perfectly opposable thumb. With such a hand humans were able to fashion increasingly sophisticated instruments, commencing with rocks chipped by design, and continuing to the electronic age. It has been said that the implements of early humans were as good (or as bad) as the hands that made them. Of course, evolution of the brain was an essential concomitant. Still, it seems impossible that any species on

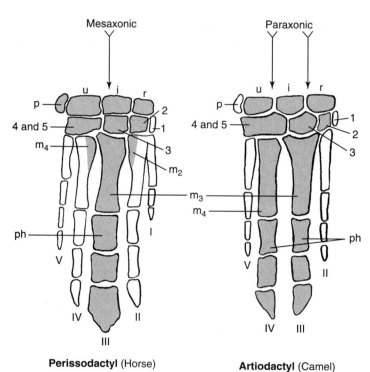

FIGURE 10.35

Mesaxonic and paraxonic manus of representative ungulates, showing bones lost (*white*) and retained (*red*) and indicating distribution of body weight through wrist and digits. The number of bony elements is correct for these animals. **i,** intermedium; **m₂** to **m₄,** second, third, and fourth metacarpals; **p,** pisiform; **ph,** first phalanx; **r,** radiale; **u,** ulnare; **1** to **5,** distal carpals; **I** to **V,** digits.

TABLE 10.3 Comparable Skeletal Elements of Manus and Pes

Manus*	Pes	Synonyms in Humans
Radiale	Tibiale	Talus or astragalus†
Intermedium	Intermedium	
Ulnare	Fibulare	Calcaneus
Pisiform		
Centralia (0 to 4)	Centralia (0 to 4)	Navicular
Distal carpal 1	Distal tarsal 1	Entocuneiform
Distal carpal 2	Distal tarsal 2	Mesocuneiform
Distal carpal 3	Distal tarsal 3	Ectocuneiform
Distal carpal 4 ⎤ Hamate	Distal tarsal 4 ⎤	Cuboid
Distal carpal 5 ⎦	Distal tarsal 5 ⎦	
Metacarpals (1 to 5)	Metatarsals (1 to 5)	
Digits (I to V)	Digits (I to V)	

*For synonyms, see Table 10.2.
†Incorporates the intermedium and a centrale.

earth lacking a prehensile hand, even having the brain, could evolve so sophisticated an existence.

Pes: The Hind Foot

A generalized pes is comparable bone for bone with the manus except that there is no consistent equivalent to the pisiform bone (table 10.3). Many early tetrapods had four centralia in the ankle (figs. 10.37 and 10.38, labyrinthodont). The number of centrale bones became smaller as diversity of modes of locomotion expanded. A glance at the centralia (dark red) of selected ankles in figures 10.38–10.40 illustrates the trend. The reduction is partly accounted for by the fact that primitive ossification centers for centralia tended to coalesce with those of the proximal or distal row of tarsals, thereby reducing the number of centralia. Some of the primitive centers still appear transitorily in the ontogeny of modern tetrapods.

The pes of three amphibians is illustrated in figure 10.38. The bones labeled **prehallux** in the labyrinthodont and anuran may be vestiges of a tarsal or metatarsal that was associated with an ancestral digit that has been lost. A pentadactyle limb was not the most primitive, so it is unlikely that these specific vestiges are homologues.

The number of ankle bones in anurans is considerably smaller than in urodeles. However, the tibiale and fibulare have elongated and united firmly at each end to articulate distally with the remaining tarsals in an intratarsal joint. This condition, although not unique, is not universal. Such an ankle, coupled with unusually long, webbed metatarsals and phalanges, provides a foot well adapted for pushing off for a leap on land, their only mode of terrestrial locomotion, or swimming in water, their alternate natural habitat. In swimming, the hind limbs are drawn forward against the body alternately, and then quickly thrust backward, toes spread apart, so that the broad fan-shaped foot pushes against the water, providing forward

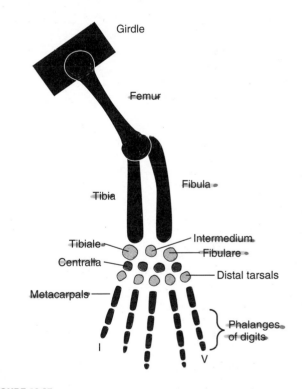

FIGURE 10.37

Skeletal organization of a primitive left hind limb. Tarsal bones (ankle) in *red,* centralia *dark red.* **I** and **V** are first and fifth digits.

thrust. The forelimbs assist in swimming to some extent, but are principally used in maneuvering.

Living reptiles display considerable loss and fusion of ankle bones, turtles to a lesser degree than others (fig. 10.39). The proximal row of tarsals is reduced by coalescence to a single bone in *Sphenodon* and many lizards and has been given the name **astragalocalcaneus.** It may incorporate all proximal tarsal bones and a centrale. A highly flexible intratarsal joint between

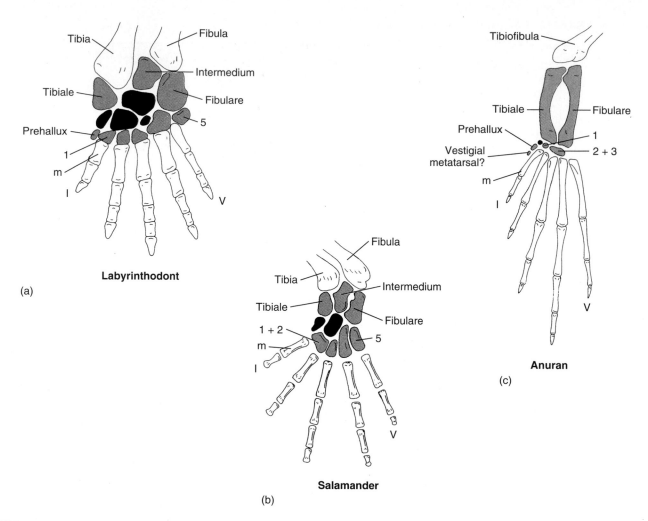

FIGURE 10.38

Left pes of amphibians. (*a*) Labyrinthodont. (*b*) Generalized plethodontid salamander. (*c*) *Rana catesbeiana.* Tarsal bones are *red,* and centralia are darker. **I** and **V**, first and fifth digits; **1** to **5**, distal tarsals; **m**, metatarsal.

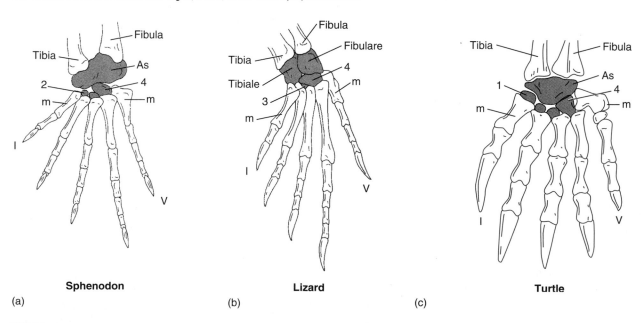

FIGURE 10.39

Left pes of reptiles. (*a*) *Sphenodon.* (*b*) The lizard *Uromastix.* (*c*) The snapping turtle *Chelydra.* Tarsal bones are *red.* **I** and **V**, first and fifth digits; **1** to **4**, distal tarsals; **As**, astragalocalcaneous; **m**, metatarsal. There are no independent centralia.

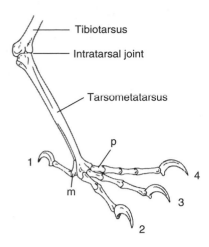

FIGURE 10.40

Left ankle and digits of a passerine bird, medial view, head to the right. **1** to **4**, digits; **m**, first metatarsal; **p**, phalanx. The claws obscure the terminal phalanges.

proximal and distal tarsals helps bipedal lizards to run rapidly in a digitigrade manner, the long tail maintaining balance. Most reptiles have five toes, although alligators and some lizards have four and some freshwater turtles have three. The phalangeal formula for *Sphenodon*, 2-3-4-5-4, is the generalized formula for reptiles. Alligators have reduced it to 2-3-4-4-0, and turtles to 2-3-3-3-2.

A bird's foot is highly modified (fig. 10.40). The proximal tarsals are united with the lower end of the tibia to form a **tibiotarsus;** there are no centralia; and the distal tarsals are united with the upper end of three fused metatarsals to form a long, rigid **tarsometatarsus.** A vestigial first metatarsal remains independent in some species. *There is an intratarsal joint between the tibiotarsus and tarsometatarsus and a joint between the tarsometatarsus and the toes.* The latter permits a digitigrade stance (fig. 4.29). This stance keeps the bird "ready" for takeoff, because extending the leg at the knee and intratarsal joints gives the initial thrust for becoming airborne. The thigh of a bird that is not in flight is not visible because it is quite short and, along with the knee, tucked away behind the wing.

Most birds have four toes, a few have three; ostriches alone have two. The phalangeal formula is the same as that for the first four toes of *Sphenodon* (thecodonts were ancestral to both). Usually three toes are directed radially forward and one comes off the back of the foot; but a few birds, including woodpeckers and parrots, have two toes at the back, the four forming an X (**zygo-dactyly**). This enables woodpeckers to obtain a firm grip on rough bark on a vertical tree trunk while drilling, braced against the trunk by stiff tail feathers. Birds can sleep while on a perch because the long tendons of the flexor muscles of the lower leg pass along the posterior aspect of the ankle to insert on the claw-bearing digits. The weight of the body pulls on the tendons, keeping the claws flexed on the perch (fig. 11.23).

Mammals, like their therapsid ancestors, lack an intratarsal joint but have a large hinge joint where the tibia and fibula meet the ankle (fig. 10.33). The tibiale is the principal weight-bearing bone of the ankle (fig. 10.41*b*). The other proximal tarsal, the fibulare, is elongated backward in plantigrade mammals, upward in digitigrades and unguligrades (fig. 10.33). It is the heel-bone of plantigrades. Except for a reduction in the number of centralia, the number of mammalian ankle bones has not changed appreciably since the first therapsids. In

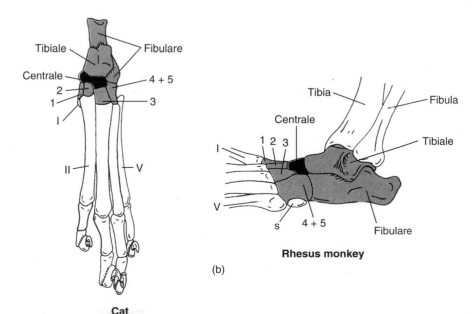

Cat

(a)

Rhesus monkey

(b)

FIGURE 10.41

(*a*) Left pes of cat, anterior view. Ankle bones are *red*. The phalangeal formula is 0-3-3-3. (*b*) Left ankle and associated bones of rhesus monkey, lateral view. **1** to **5**, distal tarsals; **I, II,** and **V**, metatarsals; **s**, sesamoid in the peroneus longus muscle.

Monkey   **Human**

FIGURE 10.42

The partially opposable big toe of an Old World monkey and the nonopposable toe of human.

hominoids, a **metatarsal arch,** or instep, distributes the body weight over four solid bases: the heel and ball of each foot. This has been compared with the four-pedestal architecture of the Eifel Tower. The arch also absorbs some of the shock generated by bipedal locomotion and provides "spring" for walking and running.

The phalangeal formula for the pes of early therapsids was 2-3-3-3-3, and this is still the typical formula for modern pentadactyl mammals and humans. The great toe, or **hallux,** is opposable in many primates but not in humans (fig. 10.42). The condition, when present, is correlated with brachiation.

Sculling and Galloping in Marine Mammals

Wriggling seals, walruses, cetaceans, and sirenians have unusual modes of swimming for mammals, dictated by the size and shape of the body and the morphology of the limbs they have inherited. In none do anterior flippers provide forward thrust for swimming, these being used chiefly for maneuvering. Cetaceans and sirenians have no pelvic flippers. How then do these marine mammals manage to survive?

Wriggling, or earless, seals swim with their foreflippers adducted against their sides, their necks retracted, and with *lateral undulations* of the posterior portion of their trunks. The tail is short, and of no use in locomotion. The posterior flippers, affixed to the side of the trunk by a narrow connecting "neck" enclosing the tibia and directed caudad, are swept alternately to the left and right with each lateral undulation cycle of the trunk. These sculling movements, in conjunction with synchronous limited flexion at the tibiofemoral and ankle joints with each sweep, propel these pinnipeds forward (fig. 10.43). Walruses also swim by sculling, but they are not strong swimmers. Sculling is a unique mode of swimming but one could say that seals and walruses make the most of what evolution has given them!

FIGURE 10.43

Sculling in a wriggling seal.

Source: Data from F.E. Fish, et al., "Kinematics and Estimated Thrust Propulsion of Swimming Harp and Ringed Seals" in *Journal of Experimental Biology,* 137:157, 1988.

Unlike seals, cetaceans derive their forward thrust from a stiff, horizontal bifluked tail, which is operated by alternate *dorsal and ventral undulations* of the caudal portion of the trunk—a sort of legless gallop—rather than by lateral undulations. This swimming mode is analogous to that in the similarly shaped tuna and other fishes of the mackerel family, and represents functional convergence resulting from analogous morphological features. Both functional modes of swimming—sculling in pinnipeds and galloping in cetaceans—have been shown by kinematic studies to represent highly efficient propulsive mechanisms (Fish and others, 1988).

Sirenians, too, derive their locomotor thrust from dorsoventral undulation of their flukelike tail; but these vegetarian, shallow-water creatures are sluggish swimmers (their food cannot flee!) and can scarcely be said to gallop.

How Pinnipeds Maneuver on Land

Fur seals, sea lions, and walruses come ashore more frequently than wriggling seals. They have an adaptation that enables the hind flippers to change readily from a caudally directed swimming position to the equivalent of a tetrapod stance. The hind flippers are simply rotated beneath the body from a swimming position to a tetrapod position or the reverse, as the animal leaves or enters the water. (Human beings can't do this!) A flexible wrist joint further adapts them for locomotion on land. Wriggling seals are not so gifted. Their hind limbs are permanently bound to the tail; consequently, movements on land consist of pulling themselves along by wriggling, which accounts for their name.

The stance of young fur seals during their first months on land is more tetrapod than pinniped (fig. 10.44). Both pairs of juvenile flippers are disproportionately large compared with those of adults. These oversized flippers, coupled with rotating pelvic appendages, enable pups to run quickly and nimbly over the rookery grounds.

Origin of Limbs

Although the problem of the origin of paired fins may never be solved, the same is not true of the origin of tetrapod limbs. The paired fins of some ancient fish must have been the precursor of tetrapod limbs. The question then arises, did the skeleton of the fin of any known Devonian fish evince the potential for becoming a limb? For an

FIGURE 10.44

Tetrapod stance of an eared seal pup.

answer we turn to the lobe-finned crossopterygians, whose skull and fin skeleton bore a remarkable resemblance to those of the first tetrapods (figs. 9.8 and 10.45).

In the pectoral fins of crossopterygians, a single basal bone, which we will call a humerus, articulates proximally with the scapula and distally with a pair of radials we will call radius and ulna (fig. 10.45). Loss of fin rays and modifications of the radials distal to the radius and ulna could have produced the skeleton of the first tetrapod limb. It has been proposed that the first two rows of labyrinthodont carpals may be modified radials and that

Crossopterygian

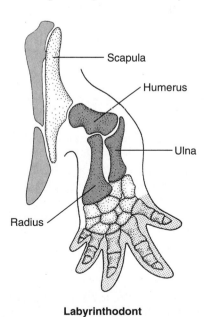

Labyrinthodont

FIGURE 10.45

Left pectoral fin of a Devonian crossopterygian, and left forelimb of an early tetrapod oriented to show similarities of endoskeleton. *Dark gray* shows dermal bones of pectoral girdle. The fin is viewed from above; the limb is viewed from in front.

the more distal bones of the manus may have evolved new ossification centers. Meanwhile, the girdles of labyrinthodonts remained fishlike.

The changes in external appearance necessitated by the transition of fins into limbs are illustrated in figure 10.46. These involved the following modifications, not necessarily in the sequence listed, not simultaneously, and certainly not exclusively, inasmuch as concomitant adaptive modifications were necessitated in the entire skeletomuscular system subserving each appendage: (1) elongation of the two bones of the epipodium; (2) formation of hinge joints between the propodia and epipodia (now elbow and knee joints), and between the epipodium and wrist or ankle; (3) rotation of the long axes of the humerus and femur to parallel the vertebral column (fig. 10.32a); and (4) emergence of a definitive manus and pes. The transition appears to have occurred sometime during the Devonian period.

It is probable that crossopterygian fins were used at times for resting on the water bottom. They would not have borne appreciable weight because of the bouyancy of water. Minor modifications would have permitted "walking" on the muddy bottoms close to shore. Several hundred living species of fishes do this today, including the Australian lungfish. Using pectoral fins, some living fishes move several feet inland nightly. Some climb inclined planes with little fringed fins that have a remarkable resemblance to hands.

The pressures that drove vertebrates onto the land must, of necessity, be conjectural. It may have been the absence of predators on land, or that there was less competition for food, or simply that food was abundant on land. Or perhaps it was simply a manifestation of the tendency of organisms to invade a contiguous environment whenever nothing prevents the invasion. Whatever the explanation, it was inevitable that a limb more suitable for life on land would evolve from the fin of a fish.

Locomotion on Land without Limbs

Limbs provide leverage for locomotion on land. Snakes and other limbless tetrapods cannot propel themselves in this manner. Nevertheless, snakes are highly successful on land because modifications of the vertebral column, ribs, body wall musculature, and skin have provided them with alternative methods of moving about.

The most common method of locomotion in snakes and limbless lizards is by forming irregular loops that become propped against, and push against, any available stationary object on their path, such as a clump of vegetation, a rock, a root, or simply an irregularity in the surface (fig. 10.47). While the head is being thrust forward, the eyes scan the substrate for the next contact point. When one looms, the head is thrust in that direction. On contact a wave of muscular contraction commences at the head and travels to the last segment of the tail, drawing the trunk and tail toward the head. A minimum of three

FIGURE 10.46

From fin to limb in the Devonian. (a) Crossopterygian. (b) Hypothetical transitional stage. Note location of future elbow joint (arrow **1**). (c) and (d) Temnospondylous labyrinthodonts. In (c), note elongation of forearm and formation of a wrist joint (arrow **2**). (e) and (f) Orientation of skeletal elements of (b) and (c). In (f), note that some reorientation of the radius and ulna may have been necessary for the manus to lie flat on the ground.

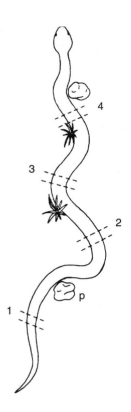

FIGURE 10.47

Serpentine locomotion. Forward progression is achieved by exerting pressure against contact points, or props, **p**, on the path. The head moves forward to a new contact point, and the rest of the body follows the established looping path, the segment at position **1** moving successively to positions **2, 3,** and **4** while the loops remain stationary with reference to the contact points. Eventually, each loop will be at the end of the tail and new loops will have formed anteriorly. Any track consists of sinuous waves.

From Carl Gans, *Biomechanics: An Approach to Vertebrate Biology*, 2nd edition. Copyright © 1980. University of Michigan Press, Ann Arbor, MI. Reprinted by permission.

contact sites are necessary for this mode of locomotion, and no appreciable force is exerted downward against the ground. This mode of locomotion is referred to as **serpentine.** A basic necessity is a body that is metameric with respect to the adult skeletomuscular system. The term **lateral undulation** has also been applied to serpentine locomotion, but there are functional aspects of serpentine locomotion that are not operative in fishes.

Some snakes glide—seemingly flow—forward on the substrate while keeping the entire body in a straight line, a spectacle that some viewers find unnerving! The ventral skin acts like a conveyor belt, sections of which stop and go. This has been termed **rectilinear locomotion,** and it depends on generating friction between sections of the ventral skin and the substrate. Unlike snakes that press laterally against contact points for leverage, these snakes press against the ground, using bunched groups of belly scutes as intermittent holdfasts. After a brief interval of holding, during which time the next more caudal group of scutes is catching up, the downward

pressure exerted by the first group is released, and the group flows forward, stretched out and elevated from the substrate, until it catches up with the group ahead. Meanwhile, the group immediately behind is holding fast. Thus rhythmic waves of forward flowing and resting scutes sweep along the body from head to tail. The skin, like that of many other snakes, is only loosely attached to the underlying tissues; there is much elastic tissue in the dermis, and the scutes, approximately one body segment long, overlap and are connected to the next by a pleat of cutaneous membrane that unfolds during movement. This combination of factors enables each scute to be released from the ground and commence its forward movement an instant before the next.

Two sets of striated muscles are responsible for rectilinear locomotion. A pair of slender costocutaneous muscles extends downward and backward from high on each rib to the dermis at the edges of each scute. When these contract they lift the scute off the substrate so that it no longer serves as a holdfast, and they draw it forward while stretching the interscutal membrane. A second, more powerful pair extends less obliquely from a scute to the lower end of a more caudal rib. Contraction of the latter maintains the forward movement of the body mass within the skin envelope. Amphisbaenians use rectilinear progression, but the entire skin moves, not just the ventral skin.

Sidewinding (fig. 10.48) enables rattlesnakes and other snakes to inhabit sandy deserts, where the ground offers too few stationary contact points for

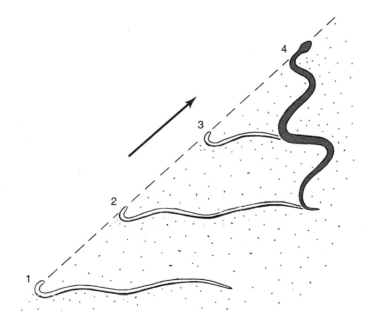

FIGURE 10.48

Locomotion in sidewinder snakes. **1** to **4,** successive tracks in sand left when the head and neck are lifted from the ground, thrust forward, and set down firmly at the next position. The body does not touch the ground between tracks. The rest of the body is "flowing" above ground to the new position. Dotted lines and arrow indicate direction of movement.

serpentine locomotion and is too unstable to provide sustained friction for rectilinear locomotion. Sidewinding is also used by many snakes in other environments when they are temporarily in a situation where, because of the nature of the surface, other methods of locomotion would be clumsy or ineffective. In sand, the body usually occupies two or three tracks at a time. More or less of the anterior quarter of the snake is thrust forward to start a new one.

Snakes propel themselves in a burrow by modified serpentine movements, bracing S-shaped loops against the burrow wall and exerting horizontal force while thrusting the head and forebody forward. The more caudal sections then advance. The process, reminiscent of the appearance of the bellows of an accordion (concertina) that is being played, has been called **concertina movement.** Snakes often combine several methods of locomotion in appropriate habitats as a common method of going from place to place.

All the foregoing methods of locomotion of snakes on land are made possible by vertebral columns consisting of up to 400 or more highly flexible intervertebral joints; by exceptional ribs that extend from the atlas to the tip of the tail and reach almost to the midventral line; by the unusually large number of muscle bundles that interconnect vertebrae and connect vertebrae with ribs; by wide, smooth, overlapping horny ventral scutes interconnected by pleated membranes; by the exceptional elasticity of the dermis; and by loose skin that allows independent movement of skin and enclosed body mass.

Of course, snakes and limbless lizards are not the only tetrapods that move about on land without limbs. Amphisbaenians use a modified serpentine movement in their burrows. Limbless amphibians on land employ basically the swimming movements of fishes, that is, lateral undulation of the body and tail. It doesn't provide much speed, but these amphibians don't venture far from water or moist habitats. The most specialized techniques and the largest number of adaptations are those in snakes.

About 15 genera of sea snakes, some of which attain a length of 3 meters, inhabit the tropical waters of the Indian and Pacific Oceans. These well-adapted aquatic snakes have a body that is flattened laterally, and their oarlike tail is used as a scull to produce the thrust for locomotion in the sea. Locomotion in marine mammals was discussed earlier.

SUMMARY

1. The appendicular skeleton consists of pectoral and pelvic girdles and the skeleton of the fins or limbs.

2. All pectoral girdles are modifications of a basic morphologic pattern seen in early bony fishes, consisting of three cartilages or endochondral bones and four or five membrane bones derived from dermal armor.

3. Coracoids, scapulae, and suprascapulae are endoskeletal. Coracoids are absent above monotremes. Scapulae are universal except in some species that lack forelimbs.

4. Clavicles, cleithra, supracleithra, posttemporals, and postcleithra are dermal bones. Cleithra and supracleithra are confined to fishes. Clavicles occur in all classes of bony vertebrates but are best developed in tetrapods. A midventral interclavicle of intramembranous origin is found in reptiles, birds, and monotremes.

5. Recent bony fishes have tended to lose the replacement bones of the pectoral girdle. Tetrapods have tended to lose dermal bones.

6. There are no dermal bones in the pelvic girdle.

7. The pelvic girdle of fishes consists of two pelvic (ischiopubic) plates. These articulate laterally with a pelvic fin, and usually meet ventrally in a symphysis. The plates unite to form a median bar in sharks and lungfishes.

8. In tetrapods, two ossification centers in each pelvic plate become ischial and pubic bones. An ilium braces the girdle against the vertebral column.

9. Ilium, ischium, and pubis unite to form an innominate bone (coxa) in mammals. It encloses an obturator foramen.

10. Associated with the pelvic girdle midventrally are prepubic (ypsiloid) cartilages in some urodeles, and epipubic and hypoischial cartilages in many reptiles, monotremes, and marsupials.

11. The sacrum and pelvic girdle of mammals and some reptiles constitute a pelvis that encompasses a pelvic cavity.

12. Paired fins are of three types: lobed (in Sarcopterygii), fin fold fins (in Chondrichthyes), and ray fins (in Actinopterygii).

13. The skeleton of a generalized fin (paired or median) consists of layered rows of basal and radial cartilages or bones.

14. Lobed fins have a central bony axis and pre- and postaxial radials or a modification thereof. Ray fins have lost basals and the radials may be reduced or absent.

15. Cartilaginous fishes have fibrous fin rays (ceratotrichia). The rays of bony fishes (lepidotrichia) consist of jointed dermal bones.

16. Dorsal and anal fins are unpaired. Caudal fins are diphycercal, heterocercal, homocercal, or hypocercal, depending on the direction taken by the notochord and vertebral column within the fin. The basal skeleton of homocercal fins consists of residual components of neural and hemal arches and spines of modified caudal vertebrae. Homocercal fins are further stiffened by a urostyle and hypural bones.

17. The origin of fins is unknown. Fin folds, gill arch, and fin spine hypotheses have been proposed.

18. Tetrapod limbs are modifications of crossopterygian fins.

19. Limb skeletons consist of a propodium (humerus or femur), epipodium (radius and ulna or tibia and fibula), and autopodium (manus or pes).

20. Structural modifications of the manus and pes involve reduction in the number of bones, disproportionate lengthening or shortening of segments, and in some aquatic tetrapods, increase in number of phalanges.

21. Digits have been reduced to four or fewer in modern amphibians, three in birds, and as few as one in some ungulates. Loss of digits is accompanied by loss or reduction of associated carpals and metacarpals or tarsals and metatarsals.

22. The most striking modifications of the manus are in flying tetrapods (pterosaurs, birds, bats), in water-adapted amniotes, and in ungulates.

23. Mammalian stances are plantigrade, digitigrade, or unguligrade. Ungulate feet are hoofed and mesaxonic (perissodactyls) or paraxonic (artiodactyls).

24. A reduced number of ankle bones in birds have fused with the tibia and metatarsals to add an additional elongated segment to the hind limbs. An intratarsal joint adds to the flexibility of the pes.

25. Excepting sesamoid bones, the bones of fins and limbs are all endochondral.

26. Most fishes swim by lateral undulation. Marine tetrapods obtain forward thrust from anterior flippers, lateral undulation, sculling, or, in cetaceans, galloping. Of the limbless terrestrial tetrapods, snakes show the largest number of adaptations for locomotion. Modifications of the manus are among the adaptations that enable true flight.

SELECTED READINGS

Additional references to locomotion will be found in Selected Readings at the end of chapter 11.

Alberch, P., and Gale, E. A.: A developmental analysis of an evolutionary trend: digital reduction in amphibians, Evolution **39**(1):8, 1985.

Blake, R. W.: Fish locomotion, New York, 1983, Cambridge University Press.

Coates, M. I., and Clack, J. A.: Polydactyly in the earliest known tetrapod limbs, Nature **347**:66, September, 1990.

Fish, F. E., Innes, S., and Ronald, K.: Kinematics and estimated thrust propulsion of swimming harp and ringed seals, Journal of Experimental Biology **137**:157, 1988.

Gans, C.: Biomechanics: an approach to vertebrate biology, Philadelphia, 1974, University of Michigan Press.

Gould, S. J.: Bent out of shape, Natural History, **99**(5):12, May 1990. A popular story of the discovery of the hypocercal tail of ichthyosaurs.

Gould, S. J.: Eight (or fewer) little piggies, Natural History, **100**(1):22, January 1991.

Jarvik, E.: Basic structure and evolution of vertebrates, vol. 1, New York, 1980, Academic Press.

Padian, K.: The flight of pterosaurs, Natural History, **97**(12):58, 1988.

Rackoff, J. S.: The origin of the tetrapod limb and the ancestry of tetrapods. In Panchen, A. L., editor: The terrestrial environment and the origin of land vertebrates, New York, 1980, Academic Press.

Romer, A. S.: Osteology of the reptiles, Chicago, 1956, University of Chicago Press.

Shubin, N., and Wake, D.: Phylogeny, variation, and morphological integration, American Zoologist **36**:51, 1996.

Strathmann, R. R.: Why life histories evolve differently in the sea, American Zoologist **30**:197, 1990.

Webb, P. W.: Body form, locomotion and foraging in aquatic vertebrates, American Zoologist **24**:107, 1984.

CHAPTER ELEVEN

MUSCLES

In this chapter we will classify muscle tissues and muscles from several viewpoints and then examine skeletal muscles of fishes and tetrapods in functional groups, noting the primitive pattern and seeing how it has been modified for life on land.

OUTLINE

uscle is a tissue; muscles are organs. The tissue, hence the organ, is specialized to perform a single function: to shorten when stimulated, and thereafter to recover. Shortening is a result of chemical changes in two muscle proteins, actin and myosin. Shortening of a sufficient number of contractile units causes a corresponding shortening and fattening of the entire muscle mass. If the muscle surrounds a lumen, the lumen is compressed. If it extends between two structures, one of these is drawn toward the other. The usual stimulus for muscle contraction is a nerve impulse, although this is not true of cardiac muscle, as will be explained shortly.

This chapter is concerned primarily with the muscles that stabilize or operate the skeleton. These muscles are the final determinants of posture, locomotion, and orientation of the body in the environment.

MUSCLE TISSUES AND MAJOR CATEGORIES OF MUSCLES

Tissues and organs may be classified in accordance with any criterion that serves a useful purpose. The major categories of muscle tissues and muscles, as employed in functional and descriptive morphology, is the subject of the paragraphs that follow.

Striated, Cardiac, and Smooth Muscle Tissues

The criterion most commonly employed in classifying muscle tissue is histology. Using this criterion, muscle tissues are either striated, cardiac, or smooth.

Striated Muscle Tissue
Striated muscle is a tissue composed of long, cylindrical, multinucleate **muscle fibers,** each with transverse bands and longitudinal striae visible with light microscopy (fig. 11.1*a*). The longitudinal striations are caused by parallel threadlike **myofibrils** within the fiber. Each myofibril consists of repeated **sarcomeres** along its length. Each sarcomere consists of two species of longitudinally oriented proteinaceous **myofilaments,** myosin and actin, not visible by light microscopy. The cross striations of the muscle fiber are the result of the perfect alignment of all the sarcomeres within the muscle fiber.

The striated muscle fiber is not a cell but a **syncytium** (a single, functional unit) formed during histogenesis when myoblasts become aligned end to end and unite, contributing nuclei and protoplasm to the elongated syncytium contained within a delicate plasma membrane, the **sarcolemma.** Muscle fibers assemble to form skeletal muscles consisting also of the connective, vascular, and neural tissues that provide mechanical support, tensile strength, metabolic needs, and stimuli for contraction. Impulses arrive at motor nerve endings and cause release

of neurotransmitter amines into a space bordered by the **motor end plate** (fig. 11.2), a portion of the sarcolemma with receptors for neurotransmitters. The nerve endings depress the sarcolemma at this point but do not pierce it. The motor end plate initiates stimuli along the sarcolemma that result in biochemical interactions between the actin and myosin that result in shortening of the sarcomeres. It is hypothesized that the reaction causes actin molecules to slide toward one another. When sarcomeres shorten, myofibrils bulge and thus the entire muscle shortens and thickens (fig. 11.1*a*).

A single nerve cell supplies motor end plates on not one, but many, muscle fibers, so that a functional group of muscle fibers known as a **motor unit** contract simultaneously. The larger the number of motor units stimulated, the greater will be the effect of the contraction.

Cardiac Muscle Tissue
Heart muscle is a unique variety of striated muscle tissue (fig. 11.1*b*). It contains myofibrils and filaments of actin and myosin arrayed similarly to those in skeletal muscle fibers. The mechanisms of contraction are essentially the same. The tissue differs in that (1) its cells are generally uninucleate, (2) cells are separated by unique boundaries—**intercalated disks,** (3) it can contract without nervous stimulation, and (4) it is innervated by fibers of the autonomic nervous system. The intercalated disks link cytoplasms of adjacent cells, facilitating ion transport and thereby the rapid passage of action potentials. The disks and branching nature of the cells thus form another type of muscle syncytium. Cardiac tissue regularly self-depolarizes (myogenic) and conducts impulses through its own system of fibers. Autonomic nerves modify the rhythmicity of this myogenic activity.

Smooth Muscle Tissue
The cells of smooth muscle tissue are fusiform, uninucleate, have myofibrils, but lack cross striations (fig. 11.1*c*). Unlike striated muscle tissue, it occurs most often in sheets as part of an organ. Like cardiac muscle, it is innervated by the autonomic nervous system.

Major Categories of Muscles

Somatic Muscles
On the basis of the major role a muscle plays in survival of the organism, it can be classified as somatic or visceral. Somatic muscles orient the body (soma) of the organism in the external environment. They are striated muscles that are attached to the ligaments, tendons, and bones of the axial skeleton, to the appendicular skeleton, and to the skeletal components of the lateral and ventral body walls. They are innervated by spinal nerves, excepting the small number that operate the eyeballs and the tetrapod tongue.

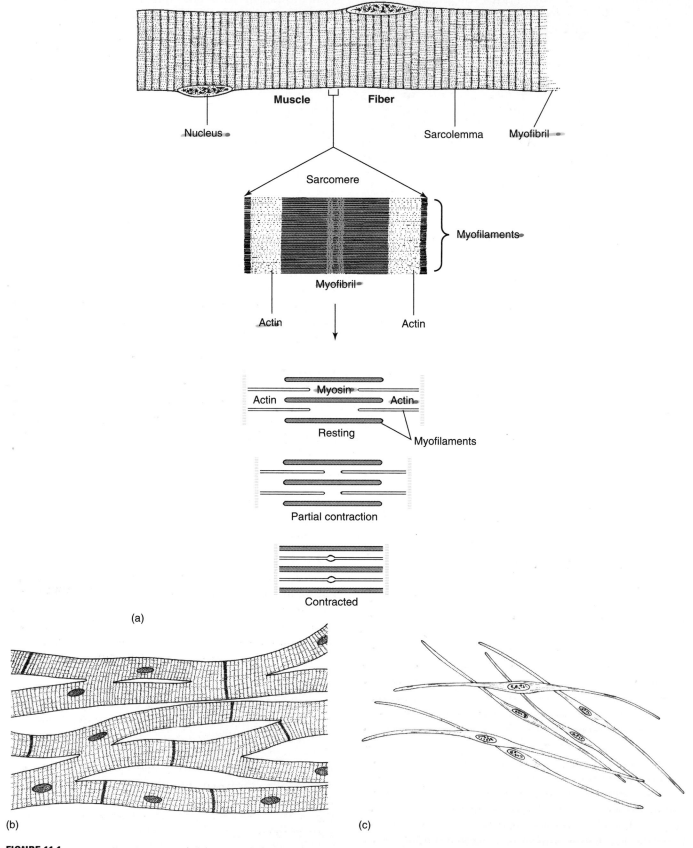

Nucleus

Muscle   Fiber

Sarcolemma   Myofibril

Sarcomere

Myofilaments

Myofibril

Actin   Actin

Myosin

Actin   Actin

Resting

Myofilaments

Partial contraction

Contracted

(a)

(b)   (c)

FIGURE 11.1

Muscle tissues. (*a*) Section of a striated muscle fiber, as seen with light microscopy (top); a schematic section of one myofibril consisting of myofilaments of myosin (*red*) and actin; and a schematic representation of muscle contraction as a function of sliding myofilaments (*lower*). The left and right boundaries of the sarcomere are too fine to be seen with light microscopy. (*b*) Cardiac muscle tissue. Dark bars are intercalated disks at cell boundaries. (*c*) Isolated smooth muscle cells from intestinal wall.

Myelinated fibers

Naked fibers

End plate

FIGURE 11.2
Innervation of skeletal muscle fibers.

Courtesy Gerrit L. Bevelander and Judith A. Ramaley.

In general, somatic muscles are said to be **voluntary** if they can be contracted at will. This doesn't preclude reflex contraction when the body is endangered, such as when the skin unexpectedly comes in contact with a pin, when holding one's breath becomes detrimental to physical welfare, or when a tendon is unavoidably stretched.

Somatic muscles are derivatives, either ontogenetically or phylogenetically, of the myotomes of mesodermal somites (figs. 5.9, 11.6, and 11.7). For this reason, they are often referred to as **myotomal,** or **somitic, muscles.**

Visceral Muscles

Visceral muscles in general maintain an appropriate internal milieu. They are the smooth muscles of hollow organs, vessels, tubes, and ducts, the intrinsic musculature of the eyeballs, and the erector muscles of feathers and hair. Visceral muscle also includes cardiac muscle. They are derived chiefly from splanchnic mesoderm (figs. 5.11 and 11.7) and are innervated by the autonomic nervous system. Figure 16.35 shows the innervation of a few visceral organs.

Visceral muscles have undergone relatively few evolutionary changes through vertebrate history because they are less subject to the vagaries of the environment. Therefore, they will receive little attention in this chapter. Some contrasts between somatic and visceral muscles are listed in table 11.1.

Branchiomeric Muscles

Branchiomeric muscles belong to the pharyngeal arches and their ontogenetic or phylogenetic derivatives from fishes to human beings. They are striated skeletal muscles, but they have features, chiefly developmental, that do not fit nicely into either the somatic or visceral category. They are innervated by cranial nerves. They will be discussed later in the chapter.

The remainder of this chapter focuses primarily on skeletal muscles. These are the ones that have diversified beginning with the earliest tetrapods. Discernment of a basic pattern of disposition of skeletal muscles from fishes to humans, discovery of the commonality of their embryonic origins, and the uncovering of the evolutionary changes that have occurred in muscle systems during geologic time have been achieved by research that began during the Renaissance and that will continue into the twenty-first century aided by increasingly sophisticated techniques of present and future scientists.

INTRODUCTION TO SKELETAL MUSCLES

Skeletal Muscles as Organs

Skeletal muscles consist of muscular and tendonous portions (fig. 11.3). Surrounding a muscle is a tough, glistening fibrous sheath, the muscle fascia, or **epimysium.** It consists chiefly of collagenous connective tissue and elastic fibers in small amounts that vary with the muscle. Major bundles of muscle fibers (fascicles) within the muscle are surrounded by a **perimysium,** which also penetrates the bundles to encapsulate smaller fascicles. The smallest bundles consist of relatively few muscle fibers and constitute the functional units of the muscle. The muscular, neural, and vascular components of each functional unit are supported by a very fine collagenous reticulum, the **endomysium,** a continuation of the perimysium. It surrounds each individual muscle fiber superficial to the sarcolemma. The epimysium, perimysium, and endomysium comprise a single continuum of high tensile strength that encompasses all contractile units of the organ.

Tendons are continuations of the muscle beyond the site where fascicles end. The collagenous bundles of the perimysium and epimysium continue into and become

TABLE 11.1   Contrasts between Somatic and Visceral Muscles

Somatic Muscles	Visceral Muscles
Striated, skeletal, voluntary	Smooth, nonskeletal, involuntary*
Primitively segmented	Unsegmented
Myotomal	Arise mostly from lateral mesoderm
Mostly in body wall and appendages	Mostly in splanchnopleure†
Primarily for orientation in external environment	Regulate internal environment
Innervated directly by spinal nerves and cranial nerves III, IV, VI, and XII	Innervated by postganglionic fibers of autonomic nervous system

*Cardiac muscle is striated.
†Those that erect hairs or feathers or constrict blood vessels are in the skin.

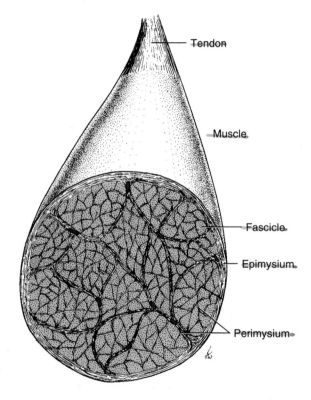

FIGURE 11.3

A mammalian fusiform skeletal muscle in cross section. All visible subdivisions are fascicles. The smallest fascicles and endomysium are too fine to be seen.

part of the tendon. Similarly, at the site of attachment of the tendon to the skeleton, the collagenous bundles of the tendon continue into and contribute to the perichondrium or periosteum of the bone to which they are attached. Consequently, tension produced by muscle contraction is transmitted throughout the entire organ and from one skeletal attachment to another.

Red (Slow) and White (Fast) Muscle Fibers

The fibers of striated muscle are of several functional varieties that can be identified histochemically on the basis of the biochemical reactions that drive the contractions and subsequent recovery of the tissue. They can be identified less precisely as red and white muscle fibers. Muscles in which either type predominates are visibly identifiable in the body wall of a shark in a dissection laboratory, in many skinned game fishes, and in fish and poultry at the dinner table.

Red fibers (slow fibers) have a richer blood supply, contain relatively large amounts of a pigmented hemoglobinlike molecule (**myoglobin**) that gives them a dark color, contract more slowly, and are relatively fatigue resistant, or in the case of fibers that simply maintain muscle tonus (tonic fibers), almost nonfatigable. These fibers enable a muscle to work for extended periods of time as, for example, in fishes that are cruising rather than swimming in fast spurts, or in mammals that are merely maintaining their posture as opposed to fleeing from an enemy or pursuing prey. White fibers (fast fibers) lack myoglobin and contract rapidly, but they fatigue more quickly.

Most muscles from fishes to humans are mixtures of red, white, and intermediate types, the proportions of each varying with the role the muscle plays in survival. For example, the muscles that move the head of a rat consist of a very high proportion of fast fibers, whereas in corresponding muscles of domestic cats the frequency of slow fibers is considerably higher. This correlates with the fact that rats (sitting on their haunches) monitor the environment more frequently and with more rapid head movements than domestic cats. On the other hand, the appendicular muscles of both domestic cats and rats are rich in fast fibers. Continued use of either fast or slow fibers results in increase in the size of the fiber, hence in the size and strength of the muscle.

During dissection of adult mammals, students frequently notice that skeletal muscles differ in size in the two sexes. Androgens, the predominant gonadal hormones of males, cause amino acids to be linked together into polypeptides and proteins. Inasmuch as muscle is 80 percent protein, androgen produces statistically demonstrably larger muscles in males.

Origins, Insertions, and Muscle Shapes

The **anatomic origin** of a muscle, as opposed to its phylogenetic or ontogenetic origin, is the site of attachment that remains fixed under most functional conditions; that is, the bone on which it originates is not displaced when the muscle contracts. For example, when the biceps muscle of the upper arm contracts (*Bi*, fig. 11.4), the forearm is flexed, that is, drawn toward the upper arm. The origin of the biceps is therefore somewhere above the elbow. The **insertion** of a muscle is the site of attachment that is usually displaced by contraction of the muscle. The biceps inserts on the forearm. A muscle may cause displacement of the bone of origin instead of the bone of insertion if the former is immobilized by other muscles. For example, the geniohyoid muscle, which extends between the hyoid bone and the lower jaw at the chin, either lowers the lower jaw or draws the hyoid forward, depending on which bone is immobilized at the time.

When thinking of muscles, one generally envisions a muscle structured like the biceps brachii of a rabbit, consisting of a belly and two tendons (fig. 11.4). This is a fusiform muscle. (The name biceps is a misnomer for this muscle in cats and rabbits, because it has only one head, or tendon of origin. In humans it has two.) Some muscles (digastrics) have two bellies; some have three heads or four. Many muscles are straplike as, for example, the geniohyoideus of tetrapods (fig. 11.15) and the sternomastoid muscles of mammals (fig. 11.25c). Less frequently seen are pinnate muscles, so named because their structure resembles that of a contour feather. Pinnate muscle slips "feather out" from a central or bordering tendon to insert at multiple sites, such as on successive vertebrae. Muscles with still other contours, such as the domed diaphragm with a central tendon (fig. 11.13), are seen in mammalian anatomy laboratories.

Tough, thin, sheetlike expanses of mammalian tendons and ligaments are known as **aponeuroses.** One of these, the **galea aponeurotica,** is the major component of the mammalian scalp, lying in intimate association with the integument and being the common tendon of insertion for a number of thin, broad integumentary muscles of the forehead, temporal, and occipital regions (fig. 11.26e and f). If you place your fingers on your own scalp and move it, you will feel the galea aponeurotica move. **Raphes** are long, seamlike tendons, such as the **linea alba** in the ventral midline of the trunk (fig. 11.14). Muscles inserting on raphes often compress a cavity and the organs within it. Fine nonmetameric myoseptalike **tendonous inscriptions** traverse many straplike muscles or broad muscular sheets, adding tensile strength to the muscle (fig. 11.15). They are prominent in the rectus abdominis of humans and in the broad oblique and transverse muscle sheets of the abdomen of anurans and many amniotes.

Actions of Skeletal Muscles

Skeletal muscles may be categorized according to their function. **Extensors** tend to straighten two segments of a limb or vertebral column at a joint. **Flexors** tend to draw one segment toward another. **Adductors** draw a part toward the midline. **Abductors** cause displacement away from the midline. **Protractors** cause a part, such as the tongue or hyoid, to be thrust forward or outward; **retractors** pull it back. **Levators** raise a part; **depressors** lower it. **Rotators** cause rotation of a part on its axis. **Supinators** are rotators that turn the palm upward, **pronators** make it prone (turn it downward). **Tensors** make a part such as the eardrum more taut. **Constrictors** compress internal parts. **Sphincters** are constrictors that make an opening smaller; **dilators** have the opposite effect. Most sphincter and dilator muscles are nonskeletal.

Although the action of any muscle is dependent on its origin, insertion, and shape, few muscles act independently. More frequently, they act in functional groups and also synergistically with other functional groups that have an opposing action. While one group is contracting, the opposing group must relax simultaneously and at the same rate; otherwise, a stalemate would be the minimum effect. More likely, a torn muscle or stretched

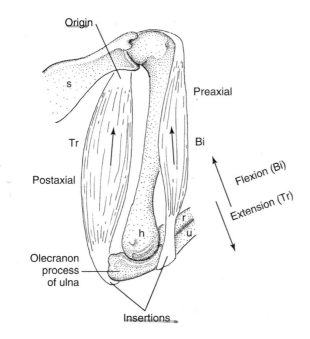

FIGURE 11.4

Origin, insertion, and direction of tensile force exerted by two muscles of the right upper arm of a rabbit, lateral view (i.e., head is to the right). Arrows within muscles indicate direction of pull on the forearm; arrows at right indicate direction of displacement of forearm when muscles contract. **Bi,** biceps brachii; **h,** humerus at elbow; **r,** radius; **s,** scapula; **Tr,** long head of triceps brachii; **u,** ulna. **Tr** provides the motive power for a first-class lever, and **Bi** for a third-class lever. The two muscles must operate synergistically, monitored by proprioceptive feedback to the central nervous system to effect smooth operation of the forearm against a load.

tendon or ligament would result. For both muscle groups to function smoothly they must be under reflex regulatory control of the cerebellum, which dispatches motor impulses to appropriate muscles on receiving sensory feedback from proprioceptive receptors located in the muscles, in their tendons, and in the bursas and capsules of affected joints (chapter 17).

Names and Homologies of Skeletal Muscles

Skeletal muscles have been named for the direction of their fibers (oblique, rectus), location or position (thoracis, supraspinatus, superficialis), number of subdivisions (quadriceps, digastric), shape (deltoid, teres, serratus), origin or insertion (xiphihumeralis, stapedius), action (levator scapulae, risorius), size (major, longissimus), and for still other features including combinations of these. Insight into the significance of a muscle's name should aid in recall of other information about the muscle.

Names were originally given to muscles in accordance with one of the preceding criteria without regard to homology. Therefore the fact that two muscles in two different species have the same name is no assurance that they are homologous. The more distantly related the two animals are, the greater is the likelihood that they are not homologous. Similarity of location, origin, and insertion is a good starting point but not a reliable basis for establishing homologies. One reason is, muscles sometimes alter their sites of attachment during evolution as they spread and produce new slips. An example is the genioglossus muscle that inserts on the mammalian tongue. It was inherited, of course, from reptiles. The probable homologue in some birds inserts on the sublingual seed pouch.

A way to test this hypothesis is through a combination of methods involving, although not exclusively, their embryogenesis and nerve supply. One method of seeking homologues, therefore, is to compare the early developmental stages of two muscles suspected of being homologous. Do the anlagen appear to be identical? If so, the two should be homologues, even though in one species the anlage expands in one direction or another to assume new relationships, whereas the anlage of the other does not do so. If the anlagen are similar, the muscles may be homologous. On this premise the coracoscapular muscle of reptiles is thought to be homologous with the supraspinatus and infraspinatus muscles of mammals. Neurologic studies often provide substantiating evidence. If the motor cell bodies that innervate two muscles suspected of being homologues are located in homologous nuclei within the cord or brain, this is substantiating evidence. Of course, there must also be some confidence that the two motor nuclei are homologous. The latter method is able to identify muscles of the mammalian head and neck, the homologues of which operated the gills of fishes, because most of the motor cell

bodies of branchiomeric muscles are in a different motor column than the motor cell bodies supplying any other group of muscles. Embryologic and neurologic evidence is the most reliable at present for establishing muscle homologies. However, such data are available for relatively few individual muscles. On the other hand, homologies between *functional groups* of muscles may be deduced with a much greater degree of reliability.

AXIAL MUSCLES

Axial muscles are the skeletal muscles of the trunk and tail. They extend forward beneath the pharynx as hypobranchial muscles and, in amniotes, as tongue muscles. They are present in the orbits as extrinsic muscles of the eyeballs. They do not include branchiomeric or appendicular muscles.

The immediately evident feature of axial muscles of fishes and generalized tetrapods is their metamerism (fig. 11.5). This primitive condition along with a flexible metameric vertebral column enables fishes and some aquatic tetrapods to propel themselves in water by lateral undulation (fig. 11.8). These muscles perform the same function for limbless amphibians living on land. The metamerism of the axial musculature was increasingly obscured as locomotion by lateral undulation was superceded by locomotion by limbs. Nevertheless, even in mammals, traces of metamerism remain in the axial musculature.

The axial muscles are inherently segmental because of their embryonic origin: they arise from segmental mesodermal somites (fig. 11.6). Mesenchyme cells from the myotome of each somite stream into the embryonic lateral body wall and migrate ventrad while undergoing repeated cell division (fig. 11.7). They cease migrating when they reach the midventral line, where the linea alba develops. These myotomal cells give rise to blastemas for body wall muscles. Because the somites are metameric, the blastemas are initially metameric. Blastemal cells, having become myoblasts, unite to form striated muscle fibers, and the body wall muscles commence to take shape. The metamerism of the somites is expressed as myomeres in those adults in which **myosepta (myocommata)** separate the muscle of one body segment from the next.

Myosepta don't form in the abdominal region of anurans and amniotes (fig. 11.5e). As a result, their abdominal musculature consists of broad sheets strengthened by tendonous inscriptions. Nevertheless, these sheets are innervated by as many spinal nerves as there were somites that contributed to them.

Trunk and Tail Muscles of Fishes

The musculature of the body wall and tail of fishes consists of myomeres separated by myosepta to which the

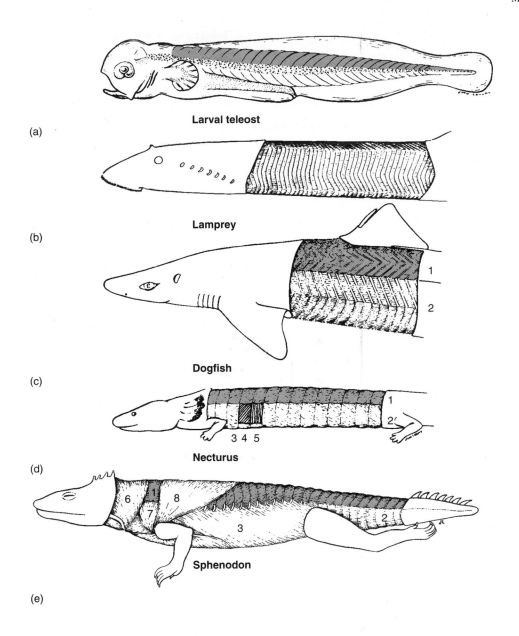

FIGURE 11.5

Trunk muscles of selected vertebrates. **1**, epaxials (*red*); **2**, hypaxials; **3**, external oblique; **4**, internal oblique; **5**, transverse muscle of abdomen. In (*e*), the appendicular muscles are **6**, trapezius; **7**, dorsalis scapulae; **8**, latissimus dorsi. In (*a*), stipple indicates position of notochord. In (*c*) and (*d*), **1** and **2** are separated by the horizontal skeletogenous septum. In (*d*) and (*e*), the epaxials are the dorsalis trunci muscles.

longitudinally directed muscle fibers attach (fig. 11.5*a* and *c*). The first and foremost role of this musculature is locomotion. Typically, there is a myomere for each vertebra and a spinal nerve for each myomere (fig. 11.6). Except in agnathans, the myomeres are divided into dorsal and ventral masses, the **epaxial** and **hypaxial muscles,** by a fibrous sheet, the **horizontal skeletogenous septum.** The septum is anchored to dorsal ribs when the latter are present and stretches between the vertebral column and the skin along the entire length of the trunk and tail (figs. 8.5 and 11.5*c*). Ventral ribs in the lateral body wall of most fishes develop within the myosepta. Brook lampreys lack not only the horizontal septum, but also myosepta.

The myosepta of fishes are seen to zigzag when the skin is removed, but deeper within the wall, the myosepta are elaborately folded, the angles of each zigzag being elongated forward or backward to form muscular cones that fit into the cones of adjacent myomeres like stacked dunce caps. The cones become longer toward the tail, and the apices of caudally directed cones near the end of the trunk are often continued as tendonous extensions that insert on caudal vertebrae. The forces exerted by contractions of the myomeres are thereby distributed over more than one body segment and become most powerful in the tail, where flexibility of the vertebral column is also the greatest. In response to waves of motor impulses that sweep caudad over

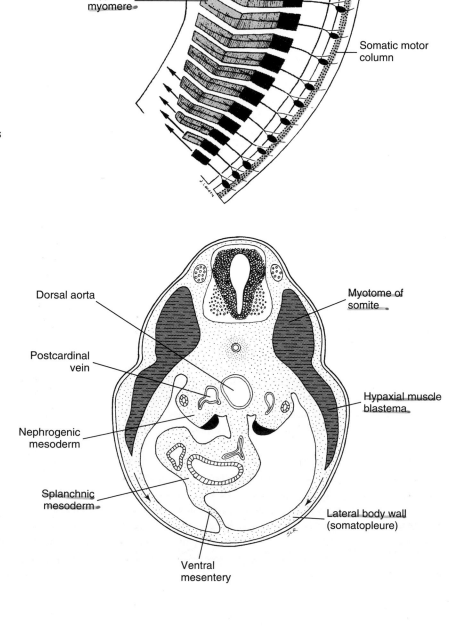

FIGURE 11.6

Axial muscle origins and innervation in a
generalized vertebrate embryo (diagrammatic).
The eyeball muscles organize from three preotic
somites at the level of the oculomotor, trochlear,
and abducens nuclei, respectively. Segmental
muscles of the trunk arise from trunk somites and
are supplied by corresponding segmental nerves.
Hypobranchial musculature migrates forward into
the floor of the pharynx accompanied by a nerve
supply. Motor fibers innervating myotomal muscles
have cell bodies in the somatic motor column
(interrupted in the brain). Nondiscrete postotic
somitomeres are indicated by dotted lines. The
central nervous system has been projected above
the embryo.

FIGURE 11.7

Cross section of mammalian embryo showing
invasion of myotomal cells (*red*) into lateral body
wall to form hypaxial muscles.

successive spinal nerves, the wave on one side being in advance of the other, successive myomeres exert a pull on the vertebral column, first on one side then the other, evolving rhythmic lateral undulations of the trunk and tail. These propel the fish forward (fig. 11.8).

A thin sheet of **oblique fibers** lies superficial to the main hypaxial mass ventrolaterally in many fishes, and a thin ribbon of still more superficial fibers parallels the linea alba on each side. The latter resembles the rectus abdominis muscle of tetrapods (fig. 11.15).

The bluefin tuna, which is among the fastest swimming fishes and weighs an average of 500 lbs (225 kg), is often said to be "warm blooded" by commercial fishermen because heat generated by contractions of the voluminous axial musculature warms its body to temperatures far exceeding that of the surrounding water.

The metamerism of the hypaxial muscles of fishes is interrupted where the pectoral and pelvic girdles are built into the body wall, and by the gills. Dorsal to the gills, the epaxial muscles continue to the skull as **epibranchial muscles** (fig. 11.25a). Beneath the gills, the hypaxials extend to the lower jaw as **hypobranchial muscles,** to be discussed shortly.

Upon emerging from the vertebral column through intervertebral foramina, spinal nerves divide into two major rami (branches) that supply, among other structures, the axial musculature. Dorsal rami supply epaxial myomeres. Ventral rami, which are larger because they innervate a greater muscle mass, innervate hypaxial myomeres. If, during embryonic development, premuscle mesenchyme from a given body segment streams beyond the limits of that segment before differentiating, branches of the spinal nerve from the original metamere will innervate the muscle in its adult location. Later in the chapter, when we study the hypobranchial muscles, we will see an example of this phenomenon. The innervation of the epaxial and hypaxial musculature and their derivatives is no different in tetrapods.

Trunk and Tail Muscles of Tetrapods

Tailed amphibians have retained the primitive metamerism of epaxial and hypaxial muscles (fig. 11.5d). This enables aquatic urodeles to swim by fishlike lateral undulation of the trunk and tail. However, a strictly metameric axial musculature unaccompanied by a well-developed appendicular musculature is ill-suited for locomotion on land. Amniotes gradually lost much of the metamerism of the axial muscles and developed, instead, a complex appendicular musculature well adapted for terrestrial life.

Disappearance of epaxial myosepta in amniotes gave rise to long, straplike or pinnate bundles disposed above the transverse processes, leaving only a vestige of metamerism in the deepest bundles (fig. 11.9). Similarly modified were the muscle bundles immediately beneath the transverse processes, that is, in the roof of the coelom. This revised architecture of the vertebral musculature, accompanied by modifications of the intervertebral joints, greatly increased the flexibility of the reptilian and mammalian vertebral columns, permitting dorsal arching, especially in mammals, and greater lateral undulation along a larger extend of the trunk, especially in reptiles other than turtles. Birds forgo these benefits for another: the rigid column of birds requires few epaxial muscles caudal to the neck but makes avian flight possible.

Changes with the advent of limbs were not confined to the vertebral musculature; they occurred also in the lateral body wall. Hypaxial myomeres in that location were gradually replaced by strata of broad muscular sheets, the fibers of which were disposed in a different

FIGURE 11.8

Propulsion by lateral undulation in a fish. *Red arrows* show progression of counterforces exerted by the surrounding water against one side commencing at the head during passage of a single contractile wave. *Black arrows* show other simultaneous counterforces. Each is at an angle that propels the fish forward. Counterforces are evoked by pressure of the undulating body against the water.

From Carl Gans, *Biomechanics: An Approach to Vertebrate Biology,* 2nd edition. Copyright © 1974. University of Michigan Press, Ann Arbor, MI. Reprinted by permission.

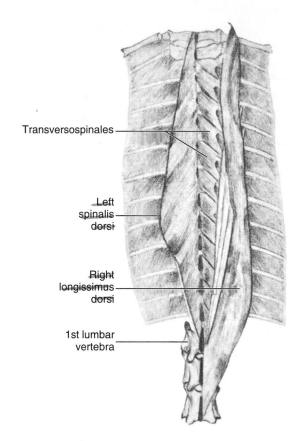

FIGURE 11.9

Three epaxial muscles of the thorax of a hamster, dorsal view. The right spinalis dorsi has been removed to show the underlying transversospinales. For the relationship of the longissimus to the spinalis, see fig. 11.12.

Labels on figure: Transversospinales; Left spinalis dorsi; Right longissimus dorsi; 1st lumbar vertebra

direction in each layer. This alteration of the lateral body wall (*parietal*) musculature was able to occur without negatively affecting survival because tetrapods had evolved paired limbs for locomotion, freeing the parietal muscles from this role. The new parietal musculature supports the trunk viscera in a muscular sling suspended well above the ground.

Loss of metamerism of the axial musculature, along with adaptations of the vertebral column and appendages, enabled tetrapods to better meet the locomotor needs imposed by competition in the terrestrial habitat. In no tetrapod is the locomotor effectiveness of this revised axial muscle better demonstrated than in the agile anole lizards, in fleet digitigrade carnivorous mammals that must "hump" (hustle) to capture food, or in jackrabbits, who must scamper to avoid being eaten (fig. 11.10).

Intermediate stages in the adaptation of trunk and tail muscles to life on land are seen in *Necturus* and *Sphenodon* (fig. 11.5*d* and *e*). *Necturus* has retained epaxial and hypaxial myomeres, but each hypaxial myomere is split into three layers. *Sphenodon* has retained epaxial myomeres only, and the lateral body wall musculature consists of broad stratified sheets. With loss of metamerism of the axial muscles, the horizontal skeletogenous septa also disappeared.

Figure 11.11 illustrates the components of the axial musculature as they are disposed in the abdominal wall of a mammal.

Epaxial Muscles of the Trunk

The epaxial muscles of tetrapods extend from the base of the skull into the tail for varying distances (fig. 11.12*a*). The most anterior ones attach to the occiput. In amphibians other than anurans, the epaxials retain their primitive metamerism, arising and inserting on myosepta and transverse processes, constituting collectively the **dorsalis trunci.** In amniotes, excepting rhynchocephalians, most of the epaxials are long bundles, some of which extend over many body segments.

Epaxial muscles collectively function in straightening (extending) the vertebral column and in lateral flexion of the body. Keep in mind that muscles do not act independently, but synergistically with others and that this dictates, within limits, their specific effect on the skeleton at any given moment. Note, too, that the action of a given appendicular muscle in a bipedal animal may not be the same as it is in one that walks on all four legs. The epaxial muscles of amniotes are divided arbitrarily into four groups: **intervertebrals,** a **longissimus, spinales,** and **iliocostales.**

Intervertebrals are the deepest epaxial muscles and the only ones to retain their primitive metamerism. They extend between two successive transverse processes **(intertransversarii),** two neural spines **(interspinales),** two neural arches **(interarcuales),** or two successive zygapophyses **(interarticulares),** and participate with

FIGURE 11.10

Extension and flexion of the vertebral column of a fleet digitigrade mammal.

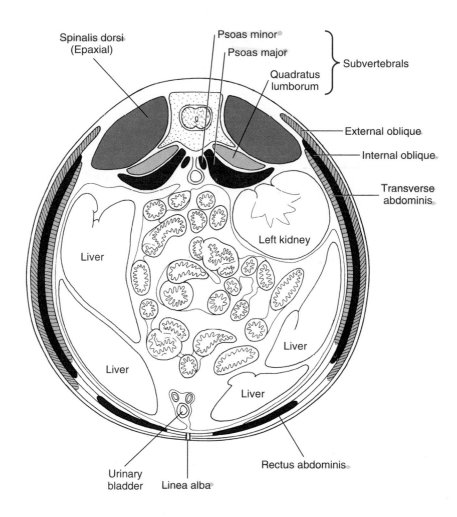

FIGURE 11.11

Disposition of axial muscles in the abdominal region of a mammal, based on cross section of a rabbit, cephalic view. The cutaneous maximus muscle has been removed.

longer epaxials in maintaining a vertebral posture appropriate to the needs of the moment.

Longissimus and spinales muscles occupy, respectively, lateral and medial positions above the transverse processes, and individual bundles are named according to their location (table 11.2). Capitis bundles insert on the skull and assist in movements of the head. Cervicis bundles are in the neck, and dorsi bundles are in the trunk. The longissimus, so named because it is the longest epaxial mass, consists of bundles that, with their tendons, may extend over many body segments. It is the dominant extensor in mammals, in which lateral undulation is minimal, and plays virtually no role in generalized mammalian locomotion. In the lumbar region of mammals, the longissimus consists of three distinct bundles (fig. 11.12*b*). The medial bundle continues into the tail.

Spinales include long and medial bundles that connect neural spines or transverse processes with neural spines several or many segments cephalad, and transversospinales that connect transverse processes with the neural spines of the second vertebra forward (fig. 11.9).

Spinales are often grouped with lumbar intervertebrals and named, collectively, **multifidus spinae.** Their chief role in mammals is to assist in maintaining stability (temporary rigidity) of the column in whatever degree of extension or flexion is imposed by other vertebral muscles. In humans the spinales assist other bundles in maintaining an upright posture while standing.

Iliocostales are lateral to the longissimus. They constitute a thin sheet arising on the ilium and passing forward to insert on the ribs and uncinate processes. They are the dominant epaxials in reptiles, because their lateral location and their insertions along the length of the column provide leverage for lateral undulation, a significant component of reptilian locomotion. Lateral undulation in crocodilians rotates the pectoral girdle and, to a lesser degree, the pelvic girdle on a vertical axis, thereby increasing the length of the stride of both pairs of limbs. The iliocostales continue forward into the neck but not caudad into the tail.

In turtles and birds, epaxials are prominent only in the neck, since the vertebral column and, in turtles, the ribs

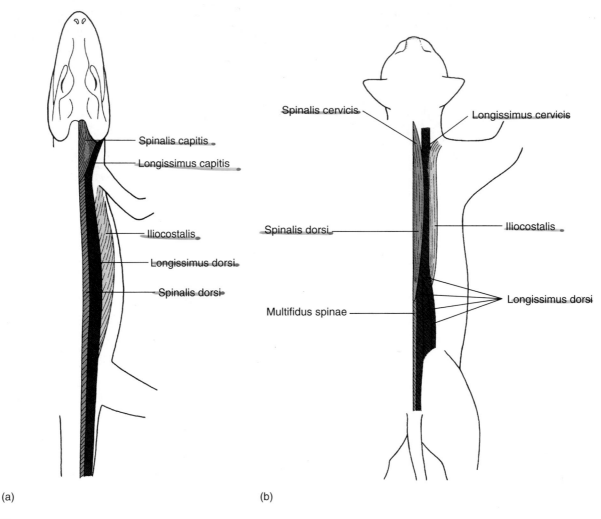

(a)   (b)

FIGURE 11.12

Disposition of the epaxial muscles. (*a*) Crocodilian. (*b*) Cat. The intervertebrals are not represented except as part of the multifidus spinae in the cat.

are immobilized by fusion with the synsacrum or cara-pace. The longissimus, for instance, is mostly ligamentous in the trunk. However, well-developed cervical epaxials participate in providing exceptional flexibility to the long necks of birds such as swans. One of the cervical epaxials of birds, the **complexus,** inserts on the interparietal bone and provides power for cracking the eggshell with the beak during hatching. In contrast to turtles and birds, snakes utilize a variety of locomotor strategies, including both dorsoventral flexion and lateral undulation. All epax-ial muscles are voluminous and complex.

The epaxial muscles became increasingly hidden by the expansion dorsad of the appendicular muscles and associated lumbodorsal aponeurosis of higher amniotes (compare *a* and *c* in fig. 11.20).

Hypaxial Muscles of the Trunk

The hypaxials of the trunk of amniotes can be divided arbitrarily into (1) **subvertebrals** (longitudinal bundles beneath the transverse processes in the roof of the

coelom), (2) **oblique** sheets, (3) **transverse** sheets in the lateral body wall (parietal muscles), and (4) **rectus abdominus** muscles (longitudinal straplike muscles on either side of the linea alba). These are listed in table 11.2 and represented schematically in figure 11.11.

Subvertebral Muscles   Subvertebral muscles form a longitudinal band of fairly powerful flexors of the verte-bral column lying *beneath* the transverse processes from the atlas to the pelvis. The portion in the neck of birds and mammals is known as the **longus colli.** Subvertebrals are meager in the thorax but become prominent again in the lumbar region, where they are represented by the **quadratus lumborum** and, in mam-mals, the **psoas minor.** The quadratus lumborum of mammals originates on the centra of several of the last thoracic vertebrae and the bases of their ribs and on the transverse processes of lumbar vertebrae, and it inserts on the ventral angle of the wing of the ilium, with varia-tions depending on the species. The psoas minor, known

TABLE 11.2 Representative Somatic (Myotomal) Muscles of Head, Trunk, and Forelimbs of Mammals

	Eyeball	Tongue	Hypobranchial
Head and neck	Superior oblique Inferior oblique Medial rectus Lateral rectus Superior rectus Inferior rectus	Genioglossus Hyoglossus Styloglossus Lingualis	Geniohyoideus Sternohyoideus Sternothyroideus Thyrohyoideus Omohyoideus

	Epaxial Muscles		Hypaxial Muscles

Trunk	Intervertebrals
	  Intertransversarii
	  Interspinales
	  Interarcuales
	  Interarticulares
	Longissimus
	  L. capitis
	  L. cervicis
	  L. dorsi
	  Extensor caudae lateralis
	Spinales
	  S. dorsi
	  S. cervicis
	  S. capitis
	  Transversopinales
	Iliocostales
			Subvertebrals
			  Longus colli
			  Quadratus lumborum*
			  Psoas minor
			Oblique group (parietals)
			  Internal and external intercostals
			  Internal and external obliques of abdomen
			  Cremaster
			  Supracostals
			    Scalenus†
			    Serratus dorsalis†
			    Levatores costarum†
			    Transversus costarum
			  Diaphragm
			Transverse group (parietals)
			  Transversus thoracis (subcostal)
			  Transversus abdominis
			Rectus muscles
			  Rectus abdominis
			  Pyramidalis

	Extrinsic		Intrinsic

Forelimb	Secondary appendicular†
	  Levator scapulae
	  Rhomboideus
	  Serratus ventralis
	Primary appendicular‡
	  Latissimus dorsi
	  Pectorales
			See table 11.4

*Secondary appendicular muscle.
†May be epaxial by derivation.
‡See discussion of these terms under "Appendicular muscles: Tetrapods."

commercially as the tenderloin, connects the lumbar vertebrae with the pelvic girdle. The action of the subvertebrals is, in general, opposite that of the epaxials.

Oblique and Transverse Muscles The parietal muscles have become stratified into superficial and deep layers, typically **external oblique, internal oblique,** and **transverse muscles of the abdomen;** and in amniotes, **external intercostal, internal intercostal,** and **transverse muscles of the thorax.** The fibers of all of these pass more or less obliquely from origin to insertion. One or another of these sheets may either be split in two or lost. In aquatic urodeles, for example, in which parietal muscles participate in locomotion, the external oblique

muscle is split into superficial and deep parts. In crocodilians and some lizards, all three layers consist of two sheets each. In anurans, the internal oblique is sometimes missing, and in birds, all sheets are thin. In turtles, they are less than thin—they are vestigial! But because of the rigid shell, a turtle couldn't use them anyway. (If you want some turtle soup, it will have to come from the neck, tail, or appendicular muscles!)

Muscle slips from the inferior border of the internal oblique of male mammals, and sometimes from the transverse abdominal muscle form a **cremaster muscle** (fig. 15.36) that loops around the spermatic cord commencing at the inguinal ring and inserts on a fibrous sheath in the wall of the scrotum below the testes. It is best developed in mammals with permanently open inguinal canals, such as rabbits, because it retracts the testes into the canals in those species.

The oblique and transverse muscles, particularly intercostals, play a major role in external respiration in most amniotes other than turtles, and an accessory role in mammals. The intercostals are assisted by subdivisions of **supracostal muscles** that differentiate on the surface of the rib cage, chiefly **scalenus, serratus dorsalis, levatores costarum,** and **transversus costarum.** On the basis of their innervation, these muscles are thought to be derivatives of the oblique sheets of primitive parietal muscles. Along with the rectus, the oblique and transverse muscles support the abdominal viscera in a muscular sling, and they compress the viscera for such functions as egg laying, delivery of mammalian young, and emptying of the digestive tract. Although they lack segmental myosepta in amniotes, their metameric history is evidenced by their embryonic origin from successive somites and by their innervation via ventral rami of successive spinal nerves.

Rectus Muscles of the Abdomen   The **rectus abdominis** extends longitudinally on either side of the linea alba between the pubic symphysis and the sternum and, in amniotes, the base of several sternal ribs. It assists in

flexing the trunk and in supporting the abdominal viscera in a muscular sling. In urodeles it is strictly segmental (fig. 11.15). In anurans and amniotes it exhibits irregular transverse tendonous inscriptions.

A pyramidalis muscle in the ventral wall of the marsupial pouch is a slip of the rectus abdominis. Higher mammals may have vestiges of the pyramidalis as a species characteristic or as an anomaly.

Mammalian Diaphragm   In mammals, mesenchyme from several embryonic somites at the level of the developing third, fourth, and fifth cervical spinal nerves, varying with the species, migrates caudad in the somatopleure to invade the embryonic septum transversum, which is thereby converted into a dome-shaped muscular diaphragm separating the thoracic and abdominal cavities (fig. 11.13). When completed, the diaphragm consists of a **central tendon** with a pair of semilunar extensions and a **muscular portion.** The latter converges on the tendonous portion from a circumferential perimeter, arising on the xiphoid process ventrally **(sternal portion),** on the caudalmost ribs or their costal cartilages laterally **(costal portion),** and from several lumbar vertebrae dorsally **(vertebral portion).** The vertebral portion consists of a pair of triangular muscular masses, the **crura,** that are firmly anchored to the lumbar vertebrae

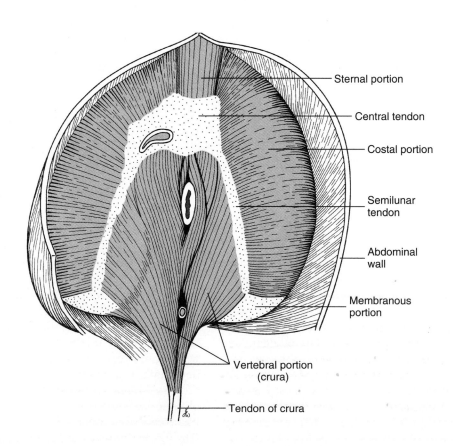

FIGURE 11.13

The domed mammalian diaphragm, viewed from abdominal cavity. The concavity is occupied by the liver. The apertures are post cava (in central tendon), esophagus (*center*), and dorsal aorta (between the crura).

by a single short, tough, cylindrical tendon. The mammalian diaphragm has become the major component of the suction pump utilized in mammalian breathing, superceding, but not totally replacing, the ribs and parietal muscles in this role. Innervation of the diaphragm by ventral rami of cervical spinal nerves indicates its hypaxial status.

Muscles of the Tail

The tail musculature of tetrapods is a continuation of the epaxial and hypaxial musculature of the trunk. This is most evident in urodeles and generalized reptiles (fig. 11.5*e*). Disruption of the continuity, especially of the hypaxial bundles, occurs to one degree or another at the level of the pelvis of amniotes generally and especially of birds and mammals. This is accounted for by the presence of the large pelvis characteristic of amniotes. (The latter is correlated with the increased volume of the extrinsic appendicular muscles of the amniote hind limb.) The volume of the caudal musculature varies, of course. Snakes, for example, employ the tail for locomotion, and epaxial and hypaxial muscles are well developed along the entire extent of the tail.

Intervertebrals and short spinales (collectively, the multifidus spinae) and the longissimus continue into the tail as medial and lateral extensors, respectively, lying on the dorsal aspect of the vertebral column. The tail section of the longissimus, the **extensor caudae lateralis**, arises from sacral and caudal vertebrae and inserts on distal tail vertebrae by many long slender tendons. In general, these epaxial muscles extend the tail and arch it upward.

Long hypaxial bundles similar to the epaxials arise from the medial surface and caudal border of the wing of the ilium, from transverse processes of the last lumbar vertebrae, or from the sacrum and pass into the tail as long bundles paralleling the vertebral column. They are supplemented by long and short flexors that arise within the tail and insert farther along by long or short tendons. Hypaxial muscles bend the tail laterad or downward.

In urodeles, reptiles, and birds, a **caudofemoralis** muscle contributes to the fleshy part of the tail. It connects several caudal vertebrae at the base of the tail with the femur, being secondarily an extrinsic appendicular muscle. It exerts a powerful backward pull on the hind limb during locomotion of lizards and crocodilians.

Caudal centra, arches, and processes become increasingly vestigial distally, especially in mammals (fig. 8.22), and all muscle bundles dwindle until only long tendons of insertion remain. The major fleshy part of the mammalian tail is therefore the proximal portion. Distally, the mammalian tail is mostly skin, stringlike tendons, and cylindrical bones united by interosseus ligaments. The meatiest oxtail soup comes from the proximal portion of the tail!

Dilator and sphincter muscles of the cloaca and the sphincter muscle of the anus are derived from the hypaxial musculature that anchors the tail to the pelvis. In turtles, for instance, they arise on sacral and proximal caudal vertebrae and insert on the skin encircling the vent. In mammals, they have lost their skeletal attachments.

Hypobranchial and Tongue Muscles

Mesenchyme from several postbranchial somites streams forward in the floor of the pharynx beneath the branchial arches to give rise to hypobranchial muscles and, when present, tongue muscles (fig. 11.6). These myotomal muscles are anterior extensions of the hypaxial musculature of the trunk. In fishes, hypobranchial muscles extend forward from the coracoid region of the pectoral girdle to insert on Meckel's cartilages (**coracomandibularis**), basihyals (**coracohyoideus**), and the ventralmost segments of the gill cartilages (**coracoarcuales**, fig. 11.14). They assist the branchiomeric muscles in respiration and feeding movements by expanding the pharynx and gill pouches, moving parts of the hyoid skeleton, and depressing the lower jaw. They perform related functions in anurans and urodeles, depending on the species. The **rectus cervicis** muscle of urodeles is a hypobranchial muscle (fig. 11.15).

With development of a neck in amniotes, hypobranchial muscles became longer and straplike. They stabilize the hyoid apparatus and larynx and draw these cephalad or caudad, depending on the concurrent actions of other muscles inserting on these same structures. They bear such names as **sternohyoid, sternothyroid, thyrohyoid, omohyoid,** and **geniohyoid.** One of these is seen in a rabbit in figure 9.44.

The tongue of amniotes is essentially a mucosal sac anchored to the hyoid skeleton and stuffed with hypobranchial muscles. Premuscle mesenchyme migrates into the developing tongue from anterior hypobranchial muscle blastemas. This accounts for the fact that, in bats, some tongue muscles extend into the tongue from as far back as the sternum. The chief extrinsic tongue muscles of mammals are the **hyoglossus, styloglossus,** and **genioglossus.** Mammals and some reptiles develop an intrinsic tongue muscle, the **lingualis.** It has no skeletal attachments.

Because hypobranchial and tongue muscles are derived from the anteriormost trunk somites, they are supplied by cervical spinal nerves or, with respect to the tongue muscles, by the last cranial nerve (hypoglossal), a cervical spinal nerve that became "trapped" within the amniote skull. The cell bodies of the motor fibers that innervate these muscles are in the same somatic motor column of the spinal cord and brain that supplies myotomal muscles elsewhere in the body, and no others (figs. 11.6 and 16.34, SE).

Extrinsic Eyeball Muscles

The extrinsic muscles that move the eyeball are striated, skeletal, voluntary muscles that arise on the wall of the orbit and insert on the fibrous sclerotic coat of the

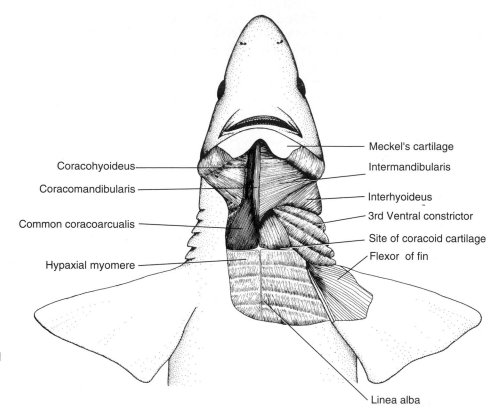

Coracohyoideus

Coracomandibularis

Common coracoarcualis

Hypaxial myomere

Meckel's cartilage

Intermandibularis

Interhyoideus

3rd Ventral constrictor

Site of coracoid cartilage

Flexor of fin

Linea alba

FIGURE 11.14

Hypobranchial muscles of a shark (*red*) and ventral branchiomeric muscles (at right of mid-line). The intermandibularis is superficial to the coracohyoideus and coracomandibularis. It has been reflected on the left. The coracobranchials are deep hypobranchial muscles that are not shown.

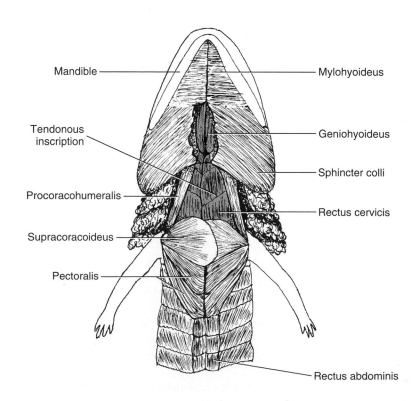

Mandible

Tendonous inscription

Procoracohumeralis

Supracoracoideus

Pectoralis

Mylohyoideus

Geniohyoideus

Sphincter colli

Rectus cervicis

Rectus abdominis

FIGURE 11.15

Anteroventral muscles of *Necturus.* The sphincter colli is a very thin muscular sheet on the surface of the powerful branchiohyoideus (see fig. 11.25*b*). Hypobranchials are shown in *red*.

eyeball. In elasmobranchs, their embryonic origin is from three pairs of preotic somites in the embryonic head (fig. 11.6). The most cephalic somite giving rise to eyeball muscles in elasmobranchs is at the level where the third cranial nerve emerges from the midbrain. This nerve innervates four eyeball muscles, the **superior, medial,** and **inferior rectus,** and the **inferior oblique** (table 11.3 and fig. 11.16). The next more caudal somite is at the level where the fourth cranial nerve emerges. This nerve innervates the **superior oblique** eyeball muscle. The last preotic somite is at the level where the sixth cranial nerve emerges. This nerve innervates the **lateral rectus** eyeball muscle.

The mesenchyme that gives rise to eyeball muscles of vertebrates other than elasmobranchs has not been traced to specific head somites, so the somites may not be myotomal by embryonic origin. However, they are preceded in ontogeny by evanescent preotic somites,

they are innervated by the same nerves as in elasmobranchs, and the cell bodies of their motor fibers are in the same motor column as the cell bodies that innervate hypobranchial muscles, tongue muscles, and the axial muscles of the trunk and tail. In view of these facts, it appears that the rectus and oblique muscles of the eyeball of all vertebrates are derivatives of an ancient segmental axial musculature that, in the head, has been conscripted to operate lateral visual organs. In other words, the eyeball muscles of vertebrates other than elasmobranchs are mytomal by phylogenetic derivation. As might be expected, vertebrates with vestigial eyeballs have vestigial eyeball muscles.

In many amniotes, muscles insert on the upper lids and nictitating membrane (**pyramidalis** of reptiles and birds, **quadratus** of birds, **levator palpebrae superioris** of reptiles and mammals). The specific source of their myoblasts has not been ascertained, but their innervation by somatic motor fibers with cell bodies in the somatic efferent column indicates that they are somatic muscles (table 11.3). **Protractors and retractors of the eyeballs** of reptiles and birds, and **depressors of the lower lids,** when present, are evidently not of axial origin, since they are innervated by the fifth cranial nerve.

APPENDICULAR MUSCLES

Appendicular muscles are those that insert on the girdles, fins, or limbs. Most fishes use axial muscles for locomotion. As a correlate, the appendicular muscles of fishes are uncomplicated, exhibit little variety, have little mass, and perform a restricted function. Those of tetrapods have become increasingly numerous and complex as a consequence of adaptations for life on land.

TABLE 11.3 Chief Extrinsic Eyeball and Eyelid Muscles and Their Innervation

Cranial Nerve Supply	Extrinsic Eyeball Muscles	Eyelid Muscles
III (oculomotor)	Superior rectus Inferior rectus* Medial (internal) rectus Inferior oblique	Levator palpebrae superioris
IV (trochlear)	Superior oblique	
VI (abducens)	External (lateral) rectus Rectractor bulbi	Pyramidalis of eye Quadratus of eye

*In lampreys the inferior rectus is innervated by cranial nerve VI.

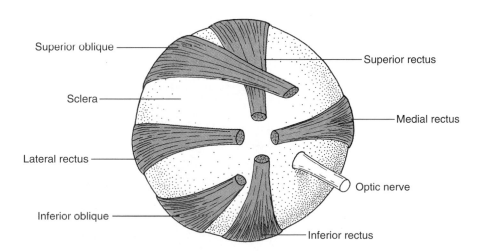

FIGURE 11.16
Extrinsic muscles of the left eyeball of a generalized vertebrate viewed from behind the eye. The rectus muscles of all vertebrates have origins on a single location on the orbital wall near the optic foramen. The origin of the obliques varies slightly.

Fishes

Paired fins arise in embryos as fin folds that protrude from the lateral body wall (fig. 11.17, shark). Thereafter, hollow **muscle buds** sprout from the lower edges of a series of embryonic myomeres near the base of each fin fold (fig. 11.18). The buds split into dorsal and ventral moities, invade the developing fin, and establish blastemas from which dorsal and ventral muscle masses are formed. Dorsal blastemas form extensors (elevators) of the fin; ventral blastemas form the flexors (depressors, fig. 11.19). The resulting musculature establishes attachments to the girdles, basalia, radialia (if any), and to the fascia overlying the base of the fin rays. Any muscle tissue that may develop distal to these masses is insignificant except muscles such as those in the claspers of male Chondrichthyes and in the undulating winglike fins of rays and skates.

The muscles of median dorsal fins organize from myotomal mesenchyme that is giving rise to epaxial myomeres. Those of median ventral fins arise from hypaxial mesenchyme. The musculature of median fins is meager, with the exception of anal fins that have become gonopodia.

Tetrapods

The appendicular muscles of tetrapods are far more complex than those of fishes because of the joints in tetrapod limbs. As in fishes, the appendicular muscles of tetrapods are disposed in opposing groups. In fishes, postaxial (posterior; dorsal) muscles extend the appendage, and preaxials (anterior; ventral) flex it. When applied to limbs that have rotated toward the long axis of the body, so that elbows point backward and knees point forward (fig. 10.46), one or another of these terms becomes meaningless to the layman. Not only have the segments and the bones within them become reoriented, but so have the muscles, which often spiral around a bone or a joint to insert on an opposite surface of the next segment. For this reason we will discuss the appendicular muscles of tetrapods with reference to their current anatomic relationships, rather than to their primitive locations.

Anatomists have found it practical for descriptive purposes to divide appendicular muscles into two groups according to their anatomic origins. **Extrinsic appendicular muscles** are those arising on the axial skeleton or fascia of the trunk and inserting on a girdle or limb.

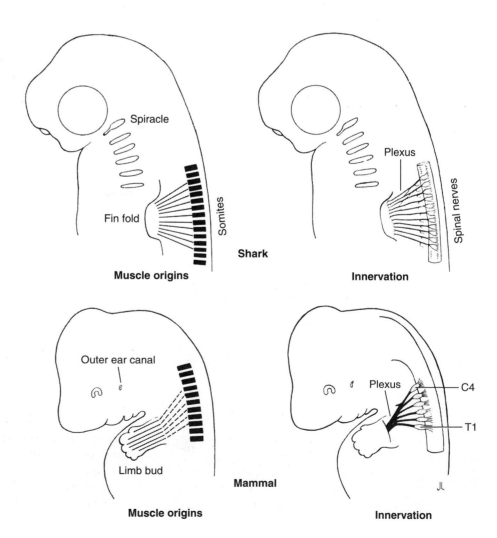

FIGURE 11.17

Top, Origin of shark appendicular muscles from somites and their innervation by corresponding spinal nerves. *Below,* Probable phylogenetic derivation of appendicular muscles of a mammal from six somites based on their innervation. **C4** and **T1,** Dorsal root ganglia of the fourth cervical and first thoracic spinal nerves of the brachial plexus.

FIGURE 11.18

Budding of fin musculature from embryonic body wall myomeres in a shark. In one electric ray 26 myomeres participate; **b,** muscle buds; **ff,** fin fold; **m,** myomere.

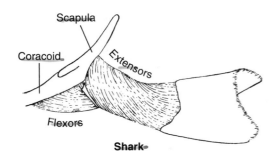

Shark

FIGURE 11.19

Muscles of pectoral fin of a shark. Extensors are dorsal (postaxial), flexors are ventral (preaxial).

Intrinsic appendicular muscles are those arising on a girdle or limb and inserting more distally on the limb. When all extrinsic muscles of the pectoral girdle and forelimbs of most mammals have been transected, the clavicle, scapula, and limb of either side can be removed from the body as a unit. The pelvic girdle cannot be similarly removed because the girdle forms an immobile joint with the sacrum.

Most extrinsic muscles of tetrapod limbs begin development from blastemas *within the embryonic body wall.* Buds from these blastemas grow toward and establish insertions on a girdle or proximal bone of the limb. Muscles that develop in this manner are also referred to as *secondary* appendicular muscles because their embryonic origin is not intrinsic to the limb. The levator scapulae, serratus ventralis, and rhomboideus (all of myotomal origin) and the trapezius, sternomastoideus, and cleidomastoideus (all of branchiomeric origin) develop in this manner. Extrinsic muscles are the most primitive, being extensions of the locomotor musculature of ancestral vertebrates.

Intrinsic appendicular muscles of tetrapods organize from blastemas *within the developing limb,* rather than as buds from hypaxial myomeres. Muscles that form in this manner are also referred to as *primary* appendicular muscles. In the latter category are a few extrinsic mus-

Necturus

(a)

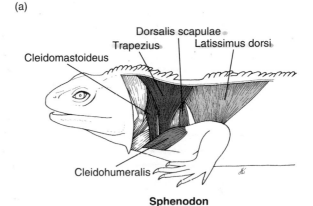

Sphenodon

(b)

Hamster

(c)

FIGURE 11.20

Superficial shoulder muscles (in color) of a urodele (*a*), generalized reptile (*b*), and rodent (*c*) illustrating the expansion of the shoulder musculature in higher tetrapods. The dorsalis scapulae and procoracohumeralis are also known as deltoid muscles.

cles that form when their blastemas spread trunkwise out of the limb and establish an attachment on the axial skeleton. Among these are the latissimus dorsi (fig. 11.20) and the iliopsoas of the posterior limb of generalized mammals.

It has not been shown that the mesenchyme contributing to these intrinsic blastemas originates in somites. Therefore it cannot be said at this time that the intrinsic muscles of the appendages are myotomal by *embryonic origin.* However, their innervation by motor fibers from the somatic motor column of the cord via ventral rami of spinal nerves is strong evidence that they are myotomal by phylogenetic derivation (fig. 11.17, mammal). Formation of intrinsic appendicular muscles in this manner, that is, in situ, skipping an ancestral ontogenetic stage of development, appears to be a time-saving modification of selective value. It is one of the many instances of developmental precocity in vertebrates.

Extrinsic Muscles of the Pectoral Girdle and Forelimbs

Dorsal Group   The most constant dorsal extrinsic appendicular muscle of tetrapods is the **latissimus dorsi,** a primary appendicular muscle that inserts on the humerus. In urodeles (fig. 11.20*a*), it is a delicate triangular muscle arising from the superficial fascia that overlies the epaxial myomeres of the shoulder region. In reptiles, it became stronger. It spread dorsad to acquire a firm attachment to the tough fascia that is anchored to neural spines, and it broadened its axial origins by spreading still farther caudad (fig. 11.20*b*). In mammals, the trend toward a broader dorsal anchorage continued. It now arises from the neural spines of most of the thoracic vertebrae caudal to the first few, and from the tough fibrous lumbodorsal fascia that overlies the lumbar vertebrae and extends all the way to the base of the tail (fig. 11.20*c*). The result of these dorsocaudad expansions was a continuing increment in the force the muscle was able to exert on the forelimb.

Deep to the latissimus dorsi (and to the trapezius, to be discussed next) in most amniotes are three extrinsic muscles that insert on the scapula: two levators of the scapula, a rhomboideus group (found only in crocodilians among living reptiles), and the serratus ventralis (serratus anterior). The mammalian levators of the scapula have their origins on the transverse processes of the atlas or on the basioccipital bone (**levator scapulae ventralis**) and on the transverse processes of the number of posterior cervical vertebrae (**levator scapulae dorsalis**). The **rhomboideus** group arises from the occiput and neural spines of a series of cervical and anterior thoracic vertebrae. The **serratus ventralis** arises by many separate prominent tendonous slips from a series of ribs near their junction with the costal cartilages (giving the muscles a serrate appearance) and, in some mammals, from a series of posterior cervical vertebrae. All the foregoing insert on the dorsal border of the scapula except the levator scapulae ventralis, which inserts on a process of the scapular spine near the glenoid fossa. (The levator scapulae dorsalis is sometimes considered part of the serratus ventralis.)

The pharyngeal arches make an important contribution to the extrinsic musculature of the pectoral girdle. The **trapezius** (fig. 11.20*b*), a superficial muscle of the shoulder region, is a survivor of the cucullaris muscle of fishes (fig. 11.25*a*). It acquired attachments to the pectoral girdle and then underwent the same expansion as the latissimus dorsi. It eventually became subdivided into several components, including **cleidotrapezius (cleidocervical), acromiotrapezius (cervical trapezius),** and **spinotrapezius (thoracic trapezius).** It receives its motor innervation via branchiomeric nerves. Two other muscles, the **cleidomastoideus** and **cleidooccipitalis** (fig. 11.25*c*) acquired an attachment to the clavicle, but they don't function as appendicular muscles. When they contract they move the head.

Ventral Group   The ventral extrinsic muscles of the forelimb, subsumed under the general term "pectoral muscles," have undergone expansive changes matching those of the dorsal group (fig. 11.21*a–e*). These fan-shaped muscles, originating primitively on the coracoid cartilages or bones and the associated midventral raphe, extended their origins to include, in one species or another, the epicoracoid, the entire length of the sternum, some of the costal cartilages, and part of the midventral raphe of the neck. They subdivided into a varying number of superficial and deep muscle masses that converge to insert on the proximal end of the humerus. The insertion sites of the more powerful pectoral muscles of amniotes are marked by prominent stress-induced ridges and tuberosities.

Two pectoral muscle masses, the **pectoralis** and **supracoracoideus,** are seen in their primitive state in a urodele (figs. 11.15 and 11.21*a*). Their different insertion sites on the humerus result in different actions on the forearm, the pectoralis being the chief adductor. In birds, in which the supracoracoideus is deep to the pectoralis (fig. 11.21*c*), these two are the powerful flight muscles that depress (pectoralis) and elevate (supracoracoideus) the wings, providing the downstroke and upstroke, respectively, when the bird is airborne. The supracoracoideus, which in birds underlies the procoracoid bone (fig. 8.28, chicken), is able to elevate the wing because its tendon of insertion passes to the dorsal surface of the humerus. (Note the caudally directed humerus on a skeleton or in fig. 4.29.) The supracoracoid has become an intrinsic muscle in mammals.

Intrinsic Muscles of the Pectoral Girdle and Forelimbs

Dorsal Group   Arising from the scapula of modern mammals are five post-axial muscles inherited from reptiles. They are the **deltoideus** or subdivisions thereof **(spinodeltoideus and acromiodeltoideus), teres major, teres minor, subscapularis** (on medial surface of the scapula), and the **long head of the triceps brachii** (fig. 11.22). With the exception of the triceps, these insert on the humerus a short distance below its head and align it with the scapula, rotate it about its long axis, or adduct it. The long head of the triceps, which has two additional heads that arise on the humerus, inserts on the olecranon process of the ulna where it exerts a

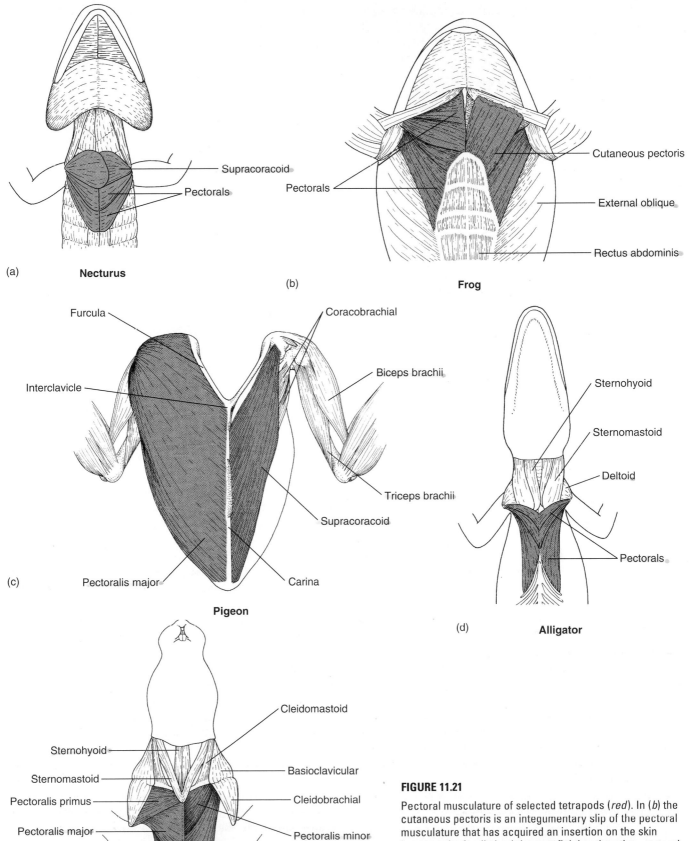

(a) **Necturus**

Supracoracoid
Pectorals

(b) **Frog**

Cutaneous pectoris
Pectorals
External oblique
Rectus abdominis

(c) **Pigeon**

Furcula
Coracobrachial
Interclavicle
Biceps brachii
Triceps brachii
Supracoracoid
Pectoralis major
Carina

(d) **Alligator**

Sternohyoid
Sternomastoid
Deltoid
Pectorals

(e) **Rabbit**

Cleidomastoid
Sternohyoid
Sternomastoid
Basioclavicular
Pectoralis primus
Cleidobrachial
Pectoralis major
Pectoralis minor
Pectoralis quartus

FIGURE 11.21

Pectoral musculature of selected tetrapods (*red*). In (*b*) the cutaneous pectoris is an integumentary slip of the pectoral musculature that has acquired an insertion on the skin between the forelimbs. It is superficial to the other pectoral muscles. In (*c*) the pectoral muscle has been removed on the right to reveal the supracoracoid. In (*e*) the pectoralis primus and pectoralis major have been removed on the right to expose deeper muscles. A pectoscapularis lies unseen deep to the cephalic border of the pectoralis minor.

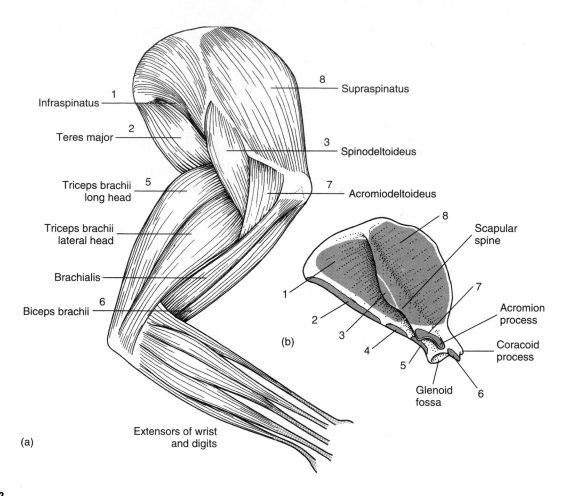

Infraspinatus — 1
Teres major — 2
Triceps brachii long head — 5
Triceps brachii lateral head
Brachialis
Biceps brachii — 6

Extensors of wrist and digits

(a)

8 — Supraspinatus
3 — Spinodeltoideus
7 — Acromiodeltoideus

8 — Scapular spine
7 — Acromion process
Coracoid process
Glenoid fossa

(b)

FIGURE 11.22

(*a*) Superficial muscles of the right scapula and proximal part of the forelimb of a cat, lateral view. (*b*) Sites of origin of correspondingly numbered muscles in (*a*); **4,** origin of teres minor, a deep muscle not visible in (*a*). Also not seen are the subscapularis on medial surface of scapula and the tiny coracobrachialis (origin, coracoid process of scapula).

powerful pull that extends the forearm (fig. 11.4). To test your own triceps, try a few pushups! This exercise extends the arm against resistance. Distal to the triceps in mammals two **supinators of the manus** connect the humerus with the radius. Finally, an assortment of **extensors of the hand and digits,** with long distal tendons, insert on the skeleton of the wrist and digits, some as far distad as the terminal phalanx (fig. 11.22). The shortest ones are intrinsic to the manus.

The mammalian deltoid muscle is probably a homologue of the dorsalis scapulae or scapular deltoid in lower tetrapods; the teres major appears to be a slip of the latissimus dorsi; and the teres minor is probably the muscle that, in reptiles, is called scapulohumeralis anterior. The mammalian subscapular is an expansion of a muscle in the same location in reptiles.

Ventral Group   Representing the primitively ventral musculature of the pectoral girdle and forelimb are two strong muscles on the lateral aspect of the scapula, the supraspinatus and infraspinatus (fig. 11.22), and a very

short, deep coracobrachialis (not illustrated). The **supraspinatus** and **infraspinatus** are derivatives of the **supracoracoid of reptiles,** a broad muscle having a ventral anatomic origin over a broad area of the procoracoid bone near the glenoid fossa. Studies on embryonic mammals have shown that a blastema for a supracoracoid muscle develops in mammals at the reptilian site, but it then spreads dorsad onto the embryonic scapula, pushing its way beneath the developing deltoideus and establishing itself on both sides of the scapular spine to become the supraspinatus and infraspinatus muscles of the adult mammal.

The **coracobrachialis** of mammals originates on the coracoid process of the scapula, the latter being the sole mammalian remnant of the coracoid bones of reptiles. The muscle is present in reptiles, passing from the coracoid bone to the shaft of the humerus. The muscle, like the bone from which it arises, has become very small in mammals.

A **biceps brachii** and a **brachialis** (fig. 11.22) are the major flexors of the forearm of reptiles and mammals.

TABLE 11.4   Chief Intrinsic Muscles of the Pectoral Girdle and Forelimbs of Mammals and Their Probable Homologues in Reptiles

	Mammals	Reptiles
Muscles of Girdle		
Girdle to humerus, proximally	Deltoideus	{ Deltoideus clavicularis
		Dorsalis scapulae
	Subscapularis	Subcoracoscapularis
	Teres minor	Scapulohumeralis anterior
	Supraspinatus }	
	Infraspinatus }	Supracoracoideus
	Coracobrachialis	Coracobrachialis
	Teres major	Slip of latissimus dorsi
Muscles of Upper Arm		
Girdle or humerus to proximal end of radius or ulna	Triceps brachii	Triceps brachii
	Biceps brachii	Biceps brachii
	Brachialis	Brachialis
	Epitrochleoanconeus	Epitrochleoanconeus
	Anconeus	Anconeus
Muscles of Forearm*		
Humerus and proximal end of radius and ulna to hand	Extensors and flexors of carpus and digits	
	Supinators and pronators of hand	
Muscles of Hand†	Extensors, flexors, abductors, adductors of digits	

*Frequently have long tendons of insertion.
†Reduced in species in which digits are reduced.

A small deep **anconeus** (not homologous with the anconeus of frogs) extends between the distal end of the humerus and the proximal end of the ulna, and a small transverse **epitrochleoanconeus** partially encircles the elbow joint medially. Distal to these, **pronators of the manus** insert on the radius and rotate this bone, and **flexors of the manus** with origins chiefly on the humerus insert by long tendons on the carpals, metacarpals, and phalanges. Intrinsic to the manus are very short **flexors of the digits.**

Present in some mammals is a **cleidobrachialis** that extends from the clavicle to the humerus or ulna. Because the clavicles of rabbits and cats are vestigial, the cleidobrachialis has the appearance of being a continuation of the **basioclavicularis (cleido-occipital)** of rabbits (fig. 11.21e) and of the cleidotrapezius of cats. However, the raphe in which the vestige is embedded is the boundary between the two. The continuity of some of their fibers is secondary to reduction of the intervening clavicle. The cleidobrachialis is innervated by spinal nerves, whereas the muscle above the raphe is supplied by branchiomeric nerves.

The musculature of birds is essentially reptilian. Much of the weight of a bird is represented by the extrinsic muscles of the forelimb and the intrinsic muscles of the hind limb. The intrinsic muscles of the wing have been reduced, the epaxial musculature of the trunk is vestigial, and the parietal muscles are thin.

Table 11.4 lists the chief intrinsic muscles of the pectoral girdle and forelimbs of mammals and their probable homologues in reptiles.

Muscles of the Pelvic Girdle and Hind Limbs

Pelvic girdles, especially in amniotes, are incapable of independent mobility. They are united with the vertebral column dorsally and the two halves meet in a symphysis ventrally, except in birds. In birds the pelvic girdle is united with the immobile synsacrum. Consequently, muscles arising from the axial skeleton and inserting on the girdle of amniotes—the extrinsic muscles of the appendages—play little or no role in locomotion.

A **caudofemoralis muscle** in urodeles, reptiles, and birds extends between some of the proximal caudal vertebrae and the femur, but it is not locomotor. It exerts its pull on the tail. However, a part of it has become a locomotor muscle, the pyriformis, in mammals.

The locomotor muscles of the hind limbs are chiefly intrinsic muscles. They extend between the bones of the girdle—ilium, ischium, pubis (one or more of these)—and the femur, tibia, or fibula. In modern reptiles, their names usually indicate their attachments—**puboischiofemoralis, puboischiotibialis, iliofemoralis, iliotibialis, iliofibularis, femorotibialis,** and others. These muscles were derived from reptilian ancestors, as were also the hind limb muscles of mammals. Consequently, many hind limb muscles of mammals and reptiles are homologues. Unfortunately, the muscles of humans were named first, and the names were often based on something other than origin and insertion. The major muscles of the hind limbs of mammals are the subject of the discussion that follows.

The **iliopsoas,** which in humans and some other mammals appears as separate **iliacus** and **psoas major** muscles, is the **puboischiofemoralis internus** of reptiles. The iliac portion arises on the ilium; the psoas major portion arises from an extensive area of fascia covering the psoas minor and from a series of lumbar vertebrae. The two portions unite and insert by a powerful tendon on the **lesser trochanter,** a protuberance on the femur near its head. The muscle rotates the femur, thereby turning the foot outward.

A functional group of three hip muscles, **gluteus** (the reptilian **iliofemoralis**), **pyriformis** (the reptilian **caudofemoralis brevis**), and **gemelli,** has, collectively, a wide origin on sacral and caudal vertebrae and on the ilium and ischium, and inserts on a large prominence, the **greater trochanter,** near the head of the femur. The group, of which the gluteus is the most powerful, abducts the thigh and, like the iliopsoas, rotates the femur to turn the foot outward. The gluteal and pyriform are primary appendicular muscles that have increased their leverage by spreading from the pelvic girdle onto the axial skeleton.

A **quadratus femoris,** consisting of four muscles (three **vasti** and a **rectus femoris**), arises on the ilium and the greater and lesser trochanters of the femur and inserts on the ligament in which the patella is embedded. The vasti extend the leg, and the rectus femoris adducts the thigh, rotating it to carry the foot inward. Other extensors or adductors of the thigh include the **semimembranosus, adductor femoris, adductor longus, pectineus, sartorius,** and **gracilis.** They arise on the ilium, ischium, or pubis, and insert on the shaft of the femur or on the patellar ligament and tibia. The sartorius, the longest muscle in the human body, is a muscular strap that arises on the ilium and passes diagonally down the medial aspect of the thigh. It is the muscle that may become irritated when an adult human who is not used to doing so sits too long on the floor with the legs crossed in the manner of oldtime tailors who, while creating sartorial splendor, sewed by hand.

Two **obturator** muscles flex, rotate, and abduct the thigh. They arise from the lip of the obturator foramen (fig. 10.12f) and from the ischium and pubis, and insert on the proximal end of the femur.

A **biceps femoris** and a **semitendinosus** are primarily flexors of the leg, although the former also abducts the thigh and the latter extends it. They arise on the ischium and insert on the patella or tibia.

The foregoing muscles commencing with the iliopsoas either insert on the femur or parallel it and insert just beyond it. Distal to these are **extensors and flexors** **of the foot and digits** similar to those of the manus. Among these are the **gastrocnemius,** an extensor of the foot made famous by Achilles. Its long, wide tendon inserts on the calcaneus bone of mammals. In birds (fig. 11.23), it inserts on the tarsometatarsus (which incorporates the homologue of the calcaneus) and also continues behind the heel to insert on the phalanges. The weight of the perching bird stretches the tendon, thereby retracting the claw-bearing toes so that a sleeping bird, muscularly relaxed and with minimal expenditure or energy, maintains its grip on the perch.

As mentioned, the pectoral girdles and forelimbs are joined to the trunk by extrinsic appendicular muscles, vessels, nerves, and skin alone, whereas the pelvic girdles are firmly ankylosed to the axial skeleton. This accounts in part for the fact that cats jump by pushing off with their hind limbs but land on their forelimbs, where much of the shock is absorbed by the extrinsic and intrinsic musculature. Among the reasons frogs can't do this is lack of a suitable axial skeletomuscular system and of a cerebellum competent to direct precision landings. The result is that frogs land awkwardly—almost abashedly—on whatever part of the body touches down first. The hind limbs of ungulates are good shock absorbers because elongated metatarsals provide what amounts to an extra shock-absorbing joint at the end of the shank.

Branches of several successive spinal nerves innervate the appendicular muscles of fishes and tetrapods (fig. 11.17). In fishes these nerves come from the same body seg-

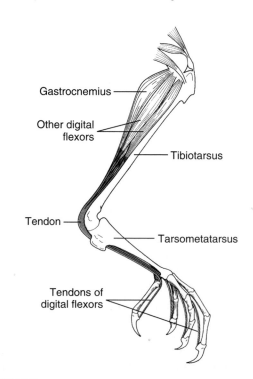

FIGURE 11.23

Perching mechanism of a bird. The weight of the body flexes the digits, causing them to grip the perch.

TABLE 11.5 Innervation of Somatic Muscles

Epaxials of trunk and tail Appendiculars of epaxial derivation	Dorsal rami of spinal nerves
Hypaxials of trunk and tail Appendiculars of hypaxial derivation Hypobranchials Mammalian diaphragm	Ventral rami of spinal nerves
Extrinsic eyeball muscles Tongue	Cranial nerves 3, 4, 6 Cranial nerve 12

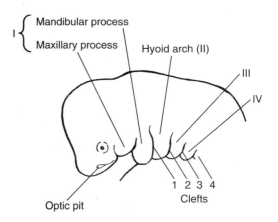

FIGURE 11.24

Embryonic pharyngeal arches of a vertebrate.

ments that contribute muscle buds to the appendage. It seems probable that the number of spinal nerves entering a tetrapod limb is indicative of the number of somites that contributed mesenchyme to the limb phylogenetically.

The pattern of innervation of somatic muscles is given in table 11.5.

BRANCHIOMERIC MUSCLES

In chapter 1 we discussed the basic architectural pattern of the vertebrate body. One of the discussions was an introduction to the pharyngeal arches (fig. 11.24). You are urged to read this discussion again, since it will have more meaning for you now at this stage of your study.

In chapter 9 we discussed how the pharyngeal skeleton was altered in tetrapods as an adaptation to life on land. Now we will examine the changes that took place in the muscles that operate this tetrapod skeleton.

Operating the skeleton derived from the embryonic pharyngeal arches, whether in fishes or tetrapods, is a series of striated, skeletal, muscles that have been termed "branchiomeric." In no living vertebrate is the basic pattern of these muscles, and their motor innervation, better exemplified than in a shark, in which they operate the jaws and gill arches. In tetrapods they still operate the jaws. However, with loss of gills, the muscles of the former gill arches have acquired new functions for life on land.

Despite the fact that branchiomeric muscles are striated and skeletal, in fishes they perform two primitive visceral functions inherited from their filter-feeding ancestors: food handling within the oropharyngeal cavity, and respiration. They did not in the past, nor do they today, orient the body (soma) in the external environment, which is the role of somatic (myotomal) muscles. The mesenchyme that provides the stem cells for these muscles comes from neural crests and, in birds, from mesenchymal blastemas in the vicinity of the nondiscrete postotic **somitomeres** (fig. 11.6), which may or may not be vestiges of ancestral somites. But the cell

bodies of the motor nerve fibers that innervate them are not in the gray motor column that innervates somatic muscles. They are in a column that operates branchiomeric muscles only (fig. 16.34, EB/SVE). Whether or not one calls them somatic muscles or visceral muscles is a matter of definition.

Muscles of the Mandibular Arch

In all vertebrates, the muscles of the first arch are chiefly jaw muscles. In *Squalus*, a **levator palatoquadrati** arises on the otic capsule and inserts on the quadrate end of the upper-jaw cartilage (fig. 11.25*a*); a powerful **adductor mandibulae** arises on the quadrate process and inserts on Meckel's cartilage; and an **intermandibularis** extends between Meckel's cartilage and a strong midventral raphe in the pharyngeal floor. A slender **craniomaxillaris** inserting on the upper jaw completes the first arch muscles of *Squalus*. The levator palatoquadrati raises the upper jaw, which is possible because *Squalus* has hyostylic jaw suspension. The craniomaxillaris assists in this. The adductor mandibulae raises the lower jaw, thereby closing the mouth during the phase of the respiratory cycle when the spiracle is closed and water is being forced over the gills by constriction of the walls of the orobranchial chamber. The adductor mandibulae also enables sharks to hold in a viselike grip any prey unlucky enough to be caught. The intermandibularis is a ventral constrictor that elevates the anterior pharyngeal floor during respiration. (The lower jaw is lowered by the coracomandibularis, a hypobranchial muscle.)

The adductor mandibulae is the most powerful muscle of the first arch in all vertebrates. In tetrapods it has become split into several muscles—many in snakes, three in mammals. The three in mammals are the **masseter** and **temporalis** (fig. 11.25*c*) and the **pterygoideus.** (The latter lies deep to the other muscles and is not illustrated.) These muscles have spread in three directions to acquire origins on the zygomatic arch (masseter), the temporal bone (temporalis), and the pterygoid fossa

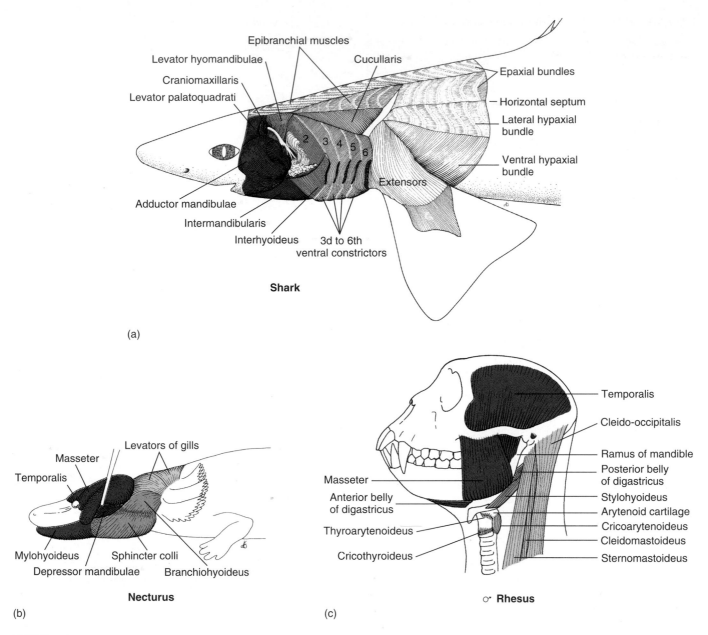

FIGURE 11.25

Selected branchiomeric muscles of three vertebrates. Muscles of the first and second pharyngeal arches are *dark* and *medium red,* respectively. Muscles of the remaining arches are *light red.* In (*a*), **2** to **6** are the dorsal constrictors of the second and successive arches. (*c*) The temporalis inserts on the coronoid process of the ramus of the mandible. The pterygoideus muscle, not illustrated, can be seen only after the mandibular symphysis has been separated and the dentary bone retracted laterad.

(fig. 9.29). The three muscles on each side constitute a muscular sling for the lower jaw and provide most of the multidirectional tensile forces that produce the side-to-side, up-and-down, forward and back, and rotary chewing movements of such different mammals as herbivores, carnivores, and rodents. The role of temporal fossae in the evolution of the amniote masseter, temporalis, and pterygoideus muscles should be reviewed at this time (chapter 9).

The intermandibular muscle of fishes is the homologue of the **mylohyoideus** of tetrapods (fig. 11.25*b*). One slip probably gave rise to the **digastricus** muscle of tetrapods

(anterior belly, when there are two bellies, fig. 11.25*c*). Also, a slip of the first arch muscle that was attached to the articular bone of therapsid reptiles remained attached when the bone became the malleus; that muscle, now the **tensor tympani,** tenses the mammalian eardrum. The mandibular arch muscles of all vertebrates are innervated by motor branches of the fifth cranial nerve.

Muscles of the Hyoid Arch

The principal muscles of the hyoid arch of *Squalus* elevate the arch or constrict the pharyngeal cavity, partly as a function of respiration and feeding. A **levator**

hyomandibulae and a **dorsal constrictor** arise on the neurocranium and insert on the hyomandibula and ceratohyal cartilage. An **interhyoideus** is a ventral constrictor (fig. 11.25a). In bony fishes the dorsal constrictor is subdivided, and part of it operates the operculum. Because of the joint between the upper jaw, lower jaw, and hyomandibula, contraction of hyoid arch muscles results in movements of the jaws.

In *tetrapods,* hyoid arch muscles continue to serve some of the functions seen in fishes, but they have also acquired some quite different roles. In *Necturus,* a **branchiohyoideus muscle** of the hyoid arch arises on the ceratohyal cartilage and inserts on the epibranchial cartilage of the first gill (fig. 11.25b). Along with levators it waves the gills back and forth in the water for respiration. A **depressor mandibulae** opens the mouth of urodeles and many reptiles; and in some mammals a **posterior belly of the digastricus** participates in chewing movements (fig. 11.25c). A slender **stylohyoideus** (anterior belly when there are two) connects the styloid or jugular process of the skull with the anterior horn or body of the hyoid in mammals (fig. 9.44).

A thin collarlike **sphincter colli,** overlying the origin of the branchiohyoideus, adheres to the skin of the neck of lower tetrapods (fig. 11.25b). In reptiles and birds, it spreads upward around the rear of the skull to insert on the skin of the head and is called the **platysma.** In mammals, the platysma spreads forward onto the face to become the muscles of facial expression, or **mimetic muscles** (fig. 11.26c–f). The sphincter colli is thought to be derived from a portion of the interhyoideus of fishes.

A small **stapedius** muscle originates on the posterior wall of the middle ear cavity of mammals and inserts on the stapes, a homologue of the hyomandibular cartilages. It contracts reflexly to impede extra loud airborne sounds that might injure the delicate hair cells of the cochlea.

All these hyoid arch muscles receive their motor innervation via the facial (seventh cranial) nerve, so named because of its distribution to the mimetic muscles. The innervation of the two muscles of the middle ear by two different branchiomeric nerves, V and VII, is not surprising to those who know the phylogenetic history of the malleus and stapes.

Muscles of the Third and Successive Pharyngeal Arches

The muscles of the arches of fishes caudal to the hyoid are **constrictors** (dorsal and ventral), **levators, adductors,** and **dorsal and lateral interarcuals,** which compress or expand the pharyngeal cavity and gill pouches during respiration. Constrictors in sharks lie just under the skin, covered by tough subcutaneous fascia that is not readily removed, and they attach to strong fascia above and below the gill pouches (fig. 11.25a). They compress the gill pouches, expelling respiratory water. The levators of these arches make up a strong muscle

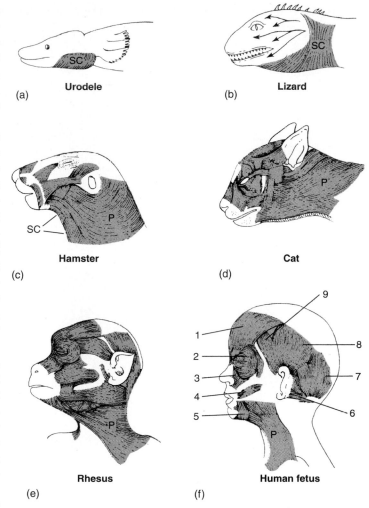

FIGURE 11.26

Evolution of mammalian mimetic muscles from hyoid arch muscles of lower tetrapods. The sphincter colli, **SC,** spreads onto the neck in reptiles and becomes the platysma, **P,** in mammals. It then spreads forward onto the head and face, as indicated by *arrows* in (b), to become muscles of facial expression. Note increasing differentiation of mimetics in the mammals shown. **1,** frontalis; **2,** orbicularis oculi; **3,** quadratus labii superioris; **4,** risorius; **5,** triangularis; **6,** posterior auricular; **7,** occipital; **8,** superior auricular; **9,** anterior auricular. In (e) and (f), the white expanse on top of the head is the galea aponeurotica.

sheet, the **cucullaris,** which raises the pharyngeal wall assisted by the levator hyomandibulae of the second arch. Adductors deep in the gill arches connect epibranchial and ceratobranchial cartilages and cause the lateral pharyngeal walls to bow outward when the muscles contract, thereby expanding the pharyngeal chamber. Two sets of interarcual muscles in the roof of the pharynx just above the mucosa connect successive pharyngobranchial cartilages with epibranchials and draw these together. Their action assists in further expanding the pharynx. *The interarcuals in the floor of the pharynx are hypobranchial, not branchiomeric muscles.* In bony fishes the branchiomeric muscles caudal to the hyoid arch are

TABLE 11.6   Chief Branchiomeric Muscles and Their Innervation in *Squalus* and in Tetrapods

Pharyngeal Arch	Pharyngeal Skeleton in *Squalus*	Chief Branchiomeric Muscles		Cranial Nerve Innervation
		Squalus	Tetrapods*	

I Mandibular arch	Meckel's cartilage	Intermandibularis
		
		Adductor mandibulae
			Intermandibularis
			  Mylohyoideus (anterior part)
			  Digastricus (anterior part)
			Adductor mandibulae
			  Masseter
			  Temporalis
			  Pterygoidei
			  Tensor tympani
				V

	Pterygoquadrate cartilage	Levator palatoquadrati Craniomaxillaris		

II Hyoid arch	Hyomandibula
	Ceratohyal
	Basihyal
		Levator hyomandibulae
		Dorsal constrictor
		Interhyoideus
			Stapedius
			Stylohyoideus (anterior part)
			Depressor mandibulae
			Digastricus (posterior belly)
			Sphincter colli
			  Platysma
			  Mimetics
				VII

III	Gill cartilages	Constrictors Levators Adductors Interarcuals	Stylopharyngeus Stylohyoideus (posterior part)	IX
IV to VI	Gill cartilages	Constrictors Levators Adductors Interarcuals	Thyroarytenoideus Cricoarytenoideus Cricothyroideus	X
		Cucullaris (derived also from dorsal constrictor 3)	Trapezius Sternomastoideus[†] Cleidomastoideus[†] Basioclavicularis[†]	Occipitospinal nerves in shark; spinal roots of XI in amniotes

*Indented muscles in this column may be derivatives of the preceding muscle.
[†]Presumed to be branchiomeric on the basis of their motor innervation.

much reduced as a consequence of the role of the operculum in moving respiratory water across the gills.

In tetrapods, branchiomeric muscle has pretty much disappeared from what had been the gill-bearing arches. Remaining from arch III are a **stylopharyngeus** muscle that is used in swallowing and, in some mammals, a **posterior belly of the stylohyoideus** (fig. 9.44). Remaining from arch IV are the intrinsic muscles of the mammalian larynx: **cricothyroideus, cricoarytenoideus,** and **thyroarytenoideus** (fig. 11.25c).

The trapezius muscles of amniotes (fig. 11.20b and c), discussed earlier with other extrinsic muscles of the pectoral girdle, are derivatives of the cucullaris muscle of fishes. Whether or not muscles with that name in urodeles are homologues is unknown. The sternocleidomastoid complex (sternomastoideus, cleidomastoideus, and, in lagomorphs and a few other mammals, a basioclavicularis, also known as the cleido-occipitalis) has been thought to be of branchiomeric origin on the basis of its innervation by the spinal accessory nerve. However, the history of the components of this nerve, discussed in chapter 16, is conjectural. The chief branchiomeric muscles and their innervation are given in table 11.6.

INTEGUMENTARY MUSCLES

In fishes and amphibians, slips of branchiomeric or myotomal muscles insert on the dermis at one site or another, attaching the skin to the underlying muscle at these locations. **Costocutaneous muscles** of snakes are hypaxial integumentary muscles used in locomotion (chapter 10). But it is only in mammals that integumentary muscles are well differentiated. A **panniculus carnosus (cutaneous maximus)** wraps around the entire trunk of some mammals, enabling armadillos to roll into a ball when endangered, forms a sphincter

Cat   Monkey

FIGURE 11.27

Panniculus carnosus (cutaneous maximus) of a cat and primate. Note difference in the extent of the muscle in the two animals.

FIGURE 11.28

Electric organs in tail of electric eel. Each nucleated horizontal disk (electroplax) is a single modified hypaxial muscle fiber. **C,** centrum; **M,** epaxial myomere.

around the entrance to the abdominal pouch of marsupials, and vigorously shakes flies off horses (fig. 11.27). It probably started as superficial slips of the pectoral musculature that became attached to the undersurface of the skin and then spread, as did the longus colli. The panniculus is poorly developed in monkeys, absent in humans. A sheet of pectoral musculature, the **cutaneous pectoris,** maintains its original attachment to the chest wall in anurans (fig. 11.21*b*); and in bats, slips of pectoral muscles insert on the skin of the wing membranes as **patagial muscles.** All these are myotomal muscles innervated by spinal nerves.

The most notable integumentary muscles are the **mimetics** mentioned earlier (fig. 11.26). More than 30 different muscles in humans depress the corners of the mouth in grief, raise them for smiling, wrinkle the forehead, raise the eyebrows quizzically, close the lids tightly, pucker the lips, draw milk when suckling, or dilate the nostrils. Other mammals have fewer of these mimetic muscles. One of them, the **caninus,** elevates the part of the upper lip that hides the spearlike canine tooth used by carnivores for ripping flesh. When a human uses the muscle, he or she is said to be sneering. **Auricular muscles** in nonhuman mammals direct the pinnas of the ears toward faint sounds. Mimetic muscles are innervated by cranial nerve VII, a branchiomeric nerve.

All the preceding are *extrinsic* integumentary muscles: their developmental and anatomical origins are away from the dermis. Their insertions are along the undersurface of the dermis. *Intrinsic* integumentary muscles develop entirely within the skin, in the dermis. They are the **arrectores plumarum** and **arrectores pilorum** that insert on feather or hair follicles and ruffle the feathers or elevate

the fur. They are smooth muscles in nearly all species and are innervated by the sympathetic nervous system.

ELECTRIC ORGANS

In several hundred species from at least seven families of cartilaginous and bony fishes, certain muscle masses are modified to produce, store, and discharge electricity. In *Torpedo,* the electric ray, an electric organ lies in each pectoral fin near the gills. It is of branchiomeric origin, being supplied by motor fibers of cranial nerves VII and IX. In *Raia,* a skate, and *Electrophorus,* a South American electric eel, electric organs lie in the tail and are modified hypaxial muscles (fig. 11.28). The potential produced by these organs in eels amounts to 600 volts and is used to paralyze or kill prey or to repel other organisms. Some other fishes have electric organs with a lower electric potential that serve as radarlike mechanisms or for communication. Certain neuromast organs (see chapter 17) are receptors for these signals.

Electric organs consist of a large number of electric disks, up to 200,000 in the tail of one ray, piled in either vertical or horizontal columns. Each disk, or **electroplax,** is a modified multinucleate muscle fiber embedded in a vascular jellylike extracellular matrix surrounded by connective tissue. Nerve endings terminating on each disk induce the discharge.

Electric organs seem to have no systematic distribution among fishes and probably evolved independently many times. The electric organ of *Malapterurus,* a freshwater catfish from the Nile, envelops nearly the entire body—head, trunk, and tail—in a jellylike mass located subcutaneously rather than among muscles. It appears to be composed of modified skin glands rather than muscle tissue.

S U M M A R Y

1. Histologically, muscle tissues are striated, smooth, or cardiac.
2. Functionally, muscles can be classified as somatic, visceral, or branchiomeric.
3. Somatic muscles orient the organism in the environment. Visceral muscles regulate the internal milieu. Branchiomeric muscles operate the pharyngeal arches and their derivatives.
4. Somatic muscles are striated and skeletal. They are myotomal by origin or derivation and, excepting eyeball muscles, they are innervated by spinal nerves.
5. Visceral muscles are mostly smooth or cardiac. Generally, they arise from the splanchnopleure and are innervated by the autonomic nervous system.
6. Branchiomeric muscles are striated and skeletal muscles of the pharyngeal arches. They receive their motor innervation from branchiomeric nerves.
7. Muscles may flex, extend, abduct, adduct, protract, retract, elevate, depress, rotate (pronate, supinate), constrict, dilate, or otherwise alter the spatial relationships or shape of a body part.
8. Skeletal muscles are named according to the direction of their fibers, location, number of parts, shape, attachments, action, size, a combination of these, and miscellaneous other considerations.
9. The trunk and tail muscles of jawed fishes are locomotor muscles arranged as myomeres divided into epaxial and hypaxial moities by a horizontal skeletogenous septum. Loss of myosepta obscures the primitive metamerism in most tetrapods.
10. The metamerism of the axial muscles of fishes and urodeles is an expression of the metamerism of embryonic somites.
11. Epaxial and hypaxial myomeres are innervated by dorsal and ventral rami, respectively, of spinal nerves.
12. The epaxial muscles of higher tetrapods are disposed as longitudinal bundles, some of which extend over many body segments. Hypaxial muscles of the lateral body wall have lost their metamerism except where there are long ribs, and they are disposed in oblique and transverse strata.
13. Hypaxial muscles immediately ventral to transverse processes in amniotes are disposed in long (subvertebral) bundles. Those paralleling the midventral line form long straplike muscles.
14. In all vertebrates, hypaxial muscles extend forward beneath the pharynx as hypobranchial muscles. In amniotes, they continue forward into the tongue.
15. The mammalian diaphragm and cremaster muscles are hypaxial muscles that have migrated accompanied by their spinal nerves.
16. The muscles of the vertebrate tail are continuations of the epaxial and hypaxial vertebral musculature of the trunk.
17. Eyeball muscles arise from three preotic somites in elasmobranchs and develop in situ in other vertebrates. They are innervated by cranial nerves III, IV, and VI and are myotomal by phylogenetic derivation.
18. Extrinsic appendicular muscles have an anatomic origin on the axial skeleton and insert on the girdle or limb. Intrinsic appendicular muscles arise on a girdle or proximal skeletal segment of the limb and insert on more distal segments.
19. The appendicular muscles of fishes are chiefly extrinsic. They differentiate from blastemas within the embryonic trunk and spread into the developing fin fold where they become disposed as ventral (preaxial) flexors and dorsal (postaxial) extensors. Reorientation of the limb in tetrapods generally has obscured these primitive relationships.
20. Appendicular muscles of tetrapods are extrinsic or intrinsic. Blastemas giving rise to extrinsic muscles develop in two different sites: (1) within the embryonic body wall, then spreading onto the girdles and into the developing limb buds, as in fishes (secondary appendicular muscles); and (2) within the limb bud, then spreading trunkward to establish an attachment on the axial skeleton (primary appendicular muscles).
21. Intrinsic appendicular muscles of tetrapods arise from blastemas that develop within the embryonic limb.
22. The muscles of the first pharyngeal arch operate the jaws. In mammals, a derivative inserts on the malleus. They are innervated by the fifth cranial nerve.
23. The muscles of the second arch attach to the hyoid skeleton, lower jaw, and, in fishes, operculum. In mammals, a derivative inserts on the stapes. A sphincter colli of lower tetrapods spreads onto the head of amniotes to become the platysma and mimetic muscles. The muscles are innervated by the seventh cranial nerve.
24. Muscles of the third and remaining arches operate gills in fishes and perform new functions in tetrapods. Muscles of the third arch are innervated by the ninth cranial nerve, those of the remaining arches by the vagus (tenth cranial nerve).
25. The cucullaris, a branchiomeric muscle of fishes, has given rise to the trapezius and sternocleidomastoid complex of amniotes.
26. Extrinsic integumentary muscles reach peak development as the panniculus carnosus (cutaneous maximus) and the mimetic muscles of primates. Erectors of feathers and hairs are smooth muscles that are intrinsic to the dermis of birds and mammals.
27. Electric organs are modified axial, appendicular, or branchiomeric muscles, the fibers (electroplaxes) of which produce, store, and discharge electric potential. A few appear to be modified skin glands.

SELECTED READINGS

Alexander, R. M.: Functional design in fishes, ed. 2, London, 1970, Hutchinson University Library.

Allen, E. R.: Development of vertebrate skeletal muscle, American Zoologist **18**:101, 1978.

Armstrong, R. B., and Phelps, R. O.: Muscle fiber type composition of the rat hind limb, American Journal of Anatomy **171**:259, 1984.

Bennett, M. V. L.: Electric organs. In Hoar, W. S., and Randall, D. J., editors: Fish physiology, vol. 5, New York, 1971, Academic Press.

Cundall, D.: Activity of head muscles during feeding by snakes: a comparative study, American Zoologist **23**:383, 1983.

Gans, C: Biomechanics: an approach to vertebrate biology, Philadelphia, 1974, J.B. Lippincott Co. (Copyright now held by University of Michigan Press.)

George, J. C., and Berger, A. J.: Avian myology, New York, 1966, Academic Press.

Jayne, B. C.: Muscular mechanisms of snake locomotion: an electromyographic study of lateral undulation of the Florida banded water snake (*Nerodia fasciata*) and the yellow rat snake (*Elaphe obsoleta*), Journal of Morphology **197**:159, 1988.

Lauder, G. V.: On the relationship of the myotome to the axial skeleton in vertebrate evolution, Paleobiology **6**:51, 1980.

Morgan, D. L., and Proske, U.: Vertebrate slow muscle: its structure, pattern of innervation, and mechanical properties, Physiological Reviews **64**:103, 1984.

Sacks, R. D., and Roy, R. R.: Architecture of the hind limb muscles of cats: functional significance, Journal of Morphology **173**:185, 1982.

Wardle, C. S., and Videler, J. J.: Fish swimming. In Elder, H. Y., and Truman, E. R., editors: Aspects of animal movement, Cambridge, England, 1980, Cambridge University Press.

CHAPTER TWELVE

DIGESTIVE SYSTEM

In this chapter we will examine the alimentary canal, the glands and other evaginations that arise from it, and some spectacular modifications associated with varying food habits, such as tongues that are stored under the scalp, tooth plates that crush molluscs, gizzards that macerate, stomachs that return cuds to the oral cavity, and ceca that are the home of cellulose-digesting bacteria. We will start by recalling some of the means employed by vertebrates for obtaining energy in the form of food from their environment.

OUTLINE

PROCURING FOOD

Ancestral chordates and ostracoderms were filter feeders. This process, which can be employed only by aquatic organisms, consists of passively filtering organic matter out of the incoming respiratory stream and propelling the food particles to the rear of the pharynx for swallowing. Filter feeding in sea squirts, amphioxus, and larval cyclostomes has been described in chapter 3. A more active version of filter feeding is employed by some fishes, such as spoonbills, tiny clupeids, and huge basking sharks. In these, plankton or small fishes are strained out of the respiratory water stream by long filamentous gill rakers that hang into the pharyngeal chamber from the gill arches. The largest mammals, baleen whales, strain tons of small fish, jellyfish, and other invertebrates from the sea each day through sieves of whalebone, or baleen, that hang into their oral cavity. However, whales take water into the oral cavity solely for the food it contains, since they don't breathe with gills, and the water spills out of the mouth and back into the sea.

With the advent of fishes with jaws and muscular body walls that can be used for locomotion and pursuit, more aggressive methods of obtaining food became possible. Jaws, at first invested with denticulated bony dermal armor, ultimately were furnished with small, often sharp, single denticles that fit our concept of teeth. Many organisms thereupon became predators, and their method of feeding, as in modern sharks, was a bite-tear-swallow technique that required no tongue or other specializations of the oral cavity.

A less energy-consuming procedure evolved as a result of further adaptations of the skull and hyoid arches. The modifications enabled some teleosts to approach close to organisms that are small enough to be swallowed, extend protrusible jaws, create suction, close the mouth, retract the jaws, and swallow. Anyone who has watched a goldfish feeding on flakes of fish food has observed this technique. Lampreys, being parasitic, have a different feeding technique. They rasp the tissues of the host with their spiny "tongue," a muscularly propelled fleshy and cartilaginous rod armed with horny teeth, and suck the debris into their pharynx.

On land, long sticky tongues are found among amphibians, squamate reptiles, and many birds and mammals. Some snakes impale prey on their upper jaw teeth. Winged tetrapods pick up grubs, seeds, and grains and perform other food-getting acts with appropriately shaped beaks (fig. 2.2) and shorebirds pierce fish with them, or scoop fish from the sea, pelican fashion. Sanguinivorous bats suck whole blood and use anticoagulants in their saliva to prevent clotting. And baby mammals suckle milk, using muscular cheeks and lips.

Herbivorous ungulates crop grasses, and carnivorous mammals use a snap-bite-tear technique that often includes the piercing effect of a saberlike tooth. Because of appropriate joints in the wrist or digits, primates can grasp food and convey it to the mouth, and rodents can hold food between their hands and nibble. Other methods of food taking will come to mind.

Food taking depends on food finding. This is accomplished by sense organs that monitor the external environment. Chemical receptors such as olfactory organs, mechanical receptors such as inner ears and lateral line organs of fishes and amphibians, thermal receptors such as the loreal pits of some snakes, capsulated touch receptors such as those on the sensitive snout of pigs, visual receptors, and electroreceptors—one or more of these alert one vertebrate or another to the presence and location of food. Once the energy-containing food is within the body, the digestive tract can process it, extract needed nutrients, and return unassimilated matter to the environment.

THE DIGESTIVE TRACT: AN OVERVIEW

The digestive tract, or alimentary canal, is a tube, seldom straight and often tortuously coiled, that commences at the mouth and empties either into a vented cloaca or directly to the exterior via an anus (fig. 12.1). It functions in the digestion and absorption of foodstuffs and in elimination of undigested wastes. Food is propelled from the pharynx to the vent or anus by smooth muscle tissue in the walls of the tract. The process is termed **peristalsis.**

Major subdivisions of the tract are an oral cavity and pharynx (oropharyngeal cavity in fishes), esophagus, stomach, and intestine, the latter being divided into small and large intestines in tetrapods. Emptying into the tract are ducts from accessory organs, principally pancreas, liver, and gallbladder. These organs and their ducts arise as evaginations from the embryonic digestive tract. Other evaginations known as **ceca,** having one function or another, are common. The digestive tract and accessory organs constitute the **digestive system.**

The digestive organs of fishes, amphibians, and most reptiles occupy the **pleuroperitoneal cavity,** a major component of the coelom, along with the lungs in tetrapods (fig. 1.12, fish and amphibian). In birds, mammals, and a few reptiles, the lungs occupy separate **pleural cavities** and the digestive organs beyond the esophagus occupy an **abdominal (peritoneal) cavity** (fig. 1.12). The coelom arises by the delamination, or splitting, of the early lateral mesoderm into two sheets, somatic and splanchnic mesoderm (fig. 5.9d). The space between the two sheets becomes the **coelom.**

Most of the embryonic digestive tract other than its endodermal lining arises in cephalochordates and vertebrates from splanchnic mesoderm (fig. 5.3c). The outer covering of the digestive tract is **visceral peritoneum** (6, in fig. 1.2). It is continuous with the **parietal peritoneum,** which lines the body wall (5, in fig. 1.2). Early

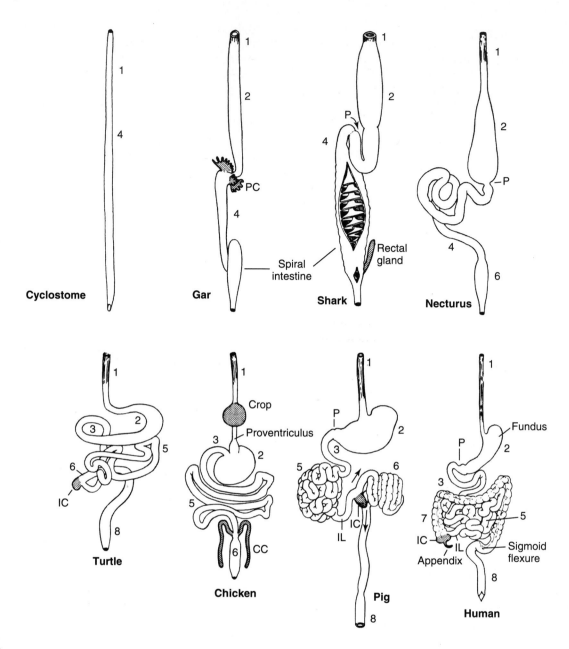

FIGURE 12.1

Digestive tracts of selected vertebrates. The intestine empties into a cloaca except in the mammals. **1,** esophagus; **2,** stomach; **3,** duodenum; **4,** intestine; **5,** small intestine; **6,** large intestine (colon in humans); **7,** colon; **8,** rectum; **CC,** paired ceca of bird; **IC,** ileocolic cecum; **IL,** ileum; **P,** pyloric sphincter; **PC,** pyloric ceca. Ceca are shown in stipple.

in embryonic life, the parietal and visceral peritonea are continuous via dorsal and ventral mesenteries, at which time the coelom is divided into separate right and left cavities. The dorsal mesentery remains essentially intact throughout life, conducting blood vessels and nerves from the roof of the coelom to the digestive organs. The ventral mesentery eventually disappears except at the level of the liver and urinary bladder. The fate of the ventral mesentery of the liver will be discussed when we study that organ.

The embryonic digestive tract consists of three regions. The part containing the yolk, when present, or

to which the yolk sac is attached, is the **midgut** (figs. 5.13*b* and 12.2). Caudal to the midgut is the **hindgut,** and anterior to the midgut is the **foregut.** The foregut elongates to form part of the oral cavity, the pharynx, esophagus, stomach, and much of the small intestine. The hindgut becomes the remainder of the intestine and the cloaca. Little of the midgut remains in adults.

The anterior portion of the oral cavity, or of the oropharynx of fishes, arises as a midventral invagination of the ectoderm of the head, the **stomodeum** (figs. 1.1 and 12.2). A thin membrane, the **oral plate,** temporarily separates the early embryonic foregut from the exterior. It

FIGURE 12.2

Sagittal section of 3.5 mm frog tadpole. An oral plate separates the stomodeal invagination from the pharynx.

From Phillips, JB: *Development of Vertebrate Anatomy* 1975 (Fig. 13-1, page 284).

soon ruptures to provide an anterior entrance to the digestive tract. A similar invagination, the **proctodeum**, provides an exit from the hindgut when the **cloacal plate** ruptures (fig. 1.1). Like all chordates, therefore, vertebrates, including humans, are deuterostomous.

Differences in the anatomy of digestive tracts caudal to the pharynx are correlated not so much with whether an animal lives on land or in the water as with the nature and abundance of food. Is it readily absorbable when ingested, as in vampire bats, or does it require extensive enzymatic activity or mechanical maceration, as in carnivores? Is the food supply constant, so that whenever an animal is hungry the food is likely to be there, or does it have to be stalked? If the latter, the meal is probably bulky and there has to be room for it in an appropriately expandable stomach until it can be digested. And what is the shape of the animal's body? If it is long, like that of a cyclostome or snake, the tract will probably be straight. If the trunk is short, as in turtles and anurans, the intestine must be coiled in order to provide a sufficient absorptive area. Some fishes have a special membrane that increases the absorptive area.

MOUTH AND ORAL CAVITY

The mouth is the entrance to the digestive tract. In gnathostome fishes, it opens into an **oropharyngeal cavity** that exhibits teeth of one variety or another, and walls perforated by gill slits. The oropharyngeal cavity terminates at a short esophagus. In tetrapods, the mouth opens into an **oral cavity (buccal cavity)** housing the teeth and tongue. The oral cavity leads to the pharynx.

The roof of the oropharyngeal cavity of fishes and of the oral cavity of amphibians is a primary palate. It is pierced anteriorly in lungfishes and amphibians by internal nares (fig. 12.3a). Most reptiles and birds have an incomplete secondary palate. This leaves in the roof of the oral cavity a deep cleft, the **palatal fissure**, that channelizes respiratory air between the choanae (internal

(a)   **Male frog**

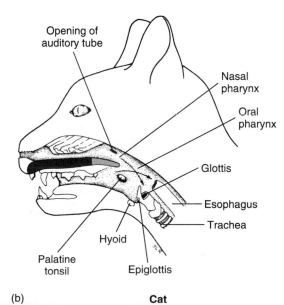

(b)   **Cat**

FIGURE 12.3

Oral cavity of an amphibian and a mammal. The roof of the oral cavity in the frog is a primary palate. In the cat it is a secondary palate (*red*). The latter consists of a hard (bony) palate (*dark red*) and a soft palate (*light red*). Arrows indicate pharyngeal chiasma where food and airstreams cross.

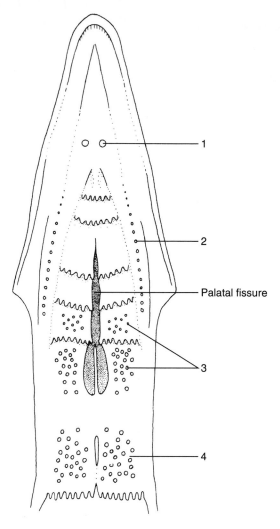

FIGURE 12.4

Palate of domestic hen. **1** to **4,** gland openings as follows: **1,** paired maxillaries; **2,** lateral palatines; **3,** medial palatines; **4,** sphenopterygoids. Internal nares (choanae) are above palatal folds at anterior end of palatal fissure.

nares of amniotes) and the pharynx (fig. 12.4). The secondary palate of crocodilians and mammals provides a cleftless roof for the oral cavity all the way from the mouth to the pharynx (figs. 9.18*c* and *d,* and 12.3*b*).

In anurans, paired **vocal sacs,** which are reverberating chambers beneath the floor of the pharynx, open into the oral cavity near the angles of the jaws. In all tetrapods, a variety of oral glands, or their ducts, empties into the oral cavity.

In mammals, a trench called the **oral vestibule,** separates the gums, or **alveolar ridges,** from the cheeks and lips. In many rodents, an opening leads from the vestibule on each side into a **cheek pouch** in which hamsters and others transport grain from the fields to the burrow for storage. The pouches of hamsters extend from the first cheek teeth caudad beneath the orbit, beneath the ears, and above the forelimb, to a position lateral to the scapula.

The collagenous connective tissue walls are lined by a moderately low keratinized squamous epithelium that protects against abrasion. Inserting on the wall is a slip of the buccinator, a mimetic muscle that acts as a retractor. The overlying skin is loose, permitting expansion of the pouches which, when full, create the impression that the hamster has a bad case of the mumps! Some seed- and grain-eating birds have a median sublingual **seed pouch** that is employed in a similar fashion. The pouch lies upon the caudal portion of the mylohyoid muscle beneath the oral cavity and is retracted by what may be a homologue of the genioglossus muscle of mammals. When full, it hangs beneath the rear of the oral cavity in a sling composed of the mylohyoid muscle. It is emptied by shaking the head vigorously. The tongue, oral glands, and teeth will be discussed in the sections that follow.

Tongue

The tongue of elasmobranchs, bony fishes, and perennibranchiate amphibians is a mere crescentic or angular elevation of the floor of the oropharyngeal cavity shaped by the underlying basihyal and ceratohyal cartilages which constitute a skeleton. This lean hyoid elevation is a **primary tongue** (fig. 12.5, adult *Necturus*). A primary tongue has no musculature, hence cannot be independently manipulated. It may assist the jaws in holding prey within the oropharyngeal cavity until it can be swallowed, especially if the surface of the tongue has teethlike denticles. It is, however, the forerunner of the tongues of tetrapods.

The tongue of terrestrial urodeles and of anurans consists of a primary tongue and of an extension that can be flipped out of the mouth. In both groups, the primary tongue develops from hyoid arch mesenchyme and the extension develops from an embryonic **glandular field** in the pharyngeal floor anterior to the hyoid arch (fig. 12.5, larval frog). The glandular field is so named because this region of the adult tongue secretes a sticky mucus that entangles insects when the tongue is suddenly thrust out of the mouth. The tip of the tongue of insectivorous amphibians usually terminates in a broad, sticky, fleshy expansion that increases the probability that the prey will be caught and then delivered into the oral cavity. The root of the tongue of terrestrial urodeles becomes anchored to the basihyal and ceratohyal cartilages in the floor of the pharynx; that of anurans, unlike other tetrapods, becomes anchored to the floor of the oral cavity immediately behind the mandibular symphysis. In the toad family *Pipidae,* no tongue develops.

The tongue of reptiles and mammals has three distinct features: (1) a pair of **lateral lingual swellings** formed by mesenchyme from the mandibular arch, a contribution not found below amniotes, (2) a primary component from the hyoid arch that develops a glandular field, and (3) mesenchyme from the third pharyngeal arch that spreads forward over some of the second-arch mes-

Digestive System

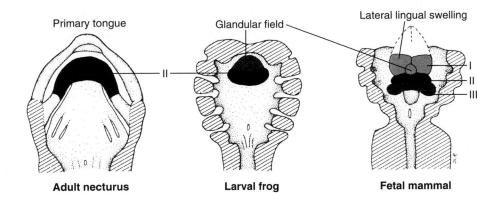

FIGURE 12.5

Floor of the oral cavity depicting stages in evolution of the tongue. **I, II, III,** derivatives of pharyngeal arches 1, 2, and 3. Dotted outline in mammal shows final extent of lateral lingual swellings.

Labels in figure: Primary tongue, Glandular field, Lateral lingual swelling, Adult necturus, Larval frog, Fetal mammal.

enchyme (fig. 12.5, fetal mammal). Sensory epithelia of the tongue (general sensation and taste) are thus formed from mesenchyme from pharyngeal arches 1, 2, and 3 and thereby innervated by three different cranial nerves—5, 7, and 9, respectively. Hypobranchial musculature invades the entire complex and receives somatic motor innervation from the twelfth cranial nerve. The innervation of the tongue of a mammal is detailed in chapter 16 under "Innervation of the Mammalian Tongue: An Anatomic Legacy." As in the lowest tetrapods, the tongue of amniotes is anchored to skeletal components of the hyoid arch. However, the tongue of turtles, crocodilians, alligators, and some baleen whales is affixed also to the floor of the oral cavity, hence is not protrusible. Nevertheless, the immobilized tongue of baleen whales plays a vital role in food handling. Garter snakes and a few other snakes have no tongue.

In birds the lateral lingual swellings are suppressed, although less so in birds of prey, and the tongue is almost lacking in intrinsic muscles. The only movement, therefore, comes from muscular operation of the hyoid skeleton to which the tongue is anchored. A unique use of the posterior horns of the hyoid of woodpeckers for "shooting" the tongue out of the mouth has been described in chapter 9 under "The Amniote Hyoid."

The "tongue" of agnathans is not homologous with any component of the tongue of gnathostomes, being simply a rodlike lingual cartilage of unknown homology capped by horny spines and operated by protractor and retractor muscles (figs. 9.35 and 13.1*b*).

The tongue is widely used for capturing or gathering food. Long, sticky tongues of insectivorous or nectar-feeding tetrapods from salamanders to bats and anteaters dart in and out at lightning speed. The sticky tongue of the four-foot-long Great Anteater is two feet in length! When the mouth of a toad is closed, the tongue lies folded back over itself so that the tip is directed toward the esophagus. Flipping the tongue out of the mouth

occurs when the long fibers of the genioglossus medialis muscle, which extend to the tip of the tongue, stiffen to form a complex of intrinsic rods, and the genioglossus basalis muscle, which forms a wedge beneath the anchored anterior end of the tongue, suddenly swells. The combination flips the rigid tongue over the mandibular symphysis. It returns to the mouth by contraction of the hypoglossal muscles (Gans and Gorniak, 1982).

Woodpeckers have a barbed tongue that shoots like an arrow into dark crevices in tree trunks, impaling grubs and carrying them to the mouth (fig. 12.6*a*). The tiny tongue of hummingbirds darts rapidly back and forth between flower and mouth, collecting droplets of nectar at the hollow frayed tip (fig. 12.6*b*). A parrot's tongue is armed with two flexible horny shields in the walls of a seed cup that is used for shucking hard-coated seeds and preparing grain for swallowing (fig. 12.7). Like fingernails, the shields are composed of keratinized epithelial cells that grow forward from a nail-like bed halfway back on the tongue. The anterior edge is constantly being worn away and replaced. Embedded within the tongue of birds and lizards is an entoglossal bone, an anteriorly directed process of the hyoid (fig. 12.6*a*). In many birds, a pair of long paraglossal bones are attached to the entoglossus near the tip of the tongue and extend caudad embedded in its edges.

The huge immobilized muscular tongue of some baleen whales directs tons of incoming seawater into huge reservoirs under the throat and chest. When these are being emptied by compression of their walls, the water is strained back into the sea through frayed sheets of baleen, and organisms strained out of the water accumulate on the deeply furrowed tongue and are then swallowed. Newborn baleen whales have only rudiments of baleen, so there is room in the neonatal oral cavity for a mobile tongue to be manipulated for suckling for about six months before it becomes immobilized. The tongues of most other mammals are protrusible, although tied to

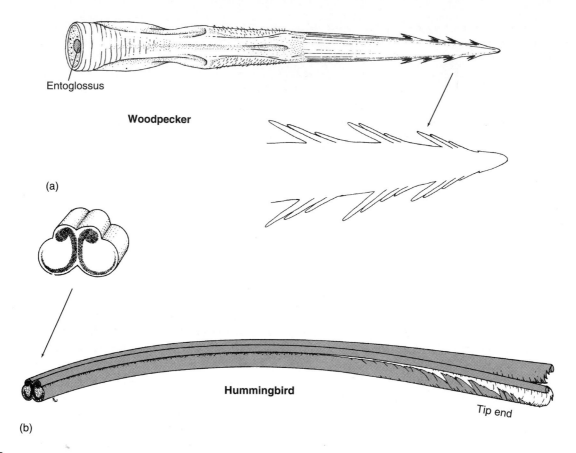

Entoglossus

Woodpecker

(a)

Hummingbird

Tip end

(b)

FIGURE 12.6

Adaptations of the tongues of two birds with different feeding habits. (*a*) Ventral view of a grub-eater. (*b*) Lateral view of a nectar-feeder.

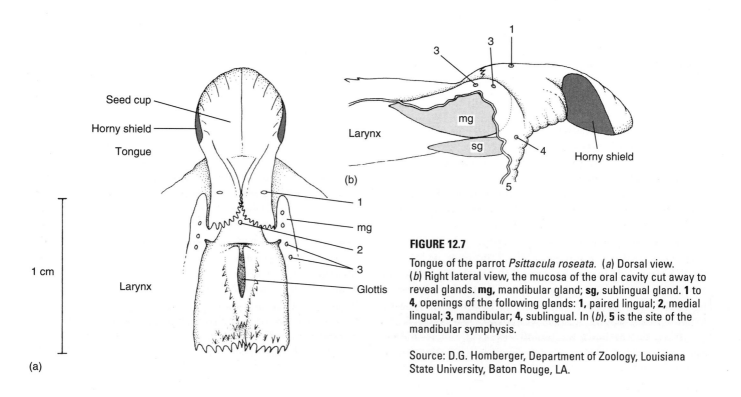

Seed cup

Horny shield

Tongue

Larynx

1 cm

Glottis

1

mg

2

3

(a)

Larynx

mg

sg

Horny shield

(b)

FIGURE 12.7

Tongue of the parrot *Psittacula roseata.* (*a*) Dorsal view.
(*b*) Right lateral view, the mucosa of the oral cavity cut away to reveal glands. **mg,** mandibular gland; **sg,** sublingual gland. **1** to **4,** openings of the following glands: **1,** paired lingual; **2,** medial lingual; **3,** mandibular; **4,** sublingual. In (*b*), **5** is the site of the mandibular symphysis.

Source: D.G. Homberger, Department of Zoology, Louisiana State University, Baton Rouge, LA.

the floor of the oral cavity by a ligament, the **frenulum linguae.** In human beings, if the frenulum inhibits movements of the tip of the tongue, making the individual "tongue-tied," it can be snipped at its anterior edge.

The mucosa of the tongue contains receptors not only for taste but for other stimuli, including, in amniotes, **stereognosis** (perception of the shape, weight, and texture of a solid body by feeling, handling, or lifting it). These encapsulated nerve endings on the tip of the tongue enable insectivores to locate food by exploring dark recesses with the tongue. Seed-eating birds use stereognostic information in manipulating a seed that is being husked in the seed cup.

Tongues not used to procure food or water manipulate fluids and solids in the oral cavity; and the tongues of most tetrapods participate in swallowing. Long tongues function in overheated mammals as a site for cooling the blood by evaporation of saliva while panting. Lizards clean their spectacles (transparent eyelids) with them. Spiny papillae on the tongue's surface are used by carnivores for rasping bones and by many mammals for grooming. The latter accounts for hairballs that often form in the stomach of domestic cats. And, of course, human speech as we know it would not be possible without a tongue.

Oral Glands

Terrestrial tetrapods have a variety of multicellular glands that secrete watery or viscous fluids into the oral cavity. The chief ingredient is usually mucus of varying viscosities and chemical composition. It moistens the food as necessary to produce a bolus that can be manipulated by the tongue, when present, and lubricates dry food for passage through the pharynx and down the esophagus. Moisture is also essential for taste buds to function because the stimulant for taste must be in solution to evoke a gustatory response. A taste bud is illustrated in figure 17.23. Other secretions include, in one species or another, serous fluids, toxins, venoms, and an enzyme that digests starch, although the latter is rare in orders below mammals. Viscous secretions keep the tongue sticky in species where this is essential, and venom tranquilizes prey or dispatches them promptly.

Oral glands are usually named according to their location. Labial glands open into the oral vestibule at the base of the lips, molar glands lie near the molar tooth, infraorbital glands are in the floor of the orbit, palatal glands open onto the palate, and sublingual and submandibular glands open via common papillae under the tongue. (To observe these papillae in your oral cavity, open your mouth while looking into a mirror, raise your tongue, and note the two papillae just behind the mandibular symphysis at the base of the tongue. A flashlight may help!) Intermaxillary (internasal) glands lie near the premaxillary bone. Frogs have up to 25 small intermaxillary glands, each with its own duct that delivers a sticky secre-

tion onto the palate. The large venom glands of the four families of venomous snakes are palatal glands. Their ducts open at the base of a maxillary tooth and the venom exudes into a groove or tube in the tooth (fang) (fig. 12.8, snake). In *Heloderma*, the only poisonous lizard, sublingual glands secrete the toxin.

Saliva is a mixture of oral secretions, but the term is usually reserved for oral secretions of mammals. The mammalian parotid is the largest tetrapod salivary gland, and ptyalin (amylase), an enzyme that initiates digestion of starch, is always one of its secretions. Mammals have several salivary glands, but not all of them produce ptyalin. The specific mix of mucus, serous secretions, and ptyalin produced by any mammalian salivary gland varies with the species, and is correlated with dietary habits. The duct of the parotid gland crosses the masseter muscle to open into the oral vestibule near one of the upper molar teeth (fig. 12.8, cat). The poison glands of reptiles resemble the parotid gland histologically. Birds do not have a copious supply of saliva, but the openings of some of the oral glands of birds are shown in figures 12.4 and 12.7.

Aquatic vertebrates have adequate moisture in the oral cavity for taste buds to function, but glands that secrete digestive enzymes would be a waste of energy. Any enzymatic secretion would be diluted and washed away. Mucus-producing **goblet cells** are about the only

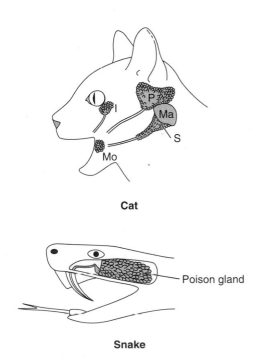

Cat

Snake

Poison gland

FIGURE 12.8

Oral glands in a cat and a snake. **P,** parotid; **Ma,** mandibular; **S,** sublingual; **Mo,** molar; **I,** infraorbital. A proximal portion of the mandibular duct (*dotted*) is surrounded by the sublingual gland. Ducts for molar and sublingual glands are small and cannot be seen. A duct from the poison gland drains into a grooved maxillary tooth.

source of oral secretions, and these are common. Mucus lubricates the esophagus. Goblet cells perform a special function in the male population of a few species of cat-fishes. Males carry fertilized eggs in temporary folds, or crypts **(brood pouches),** in the mucosa of the palate and retain the hatchlings in the crypts for about two months during which time the male does not feed. The goblet cells in the brood pouches produce a copious nutritious secretion that maintains a suitable environment for development of the eggs and later nourishes the hatch-lings. After the hatchlings leave, the brood pouches atro-phy in response to altered hormonal ratios. A rare instance of multicellular oral glands in aquatic verte-brates are the anticoagulant glands of lampreys, which open into the oral cavity beneath the tongue just internal to the buccal funnel. Their secretion inhibits blood clot-ting of the prey while the lamprey is feeding.

Teeth

Bony teeth are found among jawed fishes, amphibians, reptiles, and mammals with few exceptions, and they were present in the earliest birds. In mammals they have achieved a peak in regional specialization. Lacking bony teeth are sturgeons, numerous teleosts including sea horses, a few amphibians, all turtles, and modern birds; and among mammals, whalebone whales, South American and scaly anteaters, and the monotreme *Echidna.* But even among toothless species there are a number that develop an embryonic set that simply doesn't erupt or, after erupting, disappears. Teeth are descendants of the denticles of the dermal armor that covered the head and extended into the oropharyngeal cavity of the early fishes (fig. 6.29*a*). The relatively small number of vertebrates that lack teeth have lost the genetic code necessary either to induce their formation or to complete their development.

Bony **dental plates** of dermal origin overlay the endoskeletal components of the jaws of early fishes. Many had pointed, rounded, or jagged surface projections that prevented escape of live food from the oropharyngeal cav-ity, or that were used for crushing shellfish, biting flesh, or rasping vegetation. Placoderms that were bottom scav-engers had short jaws, each covered by a single thin, bony plate. Other placoderms had long jaws covered with one or two long bony plates with sharply notched bladelike cutting edges. The plates were braced against others that lined the inside of the cheek. In some other placoderms, the bony plates bore long prongs of dentin separate from, but affixed to, the plates. Still others had many small plates, each bearing clusters of broad-based conical or stel-late denticles, and the plates were shed at intervals. In some species, the plates were never shed. In placoderms, all these denticle-bearing bony dental plates invested the underlying endoskeleton of the jaws.

In acanthodians, each denticle, usually broadly based, some conical, some slender and curved, some with a

dozen spiny cusps arranged in whirls, was attached directly to the endoskeleton of the jaws, as in modern teeth, instead of being borne on dental plates; and the jaws often were flanked by additional denticles. Some acanthodians were toothless or had lower teeth only. These bony denticles of early fishes were the precursors of the teeth of modern vertebrates.

Until the phylogenetic relationships between placo-derms, acanthodians, and later cartilaginous and bony fishes can be ascertained, only the broadest generaliza-tion regarding the phylogeny of later vertebrate teeth can be made: *they are derivatives of bony dermal armor.* Evidence comes not only from paleontology, but also from the comparative histology of armor and teeth and from the observation that dermal denticles and placoid scales, even today, show a gradual transition to teeth as they approach the cutting edges of the jaws (figs. 12.9 and 12.10).

Teeth, like placoid scales, are composed of **dentin,** which is a variety of bone, surmounted by a crown of **enamel** (fig. 12.11). The earliest indication of the devel-opment of socketed teeth is the ingrowth into the dermis of a longitudinal ridge of ectoderm, the **dental lamina,** that extends more or less the length of the jaws (2, in fig. 12.12*c*). Beneath the lamina (or above it, in the upper jaw) a linear series of dermal papillae, each designating the site of a future tooth (1 in fig. 12.12*c*) form at inter-vals, indenting the lamina, and organizing blood vessels necessary for further development of a tooth pri-mordium. The cells at the periphery of each papilla become organized into a definitive layer of **odontoblasts** that commence to deposit dentin. As deposition proceeds, the odontoblasts slowly withdraw toward the center of the

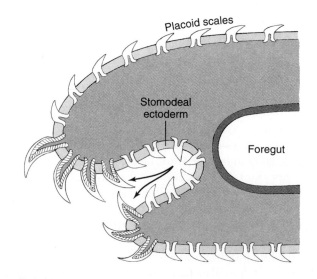

FIGURE 12.9 •

Representation of the identity of placoid scales, stomodeal denticles, and teeth in sharks. Arrows indicate direction of migration of formative to functional teeth at margins of the jaws.

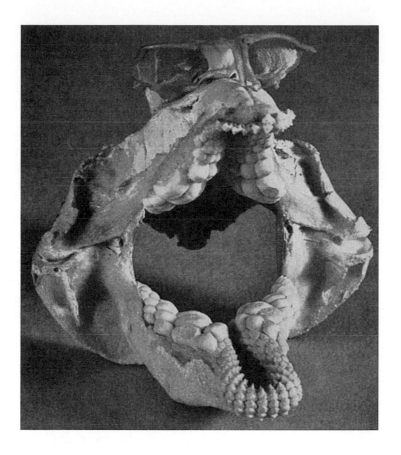

FIGURE 12.10

Jaws of the Port Jackson shark, *Heterodontus,* showing regional transition from dermal denticles to placoid scalelike teeth. The rounded teeth are used for crushing molluscs.

Courtesy Ward's Natural Science Establishment, Inc. Rochester, NY.

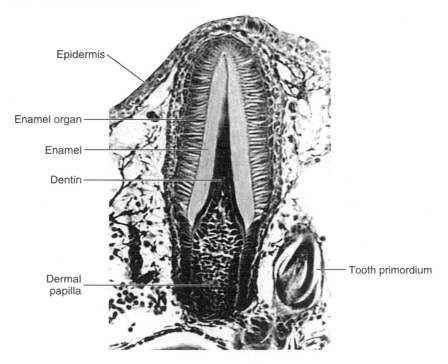

FIGURE 12.11

Unerupted tooth of garfish. Primordium is a developing replacement tooth.

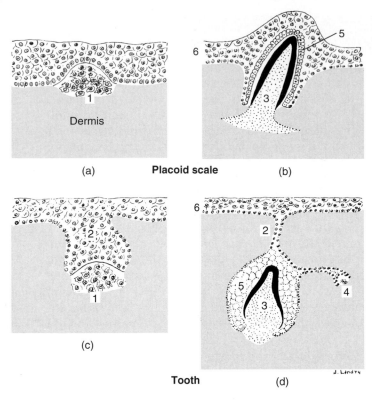

(a)   **Placoid scale**   (b)

(c)

Tooth   (d)

Morphogenesis

FIGURE 12.12

Development of a placoid scale (*a,b*) and mammalian tooth (*c,d*). *Gray* shows dermis; *red* indicates enamel. **1**, dermal papilla; **2**, (in *c* and *d*), cross section of dental lamina; **3**, dentin, produced by dermal papilla; **4**, primordium of replacement tooth; **5**, enamel organ; **6**, epidermis.

primordium, which is becoming a pulp cavity containing components of the dermal papilla. The withdrawing odontoblasts leave behind evidence of their withdrawal in the form of **dentinal tubules** that contain protoplasmic processes of the odontoblasts (fig. 7.6). The odontoblasts remain alive throughout the life of the tooth. Meanwhile, the ectoderm of the dental lamina has organized an **enamel organ,** consisting of ameloblasts, that deposits enamel on the surface of the dentin (figs. 12.11 and 12.12*d*). A thin layer of **cementum,** a type of acellular bone (chapter 7), eventually anchors the tooth to the bone of the jaw by means of collagenous fibers. Living remnants of the dermal papilla, including vessels and nerves, remain within whatever is left of the pulp cavity (root canal) throughout the life of the tooth. Vessels and nerves are both essential for maintaining the tooth in a healthy state. The details of tooth development and emergence, the time of initiation of the different stages, and the ultimate fate of erupted teeth vary with the species.

An enamel organ is present but functionless in armadillos and a few other vertebrates, hence their teeth have no enamel. Enamel in mammals, at least, is

deposited by ameloblasts of ectodermal origin. Enamels of lower vertebrates, especially fishes, differ in their physical characteristics, and the ultimate source of the scleroblasts that elaborates them (fig. 7.1) has not been ascertained. Similarities in the development of placoid scales and mammalian teeth are illustrated in figure 12.12.

Teeth vary among vertebrates in number, distribution within the oral cavity, position with reference to the summit of the jaw, degree of permanence, and shape. They are numerous and widely distributed in the oropharynx of living fishes where they develop on jaws, palatal bones, and even on the pharyngeal skeleton. For example, the blue sucker has 35 to 40 teeth on the last gill arch. In early tetrapods, too, teeth were widely distributed on the palate; and even today most amphibians and many reptiles have teeth on the vomer, palatine, and pterygoid bones and occasionally on the parasphenoid. They are confined to the jaws in crocodilians, fossil toothed birds, and mammals, and they are least numerous among mammals. Teeth, therefore, like dermal armor, have tended toward a more restricted distribution with the passage of time.

Jaw teeth may be attached to the outer surface or to the summit of the jawbone, as in many teleosts (**acrodont dentition,** fig. 12.13); they may be attached to the inner side of the jawbone, as in anurans, urodeles, and many lizards (**pleurodont dentition**); or they may occupy bone sockets, or alveoli (**thecodont dentition**). Socketed teeth are found in many fishes, in crocodilians, extinct toothed birds, and mammals. The sockets (alveoli) are deepest in mammals.

Most vertebrates through reptiles have a succession of teeth, and the number of replacements during a lifetime is indefinite but numerous (**polyphyodont dentition**). It has been estimated that an elderly crocodile may have replaced its front tooth 50 times. They and other submammalian bony vertebrates that have been studied replace teeth in waves that sweep along the jaws eliminating and replacing every other tooth. Thus in one wave, in tetrapods at least, even-numbered teeth are lost and in the next wave odd-numbered ones are lost. Meanwhile, tooth germs for the next wave of eruptions are forming. Whether the loss and replacement waves sweep from back to front or reverse is not agreed on at present. There is evidence that they sweep in different directions in different species. The waves ensure a balanced distribution of teeth along the jaws throughout life. In sharks, tooth germs form in the dermis on the oropharyngeal cavity side of the jaws and while growing migrate onto the cutting edge of the jaw (arrows in fig. 12.9) as the teeth that are being replaced move beyond the edge before falling away. The cause of the migration is not known.

Only in mammals is there a definite number of teeth in a species, with rare exceptions. Most mammals develop two sets, **deciduous,** or **milk teeth,** and perma-

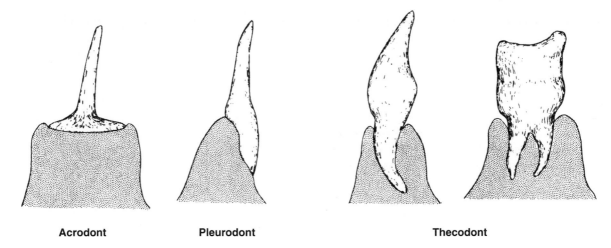

Acrodont   **Pleurodont**   **Thecodont**

FIGURE 12.13

Variations in the relationship of teeth to jaws. Acrodont teeth are attached either at the outer surface of the jaw or, as shown, at the summit. Pleurodont teeth are attached to the inner lingual surface of the jaw. Thecodont teeth occupy alveoli.

nent teeth **(diphyodont dentition)**; and there is a definite sequence in which the teeth erupt. For example, numbering the permanent set in human beings 1 to 8 from front to rear, the sequence of eruption is 6,1,2,4,5,3,7,8. Eruption of number 8, the last molar, is delayed in higher primates, and this "wisdom tooth" is sometimes imperfect, unerupted, or missing. A first set provides the constantly changing infant jaw with small temporary teeth adequate for an infant's diet until the jaws are more stabilized structurally and have elongated sufficiently to accommodate large teeth for macerating coarse foods.

A few mammals develop only a first set **(monophyodont dentition).** In a platypus, the deciduous set is replaced by **horny epidermal teeth.** In toothless whales, the first set, although formed within the jawbone, may not erupt, and if they do they are usually shed. The freshwater manatee from the Amazon River and the Australian Rock Wallaby do not have "sets," teeth being replaced throughout life by the forward migration of new teeth formed at the rear of the jaws. In the manatee, migration is at the rate of 1 or 2 mm a month. Thin, bony sockets separate the roots of successive teeth, and the bony septa are resorbed under pressure from the migrating teeth. The grasses eaten by the manatee contain abrasives that appear to be necessary for the teeth to move forward, since babies fed experimentally on milk show no forward migration of teeth. Proboscidians have a slow but constant succession of molar teeth that move forward from the rear.

Morphologic Variants in Fishes

Most sharks are fish eaters and have numerous rows of jaw teeth that are either flat, sharp, notched triangles that are used for cutting, or single or multipointed tusks that curve toward the pharynx and hold a struggling prey until it can be swallowed whole. Each shark tooth has a broad **basal plate** of dentin embedded in the dermis. A minority eat shellfish and, although the teeth at the entrance have curved caudally directed spines, the rest form batteries of rounded denticles used for crushing shells (fig. 12.10). Tiny stomodeal denticles line the pharynx in some sharks. Near the jaws these have transitional shapes between denticles and teeth.

The dental armor of today's holocephalans and modern lungfishes is like that of early jawed fishes, consisting of a few large plates of enamel-covered dentin that bear rows of various-sized rounded moundlike denticles that usually become sharp spines at the entrance to the oropharyngeal cavity. *Chimaera*, a cartilaginous fish that eats molluscs, has on each side of the upper jaw one large anterior and one small posterior dental plate that together cover the entire upper jaw. There is a single large plate on each side below. In modern lungfishes, the plates are restricted to the palate and medial aspects of the lower jaw.

The jaw teeth of actinopterygians, amphibians, and most reptiles are simple pointed cones attached to one or more membrane bones. Small teeth may be interspersed among large ones, and those in front are sometimes larger and curved slightly to the rear. Specialized shapes sometimes appear on one jaw or the other. Garfishes, for instance, have a few fanglike teeth shaped at their ends like arrows; and the fangs of venomous snakes, borne on the maxillae, are curved or bladelike, grooved on the rear surface, or tubular, for injecting venom. When all teeth are essentially similar the dentition is described as **homodont.**

Morphologic Variants in Mammals

All but a very few mammals exhibit **heterodont dentition;** that is, the teeth vary morphologically from front to rear, being **incisors, canines, premolars,** and **molars.**

Premolars and molars are "cheek teeth." Heterodont dentition arose in late synapsid reptiles. Most marine mammals of today—cetaceans, sirenians, and some marine carnivores—have reverted to homodont dentition, but their cretaceous ancestors exhibited heterodonty.

Incisors, located on either side of the mandibular symphysis, have one horizontal cutting edge and a single root. They are best developed in herbivorous mammals, which use them for holding, cropping, or gnawing. The single pair of large chisel-like incisors of rodents and the large front pair of lagomorphs have enamel on the anterior surface only. (Lagomorphs have a small second pair of incisors *behind,* not lateral to, the front pair.) Since dentin is softer than enamel, gnawing wears dentin away faster. This keeps the cutting edges of the incisors sharp. These incisors grow throughout life. Incisors may be lacking on the upper jaw only, as in bovines (fig. 12.14, ox), lacking on the lower jaw only, as in vampire bats, or lacking completely, as in sloths. Elephant and mastodon tusks are modified incisor teeth (fig. 12.14). Incisor tusks are found also in an occasional member of a few other mammalian orders. Tusks, like the incisors of rodents, grow throughout life, compensating for wear. Walrus tusks are not incisors, but canine teeth.

Canine teeth lie next to the incisors (fig. 12.14, dog). In generalized mammals, incisors and canines scarcely differ in appearance (fig. 12.14, shrew). In carnivores, the canines are spearlike and used for piercing flesh (fig. 12.15*d*). They are the tusks of a walrus (fig. 12.14).

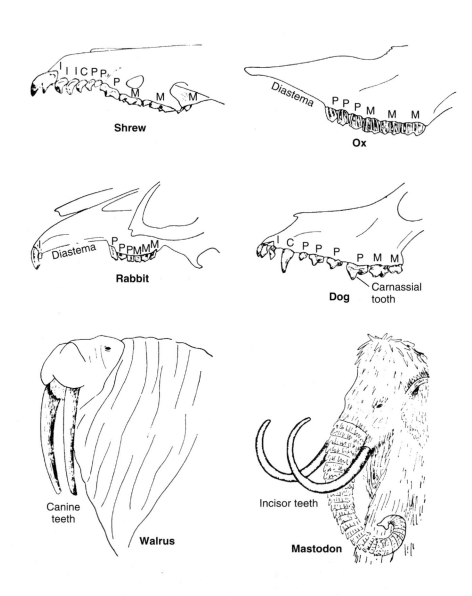

FIGURE 12.14

Mammalian upper teeth, showing generalized pattern (shrew) and specializations in a herbivore (ox), gnawing animal (rabbit), and carnivore (dog). Extreme specialization of canines is seen in the walrus and of incisors in the mastodon. **I,** incisor; **C,** canine; **P,** premolar; **M,** molar.

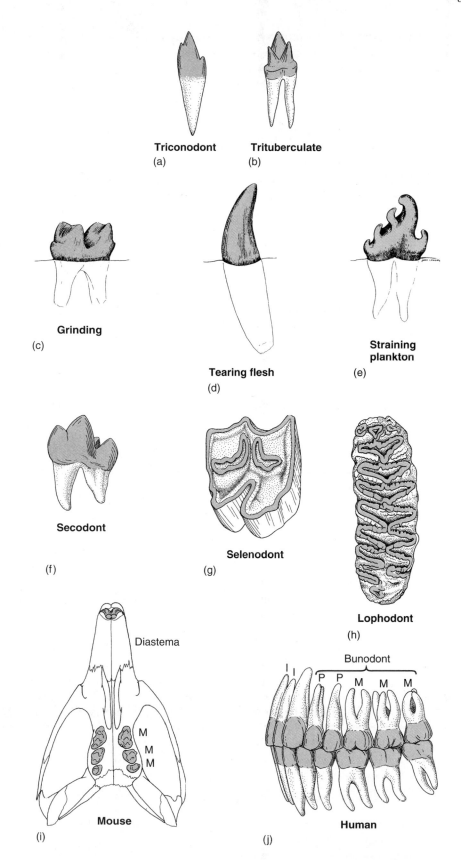

Triconodont   Trituberculate
(a)   (b)

Grinding
(c)

Tearing flesh
(d)

Straining
plankton
(e)

Secodont
(f)

Selenodont
(g)

Lophodont
(h)

Diastema

M
M
M

Mouse
(i)

Bunodont

I I P P M M M

Human
(j)

FIGURE 12.15

Varieties of mammalian teeth. (*a,b*) Early prototherian and therian teeth. (*c*) Lower molar from a young tapir whose diet is leaves and twigs. (*d*) Lower-left, flesh-tearing canine from a jaguar. (*e*) One of a series of similar lower-right molars from a crabeater seal. (*f,g,h*) Cheek teeth from a dog, horse, and elephant, respectively. (*i*) Upper-jaw dentition of a mouse. The molars are bunodont. (*j*) Complete, permanent, human dentition. *Red,* enamel.

Canine teeth are absent in lagomorphs, so there is a toothless interval, or **diastema,** between the incisors and the first cheek tooth (fig. 12.14, rabbit). In rodents, premolars are also missing, and the diastema is longer (fig. 12.15, mouse). Canine teeth attained their greatest length on the upper jaw of the now extinct saber-toothed tigers. In these cats, they extended as much as 20 cm below the lower jaw with the mouth closed. The lower canines were correspondingly reduced.

The **premolars** of most mammals other than ungulates usually have two prominent cusps, hence they are bicuspids. They also have one or two roots, and the number of roots may differ on the upper and lower jaw and among different individuals of the same population, including humans.

Molars have three or more cusps, hence they are tricuspids. They usually have three roots, but there are occasional molars with four or five. Molars are not replaced by a second set. In fact, they are late arrivals of the first set.

The **crown** of a tooth is the part above the gum line. It is covered with enamel. The crowns of the premolars and molars (cheek teeth) of carnivores and those of herbivorous ungulates show extreme morphological differences. Those of carnivores are adapted for tearing flesh and crushing bone. Those of ungulates and some other herbivores are adapted for macerating vegetation.

The crowns of carnivores' cheek teeth are laterally compressed, and the two or three cusps are interconnected by sharp ridges of enamel. The teeth also have long roots. They are known as **secodont teeth** (fig. 12.15*f*). They are spaced on the jaws in such fashion that the cusps of the teeth on the upper jaw fit between the cusps of the teeth on the lower jaw. Thus, as the jaws are closing, the sharp enamel ridges of the crowns produce the shearing effect necessary for macerating animal tissues. The last upper premolar and the first lower molar, the **carnassial teeth,** are larger and longer than adjacent cheek teeth and are employed when a particularly tough shearing problem presents itself, as when a dog is chewing on a bone.

The cheek teeth of ungulates are specialized for grinding vegetation. They are wider and longer than those of carnivores, thereby providing broad surfaces for grinding (fig. 12.15*g* and *h*); and the crowns are tall, allowing for plenty of wear. The crowns consist of crescentic columns of dentin, each column surrounded by enamel, and the columns are embedded in additional dentin that is devoid of an enamel overlay. The dentin, being softer than enamel, wears away more quickly, leaving sharp crescentic enamel ridges with a wide variety of configurations. These ridges macerate vegetation during the complex side-to-side and forward-backward chewing movements of opposing jaws. Because the cutting surfaces are crescentic, these teeth are said to be

selenodont. (For derivation, see appendix 2.) The cheek teeth of bovines are selenodont. However, there are no teeth anterior to them on the upper jaw of bovines. The cheek teeth are employed in chewing cud.

The adaptation for grinding is exaggerated in proboscidians, whose teeth are termed **lophodont** (fig. 12.15*h*). The enamel and dentin are intricately interfolded, and the enamel is disposed on ridges (lophs) on enormous plateaus of naked dentin. Lophodont teeth reach a foot or more (30 cm) in length and a third of a foot in width in the largest elephants.

The cheek teeth of the remaining mammals exhibit a diversity of molariform styles. Those of omnivores and some herbivores may lack sharp edges and pointed cusps and have, instead, low rounded cusps, as in higher primates, including humans (fig. 12.15*j*). These are **bunodont teeth.** Among other mammals with bunodont teeth are rhinos, some hogs, some primitive ruminants, and some rodents (white-footed mice, for instance).

Rodents, the largest mammalian order and with the largest variety of diets, exhibit the largest variety of teeth, some low crowned with long roots as in squirrels, some high crowned with short roots, as in wood rats. Among the most unusual mammalian teeth are those of the crabeater seal (fig. 12.15*e*), whose diet consists chiefly of crustaceans. The teeth are employed to strain small crustaceans and other plankton out of mouthfuls of seawater as it spills back into the sea.

Cheek teeth of early prototherians were **triconodont,** the crown having three conelike prominences arranged in a straight line (fig. 12.15*a*). In early therians (fig. 4.30), the cones were arranged in a triangle (fig. 12.15*b*). Those teeth are termed **trituberculate.** Whether the latter are derivatives of triconodont teeth is not known, since the fossil record of the origins of prototherians and therians is incomplete. It seems certain, however, that trituberculate teeth were the forerunners of today's tricuspids. Formation of enamel crests connecting the cones of trituberculate teeth in various configurations is thought to account for selenodont and lophodont teeth. Most of our knowledge of the evolution of mammals and much of the classification of fossil mammals is based on studies of fossil molariform teeth, and for a good reason: in many cases, these were the only recovered remnants of the extinct mammal that bore them.

The first eutherians had three incisors, one canine, four premolars, and three molars on each side of the jaw, a total of 44 teeth. This may be expressed by the formula $\frac{3-1-4-3}{3-1-4-3}$. Adults of some of the generalized mamals of today still exhibit this formula. Formulas for a few selected mammals of varied dietary habits are:

Cretaceous eutherian		$\dfrac{3-1-4-3}{3-1-4-3}$	
Insectivores	Solenodon $\dfrac{3-1-3-3}{3-1-3-3}$	Mole $\dfrac{3-1-4-3}{3-1-4-3}$	
Marsupials	American opossum $\dfrac{5-1-3-4}{4-1-3-4}$	Numbat $\dfrac{4-1-3-5}{3-1-3-6}$	
Primates	Tarsius $\dfrac{2-1-3-3}{1-1-3-3}$	Catarrhines $\dfrac{2-1-2-3}{2-1-2-3}$	
Carnivores	Canines $\dfrac{3-1-4-2}{3-1-4-3}$	Felids $\dfrac{3-1-3-1}{3-1-2-1}$	
Lagomorphs	Rabbits $\dfrac{2-0-3-3}{1-0-2-3}$	Pika $\dfrac{2-0-3-2}{1-0-2-3}$	
Rodents	Hamster $\dfrac{1-0-0-3}{1-0-0-3}$	Squirrel $\dfrac{1-0-2-3}{1-0-1-3}$	
All Bovines		$\dfrac{0-0-3-3}{3-1-3-3}$	

Teeth, along with the tongue and hyoid, constitute a functional triad that procures, manipulates, and, in mammals, masticates foodstuffs at the entrance to the digestive tract, then starts a bolus of food on its way to digestive sites.

Epidermal Teeth

Keratinized (horny) teeth sometimes function like bony ones. Agnathans have horny teeth in the buccal funnel and on the tongue, which are used for rasping. Anuran tadpoles have several rows on temporary lips perched on poorly developed jaws; they are used for rasping algae and other vegetation, which is the tadpoles' diet. At metamorphosis, the horny teeth of anurans are shed and replaced by bony ones. Before hatching, turtles, crocodilians, *Sphenodon*, birds, and monotremes have a temporary horny **egg tooth** that is used for cracking the shell; and after a baby platypus has lost its first set of bony teeth, horny teeth replace them and remain throughout life. The horny beaks of turtles and modern birds often have serrations that perform some of the functions of teeth.

PHARYNX

The adult tetrapod pharynx is the part of the digestive tract that had pharyngeal pouches in the embryo. As a functional part of the digestive tract it opens into the esophagus. The pharynx of fishes is a functional part of the respiratory system, and will be discussed in the next chapter.

The most constant features of the tetrapod pharynx are the **glottis**, a slit opening into the larynx, the **openings of the paired auditory tubes** that lead to the middle ear cavity, and the **opening into the esophagus** (fig. 12.3a and b). The pharynx of mammals exhibits additional features. It consists of a **nasal pharynx** above the soft palate and an **oral pharynx** between the oral cavity and the glottis.

Nasal passageways empty into the nasal pharynx via choanae (internal nares), and the auditory tubes, derived from the first pair of pharyngeal pouches, open into its lateral walls. A narrow **laryngeal pharynx** is located dorsal to the larynx in those mammals in which the opening to the esophagus is caudal to the glottis (fig. 13.13a).

The oral cavity in mammals leads to the oral pharynx. The transition is a narrow passageway, the **isthmus of the fauces.** The lateral walls of the isthmus each exhibit two **pillars of the fauces.** The pillars are muscular folds that arch upward from the side of the tongue to the soft palate (glossopalatine arch), and from the pharyngeal wall to the soft palate (pharyngopalatine arch). In humans and some other higher primates, a fleshy **uvula** hangs from the caudal border of the soft palate into the oral pharynx. You can readily see the arches and the uvula at the end of your own oral cavity by looking in a mirror and saying "Ah!" (Good lighting is necessary.)

In the hollow between the pillars of the fauces on each side is a lymphoid organ, the **palatine tonsil.** These tonsils develop in the walls of the embryonic second pharyngeal pouches. A remnant of a pouch often remains as a pocket-like crypt with a tonsil in its wall. **Pharyngeal tonsils (adenoids)** develop in the mucosa of the nasal pharynx and **lingual tonsils** develop on the tongue near its attachment to the hyoid bone. This encircling array of lymphoid masses is the body's first line of defense against infective agents that have entered the mouth and external nares.

In mammals, a fibrocartilaginous flap, the **epiglottis**, lies in the floor of the pharynx ventral to the pharyngeal chiasma (fig. 13.13). It is attached to the hyoid bone. In many mammals, the act of swallowing draws the larynx forward (upward, in humans) against the epiglottis, closing the glottis, and thereby momentarily preventing fluids or particles of food from entering the pathway to the lungs. In other mammals, the epiglottis and part of the larynx can be drawn into the nasopharynx to provide an uninterrupted air pathway to the lungs, while permitting foodstuffs to detour around the larynx and enter the esophagus (fig. 13.15). The advantages of these arrangements are discussed in the next chapter.

Lower tetrapods have in either the air or food pathway, or both, fleshy valves at appropriate locations that are also able to regulate air and food traffic in the pharynx. Among these, for example, are valves that open or close the entrance to the external nares in aquatic tetrapods.

In some teleosts, a pair of elongated muscular tubes, **suprabranchial organs,** evaginate from the roof of the pharynx on each side near the esophagus, extend cephalad above the membranous roof of the pharynx behind the skull, and then turn caudad to terminate as blind sacs. Elongated gill rakers from the last two gill arches form funnel-shaped baskets that extend into the entrances of the tubes, and each tube is surrounded by a cartilaginous capsule to which the striated muscle of the tube is attached. The epithelium at the blind ends has many goblet cells, and the sacs contain quantities of plankton, sometimes compressed into a bolus. It may be that a function of these is to trap plankton from the incoming respiratory water stream and concentrate it into mucified masses that are swallowed. In at least one air-gulping teleost, the cavity is filled with air, and a highly vascular epithelial lining serves as an accessory respiratory membrane.

MORPHOLOGY OF THE GUT WALL

The structure of the gut wall from the beginning of the esophagus to the cloaca or anus is essentially the same throughout its length, consisting of four histological layers: **mucosa, submucosa, muscularis externa,** and **serosa** (fig. 12.16). Differences in the detailed histology of any segment are primarily in the thickness of some of its layers and in the nature of its glands. Variations reflect the specific role of the segment.

The *mucosa* consists chiefly of (1) a glandular epithelial lining of endodermal origin; (2) an underlying layer of not very dense (areolar) connective tissue supporting the base of cryptlike epithelial glands, lymph nodules of varying sizes, and blood and lymph capillaries that service the glandular epithelium; and (3) a thin coat of smooth muscle fibers (muscularis mucosae, absent in some regions). Mucous glands are ubiquitous, providing, among other benefits, a lubricant that facilitates passage of the contents of the tract during peristalsis.

The *submucosa* is a thicker layer of connective tissue that supports the base of compound alveolar glands and, of vital importance, a rich plexus of arterioles (arteries of small diameter), venules (small veins), and lymphatics that service the capillary beds of the mucosa, bringing nutrients, oxygen, and raw materials for the synthesis of glandular secretions, and carrying away the waste products of metabolism and the absorbed products of digestion.

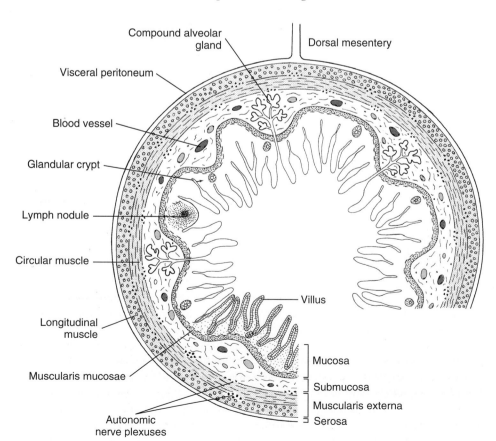

FIGURE 12.16

Cross section of mammalian digestive tract at level of small intestine. The four layers identified at bottom of figure are present from esophagus to cloaca or anus. Villi are confined to small intestine.

The *muscularis externa* consists of smooth muscle tissue arranged in two clearly defined layers, an **inner circular layer,** in which smooth muscle fibers encircle the gut, constricting the lumen on neural demand, and an **outer longitudinal layer** of muscle fibers that contracts short segments of the gut. The combined action of the two produces the macerating, peristaltic, and, in the colon of mammals, the segmenting actions of the gut. Neural stimuli for contraction are supplied from autonomic (visceral) nerve plexuses located between the longitudinal and circular muscle layers, and between the circular layer and the submucosa.

The *serosa* consists of loose connective tissue (the adventitia) and a covering of visceral peritoneum. The serosa exudes small amounts of a serous fluid that lubricates the surface of the viscera, reducing friction generated when the organs rub against one another. Inflammation of the serosa (peritonitis) results in exudation of excessive quantities of fluid.

The esophagus and the caudalmost portion of the intestine generally lie against the body wall, hence may be covered by serosa on only one surface, the one that bulges into the coelom.

The entire digestive tract is ciliated in many larval vertebrates (as in protochordates), and there are cilia in the stomach of many teleosts, in the oral cavity, pharynx, esophagus, and stomach of some adult amphibians, and in other locations in various species. Cilia are present for a time in the stomach of a human fetus. However, peristalsis is chiefly responsible for moving foodstuffs along the alimentary canal of vertebrates.

ESOPHAGUS

The esophagus is a distensible muscular tube, shortest in fishes and neckless tetrapods, and extending between the pharynx and stomach. In fishes it serves as a sphincter that closes the passageway to the stomach during the phase of respiration when water is being forced across the gills. Its only function otherwise in fishes and tetrapods is to conduct foodstuffs to the stomach. The glands in its lining secrete only mucus. Striated muscle at the cephalic end of a long esophagus is gradually replaced farther down by smooth muscle, except in ruminants. In these cud-chewing mammals, striated muscle tissue continues onto the walls of the rumen of the stomach. However, the lining of the rumen is similar to that of the esophagus, and it produces no digestive enzymes. Other features of ruminant anatomy are discussed later in the chapter.

The esophagus is lined by a stratified squamous epithelium, which, in terrestrial turtles, birds, and few mammals, is cornified, enabling the lining to withstand abrasion caused by roughage in the diet. The esophagus of marine turtles is lined by horny papillae that are directed backward, thereby preventing regurgitation while making it easy to swallow slippery seaweed, which is their preferred diet.

One of the few specializations of the esophagus of birds is the **crop** (fig. 12.20). It is a paired or unpaired membranous diverticulum, or sac, found chiefly in grain eaters, who use it for hoarding seeds and grain until there is room for them in the stomach. Under the stimulus of prolactin from the pituitary, the cells of the glandular area of the lining of the sac undergo fatty degeneration and are shed as a holocrine secretion. This is regurgitated along with partially digested food and fed to nestlings as "pigeon's milk." After the nestlings leave, prolactin levels fall and the glandular area regresses.

The esophagus of vampire bats has a very narrow lumen through which only fluids can pass. This is correlated with their sanguinivorous diet. The adaptation inspires the speculative question of whether a small esophageal lumen forced these bats to adopt a diet of mammalian blood (the only available fluid containing all the nourishment needed by most mammals), or whether the narrowing of the esophageal lumen resulted from the disuse of the part as a passageway for solid foods (see "Doctrine of Acquired Characteristics," chapter 2). Our present understanding of hereditary mechanisms provides no explanation as to how the use or disuse of a body part could alter the gamete cells necessary for reproduction. This fact makes the doctrine suspect, but does not disprove it.

STOMACH

The stomach is a muscular chamber or series of chambers that serves as a receiving site for recently ingested foods, secretes digestive enzymes and lubricatory mucus, and macerates food while mixing it with the gastric juices. The mucus and enzymes partially liquify solid foods before they are injected into the small intestine. The stomach terminates at the **pylorus,** which is the opening from the stomach leading into the duodenum. The opening is surrounded by a ring of smooth muscle, the **pyloric sphincter.**

When there is more than one chamber, examples of which are found in all vertebrate classes other than amphibians, the first chamber usually serves simply as a temporary holding site for recently ingested food. Its epithelium is no different from that of the esophagus, having many mucous glands but none of the gastric glands found farther along.

The stomach is straight when it first differentiates in embryos and remains straight throughout life in some lower vertebrates. Most often, flexures develop, producing J-shaped or U-shaped stomachs (figs. 12.17, 12.18, and 12.19). As a result, the stomach may exhibit a concave border, or **lesser curvature,** and a convex border, or **greater curvature,** the latter in mammals having been

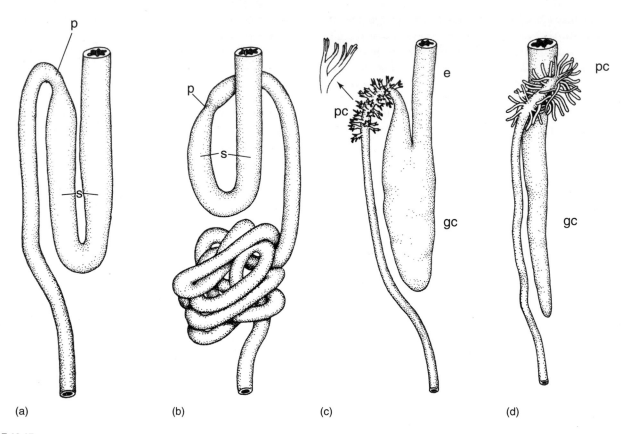

FIGURE 12.17

Digestive tracts of four teleosts. (*a*) *Fundulus.* (*b*) *Cyprinodon.* (*c*) *Elops.* (*d*) *Trichiurus.* **e,** esophagus; **gc,** cecumlike stomach; **p,** pylorus; **pc,** pyloric ceca; **s,** stomach.

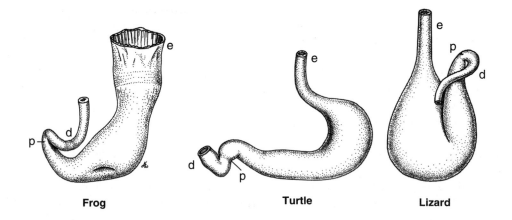

Frog   **Turtle**   **Lizard**

FIGURE 12.18

Stomach of a frog, turtle, and lizard. **d,** duodenum; **e,** esophagus; **p,** pylorus. Shape of stomachs will vary with volume of contents.

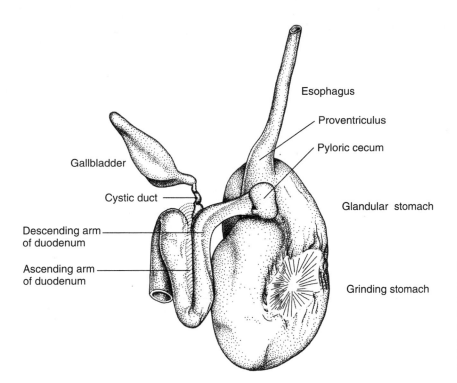

Esophagus

Proventriculus

Pyloric cecum

Gallbladder

Cystic duct

Glandular stomach

Descending arm
of duodenum

Ascending arm
of duodenum

Grinding stomach

FIGURE 12.19

Stomach and associated structures of a caiman, ventral view. The gallbladder has been lifted cephalad from its normal position between the stomach and descending arm of the duodenum. The grinding stomach has thick muscular walls, except in the midsection where there is a glistening fibrous membrane.

the embryonic dorsal border connected to the coelomic roof by the dorsal mesentery **(mesogaster).** Not only do most stomachs exhibit a flexure, but those of mammals undergo torsion along with part of their dorsal mesentery, twisting in such manner that stomach and mesentery lie more or less crosswise in the trunk. The mesentery, elongated and still attached to the greater curvature, is draped like a curtain between the ventral body wall and the intestines, where it constitutes the **greater omentum.** In humans, it covers the abdominal viscera like an apron. Within the double-walled omentum is a small part of the coelom, the **lesser peritoneal cavity,** that was trapped when the embryonic stomach and dorsal mesentery were rotating around a dorsoventral axis in the process of assuming their adult orientation in the coelom. The cavity remains continuous with the main abdominal coelom via a small passageway, the **epiploic foramen.**

Cyclostomes have no definitive stomach. The digestive tract is one long tube from mouth to vent exhibiting no gross differentiation of esophagus, stomach, or intestine (fig. 12.1). The epithelium of the agnathan tract is a single layer of cells including goblet cells that secrete mucus and flask-shaped cells that synthesize proteolytic (protein-splitting) enzymes. The base of each cell is in contact with the underlying vascularized layer of the mucosa, from which they receive nourishment.

The stomachs of fishes display a wide variety of shapes, and the epithelium is sometimes ciliated. The garfish stomach is almost straight; sharks exhibit the more common J-shape (fig. 12.1). The entire stomach of some teleosts is one large cecum (fig. 12.17c and d). Chimaeras and lungfishes have no definitive stomach, or one that is poorly differentiated and lacks digestive glands.

The stomach of a frog (fig. 12.18) is not distinguishable grossly from the esophagus. Both regions are capable of enormous distension. The stomach of a urodele is seen in figure 12.1

The stomach of crocodilians and birds is divided into two parts, **proventriculus** and **gizzard** (figs. 12.19 and 12.20). The proventriculus secrets the digestive enzymes. The gizzard, lined with a horny membrane, is simply a grinding mill that makes a mash of food mixed with gastric secretions. Pebbles that have been swallowed are usually retained in the gizzard where they assist in macerating the food. The proventriculus and gizzard of carnivorous birds are less well differentiated.

Terms derived from human anatomy have been applied to regions of the stomach of other vertebrates, sometimes inappropriately. In humans (fig. 12.21e), the region at the base of the esophagus is the **cardiac portion** because it lies close to the heart. Lateral to the cardiac region in humans is the **fundus.** It is characterized by a specific array of gastric glands. The region between the greater and lesser curvatures is the **body** of the stomach and the region preceding the pylorous is the **pyloric portion.** It houses the **pyloric canal.** X-ray films reveal that the shape and orientation of a human stomach in life are less transverse than in cadavers, and that the shape varies with posture, volume of the contents, and the activity of the intestines on which the stomach rests.

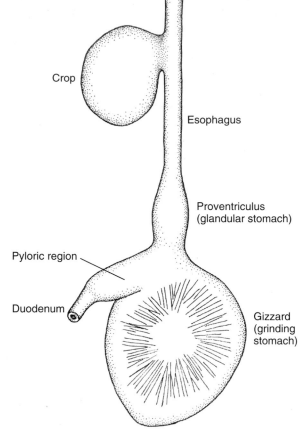

FIGURE 12.20

Esophagus and stomach of a grain-eating bird.

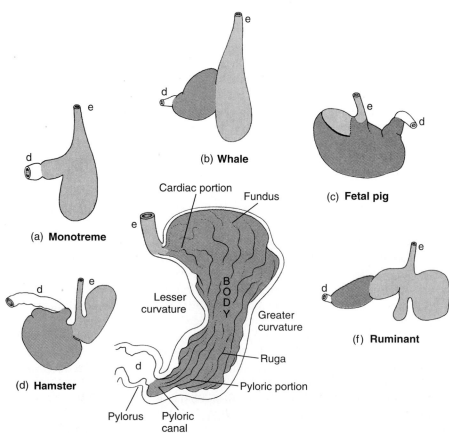

FIGURE 12.21

Distribution of esophageal-like epithelia
(*gray*) and glandular epithelia (*red*) in
stomach of selected mammals, ventral views.
d, duodenum; **e,** esophagus. The fetal pig
stomach is a dorsal view to show the
esophageal diverticulum that evaginates
near the base of the esophagus.

Fundic glands are elongated *simple tubular* glands, many with two tubules that open into a single terminal duct. There are several types of cells in their epithelium, among them zymogenic cells (**chief cells**) that synthesize and release the pre-enzyme **pepsinogen; parietal cells,** which secrete **hydrochloric acid**; and **goblet cells,** found mostly in the neck of the tubule. When released into the lumen of the stomach, hydrochloric acid splits the active enzyme pepsin (stomach protease) from a larger pepsinogen molecule, and pepsin initiates the process of protein digestion that eventually yields absorbable amino acids. Mucous cells lubricate the lining and provide a vehicle for the accumulating products of digestion.

The cardiac and pyloric regions of the human stomach can be distinguished from other regions on the basis of the histology of their glands. Neither have any zymogenic cells. The mucosa in the cardiac region resembles that of the lower end of the esophagus, the glands being *compound tubular* (fig. 6.10*d*) with many goblet cells. A few parietal cells are also present. Pyloric glands are *simple branched tubular* glands (fig. 6.10*c*), and they extend deeper into the mucosa than other types. They have many goblet cells and relatively few parietal cells. Between the fundus and the pyloric region, the mucosa exhibits typical gastric glands like those of the fundus, with a gradual shift toward a pyloric mucosa in the lower third of the organ. The pattern of distribution of these different varieties of gastric mucosas is not necessarily the same in other mammals.

As mentioned earlier, stomachs are often divided into several distinct chambers in animals whose food requires prolonged processing, as in mammals that consume grasses and grain. The stomach of a ruminant, adapted for processing cellulose, is an extreme example (fig. 12.22). Grasses are taken into the mouth, manipulated, and swallowed without processing. They pass down the esophagus and into the **rumen.** There the food becomes mixed with mucus and with **cellulase,** an enzyme that digests cellulose, a major component of the cell walls of plants. Cellulase is not secreted by the rumen, but by a multitude of anaerobic bacteria that live there. The enzyme cleaves cellulose molecules into simpler carbohydrates that can then be further digested by the animal's own digestive juices. The muscular walls of the rumen knead the contents—mucus, enzyme, and vegetation. At intervals, some of contents pass to the **reticulum,** so named because the lining is honeycombed (reticulated) by ridges and deep pits. Here the process of cellulose fermentation continues, and small boluses, or cuds, of fermenting pulp are regurgitated for further maceration by the teeth. The process of mastication, swallowing, and regurgitation may occur several times. Eventually, thoroughly masticated mash is swallowed and it passes to the **omasum,** a temporary holding site, before it is moved to

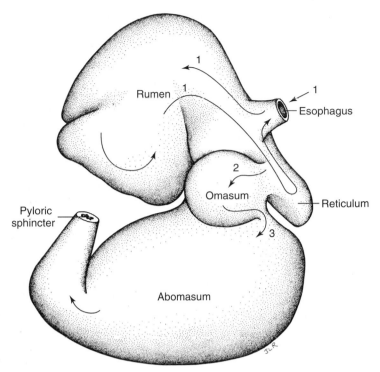

FIGURE 12.22

Stomach of a calf (ruminant). **1,** route of unmacerated food and of regurgitated cud; **2,** route of fermenting mash; **3,** entrance of mash into glandular stomach.

the **abomasum.** The latter is the true glandular stomach where gastric enzymes are added to the mash. The lining of the other chambers is much like that of the esophagus.

Other than water, few components of the diet of vertebrates are absorbed in the stomach. The accumulating product of stomach activity is a soupy mixture, **chyme.** When the pH of chyme in the pyloric canal is appropriately acidic, visceral receptors sense this and trigger a reflex arc that relaxes the pyloric sphincter for the injection of chyme into the duodenum. Figure 12.21 shows the location of gastric glands in the stomachs of mammals that have a variety of diets.

INTESTINE

The intestine begins at the pyloric sphincter and ends at its entrance into the cloaca or, if there is no cloaca, at the anus. The morphology of the intestine from fish to human is as different as the logistics attendant on digesting and absorbing the components of the diet, whether microscopic or gross, plant or animal, and on the frequency of meals and their volume. The intestine of tetrapods consists of small and large segments, each with a distinctive epithelium. That of fishes does not exhibit this differentiation.

Fishes

The intestine of cyclostomes, chondrichthyes, and lower bony fishes is quite straight (fig. 12.1, cyclostomes, garfish, shark). A few teleosts—top minnows, for example—exhibit intestinal coiling, but this is rare (fig. 12.17). A **spiral intestine**, so named because a **spiral valve**, or **typhlosole**, is suspended within its lumen, is present in many fishes other than teleosts. (fig. 12.1, shark). It serves the same function as intestinal coils, increasing the epithelial area available for absorption. Beyond the spiral intestine, a short **postvalvular intestine** leads to the cloaca. Cyclostomes have a typhlosole, but it is a mere ridge projecting into the intestinal lumen. It makes only a few spirals throughout its length.

Intestinal ceca are the major adaptations for increasing the absorptive area of the teleost intestine. Most common are **pyloric ceca,** which are diverticula near the pylorus (fig. 12.17c and d). Mackerel have up to 200 pyloric ceca.

The duct of a cecumlike, finger-shaped **rectal gland** empties into the short postvalvular intestine of sharks, but it has no digestive function. It extracts and excretes excess sodium chloride from the blood.

Tetrapods: The Small Intestine

Tetrapods have a small and large intestine. The short curved duodenum is the first segment of the small intestine. The remainder is coiled to one degree or another, except in urodeles and apodans. In lizards, birds, and mammals, it is lined with fingerlike or leaflike **villi,** low in birds, tall in mammals, so densely crowded that they give the lining a velvety appearance (fig. 12.16). (A hand lens or dissecting microscope will enable you to observe this effect if you are dissecting a mammal.) Villi greatly increase the absorptive surface of the intestine. Digested (hydrolyzed) lipids absorbed through their epithelium enter dead-end lymph vessels, or **lacteals,** within the villi (fig. 12.23). By suddenly shortening, lacteals pump the milky fluid, **chyle,** into larger lymphatics that eventually empty into a major vein near the heart (see thoracic duct of a mammal, fig. 14.40). The lipids are then carried by the bloodstream to the liver, the chief site of synthesis of cholesterol and other sterols and lipids, and to other tissues primarily for storage.

In mammals, the small intestine beyond the duodenum is divided into **jejunum** and **ileum** on the basis of the shape of the villi, the nature of the epithelial lining, and the size of the lymph nodules in the mucosa. Small nodules are common throughout the length of the intestine (fig. 12.16), but in the ileum they aggregate into large masses known as **Peyer's patches.** The small intestine of amniotes terminates at an **ileocolic sphincter** (fig. 12.24), which regulates the ejection of the contents of the ileum into the large intestine.

The small intestine is the chief site of digestion and absorption of nutrients, although, as we have seen,

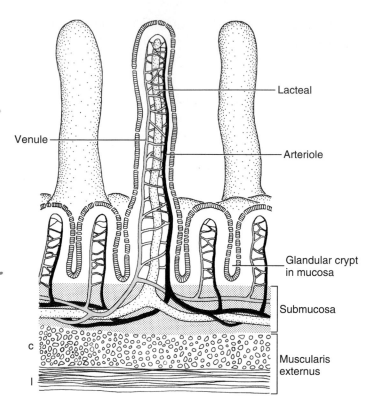

FIGURE 12.23

Blood supply and lymph drainage of a villus. **c,** circular muscle; **l,** longitudinal muscle.

ptyalin, pepsin, and cellulase (the latter from commensal bacteria), split starch, proteins, and cellulose in the stomachs of one species or another. However, the final stage of digestion that results in absorbable nutrients takes place in the small intestine in the presence of intestinal juice and pancreatic enzymes. The enzymes of **intestinal juice** are secreted by glands in the epithelial lining of the crypts and by compound alveolar glands near the pylorus (fig. 12.16). The enzymes split polypeptides into amino acids and they split disaccharides into monosaccharides (simple sugars). Amino acids and simple sugars are absorbable molecules. **Pancreatic juices** contribute an **amylase** that acts on carbohydrates, a **lipase** that digests lipids yielding absorbable fatty acids and glycerol, and **proteolytic enzymes** that continue the digestion initiated by pepsin. The digestive process reaches a crescendo in the small intestine. By the time the contents of the ileum arrive at the ileocolic sphincter, the recoverable nutrients that were ingested earlier have been absorbed into the bloodstream. Except in some herbivores, what is left is chiefly water and undigestible roughage. Most of the water is absorbed in the colon.

The Large Intestine

The large intestine is rarely coiled, but ceca are common. In mammals and in some reptiles and birds, the large intestine is divisible into **colon,** which commences at the

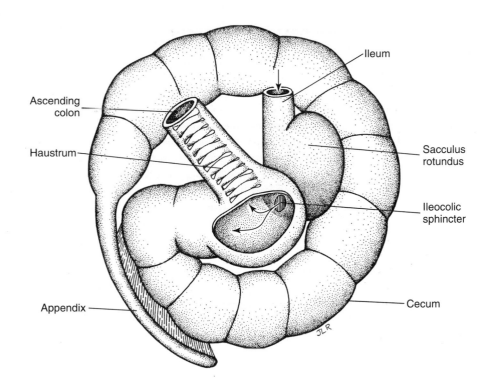

FIGURE 12.24

Ileocolic junction, cecum, and vermiform appendix of rabbit.

ileocolic sphincter, and **rectum,** a straight terminal portion in the pelvic cavity. The colon of some mammals has ascending, transverse, and descending portions with pronounced flexures between them; and in humans the descending colon ends in a **sigmoid flexure** (fig. 12.1, human). Although ceca are rare beyond the duodenum in fishes and amphibians, ceca just beyond the ileocolic sphincter **(ileocolic ceca)** are common in amniotes. Birds typically have two (fig. 12.1, chicken). In mammals that feed exclusively on fibrous vegetation, fruits, grains, and seeds, the ileocolic cecum may exceed the large intestine in capacity. An **appendix** with a lumen and histology the same as that of the cecum terminates the cecum in anthropoids, rodents, and many mammals. The cecum and appendix of a rabbit are seen in figure 12.24. As one might expect, the ileocolic cecum of a carnivore is quite abbreviated, as is that of humans.

The koala, a marsupial 2 to 2½ ft (60 to 75 cm) long that feeds preferentially on eucalyptus leaves, less frequently on eucalyptus bark, buds, and unripened woody fruits, has a cecum that may be up to 6 ft long. This is correlated with the fact that the diet has a low nutritional value, which mandates a large gut capacity. Thoroughly chewed leaves pass to the small intestine where any cytoplasmic nutrients are absorbed, then through the ileocolic sphincter and into the cecum. There, during the course of more than a week of fermentation by enzymes produced by commensal anaerobic bacteria, cellulose (from the cell walls of plants) is converted into absorbable carbohydrates, as in the rumen of ungulates.

A *Hyrax,* whose diet also consists of seeds, fruits, and leaves, has an ileocolic cecum, colic flexures, a bicornu-

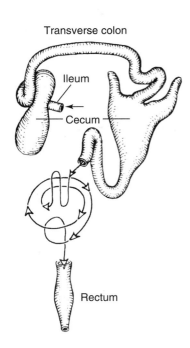

FIGURE 12.25

Large intestine of *Hyrax.* Arrows indicate direction taken by missing segment.

ate cecum farther along on the intestine, and then an intestinal coil before reaching the rectum (fig. 12.25).

The colon in all tetrapods recovers water from the residual contents (feces) of the intestine. Water reclamation in terrestrial animals is necessary to replace that "borrowed" earlier from tissues for the digestive process. Failure to recover water would lead to dehydration and possible death.

LIVER AND GALLBLADDER

The liver arises from the midventral aspect of the midgut as a hollow cecumlike diverticulum, the **liver bud** (fig. 12.26*a*). The bud grows cephalad in the ventral mesentery of the stomach. The growing tip of the bud gives rise to numerous sprouts that become the liver and the gallbladder. The cephalic pole of the liver finally becomes anchored to the embryonic transverse septum (fig. 1.12) by a **coronary ligament.**

The lobes of the adult liver are drained by several **hepatic ducts** and the gallbladder is drained by a **cystic duct.** These ducts converge very close to the base of the gallbladder at such an angle that bile from the liver can pass via the cystic duct into the gallbladder, where it is stored. The site where the hepatic and cystic ducts converge is the beginning of a **common bile duct** that empties into the duodenum (8 in fig. 12.26). A short terminal segment of the common bile duct is embedded in the wall of the duodenum, where it constitutes an **ampulla of Vater.**

Although most of the embryonic ventral mesentery disappears during subsequent development, the mesentery ventral to the duodenum and stomach that was invaded by the liver bud remains as the **hepatoduodenal ligament** connecting the duodenum and liver, and as the **gastrohepatic ligament** that connects the pyloric stomach and the liver. These two ligamentous mesenteries constitute the **lesser omentum.** It serves as a bridge that conducts the common bile duct to the duodenum and the hepatic artery and hepatic portal vein to the liver. The embryonic mesentery ventral to the liver remains in

adults as the **falciform ligament.** (The term describes its shape in humans.) The shape of the liver conforms to the space available in the coelom. In animals with eel-like bodies, the liver is elongate. In those with short trunks, it is short and wide, as in humans.

The liver has many roles. It produces **bile,** an alkaline fluid containing bile salts that emulsify lipids in the small intestine and impart to them an alkalinity necessary for digestion. Without emulsification, only a small proportion of dietary fat would be digested. Lipids are stored in liver cells and in adipose tissue throughout the body.

Some liver cells phagocytose aging red blood cells, splitting the hemoglobin molecule, and freeing the iron. The remainder of the molecule is converted into red and green pigments (**bilirubin** and **biliverdin)** that are excreted as part of the bile.* (Red blood cells are phagocytosed also in the spleen and red bone marrow. The pigments from those sites are extracted from the bloodstream in the liver and contribute to the bile.) The fetal liver is a source of red blood cells, but that role is taken over later by other tissues.

The liver removes from the circulation glucose in excess of immediate tissue needs. The glucose is converted to glucose-6-phosphate and then stored in the liver as glycogen. The latter is converted back to glucose as needed to maintain an appropriate level in the circulation.

The liver removes dietary amino acids from the hepatic portal system and deaminates them. The byproducts of deamination include ammonia, uric acid,

---

*The "black and blue" disfiguration of bruised skin is caused by the pigments released from traumatized red blood cells being phagocytosed at the site of trauma.

(a)

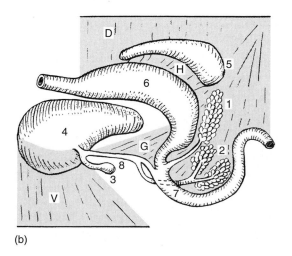

(b)

FIGURE 12.26

Development of liver, pancreas, spleen, stomach, and associated mesenteries. (*a*) Early stage. (*b*) A later stage. **1,** dorsal pancreatic bud from duodenum; **2,** ventral pancreatic bud; **3,** gallbladder; **4,** liver bud in (*a*) and liver in (*b*); **5,** spleen; **6,** stomach; **7,** duodenum; **8,** common bile duct. **D,** dorsal mesentery; **G,** lesser omentum; **H,** gastrosplenic ligament; **V,** ventral mesentery in (*a*), falciform ligament in (*b*).

and urea. These are eventually excreted in the urine in one form or another, depending on the species (see chapter 15, "Excretion of Nitrogenous Wastes").

Several blood proteins are manufactured in the liver, including **fibrinogen** and **prothrombin,** which are essential for blood clotting. The liver has still other roles that are discussed in physiology textbooks.

A gallbladder develops in most vertebrates. None develops in lampreys, a few teleosts, many birds, perissodactyls, whales, and some rodents, including rats. Vertebrates lacking a gallbladder are among those that have little fat in their diet. Human beings can live without a gallbladder but they must be careful about the amount of lipids they consume. The gastrointestinal hormones associated with the digestive process are discussed in chapter 18.

EXOCRINE PANCREAS

The pancreas consists of two histologically distinct and functionally independent components. An *exocrine* portion produces digestive enzymes in **alveoli (acini).** The enzymes are transported via **pancreatic ducts** to the duodenum. An *endocrine* portion bears **pancreatic islets (islands of Langerhans),** lacks ducts, and thus secretes its hormonal products, insulin and glucagon, into the blood-stream (fig. 12.27). These two components are not always part of the same organ (see chapter 18, "Endocrine Pancreas").

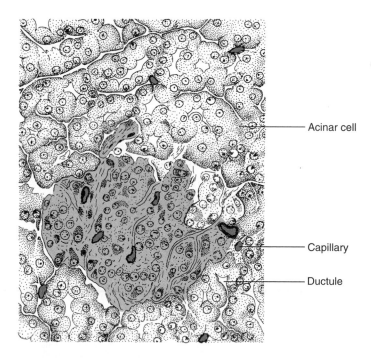

Acinar cell

Capillary

Ductule

FIGURE 12.27

Island of Langerhans (*light red*)and surrounding acini (alveoli) in a mammalian pancreas.

The pancreas varies from diffuse to compact. When diffuse, as in teleosts and many higher vertebrates, pancreatic tissue is distributed along the blood vessels in the ventral mesentery of the stomach and duodenum. When compact it may consist of several discrete lobes. In cyclostomes there is no definitive pancreas, the exocrine and endocrine components are spatially separated, and many exocrine cells remain in the intestinal epithelium.

The pancreas arises typically as one or two ventral **pancreatic buds** off the liver bud and as a single dorsal pancreatic bud from the foregut immediately caudal to the stomach (fig. 12.26*a*). The ventral buds invade the ventral mesentery of the duodenum and stomach and generally coalesce to form a ventral lobe (**body**) of the pancreas (fig. 12.26*b*, 2). The dorsal bud becomes the dorsal lobe (**tail**) of the pancreas. There are variants of this pattern. In sharks, the entire pancreas develops from the dorsal bud, and in most mammals it develops from one ventral and one dorsal bud.

Vertebrates may have as many pancreatic ducts as there were embryonic pancreatic buds, but more often one or more of the ducts loses its connection with the bile duct or gut from which it evaginated, and the pancreas is drained by the remaining duct or ducts. For example, in sheep the dorsal lobe loses its connection with the gut and the entire pancreas drains into the common bile duct. In pigs and oxen, the duct of the ventral lobe loses its connection with the common bile duct and the entire pancreas drains directly into the duodenum. In still other mammals, both ducts remain. When one duct is larger, as in cats and humans, the other is relegated to the status of **accessory pancreatic duct.** The variations in the drainage of the mammalian pancreas illustrate, once again, that a primitive pattern may be modified in any of several directions during the course of phylogenesis.

CLOACA

In many fishes and in most tetrapods other than placental mammals, the digestive tract terminates in a common chamber, the cloaca (Latin, sewer), into which the urinary and genital tracts also empty. The cloaca is then vented to the exterior. In lampreys, chimaeras, living female coelacanths, and ray-finned fishes, the embryonic cloaca becomes increasingly shallow or disappears as development proceeds, and the digestive tract eventually opens independently to the exterior. In placental mammals, the cloaca becomes partitioned during embryonic development into two or three separate passageways, one of which becomes the rectum leading to an anus (fig. 15.49). Figure 15.48 illustrates the exit from the digestive tract in two anatomically diverse mammals, a monotreme and a female rodent.

S U M M A R Y

1. Filter feeding in an aquatic environment is the oldest vertebrate method of acquiring food. It is still employed by larval lampreys and, with modifications, by a few jawed fishes and baleen whales.

2. The early embryonic digestive tract consists of foregut, midgut, and hindgut. Excepting its epithelial lining, the gut forms from splanchnic mesoderm. The epithelium is endodermal in origin.

3. Stomodeal and proctodeal invaginations establish an entrance (mouth) and exit (vent or anus).

4. The gut is suspended within the coelom by a dorsal mesentery that is continuous with the parietal peritoneum lining the body wall (somatopleure) and with the visceral peritoneum on the surface of the organs.

5. An embryonic ventral mesentery disappears except at two locations, one of which is the falciform ligament of the liver.

6. The mouth is the anterior opening to the digestive tract. Major subdivisions of the adult tract are oropharyngeal or oral cavity, pharynx, esophagus, stomach, and intestine. Chief accessory organs are tongue, teeth, oral glands, pancreas, liver, and gallbladder.

7. Nasal passageways open into the oropharyngeal cavity in lobe-finned fishes and into the oral cavity in tetrapods with primary palates. Multicellular oral glands open onto the roof, walls, and floor of these cavities but are scarce in fishes.

8. Jawed fishes and perennibranchiate amphibians have a primary tongue overlying the ventralmost component of the hyoid skeleton. In higher amphibians, a glandular field contributes to the tongue, and in amniotes paired lateral lingual swellings contribute. An entoglossal bone stiffens the tongue of lizards and birds.

9. Teeth are vestiges of dermal armor. They consist of dentin formed by odontoblasts in dermal papillae, enamel formed by ameloblasts of enamel organs, and cementum. In lower vertebrates, teeth are more numerous, more widely distributed in the oral cavity, more frequently replaced, and more alike throughout the oral cavity. A few vertebrates in every class are toothless. A few have horny teeth.

10. Dentition may be monophyodont, diphyodont, or polyphyodont; acrodont, pleurodont, or thecodont; homodont or heterodont; secodont, selenodont, lophodont, or bunodont. Primitive mammalian cheek teeth were triconodont and trituberculate.

11. The pharynx is the part of the foregut that has pharyngeal pouches in the embryo. The tetrapod pharynx has openings to the middle ear cavity, larynx, and esophagus. Amniotes have nasal and oral pharynges. Some mammals have a laryngeal pharynx.

12. The wall of the digestive tract caudal to the pharynx consists of a mucosa, submucosa, muscularis externa, and serosa (visceral peritoneum overlying a loose connective tissue membrane, the adventitia).

13. The esophagus connects the pharynx with the stomach. In birds it has a diverticulum, the crop. The esophagus is lined with mucous glands.

14. The stomach is compartmentalized in birds (proventriculus and gizzard), in cud-chewing ungulates (rumen, reticulum, omasum, and abomasum), and to a lesser degree in other vertebrates. It terminates at the pyloric sphincter.

15. The intestine is the major site of digestion and absorption. Typhlosoles, coils, ceca, and villi increase the absorptive area. Amniotes have small and large intestines, the latter consisting of colon and rectum. In mammals, the small intestine consists of duodenum, jejunum, and ileum.

16. Liver and pancreatic buds arise as evaginations from the foregut. The gallbladder is an evagination from the liver bud.

17. The intestine opens into a cloaca in most vertebrates. The cloaca opens to the exterior via a vent. In the absence of an adult cloaca, the large intestine opens to the exterior via an anus.

18. Digestive enzymes are synthesized in one species or another by salivary glands, stomach, intestine (small intestine in amniotes), and pancreas. Bile salts from the liver emulsify ingested lipids in preparation for digestion.

19. Vertebrates produce no known enzyme for digesting the complex carbohydrate cellulose. The enzyme is provided by commensal bacteria in the stomach of ungulates and in the colic ceca in most other herbivores.

SELECTED READINGS

Bramble, D. M., and Wake, D. B.: Feeding mechanisms in lower tetrapods. In Hildebrand, M., and others, editors: Functional vertebrate morphology, p. 159, Cambridge, Massachusetts, 1985, Harvard University Press.

Butler, P. M., and Joysey, K. A., editors: Development, function, and evolution of teeth. New York, 1978, Academic Press.

Domning, D. P.: Marching teeth of the manatee, Natural History 92:5 (May), 1983.

Gans, C., and Gorniak, G. C.: How does the toad flip its tongue? Test of two hypotheses, Science 216:1335, 1982.

Gorniak, G. C.: Trends in actions of mammalian masticatory muscles, American Zoologist 25:331, 1985.

Lauder, G. V.: Patterns of evolution in the feeding mechanisms of actinopterygian fishes, American Zoologist 22:275, 1982.

Moss, M.: Enamel and bone in shark teeth; with a note on fibrous enamel in fishes, Acta Anatomica 77:161, 1970.

Osborn, J. W.: The evolution of dentitions, American Scientist 61:548, 1973.

Pivorunas, A.: The feeding mechanisms of baleen whales, American Scientist 67:432, 1979.

Pough, F. H.: Feeding mechanisms, body size, and the ecology and evolution of snakes, American Zoologist 23:339, 1983.

CHAPTER THIRTEEN

RESPIRATORY SYSTEM

In this chapter we look at some of the methods and mechanisms that embryonic, larval, and adult vertebrates have developed for obtaining oxygen and eliminating carbon dioxide in aquatic and terrestrial environments. We will find that many fishes are aerial respirators, and that lungs may be older than tetrapod limbs. Along the way we will see how a crocodilian or a whale can breathe with a mouth full of water, why a nursing kitten does not have to interrupt swallowing during inhalation, and how the position of the human larynx affects human speech.

OUTLINE

he process of obtaining oxygen from the environment and eliminating carbon dioxide is **external respiration.** It is accomplished via respiratory membranes that, except in embryos, are part of some organ. Organs that are essential for external respiration constitute, collectively, the respiratory system.

External respiration precedes **internal respiration,** which is usually defined as the exchange of oxygen and carbon dioxide between capillary blood and tissue fluids. Because carbon dioxide quickly inhibits cellular activity, the continual elimination of this gas from the vicinity of the cell, and then from the organism, is essential. The role of the circulatory system in the total process of respiration is therefore vital.

External respiration is carried on through respiratory membranes. Except in very early embryos, these must be highly vascular, the epithelium must be thin, the surface should be moist to facilitate diffusion of gases through semipermeable membranes, and it must be in contact with the environment, as in external gills; or else the environment must be brought into contact with the respiratory surface, as in lungs.

The chief organs of external respiration in adult vertebrates are external and internal gills, the oropharyngeal mucosa, swim bladders or lungs, and the skin. Less common adult respiratory devices include bushy or filamentous outgrowths of the pectoral fins, as in the male *Lepidosiren,* or of the posterior trunk region and thigh, as in the African hairy frog; the cloacal, rectal, or anal lining; and the lining of the esophagus, stomach, or even of the intestine in one species or another. Embryos employ a variety of respiratory surfaces including extraembryonic membranes.

In both water and air, respiration through the skin is employed extensively by modern amphibians. It is also used by some fishes, especially those that lack scales and therefore have capillary beds close to the surface. Aquatic urodeles, lacking dermal armor and a significant stratum corneum, acquire as much as three-fourths of their oxygen from the water through the skin. Tree frogs acquire only one-fourth by that route and terrestrial species of *Rana* acquire one-third. Regardless of the proportionate role of skin and lungs in oxygen uptake in amphibians, up to nearly 90 percent of the carbon dioxide is excreted through the skin. Cutaneous respiration is not significant in amniotes because the thick stratum corneum insulates the capillaries from the atmosphere.

GILLS

The internal gills of fishes arise in the walls of pharyngeal pouches, which are paired evaginations of the embryonic pharynx (fig. 1.6). These pouches meet ectodermal grooves, which are invaginations of the ectoderm (fig. 1.7). A thin membrane, the branchial plate, separates pouch and groove, and subsequent rupture of the membrane establishes a passageway for respiratory water from the pharynx to the exterior. The gills of jawed fishes are supported by the pharyngeal skeleton. The development and fate of pharyngeal pouches and arches from fish to humans has been discussed in chapter 1.

Agnathans

Hagfishes in the genus *Myxine* have 5 or 6 pairs of well-vascularized gill pouches, rarely 7, associated with the pharynx (fig. 13.1*a*). Various species of *Bdellostoma,* another hagfish genus, have from 5 to 15 pairs. **Afferent branchial ducts** conduct respiratory water from the pharynx to the pouches, and **efferent ducts** lead from the pouches to the exterior. Each efferent duct usually has its own external aperture (fig. 1.9, *Bdellostoma*), but in myxinoids the efferent ducts unite to open via a single common external slit on each side (fig. 13.1*a*).

In hagfishes, respiratory water enters the median unpaired naris and passes via a **nasopharyngeal duct** to a **velar chamber** at the anterior end of the pharynx (fig. 13.1*a*). The wall of the chamber is a pulsating muscle, or velum, which pumps water from the velar chamber caudad into the pharynx and, according to some observers, into the gill pouches, thereby creating a vacuum in the velar chamber. The vacuum draws additional water through the naris and into the velar chamber. Muscle tissue in the walls of the gill pouches expels the water to the exterior. The velum is supported by pharyngeal cartilages and pulsates 50 to 100 times a minute in sleeping animals. On the left side only, a **pharyngocutaneous duct** of greater diameter than the afferent branchial ducts connects the pharynx with the last efferent branchial duct (*Myxine*), or, in some hagfishes, with the exterior. Periodically, debris or particles too large to enter the afferent branchial ducts are forcefully ejected through the pharyngocutaneous duct in a manner somewhat analogous to coughing. The pharyngocutaneous duct appears to be a modified last gill pouch.

The mechanism of respiration differs considerably in lampreys in comparison with hagfishes. The buccal cavity of lampreys is pressed tightly into the flesh of a host, and the duct from the unpaired naris (**nasohypophyseal duct**) does not open into the pharynx but ends blindly in a **nasohypophyseal sac** (fig. 13.1*b*). Therefore neither the nasal nor buccal route is available for the passage of respiratory water to the gills. As an adaptation, respiratory water *enters and exits* the pharynx via the external gill slits as a result of pulsations of the pharyngeal and gill pouch musculature. The external apertures are guarded by thin flaps of skin that serve as two-way valves. As a further adaptation, the lamprey pharynx becomes subdivided longitudinally at metamorphosis into an esophagus that continues to the stomach and a ventral pharynx that ends blindly (fig. 13.1*b*). The entrance to the adult pharynx is guarded by a valvelike

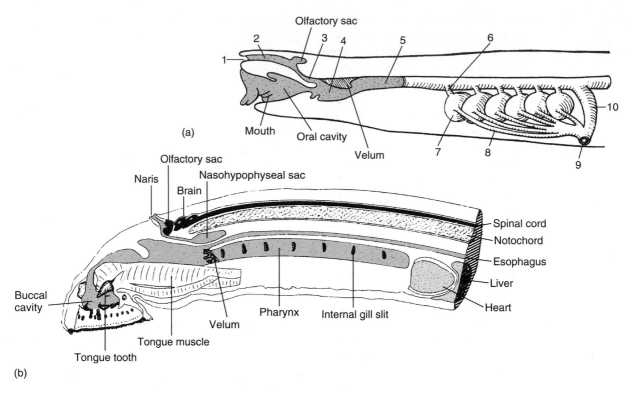

FIGURE 13.1.

Head and pharyngeal region of hagfish and lamprey. (*a*) The hagfish *Myxine glutinosa.* **1,** naris; **2, 3,** nasopharyngeal duct; **4,** velar chamber; **5,** pharynx; **6,** afferent branchial duct; **7,** gill pouch; **8,** efferent branchial duct; **9,** common external gill aperture (present on both sides); **10,** pharyngocutaneous duct (present on left side only). (*b*) The lamprey *Petromyzon,* sagittal section. Connecting naris and nasohypophyseal sac is the nasohypophyseal duct.

velum that prevents nutrients rasped from a host from entering the pharynx, becoming diluted by respiratory water, and then being lost through the external gill slits.

The seven pairs of voluminous gill pouches are lined with gill lamellae and communicate directly with the pharynx and the exterior with no intervening afferent or efferent branchial ducts. The external gill slits of *Petromyzon* are seen in figure 1.9.

The respiratory systems of hagfishes and lampreys have little in common with those of jawed fishes, structurally. The former represent two variant adaptations to parasitism. But adaptations from what preceding condition? The branchial apparatus of jawed fishes is very different. It is a pharyngeal arch mechanism. The phylogenetic history of lampreys and hagfishes, and the common ancestor of agnathans and gnathostomous fishes is unknown. However, it is highly unlikely that a system such as that of agnathans provides a glimpse into the ancestral system of jawed fishes.

Cartilaginous Fishes

Most elasmobranchs—sharks, skates, and rays—are pentanchid, having five pairs of gill pouches, each with an internal and external gill slit, and a pair of functional **spiracles** located dorsally immediately in front of the hyomandibular cartilages (fig. 13.2*a*). The shark

Hexanchus is exceptional, having a spiracle and six gill pouches; *Heptanchus*, another shark, has a spiracle and seven, this being the largest number in any gnathostome. No gill surface develops in the posterior wall of the last pouch. The external gill slits in sharks are visible on the side of the head; those in Rajiformes are on the underside (fig. 13.7). The slits are said to be "naked" because, unlike in other gnathostomes, there is no operculum. In embryos, the spiracle and gill slits are the same size and aligned in a row, but the spiracle does not keep pace in growth with the other slits (fig. 1.7). A miniature gill-like structure, or **pseudobranch,** consisting of a vascular rete (network) develops in its anterior wall. The rete may regulate blood pressure in the eyeball.

The spiracle has a one-way intake valve and is the exclusive incurrent aperture for respiratory water in Rajiformes and for much of the respiratory water in sharks except some fast-swimming voracious ones in which the spiracle becomes secondarily closed by a membrane of skin. Its dorsal position in rays minimizes the entrance of muddy debris-laden water into the pharynx of these bottom-dwelling species.

If a gill slit is probed, the instrument will enter a **gill chamber** (fig. 13.3). The anterior and posterior walls of the first four chambers each exhibit a gill surface, or **demibranch.** The last chamber lacks a demibranch in its

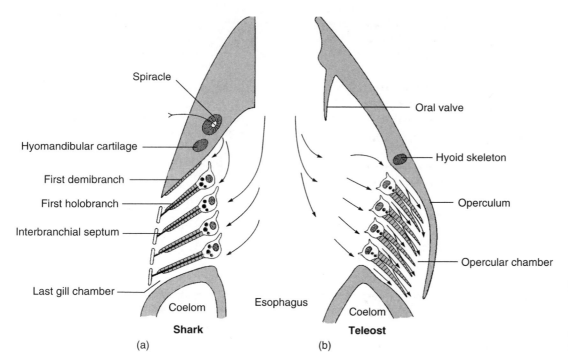

FIGURE 13.2

Gills of a pentanchid shark and teleost, frontal section, schematic. Arrows show routes of respiratory water.

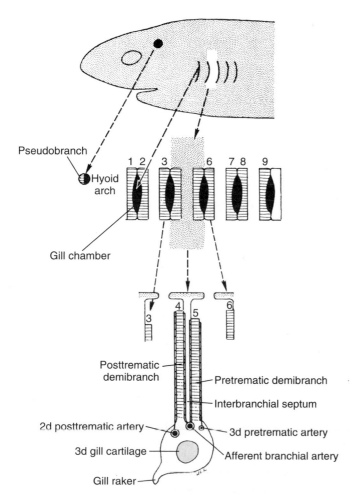

FIGURE 13.3

The gill chambers (*black*) and the nine demibranchs (*lined*) in *Squalus*. A holobranch (location indicated in *white* at top of figure) is represented at bottom in cross section.

posterior wall. Ceratohyal cartilages support the demibranch in the anterior wall of the first chamber, and epibranchial and ceratobranchial cartilages support the remaining eight demibranchs. The demibranch in the *anterior wall of a gill chamber* is a **pretrematic demibranch.** The one in the posterior wall is a **posttrematic demibranch.** Separating the two demibranchs of a single gill arch is an **interbranchial septum** that is supported by many long, tapering, and in some species branching, cartilaginous **gill rays** that radiate from the gill cartilage into the interbranchial septum along its entire extent. The two demibranchs of a single gill arch, together with the associated interbranchial septum, cartilages, blood vessels, branchiomeric muscles, nerves, and connective tissue, constitute a **holobranch** (white area in upper drawing of fig. 13.3). Stubby **gill rakers** projecting from the pharyngeal border of each gill arch guard the slitlike entrances from the pharyngeal lumen into the gill chambers, protecting the gills from mechanical injury.

The functional surface of each demibranch consists of large numbers of transverse shelflike folds, or lamellae, of gill mucosa, the epithelium of which is very thin. The folds multiply the surface area for gaseous exchange. Underlying the lamellar epithelium is a rich capillary bed supplied by afferent branchial arterioles that receive blood from the ventral aorta via afferent branchial arteries (fig. 14.17b). Since the ventral aorta carries blood low in oxygen, the blood entering the capillary bed is low in oxygen. As oxygen-laden respiratory water flows steadily across the gill lamellae, it does so in a direction opposite that of the oxygen-poor blood flowing in the capillaries. As a result, the partial pressure of oxygen in the water always exceeds that in the blood even though the blood is becoming increasingly oxygenated. This **countercurrent flow** of blood and water maximizes the efficiency of gaseous exchange, enabling the blood to "make the most" of its opportunity to acquire oxygen. Other examples of the countercurrent principle will be found elsewhere in the text. From the capillaries, oxygenated blood passes via efferent branchial arterioles to pretrematic or posttrematic arteries, then into efferent branchial arteries and the dorsal aorta (fig. 14.17b). From there it is distributed to all tissues of the body.

It has been mentioned that respiratory water enters the pharynx of most elasmobranchs via both the mouth and spiracle. Most of the water entering via the spiracle in sharks flows into the first two gill pouches, whereas that entering via the mouth enters the last three pouches. Water is sucked through the spiracle and the open mouth when branchiomeric muscles constrict the external gill slits and expand the pharyngeal chamber, creating a vacuum within the chamber. This inspiration phase slows down and then ceases when the pharynx has become filled with water. The mouth then closes, the gill chambers are expanded by the action of levator and hypobranchial muscles, and the gill chambers fill with water. In the third phase of respiration—expiration—the water is forced out of the gill chambers via the external gill slits by constrictor muscles. In the interval between expiration and the next inspiration, both the mouth and the gill slits are temporarily open. Experiments have shown that pressure is nearly always higher in the pharyngeal chamber than in the gill pouches. This assures a steady, uninterrupted flow of water over the gill lamellae, rather than tidal surges. There is a respiratory rhythm of about 35 per minute when the shark is at rest. Sharks swimming with their mouths open are utilizing the forward motion of the body to accumulate water in the pharynx, thereby reducing the cost of external respiration in terms of energy expended.

The holocephalan *Chimaera* has only four gill pouches (fig. 13.4), the spiracle closes during early ontogeny, interbranchial septa are short and do not reach the skin, and a fleshy flap, the **operculum** extends caudad from the hyoid arch, hiding the gills and deflecting excurrent water caudad. In some of these traits, *Chimaera* resembles teleosts.

Bony Fishes

The gill apparatus of cartilaginous and bony fishes exhibits the same basic pattern. A series of pharyngeal arches support holobranchs, and a stream of respiratory water flows over the demibranchs en route from the pharyngeal cavity to the exterior (fig. 13.2b). The chief differences lie in the presence of an operculum and an opercular chamber and in the shorter length of the interbranchial septa in bony fishes. The **operculum** is a bony flap that commences at the hyoid arch and extends caudad, covering the gill chambers on each side. Extending from the ventral edge of each operculum is an

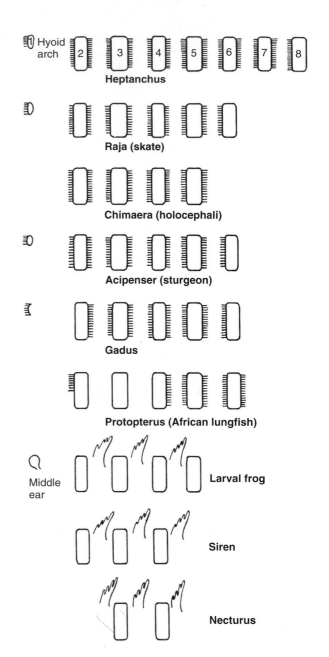

FIGURE 13.4

Open pharyngeal slits in selected aquatic vertebrates and distribution of demibranchs (*lines*) in fishes. *Heptanchus* is a primitive shark. Note that *Gadus,* a cod, retains a vascular rete in the mandibular arch even though the spiracle is closed. **1** to **8,** pharyngeal slits. The position of external gills is indicated in the three amphibians.

accordian-like **branchiostegal membrane** supported, in teleosts, by numerous long bony **branchiostegal rays.** Shorter **gular** bones, of more ancient vintage (fig. 9.10a), support the membrane found in some of the more primitive actinopterygians. The two branchiostegal membranes are united midventrally beneath the gills to enclose an **opercular chamber** that respiratory water pours into after passing over the gills and before being expelled via a cleft at the caudal end of the operculum.

The cleft varies in size. It is exceptionally small and round in eels. Infrequently in teleosts the chamber empties via a midventral aperture. The interbranchial septa are very short, with the result that the demibranchs of each holobranch are unattached distally, allowing water freer access to their capillaries (fig. 13.5).

Water is drawn into the pharynx by lowering the pharyngeal floor with the mouth open and the operculum closed. Simultaneous expansion of the opercular chamber by unfolding of the branchiostegal membrane draws the incoming water across the gills and into the chamber. The mouth is then closed and the water is forced to the exterior via the opercular cleft. This is accomplished by elevating the pharyngeal floor and compressing the opercular chamber. An **oral valve** at the anterior end of the oropharynx immediately behind the mouth prevents escape of water by mouth. Thus a suction pump and a pressure pump, operating rhythmically, keep the gills bathed in oxygenated water. A few teleosts, including mackerel and tuna, have a paucity of branchiomeric musculature and are obliged to swim with the mouth open in order to create an adequate flow of water over the gills.

Most bony fishes other than dipnoans, chondrosteans, and garfishes have four holobranchs and five gill chambers, the demibranch in the anterior wall of the first gill chamber having been lost (fig. 13.4, *Gadus*). Additional demibranchs have been lost in dipnoans (fig. 13.4,

Protopterus). In fact, the gills of *Protopterus* and *Lepidosiren* are so inefficient that these fishes suffocate when forcibly held under water. A spiracle is present in chondrosteans (fig. 13.4, *Acipenser*), but it closes during embryonic life in other bony fishes.

Larval Gills

Larval gills are of three kinds—**external gills** that are outgrowths from the external surface of one or more gill arches; **filamentous extensions of internal gills** that project through gill slits to the exterior; and the **internal gills** of late anuran tadpoles, which are hidden behind the larval operculum.

External gills usually develop before gill slits open and before any opercular fold has started to develop. They can be waved about by branchial muscles and can often be retracted. They develop in the embryonic or larval stages of most dipnoans, in all amphibians, and in a few primitive ray-finned fishes such as sturgeons and *Polypterus* (fig. 13.6a). Although external gills are larval structures, there are perennibranchiate urodeles and a perennibranchiate lungfish, *Protopterus* (fig. 14.18b). The latter has four pairs of larval gills and retains three pairs in a reduced state throughout life.

External gills develop in anuran tadpoles on pharyngeal arches III to V. Later, when pharyngeal pouches II to V rupture to the exterior, the pouch linings become folded to form a set of internal gills. Thereupon, a fleshy operculum grows backward from the hyoid arch over the gill region, enclosing external gills and internal gills in an opercular chamber that retains only a single excurrent pore, this being located at the posterior edge of the left operculum. Thereafter, the external gills gradually atrophy and the internal gills function until metamorphosis. At that time, the internal gills and operculum are absorbed along with the larval tail. All these structures are, of course, very small in a metamorphosing anuran.

Elasmobranchs develop either in a yolk-laden egg case attached to underwater vegetation, or they develop in the maternal uterus. During early development, filamentous outgrowths of the developing internal gills project through the external gill slits and into the fluids within the egg case or the uterine chamber, where they serve temporarily as respiratory organs (fig. 13.6c and d). In viviparous species, they also absorb nutrients from the uterine fluid. A few viviparous chondrostean and teleost larvae utilize filamentous external gills in a similar manner. Absorption of nutrients from uterine or other maternal fluids is known as **histotrophic nutrition.**

Excretory Role of Gills

Although usually thought of as respiratory organs, gills perform vital excretory functions. Marine fishes excrete the common marine salts chiefly via salt-secreting glands located on the lamellae of the gills. (Freshwater species secrete most of their excess chloride via the kidneys.)

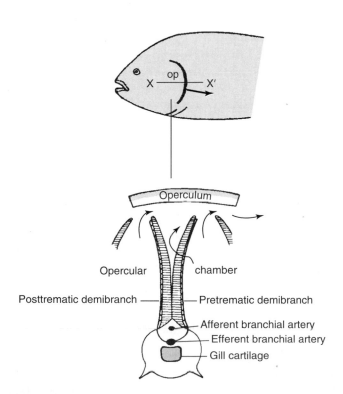

FIGURE 13.5

Operculum **(op)** and holobranch of a teleost. The holobranch, *bottom,* is seen in cross section (x—x′). Arrows indicate direction of efferent water flow.

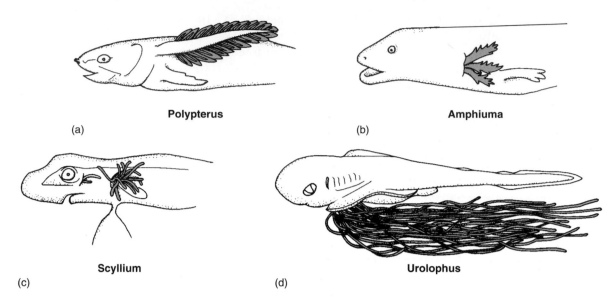

Polypterus
(a)

Amphiuma
(b)

Scyllium
(c)

Urolophus
(d)

FIGURE 13.6

Larval external gills. (*a–d*) A chondrostean, amphibian, shark, and ray, respectively. In *Amphiuma,* the gills are resorbed before the larva emerges from the egg envelope. The spiracle in *Urolophus,* presently behind the eye, will ultimately be located dorsally. Pectoral fins characteristic of Rajiformes have just appeared as limb buds.

The gills of lampreys and those of marine fishes that migrate between salt water and freshwater excrete chloride when in salt water, and they absorb chloride in freshwater, thereby assisting in maintaining homeostasis under these stressful conditions. Fishes without exception also excrete their nitrogenous wastes largely through the gills, unlike tetrapods, which excrete them via the kidneys. And, last, fishes that acquire oxygen from air in swim bladders release most of their carbon dioxide into the water flowing over the gills. Carbon dioxide dissolves more readily in water than in air, hence the partial pressure of CO_2 in the water passing over the gills is low enough to assure uninterrupted release of CO_2 from the circulation.

Aerial Respiration in Bony Fishes

Vertebrates arose in an aqueous environment, and water was the earliest source of oxygen. During the Devonian, however, if not earlier, air became a source of oxygen for many fishes. One compelling factor may have been that the water at that time was warm and swampy and, as such, low in dissolved oxygen. Air, on the contrary, contains 20 times the oxygen that oxygen-saturated water can hold. Little wonder, then, that many Devonian fishes obtained some or all of their oxygen from above the surface of the water. Today's air-breathing fishes include dipnoans, chondrosteans, garfishes, Amia, and many teleosts. The fishes snatch bubbles of air from just above the water surface, and the air then comes in contact with the oropharyngeal lining or, as we will see shortly, with the lining of swim bladders in those few fishes in which these swim bladders serve as lungs. A few teleosts swallow the bubble and extract oxygen in the stomach

or intestine. Although oxygen is acquired from the atmosphere, most of the excess carbon dioxide is eliminated through the gills into the water.

NARES AND NASAL CANALS

The paired nostrils or **external nares** of cartilaginous and ray-finned fishes open directly into blind olfactory sacs that contain the sensory epithelia for smell. Each nostril is divided into a forward-directed **incurrent aperture** through which water is driven as a fish swims forward, and a laterally or ventrally directed **excurrent aperture** through which water exits after having bathed the olfactory epithelium. In lobe-finned fishes, the nostrils are connected with the oropharyngeal cavity by a pair of nasal canals that conduct respiratory water. Their openings into the oropharyngeal cavity are **internal nares.** As we learned earlier, the nasal canals of agnathans are unpaired, and only in myxinoids do they extend to the pharynx. In no living fish, including those that obtain their oxygen from air, are nostrils used for breathing. They are part of a sensory system for monitoring one constituent of the environment—chemicals in solution in the surrounding water.

If the crossopterygian ancestors of labyrinthodonts were not using their nostrils for breathing—and whether they were is not known—labyrinthodonts were opportunists, since they commenced to use nostrils for drawing *air* into their oral cavity. With subsequent development of a secondary palate in reptiles, the internal nares opened farther caudad; and, as we have seen, the longer the secondary palate, the farther caudad the nares are located. In mammals they open into the nasopharynx above the soft palate (fig. 12.3, cat).

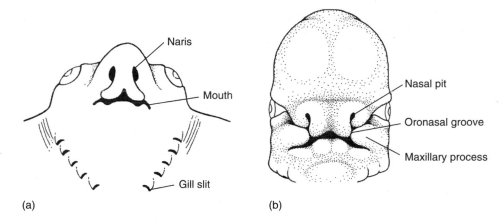

FIGURE 13.7

Oronasal relationships in (*a*), an adult skate, and (*b*), a 6-week (12 mm) human fetus.

Nasal canals (now **choanae**, appendix II), arise from paired **nasal pits** and **oronasal grooves**, the dorsolateral walls of which roll together to form a tube (fig. 13.7*b*). Although elasmobranchs lack nasal canals, oronasal grooves form and remain throughout life (fig. 13.7*a*). In mammals, the *olfactory epithelium* is restricted to the upper chambers of the nasal passageways. The lower chambers have a ciliated glandular *nasal epithelium*. Hairs at the entrance to mammalian nasal canals trap insects and coarse particulate matter in the incoming air; venous plexuses in the submucosa overlying the turbinal bones heat cold air; and air sinuses that open into the upper reaches of the nasal passageways of mammals (fig. 13.13*a*) serve partly as resonating chambers for vocalization.

A fleshy, partly cartilaginous, proboscis (nose) develops in some mammals and carries the nares to positions characteristic of the species. Compare the location of the nostrils of a catarrhine (fig. 4.39) with that of an elephant. Whales have no nose. The nostrils are far back on the top of the head and are known as **blowholes.** They are equipped with valves that close when the whale dives. (In fetal whales, the nostrils are farther forward.) In some whales the two nares unite during development to become a median blowhole.

SWIM BLADDERS AND THE ORIGIN OF LUNGS

Nearly every vertebrate from fish to human develops an unpaired evagination from the foregut that becomes one or a pair of pneumatic sacs (swim bladders or lungs) filled with gases derived directly or indirectly from the atmosphere. The only adult vertebrates that do *not* have pneumatic sacs—and their ancestors probably lost them—are lamprey, hagfish, cartilaginous fishes, a few marine teleosts, some bottom dwellers such as flounders, and a few tailed amphibians that inhabit mountain streams. Some of these develop sacs temporarily as embryos. Imprints of paired sacs with ducts have been found in a fossil placoderm. Pneumatic sacs hereafter will be referred to as swim bladders in fishes and lungs in tetrapods, regardless of their function.

After the budlike anlage for a swim bladder evaginates from the foregut, the resulting duct may retain its connection with the foregut, or the duct may close during later development. Fishes are said to be **physostomous** when the duct remains open. Chondrosteans, holosteans, the three living dipnoans, and some teleosts have open ducts. Fishes in which the duct closes, as in many teleosts, are said to be **physoclistous** (fig. 13.8).

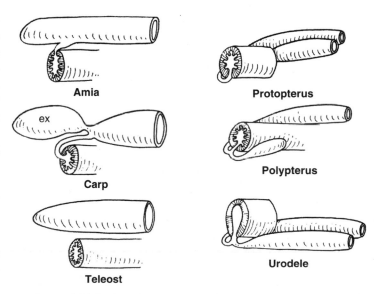

FIGURE 13.8

Swim bladders and urodele lungs. In teleosts the duct usually closes. Carp are physostomous teleosts with an anterior extension of the bladder, **ex,** connected to weberian ossicles (fig. 17.7). *Amia* is a holostean, *Protopterus* is a dipnoan, and *Polypterus* is a chondrostean.

Swim bladders may be paired or unpaired. In chondrosteans and dipnoans, the unpaired duct leads from the ventral aspect of the esophagus. It has shifted toward the dorsal side in holosteans and teleosts. Infrequently, the duct leads from the pharynx or stomach.

Swim bladders lie close to the kidneys and, like the latter, they are retroperitoneal (fig. 15.18). As the embryonic anlagen grow longer, they push their way caudad in the roof of the coelom between the embryonic parietal peritoneum and the body wall. However, adult bladders may bulge into the roof of the coelom.

Swim bladders in one species or another serve as hydrostatic organs, respiratory organs, organs that participate in sound detection, or organs of communication. Since their walls contain elastic and smooth muscle tissue, their capacity for gases can be increased or decreased as required by their roles.

Swim bladders in teleosts serve chiefly as hydrostatic organs. To maintain an appropriate depth in a body of water or to hover at a specific location, a fish must achieve, as closely as possible, a body density, or specific gravity, equal to that of the displaced water at the selected depth. This is accomplished by regulating the volume of gas in the swim bladder. Any difference in density will affect its buoyancy, causing the fish to slowly rise or sink. But since disturbances in the water preclude complete stability moment by moment, a hovering fish must also resort to frequent gentle undulations of the paired and unpaired fins to compensate for these disturbances. Some entire populations of fishes make daily changes in depth, therefore in body density, as a correlate of the time of day.

The gas in the hydrostatic swim bladder usually comes from the blood, being actively transported into the lumen of the bladder from the **red gland,** a localized rete of small arteries in the bladder lining. The blood in the rete inspired the term "red gland." The gas is eventually resorbed into the bloodstream of the bladder near its caudal end in a pocketlike area of modified epithelium. The pocket can be closed off from the main cavity by a sphincter during passage of gas into the lumen of the bladder, and relaxed when resorption of gas is taking place. In physostomes, the gas may be bubbled to the exterior through the mouth.

The gases in swim bladders differ among fishes. Some bladders contain almost pure (99 percent) nitrogen, some up to 87 percent oxygen, and all contain at least traces of carbon dioxide and argon. In deep-water fishes, nitrogen may be transported from the blood into the lumen of the bladder against a nitrogen pressure as high as 10 atmospheres.

Swim bladders function as lungs to one degree or another in physostomous fishes. Air is gulped at the water's surface and an oropharyngeal pump forces it from the oropharyngeal cavity into the swim bladder. Air depleted of oxygen is expelled from the bladder into the oropharynx as a result of a vacuum created by lowering the oropharyngeal floor while the mouth and nares are closed. Expulsion is facilitated by the elasticity of the

bladder wall, its smooth musculature, and the pressure of the surrounding water on the body wall. Oxygen-depleted air in the oropharynx is then bubbled through the mouth into the water.

Two of the three genera of true lungfishes, *Protopterus* and *Lepidosiren,* require continual gulping of air for survival, since their gills (three pairs of external ones in *Protopterus,* fig. 14.18*b*) are incapable of providing sufficient oxygen for ordinary metabolic needs. These fishes drown if held under water for a sufficient length of time. Their ability to survive tropical summers in burrows when the swamps dry up is attributable in part to the fact that their swim bladders serve as lungs. However, these fish also absorb oxygen through the skin at those times.

The swim bladder of the only other living dipnoan, *Neoceratodus,* those of two genera of ray-finned African freshwater fishes, *Polypterus* and *Calamoichthys* (sometimes inappropriately called "lungfishes"), and the swim bladder of the relict holosteans (*Amia* and garfishes) function as a lung only when the oxygen content of the water is so low that the gills alone cannot meet the respiratory needs of the organism. The lining of the swim bladders of *Polypterus* and *Calamoichthys* is almost smooth; that of true lungfishes is lined with low septa and may exhibit thousands of tiny air sacs (fig. 13.10, a dipnoan). Unlike those in other fishes, the swim bladders of chondrosteans and dipnoans are supplied by arteries arising from the sixth embryonic aortic arch, a tetrapod condition (fig. 14.18*b*); and, as in tetrapods, the venous return in dipnoans is directly to the left atrium of the heart. This is not true in any other fish.

In addition to serving as hydrostatic organs, the swim bladders of some teleosts participate in sound detection. In Cypriniformes (goldfish, carp, North American catfish, minnows, others), a series of small bones, **weberian ossicles,** connects the anterior end of the swim bladder with the sinus impar, an extension of the perilymphatic space of the inner ear. In another order of teleosts, Clupeiformes, a thin-walled anterior extension of the swim bladder makes direct contact with the inner ear (fig. 13.9). These fish can hear.

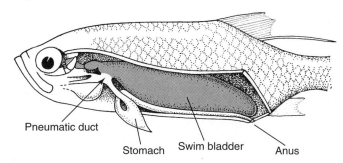

FIGURE 13.9

Unpaired swim bladder of a tarpon, a physostomous teleost. The extension into the skull contacts the inner ear.

Source: Data from J.S. Kingsley, *Outline of Comparative Anatomy of Vertebrates,* 1920, The Blakiston Company, Philadelphia, after de Beaufort.

Pneumatic duct
Pulmonary arteries

FIGURE 13.10

Swim bladders of the lungfish *Protopterus*.

In a few fishes, contraction of striated muscles attached to the swim bladder cause it to emit thumping sounds, or they force air back and forth between chambers separated by muscular sphincters within the bladder to produce croaking or grunting sounds for communication.

Swim bladders usually degenerate in fishes that are bottom feeders. Elimination of this hydrostatic organ has optimized their body density, enabling them to hover close to their food supply with a minimal expenditure of energy. Although degeneration of the swim bladder may be an adaptation for bottom feeding, we will never know the nature of the original selective pressures responsible for the advantage. An analogous condition is seen in salamanders who live in swift mountain streams and whose lungs are only a few millimeters long, or absent. Any buoyancy that might be conferred by lungs would be detrimental. Optimal body density and the saving of energy are the benefits. It should be noted, however, that *no* member of the plethodont family has lungs, and not all inhabit swift streams. These urodeles breathe through their skin.

The striking similarities between swim bladders and lungs, especially bladders that are supplied via the sixth aortic arches, suggest that these may be the same organs. In the Devonian, when the freshwaters of the swamps were warm and periodically stagnant, and therefore low in dissolved oxygen, aerial respiration may have made the difference between survival and extinction. At that time, placoderms and most crossopterygians had pneumatic sacs. We can only speculate on which came first, respiratory or hydrostatic bladders, but these two conclusions seem warranted: a pneumatic sac was functioning in aerial respiration long before vertebrates ventured onto land, and closure of the pneumatic duct in physoclistous fishes is probably a mutation from a more primitive open-duct condition.

LUNGS AND THEIR DUCTS

Tetrapod lungs arise as an unpaired evagination, the **lung bud,** from the caudal floor of the pharynx just beyond the last pharyngeal pouch (fig. 1.6). The opening in the pharyngeal floor becomes a longitudinal slit, the **glottis** (fig. 13.13*a*). The unpaired lung bud elongates only slightly before bifurcating to form bronchi and lungs (fig. 13.22). The lung primordia push caudad underneath the foregut until they bulge into the coelom lateral to the heart. As they grow into the coelom, they carry along an investment of peritoneum, which becomes the visceral pleura. The part of the lung bud between glottis and lungs develops into larynx, trachea, and bronchi.

Larynx and Vocalization

The larynx is a short air passageway between the glottis and the upper end of the trachea of tetrapods, the walls of which are supported by cartilaginous derivatives of the caudalmost pharyngeal arches (table 9.4). The glottis and associated cartilages have become an instrument for vocal communication in species in which vocal cords develop. In urodeles, the larynx is a primitive structure consisting of a single pair of lateral cartilages that surround the glottis and support it as their sole role. Most other tetrapods below mammals have two pairs of laryngeal cartilages, **arytenoids** and **cricoids** (fig. 13.11), and mammals have a third pair, the **thyroid** cartilages. Additional small cartilages—**cuneiforms, corniculates, procricoid,** and others—develop in some mammals (fig. 13.12). The paired cricoids and thyroids of mammalian embryos tend to unite across the midventral line during ontogeny (fig. 13.13*b*). The laryngeal cartilages are interconnected by ligaments and by intrinsic laryngeal muscles. Extrinsic laryngeal muscles, particularly the sternothyroid and thyrohyoid of mammals, provide the larynx with the mobility that is necessary during the act of swallowing.

Folded within or stretched across the laryngeal chamber in anurans, some lizards, and most mammals are **vocal folds,** more often referred to as **vocal cords,** which endow these species with the ability to communicate vocally to one degree or another with other members of the species. The cords vary from mere fleshy folds surrounding the glottis and attached to the arytenoid and cricoid cartilages, as in anurans, to a pair of mammalian folds containing strong bandlike elastic ligaments

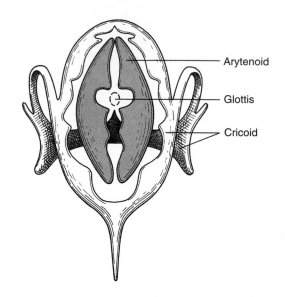

FIGURE 13.11
Laryngeal skeleton of a frog.

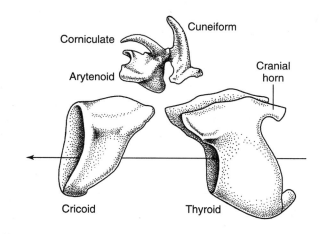

FIGURE 13.12

Laryngeal cartilages of a dog, disarticulated. The arytenoid has been elevated above its articulation sites. Cranial horns of thyroid articulate with caudal horns of hyoid seen in fig. 9.43. Arrow is in respiratory passageway to trachea.

that are stretched between the arytenoid and thyroid cartilage on each side of the sagittal plane. Air expelled from the lungs passes through a narrow slit, the glottis, between the folds. When not in use for vocalization the cords are relaxed, and exhaled air passes between them silently. When under tension the cords vibrate, giving rise to sounds. During vocalization, *intrinsic* muscles of the larynx alter the position of the thyroid and arytenoid cartilages with respect to one another, thereby regulating the tension on the cords. The pitch (frequency of vibration) of the human voice is a function of the amount of tension within the cords. Among mammals, hippos and a few others lack vocal cords and one breed of dogs,

basenjis, have poorly developed ones and cannot bark. By contrast, the thyroid and associated hyoid bones of howler monkeys are enormous, causing a goiterlike bulge in the neck (fig. 13.14). They surround a voluminous resonating chamber that enables the weird howl of this species to penetrate deep into the jungle.

Except in birds and mammals, vocalization is not a significant means of communication among tetrapods. Apodan and urodele amphibians are mute as are most reptiles. Male anurans force air out of the lungs, over vocal folds, and into the oropharyngeal cavity with the mouth and nares closed, and from there into **vocal sacs,** which produce a croak each time they refill and deflate (fig. 12.3*a*). The vociferous croaking of male toads immediately after a rainstorm is a mating call. (The eggs of oviparous females are laid after a good rain has filled the ponds and ditches in which eggs are deposited. Sperm is deposited over the eggs while they are being extruded.) Each species has its own breeding call. There may be other calls also. Birds lack vocal cords; their calls are produced by an avian voice box, the syrinx, at the base of the trachea.

Most mammals have "false" vocal cords, as well as true ones. In some species these produce unusual sounds such as purring in kittens. They are fleshy folds located at the entrance to the **vestibule of the larynx,** an antechamber located just above the true cords. The vestibule in howler monkeys has become the enormous resonating chamber referred to earlier.

Anatomic adaptations prevent fluids from entering the glottis in submerged aquatic tetrapods. Chief among these are valves at the entrance to the external nares. Crocodilians have, in addition, a fleshy valve, the **palatine velum,** which consists of two large transverse folds suspended from the caudal end of the long palate just anterior to the internal nares. When a matching fold at the base of the tongue is drawn up against the velum, the resulting partition prevents the passage of water to the pharyngeal chiasma and provides a continuous passageway for air from the nostril to the lungs. Consequently, a crocodile or alligator can be submerged with only two moundlike nostrils and two elevated eyes protruding above the water, breathing (with a mouth full of water), and eyeing the environment above the water for prey. This is especially useful after an air-breathing victim is caught—a rat, for example. The victim can be carried or dragged into the water and held under by clenched jaws until it drowns. Meanwhile, the crocodile continues to breathe without water getting into the lungs. When the crocodilian is totally submerged, the valves at the entrance to the external nares prevent flooding of the respiratory tract.

In most mammals, terrestrial and aquatic, the larynx is high in the oropharynx where it is able to lock into the nasopharynx during the act of swallowing liquids. This allows adult cats, monkeys, and many other mammals to

(a)

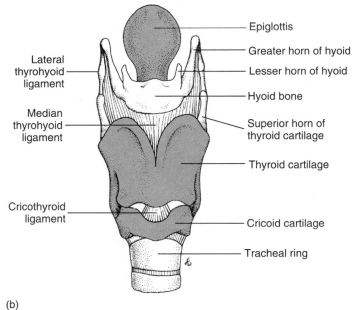

(b)

FIGURE 13.13

(*a*) Human upper respiratory tract, sagittal section. (*b*) Hyoid and larynx, frontal view. Arrows in (*a*) designate pharyngeal chiasma.

breathe while simultaneously drinking, since a pair of pharyngeal recesses lateral to the larynx serve as bypasses to the esophagus for liquids. It also allows their babies to breathe while suckling. The larynx of human babies is high in the throat until about 18 months of age, at which time it begins to descend in the throat, "altering the way in which a child breathes, swallows, and vocalizes" (Laitman, 1984). It remains high—does not descend—in monkeys or apes. In marsupials, in which the baby's mouth cannot be withdrawn from the nipple, the high larynx permits milk that is being pumped by the mother's mammary glands into the baby's esophagus from being sucked into the lungs when the baby breathes. In whales that have surfaced to breathe, the

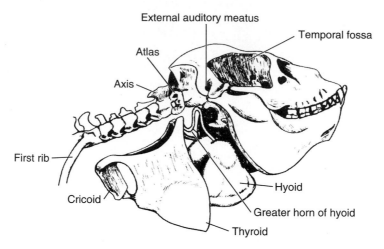

FIGURE 13.14

Hyoid bone and larynx of a howler monkey.

beaked cephalic end of the larynx (fig. 13.15) can be extended into the nasopharynx, permitting prolonged inhalation of fresh air through the blowhole while the cavernous mouth may be running over with seawater.

In humans and other mammals with a low larynx, the act of swallowing draws the larynx forward (upward in humans) against the epiglottis, momentarily blocking the entrance to the larynx and thereby preventing food from entering the lower respiratory tract. But each blockage can be only momentary. You can't eat a hamburger while holding your nose closed!

Of course, a mammal cannot breathe through the mouth if the larynx protrudes into the nasopharynx, nor could air returning from the lungs be used for articulate speech. The latter requires participation of the lips, tongue, and the cavity acquired between the root of the tongue and the glottis when the larynx descends in the course of postnatal development. Thus, the final location of the human larynx facilitates human speech.

The anterior wall of the larynx of some male lizards has a saccular evagination, the **gular pouch** or dewlap, just under the skin of the throat. When the pouch is inflated with air, it causes a ballooning of the throat; and when sunlight shines through its well-vascularized walls, the reddish pouch becomes a sex attractant. The pouch deflates noiselessly when the animal is disturbed. A process of the hyoid bone supports the inflated pouch.

Trachea, Syrinx, and Bronchi

The **trachea** is ordinarily as long as the neck. Therefore it is short in amphibians and long in amniotes. In birds and some turtles, the part of the trachea within the neck is longer than the neck as an accommodation to the stretching, twisting, and looping that these necks are capable of. When a swan's neck is relaxed, or a turtle withdraws its neck into the shell, the trachea assumes an S-shape. In crocodilians, too, the trachea is longer than the neck in some species.

The walls of the trachea are prevented from collapsing under the negative pressure of inhalation by cartilaginous or bony rings or plates that form within the wall. The rings are usually incomplete dorsally, the ends being united by smooth muscle that can change the diameter of the tube in response to the need for a greater or lesser volume of tidal air. In crocodilians and birds, however, all tracheal rings are complete. Except in lower urodeles, the trachea bifurcates to become two primary **bronchi** that are similarly strengthened.

Birds have a special voice box, the **syrinx**, at the bifurcation of the trachea (fig. 13.16). A **bronchotracheal syrinx** consists of a resonating chamber with walls strengthened by the last several tracheal rings and the first bronchial half rings. Mucosal folds project into the chamber, and a bony **pessulus** may form in a special

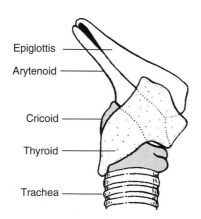

FIGURE 13.15

Laryngeal skeleton of whale. The epiglottis and arytenoid can be inserted into the nasopharynx.

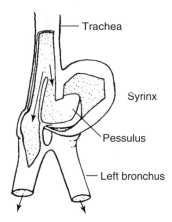

FIGURE 13.16

Asymmetrical bronchotracheal syrinx of a canvasback duck. Arrows indicate path of inhaled air.

FIGURE 13.17

Lungs of *Necturus.*

semilunar membranous fold. In a **tracheal syrinx,** parts of the last several tracheal rings are missing, enabling the membranous wall of the trachea itself to vibrate. In a **bronchial syrinx,** the membranous wall between two bronchial cartilages folds into the chamber when the cartilages are drawn together, and the fold vibrates with the flow of air.

Amphibian Lungs

The lungs of amphibians are simple sacs, long in urodeles (fig. 13.17) and bulbous in anurans, conforming to the shape of the pleuroperitoneal cavity. The internal lining may be smooth throughout, there may be simple sacculations in the proximal part, or the entire lining may be pocketed. The left lung of caecilians is rudimentary, and the lungs of salamanders that inhabit swift mountain streams may be only a few millimeters long. Plethodontid salamanders have *no* lungs.

The lungs of normally aquatic urodeles function mostly as hydrostatic organs, and respiration in species lacking gills takes place chiefly through the pharyngoesophageal lining and the skin. In water, which is the normal habitat of *Necturus,* only about 2 percent of the oxygen is acquired via lungs. Another perennibranchiate, *Siren,* a gill-breather like *Necturus* when in oxygenated water, frequently expresses air through the gill slits when obligated to breathe air. Urodeles using lungs as either hydrostatic or respiratory organs fill the lungs in the same manner as lungfishes; respiratory air enters the mouth, not the nostrils, and the muscular floor of the oropharynx pumps it to the lungs. Elasticity of the lungs probably plays a major role in emptying them.

The details of respiration in bullfrogs have been described by Gans, DeJongh, and Farber (1969). The mouth remains closed during respiration. In essence, they found that lowering the floor of the oropharyngeal cavity with the glottis closed draws air via the nostrils

into the deep inferiocaudal reaches of the oropharynx where it is put on hold. Thereupon the glottis is opened and the air already in the lungs escapes to the exterior via the nostrils. Emptying of the lungs appears to be the result of the elasticity of the lung and the smooth muscle in its walls. In the third phase, the fresh air stored in the floor of the oropharynx is forced into the lungs by raising the floor of the oropharynx with the nostrils again closed. In the fourth phase, the reservoir of air on hold is replenished. Once this sequence has been completed and the glottis closed, oscillation of the muscular floor of the oropharynx commences. The phenomenon can be observed in any frog by watching the rapid rhythmic pulsations of the throat behind the lower jaw. The oscillations maintain a steady flow of fresh air via the nostrils into and out of the oropharyngeal cavity, the lining of which is an accessory respiratory epithelium. The oscillations also probably flush out any residual air remaining from the preceding emptying of the lungs. Air is forced in from the oropharyngeal cavity by a *pressure pump* and the elasticity of the lungs pushes it out. However, a *vacuum pump* fills the oral cavity with fresh air.

Reptilian Lungs

The lungs of *Sphenodon* (fig. 13.18*a*) and of snakes are simple sacs, although the caudal third of the lining in snakes is septate and contains residual air. In lizards (fig. 13.18*b*), crocodilians, and turtles, the septa divide the lungs into numerous large chambers. These lungs are somewhat spongy because of the numerous pockets of trapped air. Although asymmetry of the two lungs is usual in tetrapods, it is pronounced in legless reptiles, in

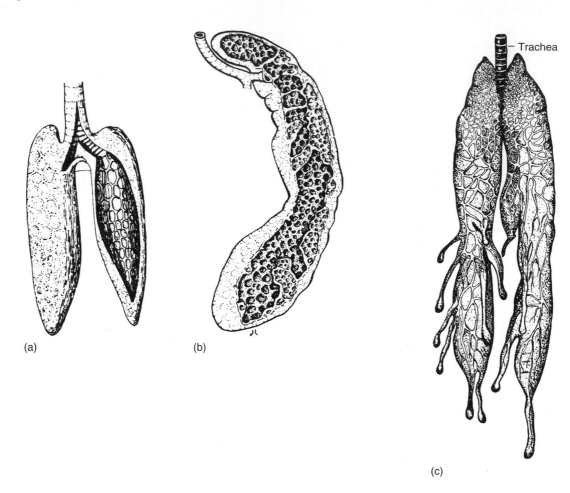

FIGURE 13.18

Lungs of three reptiles: (*a*) *Sphenodon;* (*b*) *Heloderma,* a large lizard; and (*c*) a chameleon.

which the left lung is much shorter than the other, sometimes atrophied, and occasionally totally absent.

An enormous diverticulum of the left lung of puffing adders extends into the neck. Inflation of this air sac causes the neck to balloon. Similar **air sacs** extend among the abdominal and pelvic viscera in some lizards (fig. 13.18*c*). Dinosaurs and pterosaurs had them, and in these reptiles the sacs also invaded the centra of vertebrae, as they do in today's birds. Air sacs are used by reptiles to inflate the body as a scare tactic or for other defensive purposes; in birds, as will be seen shortly, they have become a vital part of the mechanics of respiration.

Most reptilian lungs, like those of amphibians, occupy the pleuroperitoneal cavity along with the other viscera. In turtles they lie dorsal to the viscera, flattened against the carapace just caudal to the pectoral girdle on each side, and all but their caudal ends are retroperitoneal. The lungs of crocodilians and a few squamates occupy separate subdivisions of the coelom, the paired **pleural cavities.** These are isolated from the rest of the coelom by a tendonous **oblique septum.** (An oblique septum in a bird is seen in fig. 13.19.) Although the septum plays no active role in respiration, it participates passively when the

abdominal viscera that press against it are displaced cephalad or caudad by action of abdominal muscles.

Squamates inhale by employing intercostal muscles to rotate the ribs forward and outward, thereby increasing the volume of the pleuroperitoneal cavity. This action creates a vacuum around the lungs and other viscera, and atmospheric pressure forces environmental air into the nostrils and throughout the respiratory tract, causing the thin-walled elastic lungs to balloon with oxygenated air. Return of the ribs to their resting position and, in some species, contraction of the abdominal wall musculature compresses the lungs and other viscera; and this effect, aided by the resilience of the lung walls, results in exhalation.

Crocodilians may employ ribs for respiration to some degree, but they also have a unique method of getting air into the lungs which, atypically for reptiles, lie cephalad to the large bilobed liver. The liver, in turn, lies tightly against the oblique septum that separates the two organs. A pair of long striated **diaphragmatic** muscles, derived from the musculature of the hypaxial body wall, extend between the pelvic girdle and the fibrous investment of the liver. When these muscles contract (which they

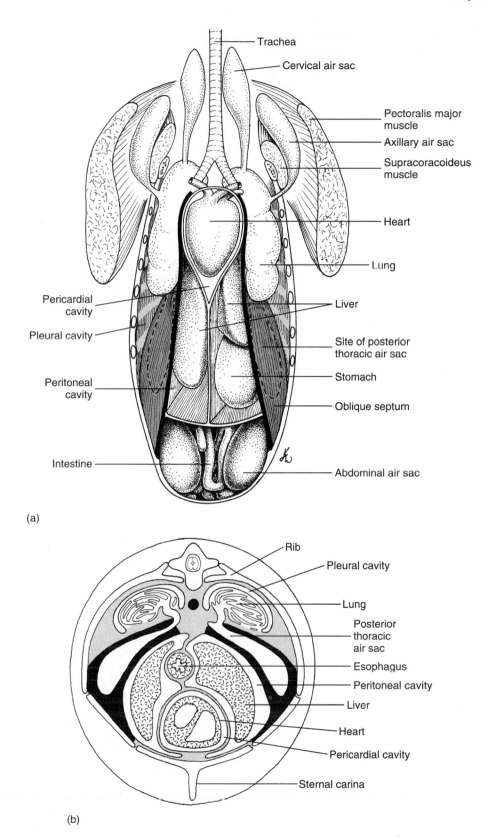

(a)

(b)

FIGURE 13.19

Oblique septum (*red*) and subdivisions of the coelom in birds. (*a*) Ventral view, interclavicular and anterior thoracic air sacs omitted to display lungs. The posterior thoracic air sacs (broken lines) are embedded within oblique septum. (*b*) Thorax in cross section.

Source: Data from E. S. Goodrich, *Studies on the Structure and Development of Vertebrates,* 1930, Macmillan and Company, Ltd., London.

presumably do in response to the same reflex stimuli that cause the mammalian diaphragm to contract), the liver and oblique septum are drawn caudad creating a negative pressure within the pleural cavities. Atmospheric pressure then expands the lungs via the respiratory tract, and air rushes into the lungs. Exhalation takes place when the transverse abdominal muscles contract, constricting the coelom and forcibly displacing the viscera. The pressure is transmitted from the coelomic viscera to the lungs, and oxygen-depleted air is forced out. Thus the body wall musculature plays the major role in respiration in crocodilians but in a unique manner.

The process of respiration in turtles is complicated by the fact that their ribs cannot participate in respiratory movements because they are fused with the carapace. And since the abdominal wall is rigid and its musculature is vestigial, it, too, cannot participate. But all is not

lost! Reflex movements of the pectoral girdle play the major role in ventilation of the lungs by altering the volume of the coelom.

Lungs and Their Ducts in Birds

The lungs of birds occupy paired **pleural cavities** separated from the rest of the coelom by a tendonous **oblique septum** (fig. 13.19*a* and *b*). Although the septum plays no active role in altering the pressure within the lungs, it participates passively, as in turtles, when the viscera within the peritoneal cavity are displaced by the action of muscles that move the ribs and sternum.

Avian lungs are unique morphologically. Incoming air enters a lung from the trachea via a **primary bronchus,** continues *nonstop* through the substance of the lung in the **mesobronchus,** and enters spacious **air sacs** before returning to the lungs (fig. 13.20*a* and *b*).

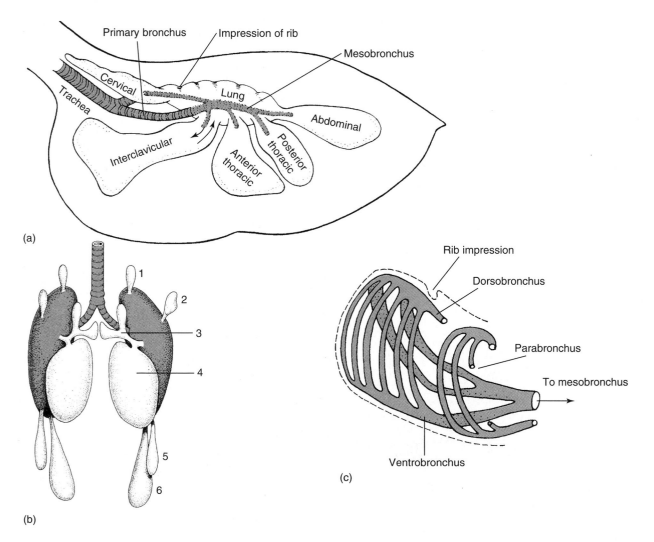

FIGURE 13.20

Lower respiratory tract of a bird. (*a*) Adult air sacs. Air leaves the sacs via recurrent bronchi (*black arrow*). (*b*) Air sacs (**1–6**) of a 12-day chick embryo. Trachea, primary bronchi, and lungs are in *red*. (*c*) Parabronchi and their connections. In (*b*): **1,** cervical; **2, 3,** lateral and medial anlagen of interclavicular; **4,** anterior thoracic; **5,** posterior thoracic; **6,** abdominal.

Source: Data from K. F. Locy and O. Larsell, "Embryology of the Bird's Lung" in *American Journal of Anatomy,* 19:447, 1916.

Air sacs are voluminous, thin-walled, distensible diverticula of the lungs. Birds have five or six pairs (fig. 13.20a): (1) cervical sacs at the base of the neck; (2) interclavicular sacs dorsal to the furcula and sometimes united across the midline; (3) anterior thoracic sacs in their own compartment of the coelom lateral to the heart; (4) posterior thoracic sacs within the double-layered oblique septum; (5) abdominal sacs among the abdominal viscera; and (6) axillary sacs (fig. 13.19a), less common, lying between the major elevator muscle of the wings (supracoracoideus) and the major depressor muscle (pectoralis major, fig. 11.21c). Except in ratites, diverticula from the sacs penetrate most of the bones of the body, including centra, entering via **pneumatic foramina.** For this reason, birds are sometimes said to have hollow bones.

Avian air sacs function in the following manner: When a bird is in flight, contraction of a set of flight muscles on the chest wall expands the coelom. This creates a vacuum around and, hence, within the air sacs, and atmospheric pressure forces fresh air via the mesobronchi into the sacs. Subsequent contraction of an antagonistic set of flight muscles compresses the sacs, forcing oxygen-rich air out of the sacs via **excurrent bronchi,** into the duct system within the lungs, across the respiratory epithelium, and on to the exterior. This muscular bellows (vacuum-compression-vacuum-compression) maintains an uninterrupted flow of air to and from the lungs as long as the bird is in flight.

When a bird is at rest the chest wall still functions to create a vacuum, but it is the intercostal and other nonappendicular muscles that expand the coelom. The thoracic wall relaxes as a result of the elasticity of the tissues of the wall, not by muscle contraction. External respiration in resting birds doesn't require the energy expenditure necessary to maintain a bird in flight.

Dorsobronchi and **ventrobronchi** within the lungs are connected by **parabronchi** (fig. 13.20c). The walls of the latter (fig. 13.21) consist of an anastomosing network of **air capillaries,** each only a few thousandths of a millimeter in diameter, surrounded by **blood capillaries.** It is the lining of the air capillaries that constitutes the respiratory epithelium.

The air capillaries are interconnected in three dimensions. Air flows freely through the network, returns to the parabronchus from which it came and, without interruption, continues to the mesobronchus as an exhalation. Further details on the flow of air through the pulmonary duct system are lacking.

FIGURE 13.21

Parabronchi and air capillaries of a bird's lung.

Avian air sacs and the open-ended duct system within the lungs result in complete replacement of the air in the lungs with every inflation-compression cycle of the bellows. Consequently, unlike the lungs of other tetrapods, those of birds contain no residual air.

Air sacs are thermoregulatory, dissipating excess heat produced by the muscles during flight. The heat is transferred directly from the tissues to the air sacs rather than via the bloodstream because air sacs have a relatively poor vascular supply. At rest, a bird's chief defense against overheating is a reflexive increase in respiratory rate. In resting doves, excess heat is transferred from an exceptionally vascular esophageal mucosa to the outgoing airstream in the adjacent trachea.

As we have seen, air sacs are not unique in birds. They are found in some lizards and snakes, and they were present in dinosaurs and pterosaurs as revealed by pneumatic foramina in their bones. There is no evidence of pneumatic foramina in *Archaeopteryx*. The extension of air sacs into bones results in energy conservation during flight by decreasing the density of the bird's body, making it more buoyant in air. The warmer the air in the sacs, the more buoyant is the bird.

Mammalian Lungs

Mammalian lungs differ from those of birds in that incoming air enters a system of passageways that branch and rebranch like the limbs of a tree. At the end of the smallest passageways, **alveoli** lined by a respiratory epithelium are the sites of gaseous exchange. Oxygen-depleted and carbon dioxide–enriched air, along with metabolic water, is exhaled via the same path by which the air entered.

Mammalian lungs usually exhibit several lobes, asymmetrically, with one more on the right (fig. 13.22, 14 mm embryo). The lungs of some aquatic and terrestrial mammals are not lobed (whales, sirenians, elephants, and horses, for example), and some are lobed on the right side only (including monotremes and rats).

Left and right lungs occupy separate pleural cavities separated in the midline by the **mediastinum,** a septum of loose connective tissue enclosing the esophagus, the heart within the pericardial sac, major ascending and descending blood vessels, lymphatics, nerves, and the lower end of the trachea.

The trachea divides into two **primary bronchi,** each of which penetrates a lung at the **hilus** and divides into one **secondary bronchus** for each lobe, when lobes are present. These give rise to **tertiary bronchi** that branch and rebranch into smaller and smaller **bronchioles,** the last of which opens into several to a dozen thin-walled **alveolar ducts** (fig. 13.22*b*). The walls of the ducts are evaginated to form clusters of alveoli, or respiratory pockets, estimated to number over 300 million in each human lung. It is in these alveoli that gaseous exchange takes place.

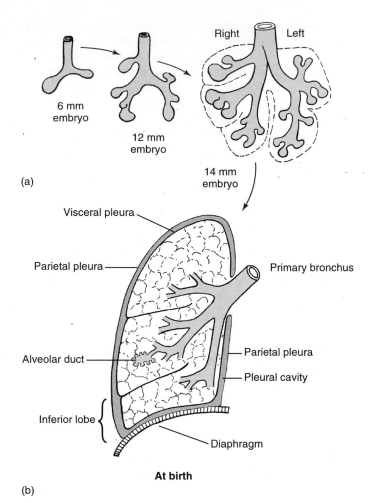

(a)

Visceral pleura

Parietal pleura

Primary bronchus

Alveolar duct

Parietal pleura

Pleural cavity

Inferior lobe {

Diaphragm

At birth

(b)

FIGURE 13.22

(*a*) Development of lung of fetal pig. Developmental stages are also applicable approximately to human embryos of the same length. The anlagen of all five adult lobes have been established by the 14 mm stage. (*b*) Relationships of parietal (*red line*) and visceral pleura, pleural cavity, and diaphragm. The width of the pleural cavity is exaggerated.

The walls of the bronchi and larger bronchioles contain smooth muscle fibers, connective tissue, and irregular cartilaginous plates, and the lining is a **ciliated pseudostratified columnar epithelium** (a single layer of tall ciliated cells with nuclei that lie at different heights, giving the impression that the epithelium is multilayered). As the branches become smaller, the cilia are lost, the epithelium becomes flatter, the cartilage disappears, and then, in some mammals at least, the smooth muscle cells disappear. Alveoli are lined by a **simple squamous epithelium** ("pavement cells") beneath which is a rich plexus of capillaries held together by a close network of reticular connective tissue fibers. Red blood cells in the capillaries take on oxygen to form oxyhemoglobin, and carbon dioxide and some metabolic water pass from the blood plasma into the alveoli to be exhaled.

The parietal peritoneum of each pleural cavity lines the chest wall as the **parietal pleura** and covers the cephalic surface of the diaphragm as the **diaphragmatic pleura** (fig. 13.22). At the **root** of each lung, where the primary bronchus and pulmonary vessels enter and leave, the parietal pleura is continuous with the visceral pleura that lies on the surface of the lung. The space between the parietal and visceral pleurae is the pleural cavity. It surrounds the lung except at the hilus. A *subatmospheric* pressure exists in the pleural cavity, and the normal atmospheric pressure exerted via the nares keeps the elastic wall of the lung in intimate contact with the thoracic wall at all times, with only a thin layer of serous lubricant intervening. The lubricant is a product of the pleurae. The fluid minimizes friction between lungs and chest wall as the latter rises and falls with each inhalation and exhalation. Inflammation of the pleura causes pleurisy, an increase of fluid in the cavity. Perforation of the thoracic wall and the parietal pleura—from a gunshot wound, for example—allows atmospheric air to enter the pleural cavity, resulting in deflation of the lung on that side, a condition known as pneumothorax.

Ventilation of mammalian lungs is accomplished primarily by a dome-shaped muscular diaphragm, unique in mammals, that functions as a suction pump (fig. 11.13). It is anchored to the xiphoid process of the sternum ventrally, to a half dozen or so of the caudalmost ribs and their costal cartilages laterally, and to several of the more anterior lumbar vertebrae dorsally. When not under tension, the dome bulges cephalad into the thoracic cavity. Contraction of the diaphragmatic muscles flattens the diaphragm. This further decreases the already subatmospheric pressure within the pleural cavity. Consequently, normal atmospheric pressure pushes more air into the lungs to fill the vacuum. This is the phenomenon known as inhalation.

Exhalation in mammals is largely a passive phenomenon attributable to the following: (1) relaxation of the diaphragm, which returns it to the domed position, decreasing the volume of the thorax and increasing the pressure around the lungs; (2) the upward pressure exerted on the diaphragm by the resilient abdominal viscera, which were under compression while the diaphragm was flattened; (3) resilience of the abdominal wall, which bulges when the abdominal viscera are compressed; and (4) return of the ribs to a resting position as the costal muscles relax; (5) elasticity of the lungs, which enables them to conform to these changes. As a result, air is squeezed out of the lungs. During forceful expiration, as in panting, roaring, coughing, or singing, the abdominal wall participates in the expulsion of air.

Marine mammals that feed in the depths of the oceans have exceptionally muscular diaphragms. The waterspout in a whale, for instance—the sign that exhalation is taking place—lasts three to five minutes. Also, deep-sea foraging mammals cannot store oxygen in their lungs in anticipation of a dive because the lungs collapse as the dive takes the animal to increasing depths. Instead, oxygen acquired prior to a dive is stored in retia mirabilia in the trunk (chapter 14). The animal exhales just before commencing a dive, and breathing does not resume until the animal resurfaces, which may be as long as two hours later.

SUMMARY

1. External respiration is the exchange of respiratory gases between organism and environment. It takes place via highly vascular membranes with thin, moist epithelia.
2. Internal respiration is the exchange of gases between capillary blood and the tissues.
3. The chief adult organs of respiration are pharyngeal gills, oropharyngeal mucosa, swim bladders or lungs, and skin.
4. Cutaneous respiration is the chief method of respiration in aquatic urodeles and some scaleless fishes.
5. Internal gills arise in the walls of pharyngeal pouches and are supported by the skeleton of the pharyngeal arches except in agnathans.
6. A typical gill, or holobranch, consists of two demibranchs and an interbranchial septum. The septum is shorter in fishes that have an operculum.
7. Respiratory water usually enters via the mouth, but it enters the spiracle in some elasmobranchs, the unpaired naris in hagfishes, and external gill slits in lampreys.
8. Elasmobranchs have naked gill slits. The slits of chimaeras, bony fishes, and larval anurans are covered by an operculum attached to the hyoid arch.
9. A branchiostegal membrane is attached to the ventral edge of each operculum in bony fishes, supported by gular bones or branchiostegal rays. The membranes enclose an opercular chamber, which receives water that has passed over the gills. The water is then expelled to the exterior.
10. The first embryonic pharyngeal slit persists as a spiracle in adult elasmobranchs and chondrosteans. In some species, it houses a pseudobranch on its anterior wall.

11. Functional hyoid arch demibranchs occur in cartilaginous fishes but tend to disappear in teleosts. The number of demibranchs is reduced still further in dipnoans.

12. Larval gills may be external or internal. They are external in dipnoans, a few primitive ray-finned fishes, and amphibians. Anuran larvae later develop internal gills. Filamentous internal gills project to the exterior in larval elasmobranchs and some bony fishes.

13. Gills in fishes function also in maintaining salt homeostasis and in excretion of nitrogenous waste and carbon dioxide.

14. External nares lead to blind olfactory sacs in jawed fishes other than dipnoans. In dipnoans and tetrapods, they are connected with the oropharyngeal cavity or pharynx via nasal canals. They are used for respiration in tetrapods only. Hagfishes have a nasopharyngeal duct that carries respiratory water. A nasohypophyseal duct in lampreys ends blindly.

15. Internal nares (choanae) open far forward in the oral cavity of dipnoans and amphibians, and farther caudad if there is a secondary palate. Despite the presence of internal nares, dipnoans and aquatic amphibians take in respiratory air through the mouth.

16. Pneumatic sacs arise as median evaginations from the embryonic floor of the foregut. They become swim bladders in fishes and lungs in tetrapods.

17. Physostomous fishes are those in which the embryonic ducts of swim bladders remain open throughout life. The ducts eventually close in physoclistous fishes, chiefly teleosts.

18. Swim bladders of fishes are primarily hydrostatic organs. They are used for respiration in dipnoans and a few physostomous ray-fins. In some teleosts, they participate in sound detection or sound production.

19. The glottis is protected against the entrance of fluids and solids by fleshy valves at one site or another, and, in mammals, by its occasional location in the nasopharynx, or by the epiglottis.

20. The larynx is supported by a single pair of lateral cartilages in urodeles and by arytenoid and cricoid cartilages in anurans, reptiles, and birds. A thyroid cartilage is added in mammals. Other small cartilages develop in mammals.

21. Vocal folds (cords) are chiefly mammalian but are found also in anurans and some lizards.

22. Tracheal walls are reinforced by cartilaginous or bony plates, rings, or half rings. The trachea bifurcates into two primary bronchi. At the base of the trachea in most birds is an avian voice box, the syrinx.

23. Paired lungs arise as an unpaired evagination from the embryonic pharyngeal floor. They occupy the pleuroperitoneal cavity in amphibians and most reptiles. Each lung occupies its own pleural cavity in crocodilians, birds, and mammals.

24. Fibrous oblique septa separate the pleural cavities from the rest of the coelom in crocodilians and birds. In mammals, a muscular diaphragm does so, and a mediastinum separates one pleural cavity from the other.

25. Lungs are saccular with shallow pocketed linings in amphibians and most snakes, more compartmentalized by septa in other reptiles, and highly spongy because of millions of blind air pockets (alveoli) in mammals. One lung in limbless tetrapods is often rudimentary.

26. Saccular diverticula (air sacs) of the lungs in some lizards and in birds extend among the viscera. In birds they also extend into hollow bones.

27. In birds, inhaled air flows nonstop through the lungs and into air sacs. It then returns to the lungs via recurrent bronchi, flows uninterruptedly through open-ended air capillaries where gaseous exchange occurs, and is then vented.

28. In addition to maintaining a steady flow of air through the air capillaries of birds, air sacs assist in thermoregulation and lighten the skeleton in flying birds.

29. Inhalation of air in most amniotes is by action of a suction pump, chiefly the thoracic wall (ribs and costal muscles). In birds in flight, certain flight muscles constitute a suction pump. In mammals, the thoracic diaphragm participates. In anurans, the oropharyngeal floor serves alternately as a suction pump and a pressure pump.

30. Unlabored expiration in amniotes results primarily from the elasticity of lungs and resilience of the thoracic wall and viscera. In birds in flight, some flight muscles act as a pressure pump.

SELECTED READINGS

Alexander, R. M.: The chordates, Cambridge, England, 1975, Cambridge University Press.

Brainerd, E. L.: The evolution of lung-gill bimodal breathing and the homology of vertebrate respiratory pumps, American Zoologist **34:**289, 1994.

Brainerd, E. L., Liem, K. F., and Samper, C. T.: Air ventilation by recoil aspiration in polypterid fishes, Science **246:**1593, 1989.

Brodal, A., and Fänge, R., editors: The biology of *Myxine,* Oslo, 1963, Universitetsforlaget (Norway).

Feder, M. E., and Burggren, W. W.: Skin breathing in vertebrates, Scientific American **235:**5: 126, November 1985.

Gans, C., DeJongh, H. J., and Farber, J.: Bullfrog *(Rana catesbeiana)* ventilation, How does the frog breathe? Science **163:**1223, 1969.

Graham, J. B.: An evolutionary perspective for bimodal respiration: a biological synthesis of fish air breathing, American Zoologist **34:**229, 1994.

Hughes, G. M., and Morgan, M.: The structure of fish gills in relation to their respiratory function, Biological Reviews **48:**419, 1973.

Laitman, J. T.: The anatomy of human speech, Natural History, August 1984, p. 20.

Laitman, J. T., and Reidenberg, J. S.: Specializations of the human upper respiratory and upper digestive systems as seen through comparative and developmental anatomy, Dysphagia **8:**318, 1993.

Liem, K. F.: Form and function of lungs: the evolution of air breathing mechanisms, American Zoologist **22:**739, 1982.

Randall, D. J., and others: The evolution of air breathing in vertebrates, New York, 1981, Cambridge University Press.

Wood, S. C., and Lenfant, C. J. M.: Respiration: mechanics, control, and gas exchange. In Gans, C., editor: Biology of the reptilia, vol. 5, New York, 1976, Academic Press.

CHAPTER FOURTEEN

CIRCULATORY SYSTEM

In this chapter we'll look at the heart, blood vessels, and lymph channels of vertebrates from fishes to human beings. We'll find that in embryonic development, structure, and function they conform to a basic pattern, and we'll see how the pattern with respect to the blood circulatory system was modified step-by-step during the evolution of modern species. Then we will examine the abrupt changes that occur when a chick or a mammalian fetus shifts from life in amniotic fluid to breathing air. Along the way we'll look at a few more examples of countercurrents, some marvelous vascular retes, and the pacemaker of the vertebrate heart.

OUTLINE

The circulatory system of vertebrates consists of the heart, arteries, veins or venous sinuses, capillaries or sinusoids, and blood **(blood vascular system)** and of lymph channels and lymph **(lymphatic system).** The blood carries oxygen from respiratory organs; nutrients from extraembryonic membranes, digestive tract, and storage sites; hormones and other substances associated with homeostasis and immunity to disease; and waste products of metabolism to the excretory organs. Blood also conducts heat to and from the skin and other surfaces where heat is exchanged, thereby regulating and equalizing internal temperatures. Lymph channels collect interstitial tissue fluids not taken up by the bloodstream and they collect lipids from the villi of the small intestine. Lymph vessels terminate in venous channels.

Arteries carry blood away from the heart. They have muscular and elastic walls (figs. 14.1a and 14.2) capable of distention with each intrusion of blood. (Feel your pulse!) The smallest arteries, 0.3 mm or less in diameter, are **arterioles.** They dilate and constrict reflexly and thereby assist in regulating blood pressure. They terminate in blood capillaries. **Veins** commence in capillaries (other than the respiratory capillaries of the gills) and carry blood toward the heart. They have proportionately less muscle and elastic tissue and more fibrous tissue than arteries and are therefore capable of less distention or constriction. The smallest are **venules.** They begin in capillaries. Arteries and veins, like most other organs, have a loose connective tissue covering, the **adventitia.**

FIGURE 14.1

Structure of a medium-sized artery and its accompanying vein. Note differences in thickness of the muscle layers (*red*) and the absence of elastic fibers in the vein.

FIGURE 14.2

Microphotograph of a medium-sized artery.

Courtesy Gerrit L. Bevelander and Judith A. Ramaley.

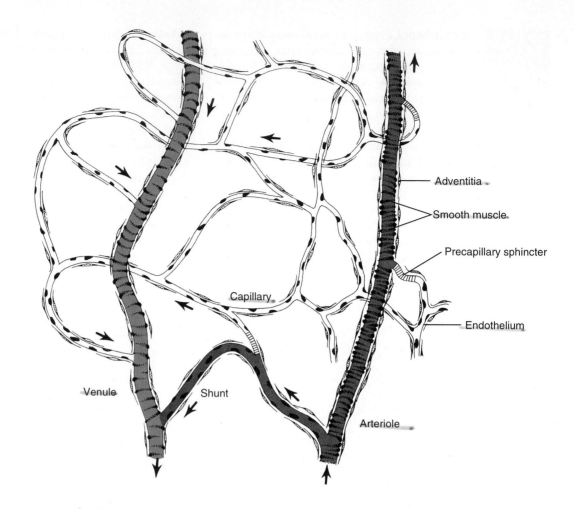

FIGURE 14.3

A capillary bed and its connections. Three crosshatched segments are sites of precapillary sphincters. Arrows indicate direction of flow.

Labels on figure: Adventitia, Smooth muscle, Precapillary sphincter, Capillary, Endothelium, Venule, Shunt, Arteriole

Capillaries (fig. 14.3) generally consist of endothelium alone, although certain capillaries are accompanied by a delicate investment of mesenchyme and a scattering of smooth muscle fibers. Their lumen is just large enough to accommodate red blood cells in single file. In fact, the cells often must "squeeze through," becoming temporarily deformed until they "pop out" into the smallest venule. A capillary "bed" or network represents all the individual capillaries served by a single arteriole. At the site where a capillary emerges from an arteriole, a short section of the capillary wall, the **precapillary sphincter,** has smooth muscle fibers that, given an adequate neural or hormonal stimulus, close off the entrance to the capillary bed at that site. Relaxation of the constrictor muscle readmits blood to the capillary. Blanching or blushing of human skin is a result of the constrictor and dilatory effects of these sphincters. The smallest arterioles and venules are connected directly by short vascular **capillary shunts** that assure uninterrupted circulation between the arterial and venous sides of the capillary beds when other capillaries are constricted.

A **portal system** is a system of veins terminating in a capillary bed (fig. 14.4). In most vertebrates, blood from the capillaries of the tail passes via a **renal portal system** to capillaries of the kidneys before continuing to the heart. Blood from the digestive tract, pancreas, and spleen passes via a **hepatic portal system** to the capillaries of the liver before continuing to the heart. Blood containing pituitary regulating hormones from the hypothalamus passes via a **hypophyseal portal system** to the adenohypophysis before continuing to the heart.

BLOOD

Blood is one of the tissues comprising the circulatory system. It consists of **plasma** and **formed elements** (fig. 14.5a). Plasma, which constitutes 55 percent of human blood, is a viscous fluid that is 90 percent water and 10 percent dissolved solids. The dissolved solids are blood proteins (serum globulin, serum albumin, and fibrinogen), substances required by living cells throughout the body (glucose, fat and fatlike substances, amino acids, ions of necessary salts), essential products synthesized by living cells (enzymes, antibodies, hormones), and some of the waste products of cellular metabolism. **Serum** is essentially plasma from

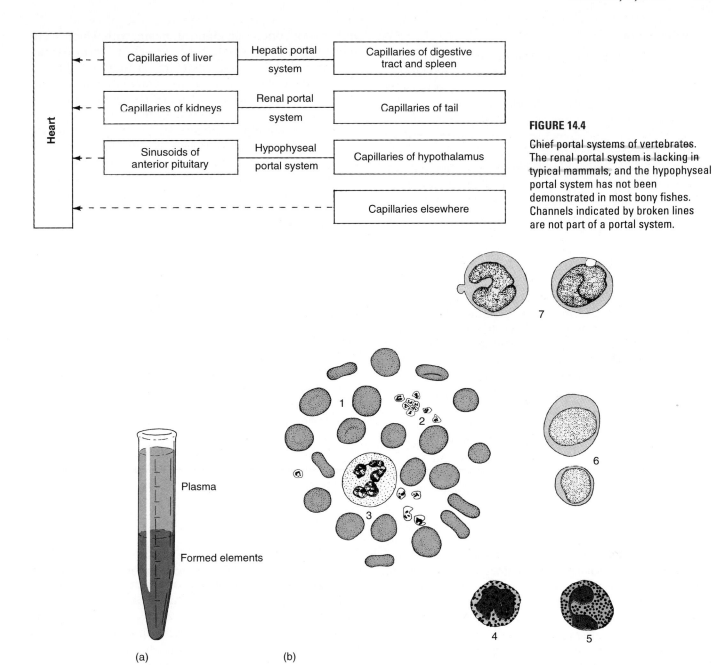

FIGURE 14.4

Chief portal systems of vertebrates. The renal portal system is lacking in typical mammals, and the hypophyseal portal system has not been demonstrated in most bony fishes. Channels indicated by broken lines are not part of a portal system.

FIGURE 14.5

Cellular constituents of blood. (*a*) Centrifuged whole blood. Plasma is a light yellow liquid; the formed elements constitute a dark sludge. (*b*) Red and white blood cells and platelets. **1,** erythrocytes; **2,** thrombocytes; **3** to **5,** neutrophilic, basophilic, and eosinophilic granulocytes; **6,** lymphocytes; **7,** monocytes.

which fibrinogen (clotting substance) has been removed.

Suspended in the plasma, and carried along in it as it flows, are **formed elements** consisting of red blood corpuscles **(erythrocytes),** white blood corpuscles **(leukocytes),** and platelets **(thrombocytes).** The formed elements can be separated from the plasma by centrifugation. With light microscopy and appropriate staining the following cell types can be identified (fig. 14.5):

Erythrocytes
Leukocytes
    Granulocytes (polymorphonuclear leukocytes)
      Eosinophils
      Basophils
      Neutrophils
    Agranular leukocytes
      Lymphocytes
      Monocytes
Thrombocytes (platelets)

Hemopoiesis

Hemopoiesis is the formation of blood cells. In nearly every vertebrate embryo from fishes to humans, the earliest sign of hemopoiesis is seen in the walls of the yolk sac, where large numbers of **blood islands** are producing the first blood cells and vessels (fig. 5.8). Beginning as concentrations of mesenchyme, blood islands produce **hemocytoblasts.** These are the stem cells from which all later blood cells arise. Blood islands also elaborate an extensive network of endothelial-lined spaces that form around the hemocytoblasts, confining them to vascular channels. Elsewhere in the embryo other blood channels and a heart begin to form, and these establish connections with the channels in the blood islands. This combined activity quickly provides an early circulatory system adequate to meet the immediate needs of the embryo, which initially are the nutrients stored in the yolk sac and oxygen from whatever extraembryonic respiratory membrane is functioning at the time. Even if the early embryo is not being nourished by yolk (and this applies to only a small minority of vertebrates), the yolk sac is still the initial source of blood cells.

With continuing development of the embryo, descendants of hemocytoblasts form in tissues within the body itself, eventually including the liver, kidneys, spleen, and bone marrow (but not confined to these) and varying with the class or order. Bone marrow is found in the spongy core of flat bones and in the spongy proximal epiphyses of long bones. Adult red blood cells of all vertebrates are produced in varying proportions in the liver, kidneys, and spleen; and in amniotes, the red bone marrow is an important source. Granular leukocytes are produced in the spleen of most vertebrates, in the intestinal submucosa of teleosts, in the bone marrow of amniotes, and elsewhere in lesser numbers. Lymphocytes are produced in the lymphatic tissues discussed later in this chapter.

Deteriorating formed elements in the blood of vertebrates in general are removed from circulation largely in the sinusoids of the spleen, liver, and bone marrow, which are lined by agranular **macrophages** that phagocytose (engulf and digest) blood cells in these locations. Blood cells and other elements are also phagocytosed by amoeboid white cells in any traumatized tissue of the body. The breakdown of hemoglobin into bile pigments was discussed in chapter 12.

Formed Elements

The red corpuscles of most vertebrates are oval nucleated cells; those of mammals are biconcave circular discs just large enough to squeeze through capillaries in a distorted shape, and without nuclei, having lost them before leaving the erythropoietic tissue where they developed. The circulating red cells of camels are an exception, resembling the oval nucleated cells of lower vertebrates. The functional constituent of red blood cells is hemoglobin, a compound protein composed of a simple protein, globin, and an iron-containing pigment, heme, with which available oxygen immediately binds loosely to form **oxyhemoglobin** (HbO_2). Iron, therefore, is indispensable for life. The oxygen is freed from the hemoglobin in any capillary where the surrounding tissue has a partial pressure of oxygen lower than that in the capillary. The result is **reduced hemoglobin.** The average life of an erythrocyte in a human being is three to four months. Since there are about 25 trillion erythrocytes in the human body, it is obvious that the erythropoietic centers are very active.

The white corpuscles are far less numerous than either red cells or platelets. Granulocytes and monocytes are amoeboid, and can pass through capillary walls between adjacent epithelial cells and enter tissue spaces where, as phagocytes, they serve as scavengers of broken-down tissues. They have additional functions. Lymphocytes are abundant in lymph nodes, in the spleen, and in other lymphoid tissues.

Platelets participate with fibrinogen in the clotting of blood. They are tiny fragments of stem cells (**megakaryocytes**) found in bone marrow. They consist of membrane-enclosed cytoplasm and lack a nucleus.

Most of the preceding discussion refers to mammalian blood. Whereas blood comprises 5 percent to 10 percent of the body weight of amniotes, it represents only about 1.5 percent to 3 percent of the body weight of fishes. The ratios of the constituents of blood differ among the vertebrate orders as adaptations to the environmental conditions with which each organism must cope.

THE HEART AND ITS EVOLUTION

The vertebrate heart is chiefly a muscular pump that occupies the pericardial cavity. The walls consist of an **endocardium, myocardium,** and **epicardium,** which correspond, respectively, to the intima, media, and adventitia of the arteries. However, as described in chapter 11, the myocardium is cardiac muscle, rather than smooth muscle as in arteries. It is especially thick in the ventricular wall. Lying on the epicardium is the **visceral pericardium,** the equivalent of the visceral peritoneum of the main coelom. The parietal and visceral pericardia are continuous with one another, being reflected over the blood vessels that enter and leave the heart. The space between these pericardia is the **pericardial cavity.**

The tissues of the heart are supplied with arterial blood by **coronary arteries** and drained by **coronary veins.** The heart muscle pulsates in response to specific electrolytes in the blood that perfuses it. The rhythmicity of the beat, which will be discussed shortly, is imposed by the autonomic nervous system.

Single- and Double-Circuit Hearts

In fishes, blood passes from the heart to the gills and from there directly to all parts of the body, after which it returns to the heart (fig. 14.6, fishes). Thus blood makes

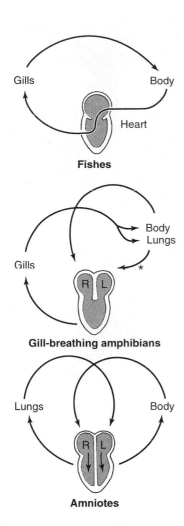

FIGURE 14.6

Usual circulatory routes in gill-breathing fishes, perennibranchiate amphibians, and amniotes. *Red* is oxygenated blood. When the gill-breathing amphibian is forced to breathe air, unoxygenated blood bypasses the gills (not shown) and is oxygenated in the lungs, making the oxygen content of the pulmonary veins (*) rise.

a *single circuit* during which it is pumped, oxygenated, distributed, and returned to the pump. No blood escapes oxygenation, and none fails to enter a capillary bed where oxygen can be released for use by the tissues.

In vertebrate classes in which all members have totally abandoned gills, namely, amniotes, a **pulmonary circuit** ordinarily carries oxygen-poor blood from the heart to the lungs and brings back oxygen-rich blood. A **systemic circuit** then carries the oxygenated blood to the other organs of the body and returns oxygen-depleted blood to the heart (fig. 14.6, amniotes). The evolution of *double-circuit* hearts was preceded by a long period of time during which natural selection produced circulatory systems adaptable for either aquatic or aerial respiration. Further mutations culminated in the double-circuit heart. It freed amniotes from the necessity of an aquatic existence but held them in bondage to atmospheric oxygen. We will have a glimpse of these transitional stages in the paragraphs that follow.

Hearts of Gill-Breathing Fishes

The hearts of fishes other than dipnoans have four chambers in a series: a **sinus venosus, atrium, ventricle,** and **conus arteriosus (truncus arteriosus).** Blood flows through these chambers in that sequence. Such a heart is seen in sharks and hagfishes (figs. 14.7 and 14.8). Designation of fish hearts as "two-chambered" recognizes only atria and ventricles as chambers. The terminology is not used in this text because it is not definitive.

The heart of sharks is typical of jawed fishes in general (fig. 14.7). The sinus venosus has thin walls, little muscle, and much fibrous tissue. Its caudal wall is anchored to the anterior face of the septum transversum. The sinus has some contractility, but is chiefly a collecting chamber for venous blood that is returning from all parts of the body. It is filled by suction each time the ventricle contracts and

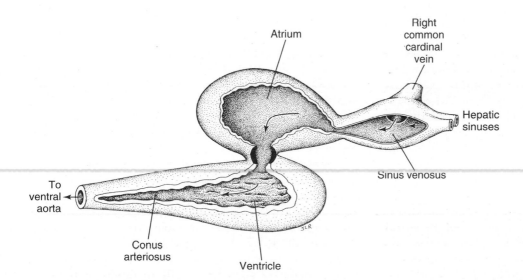

FIGURE 14.7

Heart of *Squalus acanthias* opened to show chambers and blood flow. Left lateral view.

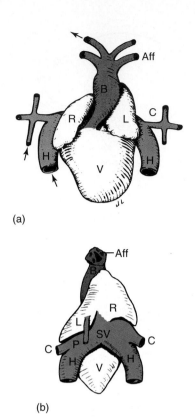

(a)

(b)

FIGURE 14.8

Heart and associated vessels of the cyclostome. *Myxine glutinosa,* ventral view. Arrows indicate direction of blood flow.

- Afferent branchial artery
- Ventral aorta
- Conus arteriosus
- Atrium
- Ventricle
- Sinus venosus

FIGURE 14.9

Heart and associated vessels of *Necturus.* (*a*) Ventral view. (*b*) Dorsal view. **Aff,** common channel leading to second and third afferent branchial arteries; **B,** bulbus arteriosus of ventral aorta; **C,** common cardinal vein; **H,** hepatic sinus; **L,** left atrium receiving, in (*b*) the pulmonary vein; **P,** pulmonary vein; **R,** right atrium; **SV,** sinus venosus; **V,** ventricle. In (*a*), a short conus arteriosus connects the ventricle with the bulbus. Arrows show direction of blood flow. Colors represent arteries (*red*) and veins (*blue*) but not necessarily oxygen content.

relaxes. Blood from the sinus gushes through the sino-atrial aperture into the atrium between two one-way valves as soon as the atrium begins to relax after emptying.

The atrium is a large thin-walled muscular sac that is a sort of staging area for blood that is about to enter the ventricle to be propelled toward the gills. From the atrium, blood pours into the relaxing ventricle through an atrioventricular aperture that is guarded by a pair of one-way valves. These prevent ventricular blood from being pumped back into the atrium when the ventricle contracts.

The ventricle has very thick muscular walls and is the actual pumping portion of the heart. The anterior end is prolonged as a muscular tube of small diameter, the **conus arteriosus,** which extends to the extreme cephalic end of the pericardial cavity, at which point it is continuous with the ventral aorta. The conus is composed chiefly of cardiac muscle and elastic connective tissue. A series of semilunar valves facing forward within the conus prevents backflow of blood into the ventricle. Because of its elasticity it balloons with each delivery of ventricular blood, and then slowly constricts, thereby maintaining a steady arterial pressure in the ventral aorta so that the flow of blood through the gill capillaries, like the flow of countercurrent respiratory water over the gill filaments, is steady despite the rhythmicity of the ventricular beat.

The conus arteriosus of teleosts is shorter than that of sharks, but a muscular swelling at the base of the teleost ventral aorta, the **bulbus arteriosus,** performs the same role. It maintains a steady flow of blood over the gills. A bulbus arteriosus is present also in a *Necturus* and some other perennibranchiate amphibians (fig. 14.9*a*).

Hearts of Dipnoans and Amphibians

Modifications in the heart and associated arteries in dipnoans and tetrapods are correlated with aerial respiration. The modifications enable oxygenated blood returning from the swim bladders or lungs to be separated in the heart from deoxygenated blood returning from other organs.

One modification in dipnoans and amphibians was the establishment of a partial or complete **interatrial septum** so that there are partial or complete **right and left atrial chambers** (figs. 14.9*a*, and 14.10, Dipnoi, urodele, anuran). The septum is complete in anurans and in some urodeles. Except in *Neoceratodus* (Dipnoi), the veins from the swim bladder or lungs empty directly into the *left* atrium; therefore the blood in this chamber is oxygen rich. The sinus venosus empties into the *right* atrium (fig. 14.9*b*); hence the blood in this chamber is low in oxygen. In lungless urodeles, the atrium remains totally undivided.

A second modification is the formation of a partial **interventricular septum** (chiefly in dipnoans but also in *Siren,* a urodele) or of **ventricular trabeculae** (in

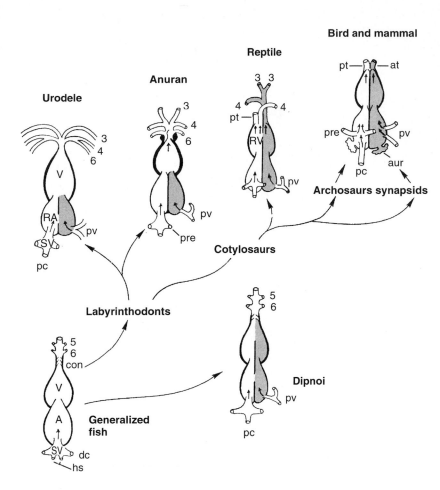

FIGURE 14.10

Heart chambers and oxygenated blood flow (*red*) in some generalized vertebrates. Distribution of oxygenated blood beyond the ventricle in dipnoan, urodele, and reptile depends on the species and the physiological needs of the organism. The parts of the heart shown are **A**, atrium; **RA**, right atrium; **V**, ventricle; **RV**, right ventricle; **SV**, sinus venosus; **con**, conus arteriosus; **aur**, auricle of mammalian heart. **3** to **6**, third to sixth aortic arches. Other vessels are **at**, aortic trunk; **dc**, common cardinal vein; **hs**, hepatic sinus; **pc**, postcava; **pre**, precava (common cardinal vein); **pt**, pulmonary trunk; **pv**, pulmonary veins.

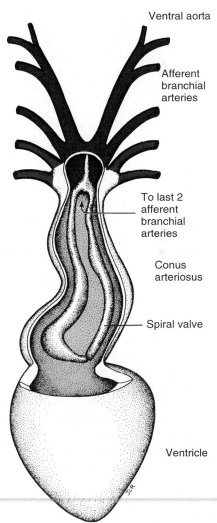

amphibians). Trabeculae are shelves or ridges projecting from the ventricular wall into the chamber and running mostly cephalocaudad. Interventricular septa and ventricular trabeculae perform identical functions: they maintain separation of oxygenated and unoxygenated blood that began in the left and right atria.

A third modification is formation of a **spiral valve** in the conus arteriosus in dipnoans and anurans. The valve in dipnoans (fig. 14.11) consists of a pair of longitudinal typhlosolelike folds of the lining of the conus. In anurans it is a single flap (fig. 14.12). The valves direct oxygen-poor blood into aortic arches that lead to gills or lungs and they channelize oxygenated blood into arches that supply other organs (figs. 14.11 and 14.18*b*, blue; and fig. 14.12).

A fourth modification shortened the ventral aorta so that it becomes practically nonexistent as embryonic development progresses (figs. 14.11 and 14.12). As a result, blood moves from the conus arteriosus directly into appropriate vessels. Urodeles, however, retain a prominent ventral aorta (fig. 14.9).

Many of the adaptations enabling aerial respiration in dipnoans and postlarval amphibians are seen also in amniotes.

FIGURE 14.11

Conus arteriosus and afferent branchial arteries of the lungfish *Protopterus*. The spiral valve distributes oxygen-rich blood (*red*) to the first three afferent branchial arteries and oxygen-poor blood (*blue*) to the last two, which supply the respiratory swim bladder and internal gills, as shown in fig. 14.18*b*.

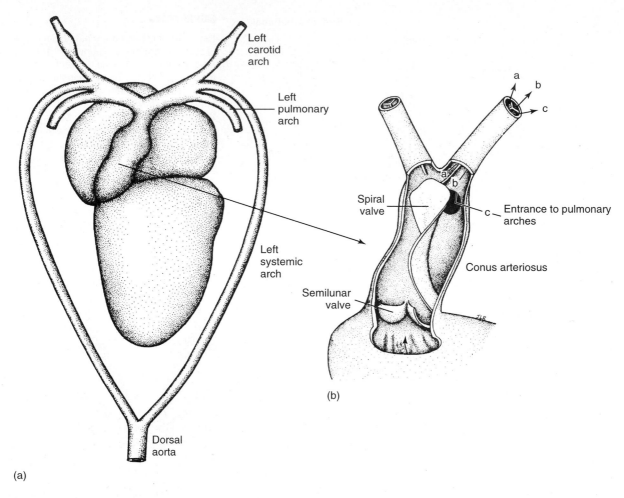

Left carotid arch

Left pulmonary arch

Left systemic arch

Dorsal aorta

(a)

a

b

c

a

b

Spiral valve

c

Entrance to pulmonary arches

Conus arteriosus

Semilunar valve

(b)

FIGURE 14.12

Heart and aortic arches of frog, ventral view. (*b*) Conus arteriosus opened to show passageways to left carotid arch (bristle **a-a**), and left systemic arch (bristle **b-b**). **c,** bristle enters common passageway to left and right pulmonary arches, then turns to enter passageway to left pulmonary arch.

Hearts of Amniotes

Amniote hearts have two atria, two ventricles (a unique third ventricular chamber in turtles and squamates), and, except in adult birds and mammals, a sinus venosus (figs. 14.13*b* and 14.24). The sinus venosus in crocodilians, unlike in other reptiles, is partially incorporated into the wall of the right atrium. Birds and mammals have a sinus venosus during early development, but it fails to keep pace with the growth of the right atrium into which it empties, and is finally indistinguishable as a separate chamber. Thereafter, the vessels that emptied into the sinus venosus empty directly into the right atrium. The embryonic location of the sinus venosus is marked in adults by a **sinoatrial node** of neuromuscular tissue. The node plays a key role in innervation of the heart, to be discussed shortly.

The right and left atria of amniotes are completely separated by an interatrial septum. Nevertheless, they are confluent during embryonic development via an **interatrial foramen**, or **foramen ovale**, which closes

near the time of hatching or at birth. The site of the obliterated foramen is marked in adult mammalian hearts by a depression, the **fossa ovale**, in the medial wall of the right atrium.

The *right* atrium of reptiles (fig. 14.24) receives blood from the sinus venosus; that of birds and mammals receives, directly, blood that earlier in phylogeny and during embryonic development emptied first into the sinus venosus (fig. 14.26). The *left* atrium receives blood from the pulmonary veins. In mammals only, each atrium has an earlike flap, or **auricle,** within which is a blind chamber (fig. 14.39). The functional advantage of the mammalian auricle has yet to be demonstrated.

The two ventricles are completely separated in crocodilians, birds, and mammals (fig. 14.26). In turtles and squamates, however, a unique third chamber, the **cavum venosum,** has developed at the upper limit of the interventricular septum (fig. 14.24). It functions to shunt oxygen-rich and oxygen-poor blood into or away from specific arteries that leave the heart. Its role can best be

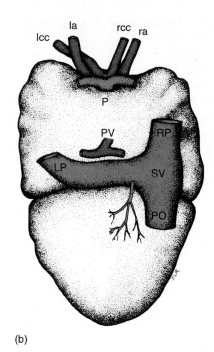

FIGURE 14.13

Heart and associated vessels of *Sphenodon.* (*a*) Ventral view. The ventricle, **V,** is divided internally into two chambers. (*b*) Dorsal view. **L,** left atrium; **R,** right atrium. *Red* vessels contain oxygenated blood. They are **la,** left aortic trunk; **lcc,** left common carotid artery; **PV,** pulmonary veins entering right atrium; **ra,** right aortic trunk; **rcc,** right common carotid artery. Blue vessels carry blood low in oxygen. They are **LP,** left precava; **P,** pulmonary trunk emerging from right ventricle; **PO,** postcava; **RP,** right precava; **SV,** sinus venosus.

understood after we have studied those arteries, which we will do shortly.

The muscular walls that line the ventricular chambers of mammals exhibit sturdy interanastomosing muscular ridges and columns, the **trabeculae carneae.** They strengthen the walls of these powerful pumps and increase the force exerted by them.

One-way valves guard the passageways from the atria into the ventricles of vertebrates. In amniotes, each valve consists of one or more fibrous flaps, or **cusps** (muscular on the right side of the heart in crocodilians and birds), connected, chiefly in mammals, by tendonous cords **(chordae tendineae)** to **papillary muscles** that project from the ventricular walls (fig. 14.26). During relaxation of the ventricles **(diastole)** blood from the atria flows freely past the cusps and into the ventricles. During ventricular contraction **(systole)** the cusps are forced forward or upward into the atrioventricular passageway, preventing reflux of blood into the atria. Each valve has one or two cusps in reptiles and birds. In most mammals, the left valve has two cusps **(bicuspid,** or **mitral valve),** and the right has three **(tricuspid valve). Semilunar valves** at the exits of the ventricles into the pulmonary and aortic trunks prevent backflow into the ventricles as the latter relax (fig. 14.26).

Innervation of the Heart

The embryonic heart begins to pulsate before any nerve fibers have reached it, and adult heart tissue, infused with an appropriate physiological solution, will continue to pulsate after its extrinsic innervation has been severed. Tissue from the heart of an embryonic chick, and its descendant cells, was kept alive and pulsating in a French laboratory for many years early in the twentieth century. The contrac-

tion of heart muscle is, therefore, autogenic, requiring no extrinsic neural stimulus for its initiation. Pulsation depends on the appropriate concentrations of certain electrolytes, especially Na, K, and Ca ions, in the tissue fluids that bathe cardiac muscle. The rate of autogenic pulsation of a denervated sinus venosus is imposed on the atria and ventricles via an intrinsic conduction system composed of atypical cardiac muscle fibers **(Purkinje fibers)** that constitute a syncytium with high conductile competence.

An extrinsic neural stimulus is necessary, however, to produce a regular beat that can be increased or slowed reflexly by the central nervous system in response to the physiological needs of the animal. Consequently, once nerve fibers have reached the sinus venosus during embryonic development, these fibers impose a rate of contraction on the cardiac muscle of that chamber. Thereafter, in fishes, amphibians, and reptiles, this beat is transmitted by the intrinsic conduction system to the atrium (or atria), then to the ventricle (or ventricles), and finally, to the conus arteriosus, if there is one. Thus, the sinus venosus is the *pacemaker* of the heart of these cold-blooded vertebrates. The sinus venosus of fishes is innervated by the vagus nerve (tenth cranial nerve) only. In anurans and amniotes, the heart is innervated also by fibers from the sympathetic trunk, the two sets of fibers having opposite effects on the heart rate (inhibitory fibers via the vagal nerve, accelerator fibers via cardiac nerves from the sympathetic trunk, fig. 16.35).

In birds and mammals, the embryonic sinus venosus eventually becomes incorporated into the wall of the right atrium, where it remains as a nodular mass of modified cardiac muscle and connective tissue known as the **sinoatrial (SA) node.** This node continues to receive the incoming fibers of the autonomic nervous system; and it is the SA node that serves as the pacemaker for the heart of birds and mammals. Impulses from the SA node are conducted via Purkinje fibers to the myocardia of both of the atria, causing them to contract. The stimuli also spread to an **atrioventricular (AV) node** embedded in the heart near the cusps of the right atrioventricular valve. The AV node is unique to birds and mammals. From the AV node an **atrioventricular bundle** of Purkinje fibers extends into the interventricular septum and splits into two parallel bundles that extend the length of the septum. From these, small bundles of Purkinje fibers ramify throughout the myocardium of both ventricles, providing stimuli that cause the ventricles to con-

tract fractions of a second after the atria have done so. The Purkinje fibers are also innervated directly by autonomic nerve fibers, as is the myocardium in general.

During the short interval between contraction of the atria and that of the ventricles, blood in the atria empties into the dilated ventricles. Contraction of the latter then propels ventricular blood into the outgoing arteries.

Morphogenesis of the Heart

When the heart is first recognizable in vertebrate embryos it is an almost straight, pulsating tube in fishes and amphibians (fig. 14.14*a*), or a pair of tubes in amniotes (fig. 14.14*d*). The tubes receive incoming

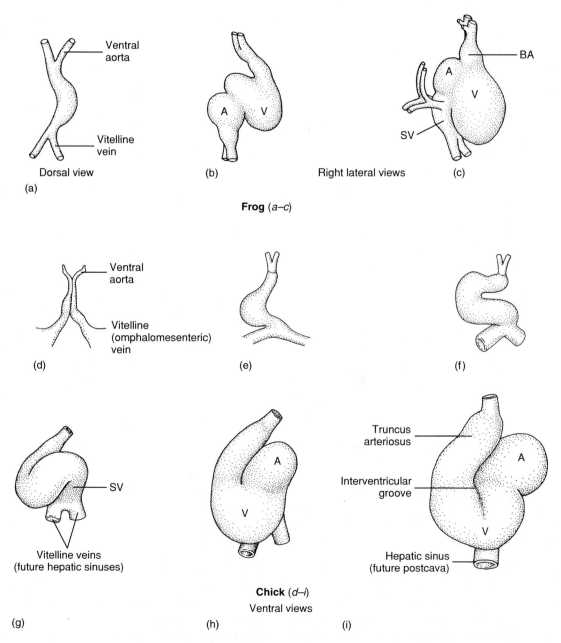

FIGURE 14.14

Development of an anuran heart (*a* to *c*, dorsal and lateral views) and an avian heart (*d* to *i*, ventral views). **A,** atrium; **BA,** bulbus arteriosus; **SV,** sinus venosus; **V,** ventricle. Oriented for optimal viewing of bending and twisting. For a view of the early chick heart in situ see fig. 16.6.

blood at their caudal ends and are continuous with the embryonic ventral aortas at their cephalic ends. The unpaired tube of sharks, lower bony fishes, and amphibians organizes from an unpaired aggregation of mesenchyme beneath the pharynx derived from lateral-plate mesoderm (fig. 14.15a). The paired tubes of amniotes organize from the same precursor tissues, differing only in that the tubes organize before the paired lateral plates have met in the midventral line (fig. 14.15, chick and rabbit). The paired tubes of amniotes eventually unite to form one (fig. 14.14e). Mesenchyme cells trapped within the tubes become hemocytoblasts (stem blood cells).

As development progresses, the tube bends to the animal's right and then twists, so that the atrial region, pre-viously at the caudal end, is carried dorsad and cephalad until it reaches its adult location. The location of the atrium in sharks is shown in figure 14.8. The twisting and bending is correlated with the confinement of the rapidly growing heart in a less expansive pericardial cavity.

In amphibians and amniotes, the twisting is carried further than in fishes, so that the atrial region finally lies cephalad to the ventricular region (fig. 14.14a to c, frog, and d to i, chick). Following this, the atrial chamber expands to form bilateral pouches, and a median dorsal fold within the atrial chamber grows ventrad, separating the pouches into left and right atria. As a final major step in amniotes, an interventricular septum commences the separation of the ventricular chamber into right and left

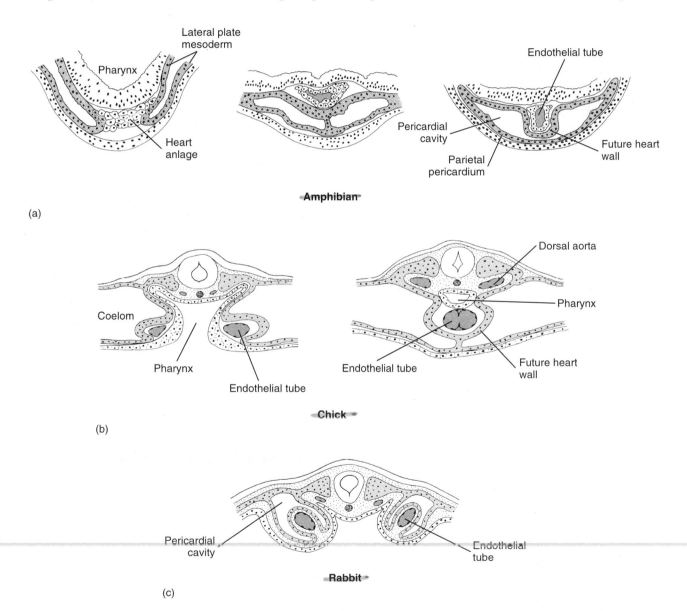

FIGURE 14.15

Origin of heart from lateral plate anlagen. In amphibians, (a), the heart forms from a median mass of mesenchyme that has a bilateral origin. In amniotes, (b) and (c), two endothelial tubes of splanchnic mesoderm are brought together to form the primordium of the heart.

sides. The site of the septum is marked externally by a longitudinal **interventricular groove** that creases the surface of the ventricle (fig. 14.14*i*).

As hatching approaches in birds, the sinus venosus becomes almost completely incorporated into the wall of the right atrium. This takes place much earlier in mammals. Because it is imperative that oxygen and nutrients be circulated as early in development as possible, the heart is the first organ to function. It does so even before any nerves have reached it to impose a rhythm.

The initial straight tube that will become the heart of sharks and amphibians organizes from *paired mesenchymal masses* of lateral-plate somatic and splanchnic mesenchyme that aggregate in the midline beneath the pharynx to form a single endothelial tube (fig. 14.15*a*). In amniotes, *a pair of already organized endothelial tubes* comes together beneath the pharynx, fuses, and forms a single tube (fig. 14.15*b* and *c*). In either case, the heart is a bilateral contribution of lateral-plate mesoderm.

ARTERIAL CHANNELS AND THEIR MODIFICATIONS

→ handout

Arterial channels supply most organs with oxygenated blood, although they carry deoxygenated blood to respiratory organs. In the primitive pattern (fig. 14.16), the major arterial channels consist of (1) a **ventral aorta** (paired early in embryogenesis) emerging from the heart and passing forward beneath the pharynx, (2) a **dorsal aorta,** paired above the pharynx only, and extending caudad in the roof of the coelom, and (3) six pairs of **aortic arches** connecting the ventral aorta with the dorsal aorta. Branches of these major channels supply all parts of the body. Modifications of this inherited basic pattern during embryonic development involve, most prominently, the aortic arches, which become modified for respiration by whatever means is employed during the life history of the species.

Aortic Arches of Fishes

Adaptive modifications of the six embryonic aortic arches for adult respiration by gills are illustrated in developing sharks (fig. 14.17). The ventral aorta in *Squalus* extends forward under the pharynx and connects with the developing aortic arches. The aortic arches in the mandibular arch are the first to develop. Shortly thereafter the other five pairs appear. Before the sixth pair is completed, the ventral segments of the first pair disappear, and the dorsal segments become the efferent pseudobranchial arteries. The second pair sprouts buds that become the first pretrematic arteries. Other buds sprout from the third, fourth, fifth, and sixth aortic arches and give rise to posttrematic arteries. The posttrematic arteries then sprout crosstrunks, which grow caudad in the holobranch and by further budding establish the last four pretrematic arteries. Aortic arches

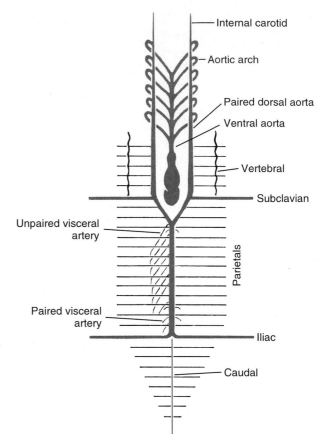

FIGURE 14.16
Basic pattern of the chief arterial channels of vertebrates.

II to VI soon become occluded at one site (broken lines in fig. 14.17*a*). The segments ventral to the occlusions become **afferent branchial arteries.** The dorsal segments become **efferent branchial arteries.** In the meantime, capillary beds are developing within the nine demibranchs. Afferent branchial arterioles connect the afferent branchial arteries with the capillaries. Efferent branchial arterioles return oxygenated blood from the capillaries to the pretrematic and posttrematic arteries. As a result of these modifications, blood entering an aortic arch from the ventral aorta must pass through gill capillaries before proceeding to the dorsal aorta.

The same developmental changes convert the embryonic aortic arches of bony fishes into afferent and efferent branchial arteries. The specific number converted determines the number of functional gills. In most teleosts, the first and second aortic arches tend to disappear (fig. 14.18*a*). In *Protopterus* (fig. 14.18*b*), the third and fourth embryonic aortic arches do not become interrupted by gill capillaries, although the fourth pharyngeal arches bear an external gill.

In dipnoans, a pulmonary artery sprouts off the left and right sixth aortic arch and vascularizes a swim bladder. This happens also in two ganoid fishes, *Amia* and *Polypterus.* (In most other actinopterygians, the swim

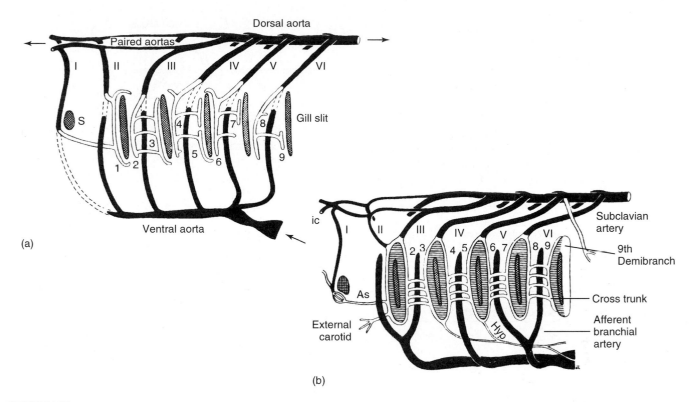

FIGURE 14.17 •

Developmental changes in embryonic aortic arches I to VI of *Squalus*, lateral view. In (*a*), buds (*white*) off aortic arches II–VI are establishing pretrematic and posttrematic arteries and cross trunks. Broken lines indicate sections of the aortic arches that become occluded, forcing blood from the ventral aorta to enter the demibranchs via afferent branchial arterioles (not shown). **1, 3, 5, 7, 9** pretrematic arteries; **2, 4, 6, 8,** posttrematic arteries. In (*b*) **I** has become the efferent pseudobranchial artery and **II** to **VI** have become efferent branchial arteries. **As,** afferent pseudobranchial artery; **Hyp,** hypobranchial; **ic,** internal carotid; **S,** spiracle.

bladders are supplied from the dorsal aorta.) This is precisely how tetrapod lungs are vascularized!

Aortic Arches of Tetrapods

Embryonic tetrapods, like fishes, construct six pairs of embryonic aortic arches (fig. 1.6). The first and second pairs are transitory, and regress fairly soon (figs. 14.19 and 14.20). After arches I and II disappear, arch III (the carotid arch) and the paired dorsal aortae anterior to arch III constitute the internal carotid arteries (figs. 14.18*c*, 14.19, and 14.20). With the exception of terrestrial urodeles and some limbless squamates, tetrapods also lose the fifth aortic arches during embryonic life (figs. 14.18*d–h* and 14.19*d–h*). Pulmonary arteries sprout off the sixth pair to vascularize the lung buds (figs. 1.6 and 14.20*d*). Other modifications of the aortic arches and associated vessels in tetrapods will be discussed in the sections that follow.

Amphibians

Most terrestrial urodeles retain four pairs of aortic arches (fig. 14.18*c*). Perennibranchiate urodeles generally retain three, since the fifth arches either disappear or unite with the fourth during embryogenesis (figs. 14.18*d* and 14.21). The larval afferent and efferent branchial arteri-

oles of perennibranchiates that carry blood from the aortic arches into the gills and back function throughout life, and a short section of the third, fourth, and sixth aortic arches becomes a **gill bypass** (fig. 14.18*d*). These bypasses are constricted while the animal is using its gills; however, when the dissolved oxygen in the pond becomes low enough to cause the animal to gulp air, the gills shrink, and the bypasses carry more blood. Similarly, when resorption of the gills of *Siren* is brought about by thyroid hormone injections, the animal gulps air, and the bypasses carry all the blood that enters the arches. On the ventral aorta of perennibranchiates, the bulbus arteriosus (fig. 14.10) maintains a steady nonpulsating arterial pressure in the gills.

Embryonic anurans, like terrestrial urodeles, have four aortic arches (III through VI) at the time they enter the larval state (6 mm stage of development). Aortic arches III, IV, and V supply the larval external gills during the five or six days these gills function; thereafter, they supply internal gills until metamorphosis. Arch VI sprouts a pulmonary artery that vascularizes the developing lung bud.

With loss of gills at metamorphosis, three changes affect the aortic arches and associated vessels of anurans (fig. 14.18*e*): (1) aortic arch V disappears, and (2) the

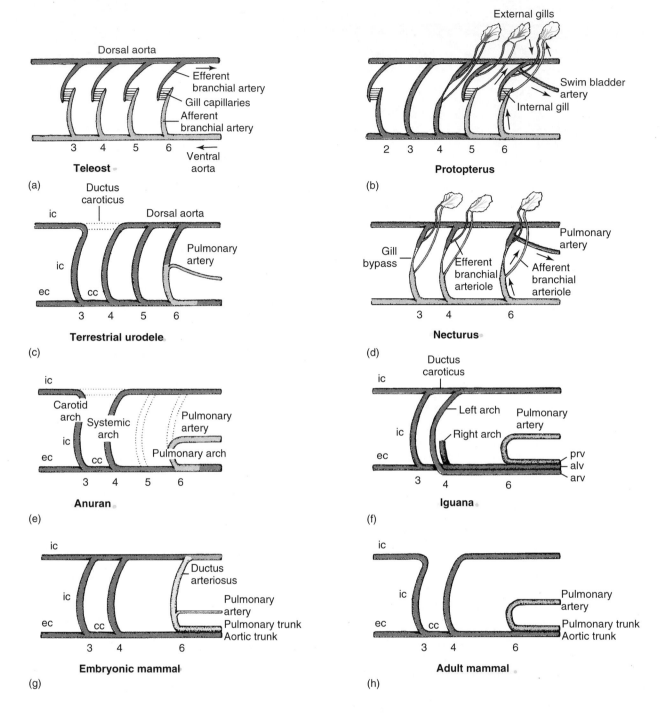

FIGURE 14.18

Persistent left aortic arches in representative vertebrates. **2** to **6,** second through sixth embryonic aortic arches. *Dotted lines* in (*c*) indicate vessel present in some species; in (*e*), vessels that are functional in larvae. *Arrows* indicate direction of blood flow. **alv,** aortic trunk from left ventricle; **arv,** aortic trunk from right ventricle; **cc,** common carotid artery, which is the paired segment of the embryonic ventral aorta; **ec,** external carotid; **ic,** internal carotid; **prv,** pulmonary trunk from right ventricle. In (*b*), (*c*), and (*e*), the ventral aorta carries venous blood (*blue*) during one phase of a single ventricular contraction, and arterial blood (*red*) during the next phase. In (*b*) the oxygen content of each vessel depends on the extent to which oxygen is being acquired via the gills as compared to the swim bladder. In (*c*) the sixth arch carries only oxygenated blood after the pulmonary artery is filled with unoxygenated blood. In (*d*), colors indicate conditions while animal is in well-oxygenated water. In (*g*) all blood in arches is mixed and colors designate predominant condition. Blood in the aortic trunk was oxygenated in the placenta.

FIGURE 14.19

Adult aortic arches (1 to 6) of representative vertebrates diagrammed from ventral aspect. The snake illustrated is a species with no left lung. **L**, aortic trunk to left systemic arch; **P**, pulmonary trunk; **R**, aortic trunk to right systemic arch. **da**, ductus arteriosus; **dc**, ductus caroticus; **ec**, external carotid artery; **ic**, internal carotid artery.

Source: Data from J. S. Kingsley, *Outline of Comparative Anatomy of Vertebrates*, 1920, The Blakiston Company, Philadelphia.

dorsal aorta between aortic arches III and IV (**ductus caroticus**) disappears. As a result of these two changes, blood entering aortic arch III (**carotid arch**) can pass only to the head. Eventually, (3) the segment of aortic arch VI dorsal to the pulmonary artery (**ductus arteriosus**) disappears. Blood entering arch VI (**pulmonary arch**) in anurans can now pass only to the lungs and skin. Aortic arch IV (**systemic arch**) on each side continues to the dorsal aorta to distribute blood to the rest of the body (figs. 14.12 and 14.19d).

As noted earlier, oxygenated blood from the left atrium and deoxygenated blood from the right are kept remarkably well separated as they pass through the ventricle in amphibians. In anurans, this is accomplished by ventricular trabeculae, by expulsion from the ventricle of right atrial blood first, and by action of the spiral valve in the conus arteriosus. At the start of ventricular systole, the valve is flipped into a position that closes off the entrance to the systemic and carotid arches, thereby diverting deoxygenated blood into the common aperture that leads to the two pulmonary arches (fig. 14.12b). Then, as back pressure builds up in the pulmonary arteries because of filling of the lung capillaries, the spiral valve flips into an alternate position that directs oxygenated blood into the systemic and carotid arches. (Late in ventricular systole a little of the left oxygenated blood may enter the pulmonary arches.) In a marine toad studied by injection of radioisotopes and using angiocardiography, the mixing of oxygenated and deoxygenated blood in the heart and aortic arches was found to be slight.

Adult apodans retain three complete aortic arches (III, IV, and V) and a ductus caroticus. However, the ductus arteriosus and ductus caroticus are of small diameter and carry little blood.

Reptiles

Generalized adult reptiles exhibit three adult aortic arches, III, IV, and the base of VI. The ductus arteriosus

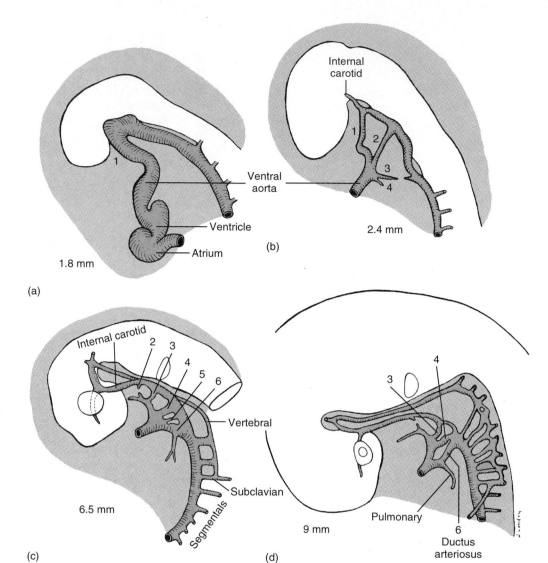

(a)

(b) 2.4 mm

(c) 6.5 mm

(d) 9 mm

FIGURE 14.20

Embryonic modifications of the left aortic arches (1 to 6) of a porcupine. In (c) and (d), **3** is the base of the internal carotid artery.

From P. H. Struthers, "The Aortic Arches and Their Derivatives in the Embryonic Porcupine" in *Journal of Morphology and Physiology,* 50:361. Copyright © 1930 John Wiley & Sons, Inc. Reprinted by permission of John Wiley & Sons, Inc.

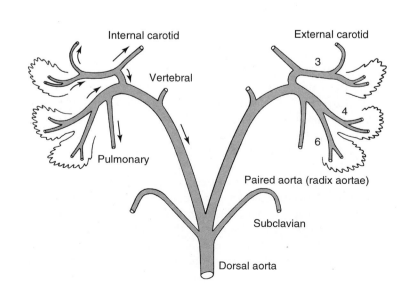

FIGURE 14.21

Efferent branchial arteries **(3, 4, 6)** of *Necturus* as seen from above the gills. Numerals designate dorsal segments of third, fourth, and sixth embryonic aortic arches. *Arrows* indicate direction of blood flow.

and the ductus caroticus are usually obliterated (fig. 14.19*e*). Exceptions are found among primitive lizards and some limbless squamates.

An innovation has been introduced in the ventral aorta of reptiles. Instead of developing a spiral valve to shunt fresh and deoxygenated blood into appropriate arches, reptiles underwent a series of mutations that had the effect of splitting the embryonic ventral aorta into three separate channels: **two aortic trunks (left and right systemic arches)** and a **pulmonary trunk** (figs. 14.18*f*, arv, alv, and prv; 14.19*e*; and 14.22, 2, 3, and 4). One result of these mutations can be demonstrated in crocodilians.

In crocodilians, (1) the pulmonary trunk emerges from the right ventricle and leads to the left and right pulmonary arches (fig. 14.19*e*, vessels in blue). Deoxygenated blood is therefore sent to the lungs. (2) One aortic trunk emerges from the left ventricle and carries oxygenated blood to the right systemic arch and the carotid arches (fig. 14.23). (3) A second aortic trunk emerges from the right ventricle and leads to the left systemic arch. One might expect that blood in the left systemic arch would be low in oxygen, since it comes from the right ventricle. However, this is not the case when the animal is breathing. An aperture, the **foramen of Panizza** (fig. 14.23*a*, *curved arrow*), connects the two aortic trunks at their base. During normal respiration, the valve at the exit from the right ventricle into the aortic trunk remains closed, and blood from the right ventricle can pass only to the lungs. Some oxygenated blood from the left ventricle is shunted through the foramen of Panizza to the left systemic arch (fig. 14.23*a*). This left-right shunt assures delivery of oxygenated blood to all parts of the body.

Crocodilians are aquatic animals that spend a proportion of their time totally submerged. In this state, they can't breathe and the oxygen content of their blood slowly falls. In response, the pulmonary arteries constrict

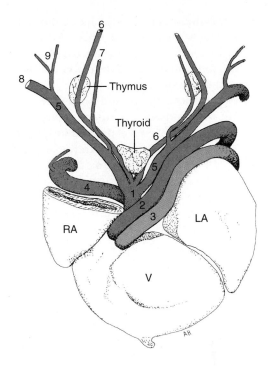

FIGURE 14.22

Turtle heart and emerging aortic arches, ventral view. **LA**, left atrium; **RA**, right atrium; **V**, ventricle. **1**, brachiocephalic artery; **2**, left fourth aortic arch; **3**, pulmonary trunk; **4**, right fourth aortic arch. **1** and **4** arise from a common trunk. **5**, subclavian artery; **6**, common carotid; **7**, ventral cervical; **8**, axillary; **9**, arteries to pectoral and shoulder muscles. *Red*, oxygenated blood; *blue*, deoxygenated blood.

reflexly, causing a backup of blood in the right ventricle. The valve between the right ventricle and its aortic trunk is thereby forced open (fig. 14.23*b*) and some right-ventricular blood is shunted via the foramen of Panizza to the aortic trunk that emerges from the left ventricle. This right-left shunt diverts considerable blood away from the lungs and into the systemic circulation.

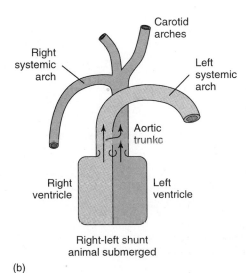

(a) Left-right shunt aerial respiration

(b) Right-left shunt animal submerged

FIGURE 14.23

Crocodilian foramen of Panizza, schematic. Curved arrows go through foramen. (*a*) Distribution of oxygenated blood during aerial respiration and, (*b*), when lungs are not ventilating.

Utilization of the shunt is facilitated by a mild reduction in blood pressure within the left ventricle. Eventually, the animal will have to come to the surface to breathe, whereupon the pulmonary circulation will revert to that of an air-breathing amniote.

As described earlier, turtles and squamates have, at the upper limit of the interventricular septum, a **cavum venosum** that receives deoxygenated blood from the right atrium and is confluent with both the left and right ventricles (fig. 14.24). How do these reptiles avoid the mixing of oxygenated and deoxygenated blood in the heart? A series of studies using cinefluoroscopy, simultaneous readings of arterial pressure in several chambers and vessels, assay of blood gases, and other cardiovascular techniques have provided the answer. During *ventricular diastole,* blood from the right atrium—venous blood—passes through the cavum venosum and enters the right ventricle, known also as the **cavum pulmonale** in these reptiles. From the cavum pulmonale, the next *ventricular systole* pumps it into the pulmonary trunk and thence to the lungs. Blood from the left atrium during *ventricular diastole* enters the left ventricle **(cavum arteriosum).** By the time the cavum venosum has discharged its unoxygenated blood into the cavum pulmonale, ventricular systole begins. The muscular activity of ventricular systole displaces the septal wall, blocking the passageways between the cavum venosum, the right atrium, and the cavum pulmonale, and opening the passage between the cavum venosum and the cavum arteriosum. Thereupon, blood from the cavum arteriosum is pumped into the cavum venosum and from there into the two aortic trunks that lead to the left and right systemic arches (fig. 14.24, 2 and 4). When this circulatory pattern is in effect, as it is when the lungs are ventilating, deoxygenated blood is sent to the lungs and oxygenated blood is sent to the rest of the body.

When physiological conditions initiate it, as when ventilation in aquatic species is halted by submersion in water, the pulmonary arteries become constricted and deoxygenated blood in the cavum venosum is shunted away from the cavum pulmonale and into the left aortic arch. A similar response has been elicited when radiant heat was applied to the reptilian body. It has been suggested that the latter response may be thermoregulatory in cold weather, enabling blood warmed by basking in the sun to be diverted from the lungs, where heat loss would occur during exhalation. When insufficient oxygen is available to these reptiles during submersion, an additional mutation provides energy for metabolic needs by glycolysis, an anaerobic process.

Primitive lizards retain the fifth aortic arch and the ductus caroticus on both sides. Limbless lizards and snakes in which the left lung is atrophied may lose the entire left sixth aortic arch, retaining only a single pulmonary artery; and in some snakes, the left third aortic arch also disappears. However, in the latter species, the ductus caroticus persists throughout life, and blood reaches the left internal carotid artery and brain by that route. Of the primitive six aortic arches, some adult snakes (fig. 14.19f) retain only the right third, the left and right fourth, the ductus caroticus, and the ventral segment of the right sixth aortic arch to the site where the right pulmonary artery arises.

Left-right and right-left shunts in reptiles, persistent carotid ducts, retention of the fifth aortic arch, and other features of the cardiovascular system that characterize one reptilian species or another enable its members to survive in the specific environment they inhabit.

Birds and Mammals
Birds and mammals became the first tetrapods in which there is no opportunity for the mixing of oxygenated

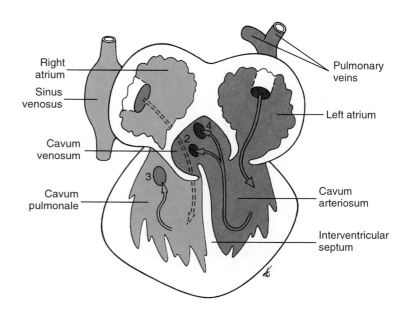

FIGURE 14.24

Turtle heart showing location of oxygenated and deoxygenated blood at start of ventricular systole when lungs are ventilating. Numerals designate vessels similarly numbered in fig. 14.22. **2,** opening to left fourth aortic (left systemic) arch; **3,** opening to pulmonary trunk; **4,** opening to right fourth aortic (right systemic) arch and brachiocephalic artery. During atrial systole, which immediately precedes ventricular systole, deoxygenated blood from the right atrium flows through the cavum venosum (broken lines) to reach the cavum pulmonale.

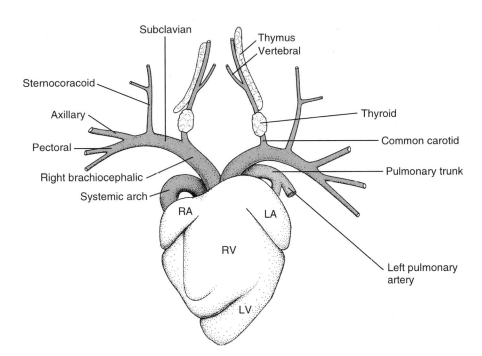

FIGURE 14.25

Heart and associated arches of a bird, ventral view. **LA, RA,** left and right atria; **LV, RV,** left and right ventricles. The systemic arch is the right fourth aortic arch. The paired brachiocephalic arteries are the embryonic ventral aortas distal to the base of the fourth arch. In adults they arise by a common stump from the aortic trunk that gives rise also to the systemic arch (fig. 14.19*g*).

and deoxygenated blood in the heart after birth or hatching (fig. 14.6, amniotes). This has come about as a result of division of the left and right ventricles by a complete interventricular septum and division of the embryonic ventral aorta into only two trunks, a pulmonary trunk from the right ventricle and a single aortic trunk from the left (fig. 14.19*g* and *h*). (The vessel carrying blood from the heart to the early aortic arches in amniotes is also referred to variously as **truncus arteriosum** and **bulbus cordis.**) The pulmonary trunk leads only to the sixth aortic arches, and the aortic trunk leads to the third and fourth. The left fourth aortic arch disappears in birds, and all but the base of the right fourth disappears in mammals. As a result, the systemic arch (arch of the aorta) of birds turns to the right (fig. 14.25), whereas in mammals it turns to the left (fig. 14.26). The part of the *right* fourth arch that remains in adult mammals becomes the proximal segment of the *right* subclavian artery.

In birds and mammals, therefore, six aortic arches develop in the embryo, and the first, second, fifth, and left fourth (in birds) or most of the right fourth (in mammals) disappear. The ductus caroticus disappears, and the ductus arteriosus closes before hatching in birds, or with the first gasp of air into the lungs of a mammal. While patent, the ductus arteriosus shunts blood away from the lungs and into the dorsal aorta (fig. 14.39, arrow). The latter supplies the embryonic respiratory membranes (allantois or placenta) with blood. Thus, a right-left shunt that diverts blood away from the lungs functions in unhatched chicks and fetal mammals just as it does in adult reptiles. The circulatory adjustments

accompanying the closing of the shunt in mammals will be discussed later.

The left fourth aortic arch of mammals is referred to simply as *the* arch of the aorta by anatomists. The left and right common carotid and external carotid arteries develop from the paired embryonic ventral aortas, and the internal carotids develop from the embryonic third aortic arches and paired dorsal aortas (fig. 14.19*h*). A few individual and species differences in the vessels arising from the arch of the aorta in adult mammals are illustrated in figure 14.27. (Compare figs. 14.26 and 14.27*a*.)

Aortic Arches and von Baer's Law

Whenever a respiratory surface develops on a pharyngeal derivative, buds off the nearest aortic arch usually vascularize that surface. When a pharyngeal arch develops an external or internal gill, the aortic arch in that pharyngeal arch vascularizes that gill. If the pharyngeal floor evaginates to form a lung bud, a sprout from the nearest aortic arch vascularizes the bud.

Vascularization of pharyngeal derivatives for respiration has phylogenetic roots that extend back to the first chordates. The external environment enters the pharynx, water devoid of foodstuffs passes between two pharyngeal arches, and oxygen is acquired by buds that sprout from the aortic arches. Development of six aortic arches in all vertebrate embryos and the subsequent modification or elimination of one vessel or another as an adaptation to new environmental niches is an expression of a nineteenth-century generalization known as *von Baer's law.* You may be interested in examining that concept. It is discussed briefly in chapter 2.

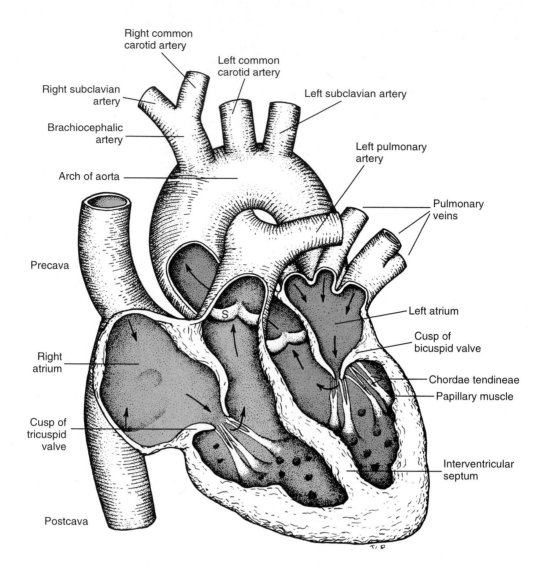

FIGURE 14.26

Human heart, auricles removed. **S,** semilunar valve at entrance to pulmonary trunk. A similar valve is seen at entrance to arch of the aorta. The third cusp of the tricuspid valve has been removed. The sinus venosus has been incorporated into the wall of the right atrium.

Dorsal Aorta

The dorsal aorta is paired in the head and above the pharynx in embryos and, to some extent, in adult fishes and gill-breathing amphibians. It remains paired in the head in all adults as the internal carotid arteries. It is unpaired in the trunk, where it gives off a series of *paired* somatic branches to the body wall and appendages, and a series of *paired and unpaired* visceral branches. It continues into the tail as the caudal artery.

Somatic Branches

The **subclavian arteries** are enlarged segmental arteries (figs. 14.16 and 14.20). They arise in embryos as branches of the paired or unpaired dorsal aortas or from the third aortic arches (some birds) or fourth aortic arches (some mammals) close to the aorta, but these embryonic relationships become obscured during subse-

quent development. In tetrapods, the subclavian artery traverses the axilla as the **axillary,** parallels the humerus as the **brachial,** and divides in the forearm into **ulnar** and **radial** arteries. In many amniotes, one branch on each side, the **vertebral** artery, passes cephalad in the neck to contribute blood to the **circle of Willis** (fig. 16.27). The vertebral artery is not well developed in birds and some reptiles.

A series of paired segmental arteries arises from the aorta along the length of the trunk. These give off short dorsally directed **vertebromuscular** branches to the epaxial muscle, skin, and vertebral column, and long **parietal** branches (appendix 2) that encircle the body wall to the midventral line. Where there are long ribs in amniotes the parietals are called **intercostal** arteries. **Lumbar** and **sacral** arteries are parietals in those regions. Segmentals in the neck may come off the vertebral arteries.

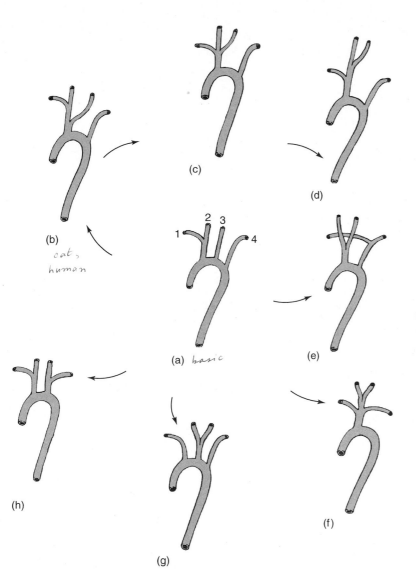

FIGURE 14.27

Individual and species differences in the relationships of the common carotid and subclavian arteries in selected mammals. (*a*) Basic pattern, seen also in human,* porcupine,* rabbit,* pig. (*b*) Cat,* dog,* pig,* human, and rabbit. (*c*) and (*d*) Rabbit, domestic cat, with (*d*) occurring with lower frequency. (*e*) Anomalous right subclavian artery in cat, human, and rat. (*f*) Many perissodactyls. (*g*) Walrus. (*h*) Human, porcupine, and rabbit. An asterisk indicates a predominant condition in the populations examined. If there is no asterisk, the condition is common but not predominant. **1,** right subclavian; **2,** right common carotid; **3,** left common carotid; **4,** left subclavian. The anomalous condition shown in (*e*) has been induced experimentally in rats by irradiation of the fetus. Any of the variants might occur in any mammalian species.

The **iliacs** are segmental arteries that supply the pelvic fins or limbs. In tetrapods, the iliac becomes the **femoral** in the thigh, **popliteal** in the knee, and **tibial** in the shank.

Branches of some of the anterior and posterior segmentals, especially of the subclavian and iliac, pass longitudinally in the body wall to anastomose (unite end to end). **Anastomoses** ensure that if one of the anastomosing vessels supplying a region becomes occluded, the vessel approaching from the opposite direction will fill the affected arterial tree beyond the occlusion. Anastomoses are common throughout the body.

Visceral Branches

A series of **unpaired visceral branches** (splanchnic vessels) pass via dorsal mesenteries to the unpaired viscera, chiefly digestive organs, suspended in the coelom. The number of such vessels is largest in generalized species such as *Necturus*. As few as three unpaired trunks— frequently **celiac, superior mesenteric,** and **inferior mesenteric**—may occur in higher vertebrates. Anastomoses between two successive visceral branches

occur along the entire length of the gut. Among anastomosing visceral branches in mammals are a superior pancreaticoduodenal branch of the celiac, which anastomoses with an inferior pancreaticoduodenal branch of the superior mesenteric; a middle colic branch of the superior mesenteric, which anastomoses with a left colic branch of the inferior mesenteric; and a superior rectal branch of the inferior mesenteric, which anastomoses with a middle rectal branch of the internal iliac. Anastomoses are also common on the greater and lesser curvatures of the stomach.

Paired visceral branches of the aorta include arteries to the urinary bladder, reproductive tract, gonads, kidneys, and adrenals. A *series* of gonadal and renal arteries occur in lower vertebrates, several pairs in reptiles and birds, and usually a single pair in mammals.

Allantoic Arteries of Amniotes

The early embryonic dorsal aorta of amniotes ends at the level of the *future* hind limbs by bifurcating into right and left **allantoic (umbilical) arteries** that carry blood

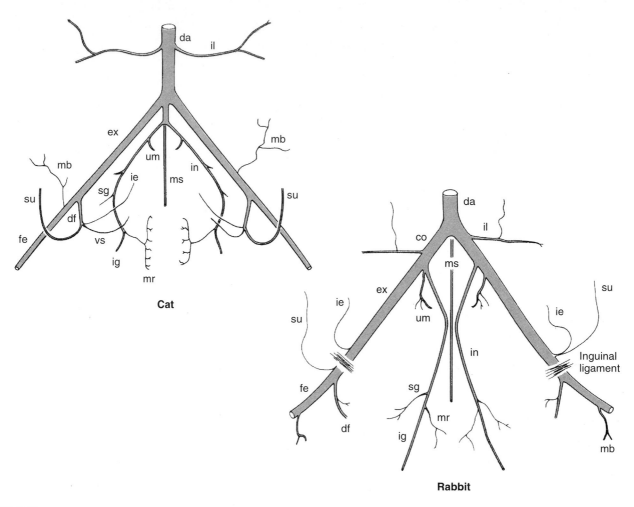

FIGURE 14.28

Contrasting termination of the dorsal aorta in two mammalian orders. There is no inguinal ligament in cats. The sources of paired vessels are frequently asymmetrical (compare the left and right superior epigastrics in the rabbit illustrated here). **co**, common iliac; **da**, dorsal aorta; **df**, deep femoral; **ex**, external iliac; **fe**, femoral; **ie**, inferior epigastric; **ig**, inferior gluteal; **il**, iliolumbar; **in**, internal iliac; **mb**, muscular branch; **mr**, middle rectal; **ms**, median sacral; **sg**, superior gluteal; **su**, superficial epigastric; **um**, umbilical; **vs**, vesicular (absent in rabbits).

to the allantois (fig. 14.38). **Internal iliacs** sprout off the umbilical arteries as limb buds develop, and the umbilicals become branches of the external or internal iliacs. Species-specific differential growth of the embryonic external iliac, internal iliac, and umbilical arteries results in species variations in the adult relations of these vessels (fig. 14.28).

Coronary Arteries

The walls of all arteries and veins except the smallest are supplied with **vasa vasorum** ("vessels of the vessels"). The heart is no exception, and the vessels here are **coronary arteries** and **veins.** In elasmobranchs, the coronary arteries arise from hypobranchial arteries that receive aerated blood from several arterial loops around the gill chambers (fig. 14.17*b*). In frogs, they arise from the carotid arch. In reptiles and birds, they arise from the aortic trunk leading to the right fourth arch, or from the brachiocephalic. In mammals, they arise from

the base of the ascending aorta just beyond the semilunar valves. In urodeles, the coronary supply consists of many small arteries.

Retia Mirabilia

Some arteries along their course develop segments that are highly tortuous, beyond which they straighten out again. These segments are **retia mirabilia** (singular, rete mirabile), or "wonderful networks." They are present in the head, neck, or trunk of one vertebrate group or another. The glomeruli in vertebrate kidneys (fig. 15.15) are retia, as are the arterial networks in the pseudobranchs of sharks. Retia serve different functions in different locations.

Cetaceans and pinnipeds have extensive retia on segmental arteries of the trunk in protected positions beneath the transverse processes of vertebrae and within the bony vertebral canal that houses the spinal cord. The retia are confluent and are drained by vertebral arteries that are en route to the brain. When the animal dives, the viscera are compressed by the weight of the surrounding water and blood is forced out of the viscera into the retia, which are protected from compression by their bony surroundings. The retia constitute a reservoir of red blood cells that were oxygenated just before a dive. The oxyhemoglobin supplies the brain with oxygen while the animal is submerged.

Often the tortuous artery is associated with an equally tortuous vein in such a manner that artery and vein lie side by side, with the blood flowing in opposite directions. In some birds that wade in icy waters, these countercurrents in retia above the thigh result in transfer of heat from warm arterial blood entering the leg to cold venous blood returning to the trunk. This conserves body heat, therefore energy, by preventing heat loss to the icy water while also warming the returning blood to body temperature. Polar bears and arctic seals have retia that perform the same function. On the other hand, retia do not occur in arteries en route to the limbs of penguins and some other arctic animals in which swimming generates excessive heat that must be dissipated. Mammals with testes in scrotal sacs have a rete, the **pampiniform plexus,** in each inguinal canal. Heat is transferred from spermatic artery to spermatic vein, assuring that the temperature within the scrotal sacs is lower than body temperature, a necessity for the viability of sperm in those species.

The **red glands** of the swim bladders of fishes are retia. Swim bladders are supplied by branches of the celiac artery and drained partly by the hepatic portal system. Venous blood leaving a rete carries away oxygen that has entered the red cells of the vein because of the high partial pressure of oxygen gas within the lumen of the bladder. This oxygen is reclaimed by arteries that are approaching the gas-secreting area of the rete. The retia therefore provide countercurrents that help to maintain an appropriate level of gaseous oxygen in the bladder.

VENOUS CHANNELS AND THEIR MODIFICATIONS

A generalized venous system consists of the following major streams: **cardinals** (anterior, posterior, and common cardinals), **renal portal, lateral abdominal, hepatic portal, hepatic sinuses,** and **coronary veins** (fig. 14.29a). The hepatic portal system is derived from the embryonic subintestinal vein and the distal portion of the embryonic vitelline veins, and the hepatic sinuses are derived from the vitellines between liver and heart. Two additional streams develop in lungfishes and tetrapods: a **pulmonary stream** from the lungs and a **postcava** from the kidneys. These eight channels and their tributaries drain the entire body—head, trunk, tail, and appendages. As development progresses they are

slowly modified by deletion of some vessels and addition of others. Modifications are few in lower vertebrates, more numerous in higher vertebrates.

Basic Pattern: Sharks

An adult shark is an ideal living blueprint of the basic venous channels of vertebrates. Inasmuch as very few changes occur in the embryonic channels of sharks during later development, knowledge of the venous channels of sharks and how they develop is a good introduction to the venous channels of other vertebrates.

Cardinal Streams
The sinus venosus receives all blood returning to the heart of sharks. Most of this blood, except that from the digestive organs, enters the sinus by a pair of **common cardinal veins** (fig. 14.29b). These use the transverse septum as a bridge from the lateral body walls to the heart. They appear early in development, and remain essentially unchanged thereafter.

Blood from the head other than the lower jaw is collected by a pair of **anterior cardinal** (precardinal) **veins** lying dorsal to the gills. The anterior cardinals pass caudad and empty into the common cardinals. The embryonic anterior cardinals, like common cardinals, remain essentially unchanged throughout life.

The earliest *embryonic* **posterior cardinal** (postcardinal) **veins** are continuous with the caudal vein (fig. 14.29a). These embryonic postcardinals pass cephalad lateral to the developing kidneys and receive a series of renal veins from the kidneys en route. They then empty into the common cardinals. Their cephalic ends in adult sharks expand to become **posterior cardinal sinuses.**

While these embryonic postcardinals are functioning, a new pair of postcardinal veins is forming *between* the kidneys from a subcardinal plexus. (A similar plexus in turtles is illustrated in fig. 14.35.) These new veins become confluent with the old postcardinals at the anterior end of the kidneys, and they, too, drain the kidneys. As more and more blood from the kidneys flows into the new veins the older postcardinals are lost anterior to the kidneys (fig. 14.29b). Thereafter, the name *posterior cardinal* is applied to the newer vessels. In adults they drain chiefly the kidneys, body wall, and gonads.

Renal Portal Stream
At an early stage in development in sharks, some of the blood from the caudal vein continues forward beneath the gut as a subintestinal vein that drains the digestive tract (fig. 14.29a). Later the connection of the caudal vein with the subintestinal is lost (fig. 14.29b). When the old posterior cardinals are lost anterior to the kidneys, all blood from the tail thereafter enters the capillaries surrounding the kidney tubules (**peritubular capillaries**). The result is a **renal portal system.**

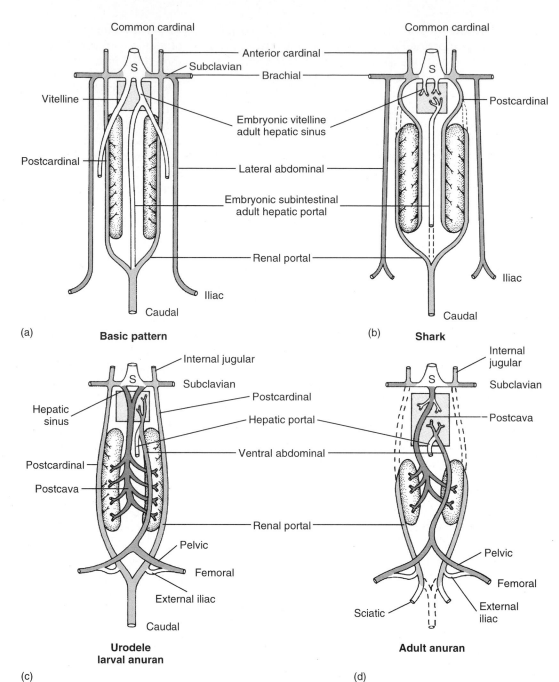

FIGURE 14.29

Modifications of the basic venous pattern in sharks and amphibians. *Broken lines* indicate lost segments. *Light blue,* cardinal and caudal (renal portal) streams; *dark blue,* postcava, receiving renal veins; *red,* abdominal stream. **S,** sinus venosus. The kidneys are *stippled,* and the liver is *gray.* The venous channels medial to the kidneys in (*b*), (*c*), and (*d*) develop from an embryonic subcardinal plexus.

Lateral Abdominal Stream

Commencing at the pelvic fin from which it receives an **iliac vein** and passing forward in the lateral body wall on each side is a **lateral abdominal vein** (fig. 14.29*b*). At the level of the pectoral fin, it receives a **brachial vein,** after which the vessel turns abruptly toward the heart to enter the common cardinal vein. The part of the abdominal stream between brachial and common cardinal is the **subclavian vein.** In addition to collecting blood from the paired fins, the abdominal stream also receives a **cloacal vein,** a metameric series of **parietal veins** from the lateral body wall, and minor tributaries. This basic channel remains unmodified during subsequent development.

Hepatic Portal Stream and Hepatic Sinuses

Among the first vessels to appear in any vertebrate embryo are paired **vitelline,** or **omphalomesenteric, veins** from the yolk sac to the heart, whether the yolk sac is functional or not (figs. 5.12 and 16.6). One of the vitelline veins is soon joined by the embryonic **subintestinal vein** that drains the digestive tract (fig. 14.29*a*). As the developing liver enlarges, it encompasses the vitelline veins, causing them to be broken into many sinusoidal channels. Caudal to the liver, one vitelline vein disappears, and the other, with the subintestinal vein, becomes the **hepatic portal system** that drains the digestive organs of the coelom, and the spleen. Between liver and sinus venosus, the two vitelline veins become **hepatic sinuses** (fig. 14.29*b*).

Other Fishes

The venous channels of other fishes are much like those of sharks. Cyclostomes have no renal portal system and no left common cardinals, although two common cardinals develop in embryos. Abdominals are lacking in most ray-finned fishes, and the pelvic fins are drained by the postcardinals. In dipnoans, the pelvic fins are drained by an unpaired ventral abdominal vein that ends in the sinus venosus, and the right postcardinal is missing. Blood from the swim bladders of all ray-finned fishes empties into the hepatic or common cardinal veins; in dipnoans it empties into the left atrium. Coronary veins in all fishes empty into the sinus venosus.

Tetrapods

The early embryonic venous channels of tetrapods are basically the same as those of embryonic sharks. We have just seen how, by adding a vessel here and dropping one there during development, the basic pattern of a shark embryo is converted into the veins of an adult shark. We will now see how the same embryonic pattern is converted into the veins of adult tetrapods.

Cardinal Veins and the Precavae

Embryonic tetrapods have postcardinals, precardinals, and common cardinals. In urodeles the embryonic postcardinals persist between the caudal vein and common cardinals throughout life (fig. 14.29c). In this respect a *Necturus* is less modified than a shark. In anurans the segment of the embryonic postcardinals anterior to the kidneys disappears as development progresses (fig. 14.29), and the connection of the postcardinals with the caudal vein is lost at metamorphosis. In amniotes, postcardinals anterior to the kidneys generally disappear during embryonic development as a newer vessel, the postcava, gradually takes on more and more tributaries (figs. 14.30 and 14.31). In mammals, however, the anterior segment of the right postcardinal persists under the name **azygos,** and part of the left postcardinal persists as the **hemiazygos** (figs. 14.32 and 14.33). These two veins drain intercostal spaces (fig. 14.34). The azygos empties into the old right common cardinal vein (precava). It receives transverse shunts from the hemiazygos.

Common cardinal veins in amniotes are known as **precavae** and anterior cardinals are known as **internal jugular** veins. Although most mammals retain both the left and right precavae (fig. 14.33), some, including cats and humans, lose most of the left precava (left common cardinal) during embryonic life (fig. 14.32). A transverse vessel, the left brachiocephalic vein, then shunts venous blood from the left side of the head and left anterior limb to the right precava (fig. 14.32). A remnant of the left precava remains as the **coronary sinus,** a venous sinus on the surface of the heart that receives several coronary veins and empties into the right atrium. The persisting

FIGURE 14.30

Systemic veins of two reptiles. Only vessels of the basic pattern illustrated in fig. 14.29a are shown. A strong branch of the renal portal vein of crocodilians continues through each kidney without ending in capillaries. **sv,** sinus venosus.

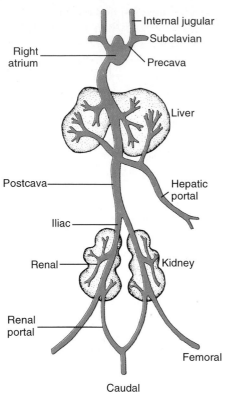

FIGURE 14.31

Major systemic veins of a bird.

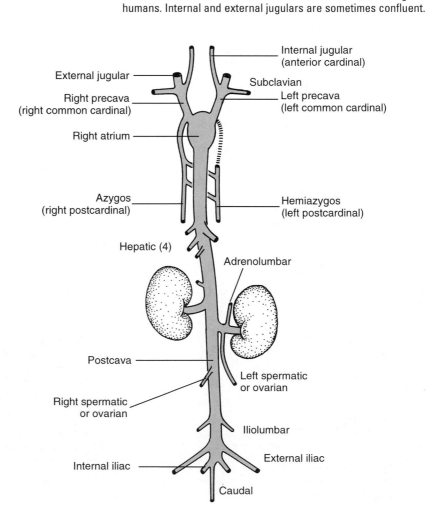

FIGURE 14.32

Basic anterior venous channels of cat and human, ventral view. Broken lines indicate vessels obliterated during ontogeny. These sometimes remain as anomalies in adult mammals, including cats and humans. Internal and external jugulars are sometimes confluent.

FIGURE 14.33

Major systemic venous channels of a rabbit, ventral view. Broken line indicates obliterated segment of left posterior cardinal vein. Entrance of one spermatic or ovarian vein into a renal vein is uncommon in rabbits, common in cats.

FIGURE 14.34

The azygos (on the animal's right) and the hemiazygos (on the animal's left) in a rhesus monkey, ventral view. This condition is one of many variants in this species. Similar variants are seen in humans.

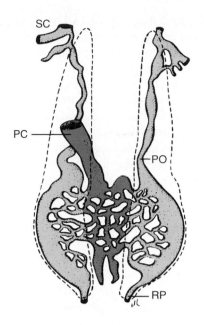

FIGURE 14.35

Embryonic subcardinal venous plexus and origin of the postcava (*dark blue*) in the turtle *Chrysemys*, ventral view. The mesonephros is outlined by broken lines. **PC,** postcava; **PO,** left postcardinal; **RP,** renal portal vein; **SC,** subclavian vein.

right precava in humans is known also as the **superior vena cava.** The precavae empty into the sinus venosus in reptiles and into the right atrium in birds and mammals.

Postcava

The postcava arises during embryonic life in a subcardinal venous plexus that receives renal veins from the kidneys (fig. 14.35). One subcardinal channel (usually the right) predominates, grows into the mesentery in which the liver is developing, and becomes confluent with the hepatic sinuses. This vessel is the **postcava.** The enlarging liver envelops it but does not break it into capillaries. Thus the postcava becomes an expressway from kidneys to heart via the hepatic sinuses (fig. 14.29a and c). In most tetrapods, the two hepatic sinuses ultimately fuse to form a median vessel that becomes part of the postcava (figs. 14.29d, 14.30, and 14.31). In mammals, the postcava is also called the **inferior vena cava.**

In crocodilians, some of the blood from the hind limbs that enters a renal portal vein bypasses the kidney capillaries and flows directly into the postcava (fig. 14.30, crocodilian). In birds most, if not all, and in mammals, all blood from the hind limbs flows directly into the postcava (figs. 14.31 and 14.33).

With establishment of a postcava, blood that previously passed from the kidneys to the heart via postcardinal veins now uses the postcava, and the postcardinals are reduced to draining the intercostal spaces.

Abdominal Stream

In early tetrapod embryos, paired lateral abdominal veins commence in the body wall at the level of the future hind limbs, pass cephalad in the lateral body wall, receive veins from the developing forelimbs, and terminate in the common cardinal veins or sinus venosus

(fig. 14.29a). As development progresses in tetrapods, this stream alters its course.

In amphibians, the two embryonic lateral abdominal veins, which lie parallel and close together on either side of the midventral line, unite in the midventral body wall as far forward as the level of the embryonic falciform ligament that connects the ventral body wall with the developing liver (fig. 14.29c and d). Blood in this unpaired segment, now known as the **ventral abdominal vein,** establishes connections with venous channels in the falciform ligament, one of these channels enlarges, and soon all the blood in the ventral abdominal vein is crossing the falciform ligament and emptying into the liver (fig. 14.29c and d). Thereafter, the abandoned segments of the lateral abdominal veins anterior to the liver disappear. The result is establishment in amphibians of a *portal stream* between the capillaries of the developing hind limbs and those of the liver. This is an addition to the already-established hepatic portal system that drains the digestive organs and spleen in all vertebrates. Rerouting of the abdominal stream dissociates it from the drainage of the anterior limbs (fig. 14.29d).

In reptiles, the two lateral abdominals do not unite (fig. 14.30). Otherwise, they undergo the same modification as in amphibians, using the falciform ligament as a bridge across the coelom to the capillaries of the liver, and losing their connection with the common cardinal veins. They acquire temporary tributaries, the two **allantoic veins,** as they pass along the ventral body wall. (The allantois in fig. 5.13b is part of the respiratory membrane of unhatched reptiles.) The allantoic vessels regress when the extraembryonic portion of the allantois is lost prior to hatching. In birds none of the embryonic abdominal stream remains in adults (fig. 14.31).

The abdominal stream serves mammals, but during fetal life only. The portion of the **umbilical (allantoic) vein** within the fetal body beginning at the umbilicus, crossing the falciform ligament, and entering the liver is all that is left of the abdominal stream (fig. 14.38). Like the lateral abdominal of lower vertebrates, it commences early in embryogenesis as a *pair* of vessels in the ventro-lateral body wall; but no connection with the drainage of the hind limbs develops. Branches from this very early vessel grow out the umbilical cord and vascularize the placenta. Soon, the paired embryonic vessels unite, forming a single umbilical vein. The abdominal stream in mammals has no function other than to drain the placenta; and when, at birth, the umbilical vein is severed along with the umbilical cord, blood no longer flows through the part that remains between the navel and the liver, and this segment becomes the **round ligament of the liver,** visible as a fibrous thickening in the free border of the falciform ligament extending between the umbilicus and the liver. Thus a history of mutations in ancestral mammals has transformed the very ancient abdominal stream, which at one time drained the posterior appendages, body wall, and anterior appendages. It is now strictly an embryonic vessel that drains the placenta.

The umbilical vein in embryonic mammals carries a heavy load of blood and erodes a broad channel, the **ductus venosus,** directly through the substance of the liver and into the postcava (fig. 14.38, 3). After birth, the channel becomes a ligament, the **ligamentum venosum.**

Renal Portal System

The renal portal system of amphibians has acquired a tributary, the **external (transverse) iliac vein** (*not homologous with the iliac veins of amniotes*), which carries some blood from the hind limbs to the renal portal vein (fig. 14.29c and d). This channel provides an alternate route from the hind limbs to the heart. The connection exists also in reptiles (fig. 14.30). It may have been one of the factors responsible for the eventual loss of the abdominal stream beyond the allantoic veins in mammals.

Snakes have no hind limbs, so the renal portal system is seen in its primitive relationship (fig. 14.36). In crocodilians, some blood from the hind limbs that enters the kidneys via the renal portal system bypasses the kidney capillaries, going straight through the kidneys and into the postcava (fig. 14.30, crocodilian). In birds, this has become the principal route from the hind limbs to the heart (fig. 14.31). In mammals above monotremes, the renal portal system disappears.

From the foregoing descriptions it can be seen that the posterior appendages of tetrapods have been drained by a succession of vessels (fig. 14.37): the abdominal stream,

FIGURE 14.36

Renal portal system and venous drainage of the kidneys of a snake.

the renal portal, and finally the postcava. Displacement of the caudal end of the nephrogenic mesoderm during embryonic development of the mammalian kidney (chapter 15) may have been a factor in the evolutionary loss of the renal portal system in mammals.

Hepatic Portal System

The hepatic portal system is similar in all vertebrates. It drains chiefly the stomach, pancreas, intestine, and spleen and terminates in the capillaries of the liver. Its origin in sharks from the embryonic vitelline and subintestinal veins has been described earlier. It arises in a similar manner in tetrapods. The abdominal stream (umbilical in mammals) becomes a tributary of the hepatic portal system, commencing with amphibians. Veins from the swim bladders are usually tributaries of the hepatic portal system in bony fishes.

Coronary Veins

Many amphibians seem to lack a definitive coronary system. In frogs, one coronary vessel enters the left precava,

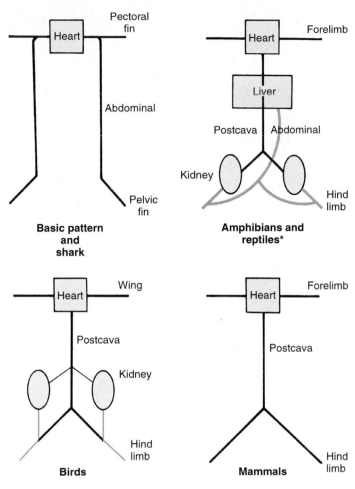

FIGURE 14.37

Venous routes from paired fins and limbs, diagrammatic. Vessels represented by thin lines in birds carry relatively less blood. *Blue* in a vessel indicates a portal stream. (*The abdominal stream in reptiles is paired.)

and another empties into the ventral abdominal vein near the liver. In reptiles, birds, and mammals, the several coronary veins empty into the coronary sinus or directly into the right atrium. The coronary sinus lies on the surface of the heart in the **coronary sulcus,** a groove between the left atrium and ventricle.

CIRCULATION IN THE MAMMALIAN FETUS, AND CHANGES AT BIRTH

Blood in the mammalian fetus passes from the caudal end of the dorsal aorta into the umbilical arteries (fig. 14.38). These extend out the umbilical cord to the placenta. From the placenta, oxygenated blood returns to the fetus via an umbilical vein that traverses the falciform ligament to enter the liver. Some of this blood, which contains nutrients as well as oxygen, enters liver capillaries for processing. Most of it continues nonstop via the ductus venosus into the postcava and finally into the right atrium. From

the right atrium it passes through an **interatrial foramen (foramen ovale of the heart)** into the left atrium, bypassing a trip to the lungs. The foramen is guarded by a one-way, flaplike valve. From the left atrium it goes to the left ventricle and is expelled at the next systole into the systemic arch. Thus, oxygenated blood is sent to the fetal head, trunk, and limbs (fig. 14.38).

The unoxygenated blood returning to the left atrium from the lungs is a very small quantity compared to that pouring in through the foramen ovale, except near the time for delivery of the fetus; and the mixing of the two only slightly dilutes the oxygenated blood.

Most of the blood returning to the right atrium from the major venous channels enters the right ventricle and, at the next systole, is pumped into the pulmonary trunk. However, the ductus arteriosus is open and functioning, and all but a small amount of this blood is shunted away from the lungs and into the dorsal aorta (fig. 14.39, blue). A quantity sufficient to satisfy the oxygen needs of the fetus reaches the placenta via the umbilical arteries.

From this account it may be seen that blood in the fetal circulation is either low in oxygen or mixed to one degree or another, except in the umbilical veins, in which it is rich in oxygen; that the umbilical arteries receive blood in need of oxidation; and that the brain and trunk receive the blood that is richest in oxygen. An essentially identical allantoic circulation is seen in unhatched chicks.

It should be pointed out that fetal circulation cannot be exactly as described until the embryo has developed two independent ventricles. In humans, this is not until the beginning of the eighth week of embryonic life.

At birth, four major circulatory changes adapt the organism for pulmonary respiration:

1. The ductus arteriosus closes as a result of nerve impulses passing to its muscular wall. These impulses are initiated reflexly when the lungs are filled with air with the first gasp after delivery. In birds this is usually the day before hatching, when the imprisoned chick pecks a hole in its extraembryonic membranes and starts breathing the air entrapped between these membranes and the shell. When the chick inside the shell starts to peep, it already has air in its lungs! Shortly thereafter, all blood entering the pulmonary trunk goes to the lungs, and the ductus arteriosus of birds and mammals becomes converted into an **arterial ligament (ligamentum arteriosum).**

2. The flaplike interatrial valve is pressed against the interatrial foramen by the sudden increase in pressure in the left atrium that results from the greatly increased volume of blood entering from the lungs. This valve prevents the unoxygenated blood in the right atrium from entering the left atrium, which now contains only oxygenated blood from the lungs. Within a few days, the foramen ovale is permanently sealed and only a scar, the **fossa ovalis,** remains.

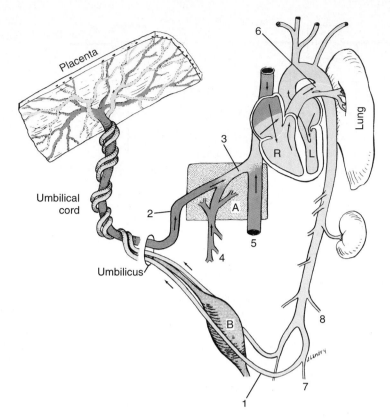

FIGURE 14.38

Circulation in a mammalian fetus. *Dark red* indicates oxygen-rich blood; *dark blue,* blood low in oxygen; *light red,* mixed, considerable oxygen; *light blue,* mixed. **1,** umbilical artery; **2,** umbilical vein in the falciform ligament; **3,** ductus venosus; **4,** hepatic portal vein; **5,** postcava (inferior vena cava); **6,** ductus arteriosus; **7,** internal iliac artery; **8,** external iliac artery growing into limb bud; **9,** umbilicus. *Gray,* allantois. **A,** liver; **B,** base of allantois, which is developing into a urinary bladder; **L,** left ventricle; **R,** right ventricle. After birth, the constricted part of the allantois distal to the bladder becomes the urachus. Much of the blood returning to the right atrium from the placenta passes through a foramen ovale (not illustrated) into the left atrium to be distributed via the left ventricle to the head and anterior limbs.

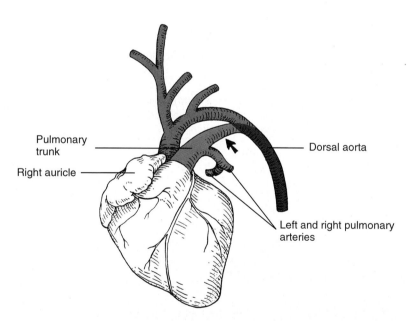

FIGURE 14.39

Ductus arteriosus (*arrow*) of fetal pig. The left auricle has been removed to show the pulmonary arteries. Blood in the dorsal aorta is mixed.

3. At birth, the umbilical arteries and vein are severed at the umbilicus. Thereafter, no blood passes through the umbilical arteries distal to the urinary bladder, which they continue to supply. Eventually, the umbilical arteries from bladder to navel are converted into **lateral umbilical ligaments** in the free border of the ventral mesentery of the bladder.

4. Blood no longer flows through the umbilical vein because the source has been cut off. Eventually, this vessel becomes converted into the **round ligament of the liver** and the ductus venosus becomes the **ligamentum venosum.** (The latter occurs halfway through gestation in whales, forcing umbilical vein blood through the liver capillaries en route to the right atrium.) As a result of these changes, the fetal mammal (or bird) is changed from an allantoic-respiring or placenta-respiring organism to one capable of breathing air and, indeed, forced to do so.

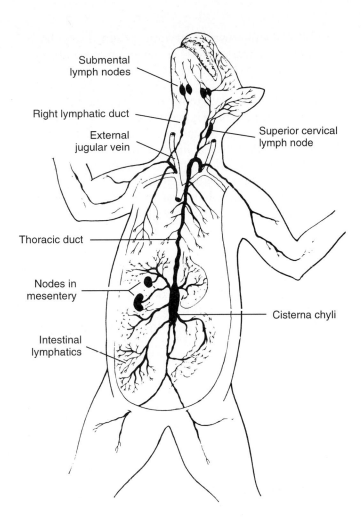

FIGURE 14.40

A few of the superficial and deep lymphatics of a cat.

Failure of the foramen ovale to close or of the ductus arteriosus to fully constrict may result in cyanosis (blueness of the skin, lips, and nail bed in humans) because of the amount of reduced hemoglobin in the red cells in the integumentary capillaries. The condition may be ameliorated if the fluid pressure in the left atrium is sufficient to keep the valve of the foramen ovale pressed against the aperture, thereby blocking the shunt.

LYMPHATIC SYSTEM

Lymph and lymph channels are found in all vertebrates (figs. 14.40 and 14.41). A lymphatic system consists of thin-walled **lymph channels, lymph** (a fluid in transit), **lymph hearts** as far up the phylogenetic scale as embryonic birds, **lymph nodes** in birds and mammals, and **solitary or aggregated masses of lymph nodules,** the largest of which is the spleen. In contrast to blood, lymph moves in one direction only: from the tissues toward the heart. Lymph channels are seldom dissected in the anatomy laboratory because of their delicate and collapsible nature. Lymph nodes are more readily observable (fig. 14.40).

FIGURE 14.41

Superficial lymph channels of a frog tadpole.

Source: Data from J. S. Kingsley, *Outline of Comparative Anatomy of Vertebrates,* 1920, The Blakiston Company, Philadelphia, after Hoyer and Udziela.

The system begins in **lymph capillaries,** the walls of which are single-layered endothelial tubes, or in **lymph sinusoids,** which are expansions of capillaries. The capillaries are branching and anastomosing tubes of slightly greater diameter than blood capillaries, and they exhibit constrictions and expansions rather than being of a standard diameter. Capillaries or sinusoids penetrate most of the soft tissues of the body other than the liver and the nervous system. They are not present in skeletal tissues. Capillaries and sinusoids collect interstitial fluids. Once inside the tubes, the fluid is colorless or pale yellow, and is called lymph. This fluid passes from one endothelial-lined channel to the next, and finally empties into a vein. Valves at these exits prevent the influx of venous blood into the lymph channels. A lymphatic network consisting of long, narrow, discrete tubular vessels with a modicum of smooth muscle in the walls is found only in birds and mammals. A series of valves line these tubes and help counteract the effect of gravity, especially in the limbs.

Lymphatics in intestinal villi of higher vertebrates collect fats absorbed from the small intestine after a meal (fig. 12.23). If the meal was particularly fatty, the lymph in these vessels has a milky appearance. For this reason these specific lymphatics are called **lacteals.** The lymph within them is **chyle.** Certain lymphatics in cyclostomes, cartilaginous fishes, and even humans contain some red blood cells. The fluid in these vessels is **hemolymph.**

Lymph channels that drain the body wall, limbs, and tail of lower vertebrates empty into nearby veins at the base of the tail, in the trunk, or in the neck. Those draining viscera are often paired in lower vertebrates, but in mammals, a single **thoracic duct** commences in a large abdominal lymph sinus, the **cisterna chyli** (fig. 14.40), and empties into the brachiocephalic or left subclavian vein, or into the external or internal jugular vein (fig. 14.40). The mammalian thoracic duct also receives lymphatics from the left side of the head and neck and from the left forelimb. One or more additional major lymphatics drain the right side of the body anteriorly.

The lymphatic system of anurans is notable because of the numerous sinusoids that take the place of capillaries at some locations. They are especially prominent just under the skin over almost the entire body, where they form huge lymph reservoirs separated from each other by connective tissue septa that attach the skin to the underlying muscles at these locations only (fig. 14.42). The skin is very thin, and the sinusoids may buffer the underlying muscles from the drying effect of air when a frog is out of water. Because of these subcutaneous lymph sinuses, the skin of a frog is easily removed from the body. Anyone who has dissected a fluid-preserved frog has seen how the skin balloons when the sacs are distended with fluid.

The flow of lymph results from a number of factors, including lymph hearts situated at advantageous locations along lymph pathways in fishes, amphibians, and reptiles. These are pulsating sinusoidal swellings on lymph pathways with thin walls containing striated mus-

cle fibers of unknown embryonic origin. Frogs have 2 pairs of lymph hearts, urodeles have as many as 16 pairs, and caecilians as many as 100. The anterior pair of lymph hearts in frogs, located just behind the transverse processes of the third vertebra, pump lymph into the vertebral veins, from which it is carried to the internal jugulars. A posterior lymph heart on each side of the urostyle empties into an external iliac vein. Amphibians, especially aquatic and semiaquatic amphibians, have more tissue fluids to manipulate than do other vertebrates, and as a result their lymph hearts move a proportionately larger volume of fluid hourly than the hearts of most other vertebrates. Semilunar valves at the exit of the hearts prevent backflow.

Lymph hearts are not present in birds after hatching, although embryonic birds have one on each side of the last sacral vertebra. None has been described in mammals. In these vertebrates, lymph flow is maintained by activity of the skeletal muscles as they contract and relax, by movements of the viscera, and by rhythmical changes in intrathoracic pressure that result from breathing. Some of these factors are also operative in lower vertebrates.

Lymph nodes are masses of hemopoietic tissue interposed along the course of lymph channels of birds and mammals. They may be no larger than a pinhead, or they may be several centimeters in diameter. They are the "swollen glands" that you can feel in the neck, axilla, and groin of humans when there is inflammation in the areas drained. Lymph enters a node via several afferent lymphatics, filters through the node, and leaves via a single large efferent lymphatic. The endothelium of their sinusoidal passageways includes many phagocytes that ingest bacteria and other foreign particles. Enmeshed in their connective tissue stroma are large numbers of small lymphocytes. The nodes are therefore the second line of defense against bacterial infections acquired through the skin, the first line being granulocytes that assemble at the invaded area.

Other **lymphoid masses** develop at one site or another. The largest of these is the **spleen,** which is unique because it is interposed into the blood circulatory stream. Its contribution to that system was discussed earlier. Other masses include the **thymus** (absent in hagfishes), the **bursa of Fabricius** of very young birds, solitary nodules or aggregates of nodules **(Peyer's patches)** in the wall of the small intestine of amniotes, and mammalian **tonsils** and **adenoids.**

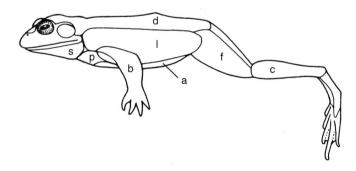

FIGURE 14.42

Subcutaneous lymph sacs of a frog. The outlines of the sacs are the sites of septa that extend between the skin and the musculature. **a,** abdominal; **b,** lateral brachial; **c,** crural; **d,** dorsal; **f,** femoral; **l,** lateral; **p,** pectoral; **s,** submaxillary.

SUMMARY

1. The circulatory system consists of a blood vascular system and a lymphatic system.

2. The blood vascular system consists of whole blood, heart, arteries, capillaries, and veins.

3. Blood consists of plasma and formed elements. The formed elements are erythrocytes, granular and agranular leukocytes (the latter includes lymphocytes), and thrombocytes (platelets).

4. Hemocytoblasts are precursors of all blood cells. Their initial source is blood islands of the yolk sac. Eventually they form in liver, kidney, spleen, and bone marrow.

5. The lymphatic system consists of lymph, lymphatic capillaries and veins, lymph hearts, lymph nodes, and miscellaneous lymphoid masses in one major taxon or another. Chyle is lymph collected in intestinal villi. Lymphatics terminate in a systemic vein.

6. Fishes have a single-circuit heart. Venous blood enters a sinus venosus, traverses an atrium, ventricle, and conus arteriosus. The last two are pumps that discharge into a ventral aorta. The latter carries blood to aortic arches that supply gills, where blood is oxygenated. It then passes via arteries to capillaries everywhere in the body, gives up oxygen, takes on carbon dioxide, and returns to the sinus venosus.

7. In vertebrates that breathe solely by lungs, a pulmonary circuit carries blood to the lungs and back and a systemic circuit carries oxygenated blood elsewhere and returns deoxygenated blood to the heart. This is a double-circuit heart.

8. Double-circuit hearts have two atria and one or two ventricles. The right atrium receives deoxygenated blood from the systemic circuit, the left receives oxygenated blood from the pulmonary circuit.

9. Mixing of oxygenated and deoxygenated blood in the ventricle of a double-circuit heart is avoided by one of several adaptations, including spiral valves when there is a single ventricle, and trabeculae and ventricular septa. A complete interventricular septum separates the blood in crocodilians, birds, and mammals.

10. Right-left shunts of blood (away from the lungs) exist in reptiles to meet metabolic needs imposed by certain behavioral patterns, including remaining underwater for long intervals without breathing.

11. A sinus venosus is present in fishes, amphibians, and reptiles. It becomes partially incorporated into the wall of the right atrium in crocodilians. In birds and mammals, it is not a sinus, but a local collection of cells in the right atrium known as the sinoatrial (SA) node.

12. The pulsations of the heart are autogenic, requiring no extrinsic neural stimulus. However, the rate of beat is imposed by the autonomic nervous system. The stimulus in fishes, amphibians, and reptiles is spread from the musculature of the sinus venosus. In birds and mammals, it emanates from the SA and an atrioventricular (AV) node.

13. The ventral aorta exhibits a swelling, the bulbus arteriosus, in teleosts and perennibranchiate urodeles.

14. In most reptiles, the ventral aorta is split longitudinally into two aortic trunks and a pulmonary trunk. The two aortic trunks exit from a cavum venosum in the ventricle and supply all functional aortic arches except the pulmonary. The pulmonary trunk emerges from the right ventricle (cavum pulmonale) and supplies the pulmonary circuit. The left ventricle has been termed the cavum arteriosum.

15. The ventral aorta in birds and mammals is split into only two trunks, a systemic trunk that emerges from the left ventricle and supplies all parts of the body except the lungs, and a pulmonary trunk that emerges from the right ventricle and supplies the lungs.

16. Six pairs of aortic arches develop in vertebrate embryos. During ontogeny the arches are reduced in number, the third (carotid), fourth (systemic), and sixth (pulmonary) being the most persistent.

17. The paired dorsal aorta between the third and fourth arches, when persistent, is the ductus caroticus.

18. The dorsal segment of the left sixth aortic arch is present until hatching or birth in birds and mammals. This segment, the ductus arteriosus, shunts blood away from the lungs and to the embryonic respiratory membranes (placenta in mammals). It becomes a ligament after birth.

19. The unpaired dorsal aorta of the trunk has paired segmental branches (somatic) to the body wall and appendages, unpaired visceral branches to the digestive organs within the coelom and to the spleen, and paired visceral branches to the urogenital organs and adrenal glands. The embryonic dorsal aorta of amniotes supplies an allantoic (umbilical) artery to the allantois or placenta.

20. Retia mirabilia are localized networks of arteries, or sites of close proximity of arterioles and venules, with functions that vary with their location and with the species. Local heat or gas exchange are two such functions.

21. Basic venous channels are anterior, posterior, and common cardinals, abdominals, renal and hepatic portals, hepatic sinuses, and coronary veins. Pulmonary veins and a postcava are added in tetrapods. Embryonic subintestinal and vitelline veins contribute to the hepatic portal system. Allantoic (umbilical) veins drain the allantois or placenta in embryonic amniotes.

22. The renal portal system drains only the tail in fishes. It acquires connections with the hind limbs in

amphibians. In crocodilians and birds, this connection bypasses the kidneys and goes directly to the postcava. There is no renal portal system in adult mammals other than monotremes.

23. The postcava becomes increasingly important in tetrapods. Commencing as an alternate route from kidneys to heart, it eventually drains the tail, hind limbs, and most of the trunk.

24. Remnants of embryonic vascular channels in adult mammals include the round ligament of the liver (remnant of left umbilical vein), ligamentum venosum (remnant of ductus venosus), ligamentum arteriosum (remnant of left ductus arteriosus), lateral umbilical ligaments (remnants of paired umbilical arteries from urinary bladder to umbilicus), and fossa ovalis (occluded interatrial foramen ovale of fetus).

SELECTED READINGS

Barker, M. A.: A wonderful safety net for mammals, Natural History **102**:8, August 1993.

Berg, T., and Steen, J. B.: The mechanism of oxygen concentration in the swim bladder of the eel. Journal of Physiology **195**:631, 1968.

Burggren, W. W.: Form and function in reptilian circulation, American Zoologist **27**:5, 1987.

Gatten, R. E., Jr. Cardiovascular and other physiological correlates of hibernation in aquatic and terrestrial turtles, American Zoologist **27**:59, 1987.

Heatwole, H.: Adaptations of marine snakes, American Scientist **66**(5):594, 1978.

Kampmeier, O. F.: Evolution and comparative morphology of the lymphatic system, Springfield, Ill., 1969, Charles C. Thomas, Publisher.

Nandy, K., and Blair, C. B.: Double superior venae cavae with completely paired azygos veins, Anatomical Record **151**:1, 1965.

Ottaviani, G., and Tazzi, A.: The lymphatic system. In Gans, C., and Parsons, T. S., editors: Biology of the reptilia, vol. 6, New York, 1977, Academic Press.

Simons, J. R.: The heart of the tuatara, *Sphenodon punctatus*, Journal of Zoology **146**:451, 1965.

Struthers, P. H.: The aortic arches and their derivatives in the embryo porcupine (*Erethizon dorsatus*), Journal of Morphology and Physiology **50**:361, 1930.

CHAPTER FIFTEEN

UROGENITAL SYSTEM

In this chapter we will be studying the urinary and reproductive systems. We will see that these organs, from fishes to humans, develop in accordance with a single, basic architectural pattern. We will see how marine fishes and tetrapods conserve water and how freshwater vertebrates get rid of it, and that urinary organs assist in saving or excreting blood-borne solutes. We will see how reproductive and urinary passageways are developmentally associated with each other and witness the diversity of reproductive strategies in vertebrates. The chapter ends describing the cloaca, a simple posterior embryonic chamber and how it may, depending on the vertebrate, subdivide into separate urinary, reproductive, and digestive passageways.

OUTLINE

lthough the function of kidneys is different from that of gonads, the urinary and genital systems of gnathostomes are so interrelated, both developmentally and structurally, that neither can be discussed without frequent reference to the other. For this reason, urinary and genital organs are discussed in a single chapter. The term urogenital is a combination of the Greek combining form *ouro-*, pertaining to urine, and an English derivative of the Latin word *genitalis*, which pertains to reproduction.

KIDNEYS AND THEIR DUCTS

Osmoregulatory Role of Kidneys

Vertebrate life began in water, and the early stages of the evolution of kidneys took place in that medium. Vertebrates that are submerged in freshwater inevitably acquire water by absorbing it through the gills, oropharyngeal membranes, or skin and by unavoidably swallowing it with their food. Therefore, to maintain their tissue fluids in an osmotic state compatible with existence in this medium, it is essential that freshwater organisms *excrete water and conserve salt.* Fishes that are submerged in salt water face a different osmotic problem. Instead of accumulating too much water, they are in danger of accumulating too much salt. Survival in salt water depends on *conserving water and excreting salt.* Vertebrates living in some environments on land face the possibility of becoming dehydrated. The major role of kidneys is **osmoregulation,** the maintenance of a ratio between the quantity of salt and water in the body fluids such that the organism can survive in its natural habitat, whether it be in water or on land. Osmoregulation is one aspect of a physiological function, **homeostasis**—the involuntary control of the chemical environment of cells.

Kidneys maintain appropriate concentrations of water and salt in body tissues by eliminating any excess water, by preventing the escape of water when necessary, and by regulating the excretion of certain salts. In the latter role, they have the assistance of extrarenal osmoregulatory organs that will be discussed later. Tetrapod kidneys have an additional role, the elimination of nitrogenous wastes, which are the product of the metabolic breakdown of proteins. In no fish is the kidney of primary importance in the excretion of nitrogenous wastes. Most of these are eliminated by gills and sometimes via the skin. The functional components of vertebrate kidneys and their adaptations is the next topic.

Whether the first vertebrates, hence vertebrate kidneys, evolved in fresh or salt water is not settled. Advocates of a freshwater origin point to the lack of ostracoderm fossils in early marine deposits. Proponents of the saltwater origin are convinced that ostracoderm fossils exist in marine deposits and they have not been discovered because ostracoderms were not abundant when these sediments were being deposited. Stahl presents both sides of the debate, pointing out that negative evidence—in this instance, absence of fossils—is not conclusive. Regardless of which view is correct, the functional components of vertebrate kidneys have been such that, when subjected to changing environments during geologic time, they were adaptable for life in salt water, in freshwater, on land, and even in a desert.

Basic Pattern and the Archinephros

Vertebrate kidneys are built in accordance with a basic architectural pattern consisting of **glomeruli, renal tubules,** and a pair of **longitudinal excretory ducts** (fig. 15.1a–c). Variations in details from fishes to humans are principally in the number and arrangement of glomeruli and in the relative complexity of the tubules.

Glomeruli are tufts of microscopic capillarylike arterial loops, or retia mirabilia, on the pathway of an arteriole, where water, ions, metabolic waste products, and certain other constituents are removed from the bloodstream. In some fishes, they are large enough to be seen with the naked eye or a hand lens. In others they are microscopic. The most primitive glomeruli are suspended in the coelom surrounded by peritoneum. They discharge their filtrate into the coelomic fluid, which is then swept into a peritoneal funnel, or **nephrostome,** leading to a tubule (fig. 15.1a). These are **external glomeruli.** In today's vertebrates, external glomeruli are confined to embryos and larvae. Glomeruli in adults are embedded within the dorsal body wall (hence said to be retroperitoneal) and are ensheathed by **Bowman's capsule,** a delicate double-walled outgrowth from a kidney tubule (figs. 15.1b and 15.2). Its inner wall adheres to the surfaces of the vascular loops. The capsular cavity collects the glomerular filtrate, which then passes into a renal tubule. These are **internal glomeruli.** A glomerulus and the surrounding capsule constitute a **renal corpuscle** (fig. 15.2). A renal corpuscle, renal tubule, and the associated **peritubular capillaries** constitute a **nephron,** the functional unit of a gnathostome kidney (fig. 15.3).

Supplying a glomerulus is an **afferent glomerular arteriole,** and emerging from the glomerulus is an **efferent glomerular arteriole** of *lesser diameter.* This has the effect of increasing the blood pressure within the glomerulus. The afferent arterioles of agnathans are supplied directly from the dorsal aorta by segmental arteries. In gnathostomes, they are terminal branches of renal arteries. Efferent glomerular arterioles supply **peritubular capillaries** along with blood from the renal portal system (figs. 15.3a and 15.4). Efferent glomerular arterioles are the sole source of blood in the peritubular capillaries of placental mammals, since a renal portal system is lacking. Peritubular capillaries are drained by venules that lead to renal veins. Agnathans alone lack peritubular capillaries, but as compensation, the longitudinal

External glomerulus

(a)

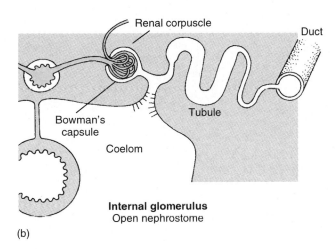

Internal glomerulus
Open nephrostome

(b)

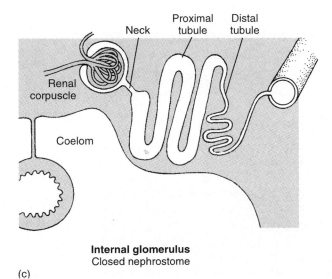

Internal glomerulus
Closed nephrostome

(c)

FIGURE 15.1

Basic components of a vertebrate kidney. (*a*) External glomerulus and tubule in the cephalic end of the segmental kidney of a larval hagfish. (*b*) Tubule with internal glomerulus and open nephrostome, as in an apodan kidney. (*c*) Generalized nonsegmental tubule of freshwater fishes, urodeles, and anurans. The tubules of marine fishes often lack glomeruli.

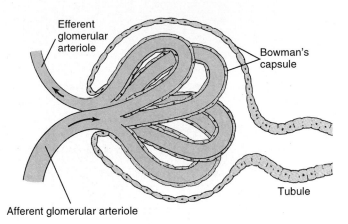

Renal corpuscle

FIGURE 15.2

Renal corpuscle consisting of glomerulus (*red*) and Bowman's capsule (*gray*).

kidney ducts are surrounded by unusually rich capillary beds that take their place functionally.

Renal tubules differentiate from a ribbon of embryonic **nephrogenic (intermediate) mesoderm** that lies lateral to the mesodermal somites and extends the length of the embryonic trunk from immediately behind the head to the cloaca (figs. 5.9*d* and 15.8). The earliest tubules appear at the anterior end of the ribbon, and additional tubules are added as the embryonic trunk elongates. The paired longitudinal ducts begin development as caudally directed extensions of the first tubules (fig. 15.5). The two ducts grow caudad until they reach the cloaca, into which they establish an opening. The ducts participate, although not exclusively, in the induction of additional tubules as the body elongates. Adult kidneys remain retroperitoneal throughout life, although in mammals they bulge into the roof of the coelom.

Except in primitive metameric kidneys, several renal tubules empty into a single **common collecting tubule** and this, in turn, empties into one of the two longitudinal kidney ducts (fig. 15.6). The longitudinal duct collects fluids from all the common collecting tubules and empties into the cloaca or a derivative thereof.

Renal tubules increase in complexity among the vertebrate classes, exhibiting increasing degrees of *histologic and functional* specialization along the course of the tubule. They are short and straight in agnathans, longest and with the greatest regional specialization in mammals. Compare the tubules in figure 15.1 with those in figure 15.3.

Many marine teleosts have renal corpuscles that are poorly vascularized, cystic, or vestigial. In some others the distal segment of their tubules is abbreviated or lost. Loss of glomeruli, the major site of water removal, results in increased water retention. Shortening of the

(a) **Aquatic urodele nephron**

(b) **Mammalian nephron**

FIGURE 15.3

Anamniote and mammalian nephrons, peritubular capillaries omitted in (*b*). In (*a*): **1,** neck of renal tubule; **2,** proximal segment; **3,** intermediate segment; **4,** distal segment. In (*b*): **1,** Neck of tubule; **6,** proximal convoluted tubule; **7,** descending arm of loop of Henle; **8,** ascending arm; **9,** distal convoluted tubule. **aff,** afferent glomerular arteriole; **eff,** efferent glomerular arteriole. The peritubular capillaries of placental mammals receive no portal blood. (Not to scale.)

FIGURE 15.4

Schematic representation of blood supply to and from the peritubular capillaries of a generalized nephron. *Red,* arterial blood; *blue,* venous blood.

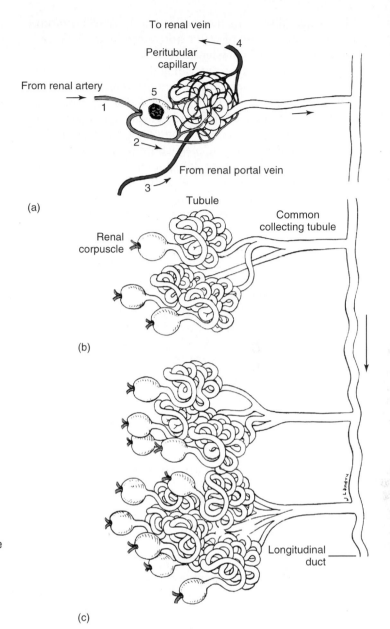

FIGURE 15.5

Pronephric tubules of a 15 mm larval frog. **II, III,** and **IV,** location of the second, third, and fourth somites. The glomus is three fused external glomeruli; the pronephric chamber is a coelomic pocket. The next tubule to form will be a mesonephric tubule at the level of somite VII.

FIGURE 15.6

Effect of multiple nephrons in a single body segment. (*a*) Segmental nephron consisting of renal corpuscle, renal tubule, and peritubular capillary. **1,** afferent glomerular arteriole; **2,** efferent glomerular arteriole; **3,** venule from renal portal system; **4,** efferent venule to renal vein; **5,** renal corpuscle, Bowman's capsule opened to show glomerulus. (*b*) Secondary and tertiary nephrons in a single body metamere. (*c*) Increased number of nephrons in a metamere disrupts the primitive metamerism of the kidney.

distal segment of the tubules eliminates the major site for reabsorption of salt. Both effects have survival value in salt water. There are, however, freshwater teleosts with the same mutations. In these, water excretion is chiefly tubular, and these freshwater fishes seem to compensate for any salt deficiency by active uptake of salt via the gills. There is reason to believe that these are former saltwater species that were seasonally migratory into freshwater streams (anadromous, appendix 2), and eventually became fully adapted to a freshwater habitat.

Elasmobranchs are exceptional. They have unusually large glomeruli, but they also have tubules that are unusually long for a fish. Those living permanently in salt water produce very small amounts of highly concentrated urine in any unit of time, thereby conserving water. They also have **rectal glands** that secrete large amounts of chlorides into the caudal end of the intestine. The total effect of the activity of their glomeruli, tubules,

gills, and rectal gland is to enable sharks to maintain their body fluids in an osmotic state that assures their survival, whether in freshwater or in the sea.

Like marine fishes, terrestrial reptiles and anurans that inhabit arid environments have very small glomeruli or none at all, a water conservation adaptation. Dehydration is the principal stimulus for release of pituitary hormones that induce water reabsorption from glomerular filtrate in tetrapods. These same hormones induce reabsorption of water from amphibian and reptilian urinary bladders, when necessary.

Several of the most anterior kidney tubules of some adult fishes and many tetrapod embryos and larvae have nephrostomes (fig. 15.1*a,b*), and vestigial nephrostomes lacking a lumen are often found in avian and mammalian embryos. Nephrostomes may be vestiges of the kidneys of a postulated ancestral protochordate in which there may have been one external glomerulus, one nephrostome, and one unconvoluted tubule in each body segment along the entire length of the coelom. This hypothetical ancestral kidney has been termed an **archinephros** (fig. 15.7). The nearest approach to such a kidney in living vertebrates is seen in larval hagfishes, in which a transient series of segmental external glomeruli, nephrostomes, and tubules is formed throughout much of the extent of the nephrogenic mesoderm (fig. 15.1*a*). However, at this stage of body elongation the kidney is still elongating. Segmental tubules with closed nephrostomes and renal corpuscles develop farther caudad. This transient larval hagfish kidney has been termed a **holonephros.** The adult hagfish kidney is a mesonephros (described shortly).

FIGURE 15.7

Hypothetical archinephros with one external glomerulus, nephrostome, and tubule per body segment. See also figure 15.1*a*.

Gathering coelomic fluid for elimination of liquid wastes is not confined to vertebrates. Nephrostomes in every body segment of marine annelids gather coelomic fluid for subsequent excretion.

Role of Glomeruli and Tubules

Three major processes are at work in urine formation: glomerular filtration, tubular reabsorption, and tubular secretion. Glomeruli function much like a simple filter. Because of an elevated blood pressure within the glomerulus resulting partly from the smaller size of the efferent glomerular arteriole, water, certain salts, glucose, and other solutes in the blood plasma are filtered into the space between the outer and inner walls of Bowman's capsule. Some constituents of the glomerular filtrate may not be expendable, and those are selectively reabsorbed during passage through specific segments of the tubule. As a result, the composition of the filtrate becomes altered in passage. All the glucose, for example, is ordinarily reabsorbed. It is the sole immediately available circulating source of energy, it is not readily acquired, and it is not ordinarily in excess in the bloodstream because, once acquired, it is quickly stored in the liver. Water and certain salts may or may not be in excess, depending on the environment. When in excess, water is allowed to pass through the length of the tubules and to enter the kidney ducts without having been reabsorbed. Under conditions that tend to lead to dehydration, water is reabsorbed from segments of the tubule. Certain salts also are selectively reabsorbed when appropriate. Finally, tubular secretion removes from the circulation useless or harmful substances that were not removed by filtration. Among these are, for instance, wastes from protein breakdown (nitrogenous wastes) in tetrapods. (Nitrogenous wastes, Na^+ and Cl^- in most fishes are eliminated by extrarenal means.) In marine fishes, certain salts (Mg^{++}, Ca^{++}, SO^-_4, phosphates) are tubular secretions, glomerular filtration being low. Glomerular filtration, tubular reabsorption, and tubular secretion produce the final excretion, urine, which varies in the amount of water, salts, and other constituents depending on the environment and the species. The constituents of urine may vary in a single organism from hour to hour, if not from minute to minute.

By varying the size and number of glomeruli in species that occupy different habitats, water can be excreted abundantly or sparingly. By varying the length of the kidney tubules, either salt or water, as necessary, can be recovered from the glomerular filtrate, or salt can be excreted abundantly. In general, glomeruli are larger in freshwater fishes and aquatic amphibians, and smaller in marine fishes and tetrapods, especially tetrapods living in arid environments.

Excretion of Nitrogenous Wastes

Nitrogenous wastes are excreted chiefly as *ammonia* by freshwater teleosts and aquatic and semiaquatic amphibians, because ammonia is highly soluble in water, and water is plentiful. These are therefore ammonotelic animals. These wastes are also excreted as ammonia in most marine fishes other than elasmobranchs, but the water that serves as a solvent in these species is provided by drinking seawater and rapidly excreting the salt. Nitrogen is excreted as *urea* by elasmobranchs and mammals (ureotelic animals). It is excreted in a semisolid urine as *uric acid* in species in which water is not abundant, as in reptiles and birds (uricotelic animals). The nature of the excreted nitrogenous compound in each instance is dictated in part by the availability of the solvent (water) that serves as a vehicle for elimination.

Pronephros

The earliest embryonic tubules arise from the cephalic end of the nephrogenic mesoderm. Because they are the first tubules to appear and are anteriorly located, they are called **pronephric tubules,** and this region of the nephrogenic mesoderm is the **pronephros** (fig. 15.8, embryo). The tubules are segmental, one pair at the level of each of the more-anterior somites. Pronephric tubules arise as solid aggregates of nephrogenic mesodermal cells that organize a lumen and frequently, except in birds and mammals, a nephrostome. Associated with each pronephric tubule in anamniotes there may be an external glomerulus. In some dipnoans and in larval urodeles and anurans, the first two or three external glomeruli coalesce to form a single large glomus that lies in a pronephric chamber, an isolated pocket of the coelom (fig. 15.5).

The number of fairly well-differentiated pronephric tubules is never large—3 in larval frogs at the level of somites II, III, and IV; 7 in human embryos at the level of somites VII to XIII; and about 12 in chicks commencing at somite V. Additional rudimentary segmental tubules usually appear and regress without reaching full development.

The distal ends of the first several pronephric tubules turn caudad and unite end to end to form a caudally directed duct that eventually establishes an opening into the cloaca. This is the **pronephric duct** (fig. 15.5).

Pronephric tubules are temporary. As additional, more complicated tubules form farther caudad, the pronephric glomeruli lose their connection with the dorsal aorta and regress. Tubules regress more slowly, and traces may remain in some adult fishes. In teleosts, the site of the embryonic pronephros is converted into a lymphoid organ housing components of the adrenal complex (fig. 18.9, teleost). *The pronephric duct does not regress. It continues to drain the mesonephric tubules that are forming farther back.*

Although pronephric tubules develop in all vertebrates, they function only in embryonic ray-finned fishes and in larval agnathans, dipnoans, and amphibians, and they atrophy during the equivalent of a late larval stage. They were also reported as being functional for a time in turtle and crocodilian embryos; otherwise, in amniote embryos they are vestigial or lost.

Mesonephros

Under the stimulus of the pronephric duct acting as an inductor, additional tubules develop sequentially in the

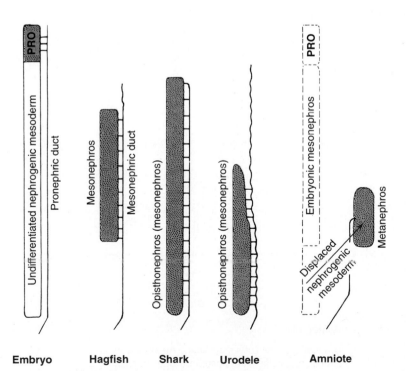

FIGURE 15.8

Fate of the nephrogenic mesoderm (*red*) in representative vertebrates. *Pro,* Pronephros. The pronephric duct persists in adult anamniotes to drain the adult kidney.

nephrogenic mesoderm behind the pronephric region and establish an entrance into the existing pronephric duct. These are destined to give rise to the **mesonephros.** With regression of the pronephric kidney and development of a mesonephros, the pronephric duct is thereafter called the **mesonephric duct.**

There is usually a gradual transition from tubules characteristic of the embryonic pronephric region to the more complicated nephrons characteristic of a definitive mesonephros. In the transitional zone, the tubules may be segmentally disposed, and in fishes, amphibians, and reptiles, they have open nephrostomes that usually close later in development. However, nephrostomes remain open in some adult fishes throughout life. Nephrostomes that do not open are common in this transitional region in the embryos of birds and mammals.

Little basis exists for drawing a definite boundary between the pronephros and the embryonic mesonephros, unless it is where one first finds secondary and tertiary tubules forming as buds from the initial segmental tubule in each body segment (fig. 15.6). As these secondary and tertiary tubules enlarge and encroach on one another, the metamerism of the developing kidney is at first obscured and then lost altogether. The mesonephric tubules remain throughout life to function as the adult kidney **(mesonephros)** of fishes and amphibians, and it is the functional embryonic kidney of amniotes.

The adult kidney of gnathostome fishes and of amphibians is sometimes termed an **opisthonephros** in recognition of the fact that kidney tubules in anamniotes may develop as far caudad as the cloaca, whereas in amniotes a section of the undifferentiated nephrogenic mesoderm anterior to the cloaca is displaced to give rise to the final (metanephric) kidney of amniotes. (For derivation of the prefix *opistho*, consult appendix 2.)

Agnathans
The kidney of the adult hagfish *Myxine* is an expression of the larval holonephros, with modifications. All nephrostomes have closed. The glomerular part occupies a 10-cm segment of the nephrogenic mesoderm commencing some distance behind the regressed pronephros and terminating some distance anterior to the cloaca. This segment consists of 30 to 35 large renal corpuscles up to 1.5 mm in diameter, strictly segmental, and connected to the longitudinal duct by very short tubules. There are no peritubular capillaries and no renal portal system. Between the regressed pronephros and the glomerular segment are a variable number of corpuscles that lack glomeruli or have lost their connection with the longitudinal duct. Caudal to the glomerular segment are additional aglomerular corpuscles. The total number of corpuscles, typical and atypical, is about 70, and they are strictly segmental.

The adult lamprey kidney is a long, thin fold extending about half the length of the coelom and, unlike other kidneys, hanging into it. Along its free edge is the mesonephric duct. There are large renal corpuscles and more tubules than body segments, but no peritubular capillaries and no nephrostomes.

Gnathostome Fishes and Amphibians
Some of the more anterior mesonephric tubules in male fishes and amphibians become utilized to transport sperm from the testis to the mesonephric duct via the substance of the kidney. This comes about because, instead of associating with a glomerulus, these tubules grow into the immediately adjacent embryonic genital ridge that gives rise to the testis (fig. 15.9). There they establish continuity with sperm channels within the developing testes, from which they eventually collect sperm. These mesonephric tubules are **vasa efferentia.** This region of the adult male kidney is therefore preempted for sperm transport and is known as the **sexual kidney.** The usually coiled part of the mesonephric duct that drains it is the **epididymis** (figs. 15.9 and 15.10a). The more caudal portion of the male kidney is the major uriniferous portion. There is no sexual kidney in male teleosts because sperm are not conveyed to the kidney (fig. 15.25f).

Shark

FIGURE 15.9

Urogenital system of a male shark. An accessory urinary duct, not shown, drains the more posterior kidney tubules.

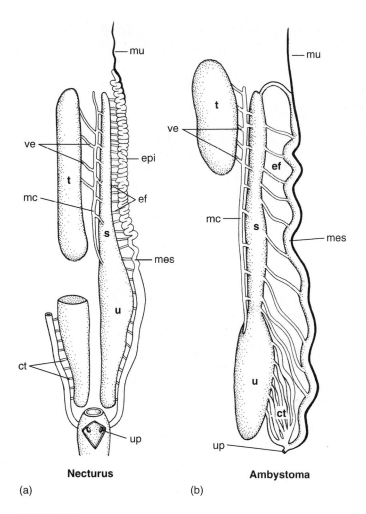

FIGURE 15.10

Left mesonephroi (opisthonephroi) and associated reproductive structures of two male urodeles, ventral view. The testes have been displaced mediad and the mesonephric ducts laterad. The sexual portion of the kidney, **s,** serves solely for sperm transport between testis and mesonephric duct. **ct,** collecting tubules; **ef,** efferent epididymal ducts; **epi,** epididymis (not highly coiled in *Ambystoma*); **mc,** marginal canal, for sperm collection; **mes,** mesonephric duct; **mu,** rudimentary oviduct (muellerian duct); **s,** sexual portion of kidney; **t,** testis; **u,** uriniferous kidney; **up,** urogenital papilla opening into cloaca; **ve,** vasa efferentia.

Accessory urinary ducts frequently supplement, or even replace, the mesonephric duct as a carrier of urine. This is particularly true in elasmobranchs. For an interesting variation in accessory urinary ducts, see figure 15.10*b*, ct.

Amniotes

The mesonephros is the functional kidney of amniotes during part of embryonic or fetal life (fig. 15.11). In embryonic chicks, it reaches a peak on the eleventh day of incubation—halfway through embryonic life. In placental mammals, it peaks earlier (nine weeks of gestation in humans). The first mesonephric tubules in humans appear during the fourth week of embryonic life. A wave of differentiation sweeps along the nephrogenic mesoderm, establishing a mesonephros; but before the last mesonephric tubules have formed the earliest ones have already regressed (fig. 15.12). As a result, the human mesonephros at its peak consists of about 30 functioning renal corpuscles, although as many as 80 have been known to form. During the time the mesonephros is functioning in amniotes, a new kidney, the metanephros, is organizing. When the metanephros is able to function, the mesonephros regresses, except for remnants. However, it functions as late as the first hibernation in some lizards and the first molt in some snakes, and it is still functioning in monotremes and marsupials at birth.

Mesonephric Remnants in Adult Amniotes

Vestiges of the embryonic mesonephros remain in adult mammals as two groups of blind tubules, the **epoophoron** and **paroophoron,** in the dorsal mesentery of the ovary (fig. 15.13), and the **paradidymis** and **appendix of the epididymis,** both near the epididymis in the dorsal mesentery of the testis (fig. 15.24). The mesonephric ducts remain as sperm ducts in all male amniotes. In females, mesonephric duct remnants include a short, blind **Gartner's duct** in the mesentery of each mammalian oviduct (fig. 15.13). Blind vestiges of the embryonic mesonephric duct are found in association with the ovaries of some lower amniotes.

Metanephros

The **metanephros,** the adult kidney of amniotes, organizes from the caudal end of the nephrogenic mesoderm, which becomes displaced cephalad and laterad during development (fig. 15.8, amniote). This is the same mesoderm that gives rise to the caudalmost part of the adult kidney of fishes and amphibians. Differentiation of the metanephric kidney commences when a hollow **metanephric bud** sprouts from the caudal end of the mesonephric duct (fig. 15.13*a*). Surrounding the bud is nephrogenic mesoderm. The bud grows cephalad, carrying the metanephric blastema along with it. Eventually, the basic components of a metanephric kidney organize in the displaced nephrogenic mesoderm, which continues to enlarge. The hollow stalk connecting the metanephros with the embryonic mesonephric duct becomes the **ureter,** and the end of the stalk surrounded by the developing blastema gives rise to the urinary channels within the kidney up to and including the common collecting tubules. Meanwhile, S-shaped renal tubules are organizing within the blastema. One end of each renal tubule grows toward and encapsulates a glomerulus to form a renal corpuscle; the other end grows toward, and acquires an opening into, a common collecting tubule. Failure to establish this connection may result in a fluid-filled renal cyst. In mammalian

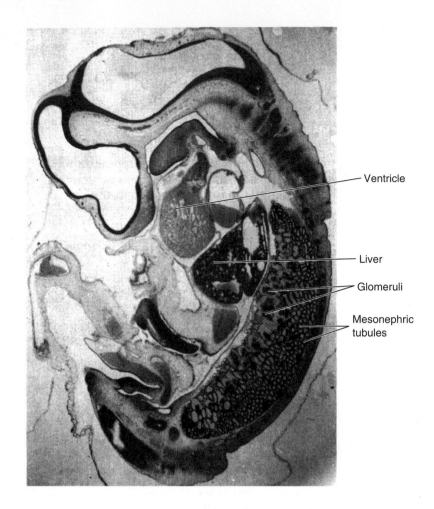

Ventricle

Liver

Glomeruli

Mesonephric
tubules

FIGURE 15.11

Mesonephros (*red*) of a 20 mm pig embryo.

From Phillips, J. B.: *Development of Vertebrate
Anatomy* 1975 (Fig. 18–7, page 440).

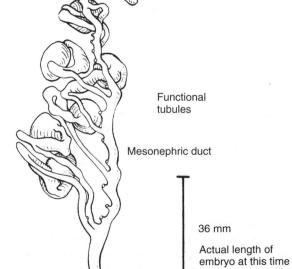

Regressing
tubules

Functional
tubules

Mesonephric duct

36 mm

Actual length of
embryo at this time

Human embryo

FIGURE 15.12

Right functional mesonephros of a 36 mm human embryo.

From M. D. Altschule, "The Change(s) in the Mesonephric Tubules of Human
Embryos Ten to Twelve Weeks Old" in *Anatomical Record,* 46:81, 1930. Copyright
© 1930 American Association of Anatomists. Reprinted by permission of John
Wiley & Sons, Inc.

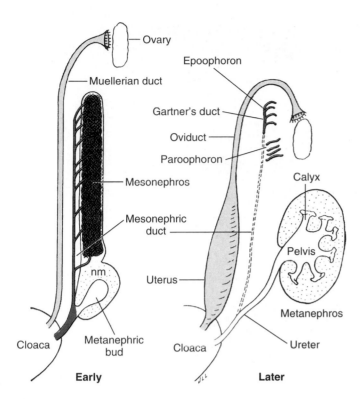

FIGURE 15.13

Developmental changes in urogenital system of female mammals. In the early stage (*left*), the mesonephric kidney and duct are present (*red*), and the metanephric bud has formed at the caudal end of the nephrogenic mesoderm (**nm**). A muellerian duct (oviduct) parallels the mesonephric duct. In the later stage (*right*), the mesonephric kidney and duct have regressed except for remnants (*red*), and the muellerian duct has differentiated to form a female reproductive tract.

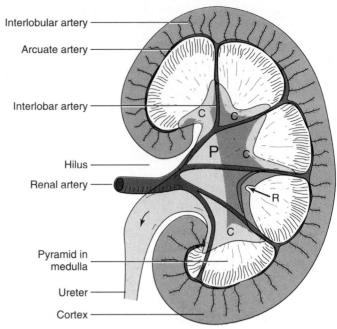

FIGURE 15.14

Mammalian kidney, frontal section, renal vein removed. Glomeruli are confined to the cortex (*red*) and are supplied by interlobular arteries. Loops of Henle, vasa recta, and common collecting tubules constitute the bulk of the pyramids. One calyx has been cut open revealing a Renal papilla (**R**). **P,** pelvis; **c,** calyx.

kidneys, a **renal pelvis** with funnel-shaped extensions (**calyxes**) collects urine from the common collecting tubules (figs. 15.13*b* and 15.14).

The tubules of mammalian kidneys have a long U-shaped **loop of Henle** of narrow diameter and with very thin epithelia-lined walls. The loop is interposed between proximal and distal convolutions of the tubule (fig. 15.3*b*). As the loops develop they grow away from the periphery of the kidney, where the glomeruli are located, and toward the renal pelvis. The kidney therefore consists of a **cortex** housing the glomeruli, convoluted tubules, and the upper ends of the loops of Henle, and of a **medulla** consisting of the loops and collecting tubules (figs. 15.14 and 15.15). Associated with the loops and paralleling each of them are similarly shaped loops of peritubular capillaries, the **vasa recta.** Blood in the vasa recta and filtrate in the loops of Henle flow in opposite directions. The adaptive value of this arrangement is explained in discussions of the countercurrent multiplier theory in textbooks on physiology. The loops of Henle, the vasa recta, and the collecting tubules are aggregated into **pyramids** to which they give a striated appearance. Each pyramid—as many as 20 in some

mammalian kidneys—tapers to a blunt tip, a **renal papilla,** that projects into a calyx along with a few other papillae. The tip of each papilla is perforated by numerous small openings through which urine from the collecting tubules enters the calyx en route to the ureter. Each collecting tubule drains a small number of renal tubules—7 to 10 in humans. Avian kidneys have loops similar to mammalian loops of Henle, but they are much shorter than those of most mammals.

Glomerular filtrate enters the loop of Henle at given concentrations of salt and water. It is the role of the loops of Henle, along with the convolutions and the collecting tubules, to so alter the osmotic pressure of the fluid within the tubule that the urine excreted is either *hyperosmotic* (higher solute concentration) or *hypoosmotic* to the blood, depending on the need to conserve or excrete salts and water. The segments of the tubule from neck to collecting tubule each has its own role. Their collective activity maintains a constant osmotic state of the circulating blood plasma within the narrow limits that will sustain life. Salt moves out of the loop of Henle and into the surrounding interstitial fluid, or it moves from the interstitial fluid into the loop. A difference in osmotic pressure between the two sites and selective tubular permeability results in *passive movement* of salt in one direction or the other. When called upon, hormones cause *active transport* of salt into or out of the loops against an osmotic gradient. Water is reclaimed from the filtrate in

FIGURE 15.15

Cortex of a mammalian kidney, frontal section. The surface of the kidney is near the top.

From G. Bevelander and J. A. Ramaley, *Essentials of Histology,* 8th edition. Copyright © 1979 Mosby-Year Book, Inc. Reprinted by permission of G. Bevelander.

Common collecting tubule

Loop of Henle

Glomerulus

Distal convoluted tubule

Proximal convoluted tubule

Bowman's capsule

FIGURE 15.16

Lobulated metanephric kidneys.

Lizard   **Newborn human**   **Adult otter**

the collecting tubules as needed. Excretion of a dilute urine rids the body of excess water.

Reptilian, avian, and some mammalian kidneys are lobulated, each lobe consisting of clusters of many tubules (figs. 15.16, 15.28, 15.29). The kidneys of snakes, legless lizards, and apodans are elongated, conforming to the slender body. Those of birds lie snugly against the contours of the sacrum and ilium. In most mammals, the kidneys are smooth and bean shaped, and the arteries, veins, nerves, and ureters enter and leave at a median notch, the **hilus.**

Because of their embryonic origin as buds off the mesonephric ducts, the ureters at first terminate in these ducts, and they do so throughout life in male sphenodons and lizards (fig. 15.26). In other reptiles and in

birds and monotremes, they ultimately open into the cloaca. In placental mammals, the ureters open into the urinary bladder because of differential growth of components of the embryonic cloaca, to be explained later (figs. 15.30 and 15.31).

Extrarenal Salt Excretion

Vertebrates that live in an environment laden with salt or that inhabit arid environments and can afford little body water for carrying away accumulated salts have extrarenal structures that supplement the kidneys in this role. Marine teleosts have chloride-secreting glands on the gills, and elasmobranchs and coelacanths have chloride-secreting rectal glands that utilize little water and empty into the caudal end of the intestine. (The rectal glands of bullsharks caught in freshwater are smaller than those caught in salt water and show regressive changes.)

Marine turtles and snakes have salt glands. So do iguanas that live in trees on the shore but feed on seaweed; and so, too, do birds that scoop fish out of seawater. Terrestrial lizards and snakes that inhabit arid environments also have them. These same lizards and snakes have atrophied glomeruli, which also conserves water. The salt glands of reptiles and birds are compact and sometimes lobulated and consist of an abundance of tubular crypts that are drained by a single duct.

The salt-excreting glands of lizards are located below the orbits just external to the olfactory capsule, and their ducts empty into the nasal canals. Whitish incrustations of sodium chloride and potassium can be seen at the nostrils. Those of marine birds are large and are located above each orbit. Their long ducts open at the base of the beak just lateral to the nostrils. A groove on the beak extends from this opening to the tip of the beak. Within 15 minutes after these birds have drunk water containing sodium and potassium salts, minute drops of fluid containing these trickle down the groove and drip or are shaken off the beak. The salt glands of sea turtles open into the orbit, and those of sea snakes open into the oral cavity in the fashion of salivary glands.

The copious mucous coat of hagfishes, which inspired the name slime eels, has been shown to be rich in salt. Hagfishes have very short kidney tubules, which deprives them of one route for salt excretion. It may be that integumentary mucous glands play a significant role in osmoregulation in those agnathans.

Sweat glands in mammals eliminate salt, but salt loss by this route is incidental to secretion of water for its evaporative cooling effect. Whereas the excretion of salt by some routes is hormone regulated, salt loss via sweat glands is unregulated. In fact, salt lost by this route, unless in excess in the tissues, must be replaced by dietary salts.

URINARY BLADDERS

Urinary bladders are adaptations to life on land. They serve as reservoirs for water that may be needed later and therefore should not be wasted. Urinary bladders arise during embryonic development as an evagination from the ventral wall of the embryonic cloaca; and in amphibians, turtles, primitive lizards, and monotremes—the only tetrapods below placental mammals that have a urinary bladder excepting a few ratite birds—the bladder continues to open into the cloaca throughout life (figs. 15.17 and 15.38, amphibians; 15.26, 15.27b, 15.40, and 15.41, reptiles; and 15.43, monotreme). Since the ureters also open into the cloaca in these tetrapods, urine backs up into the bladder from the cloaca unless the vent is open.

The urinary bladder of placental mammals, like that of all other tetrapods, develops as an evagination from the embryonic cloaca, but the evagination in placentals

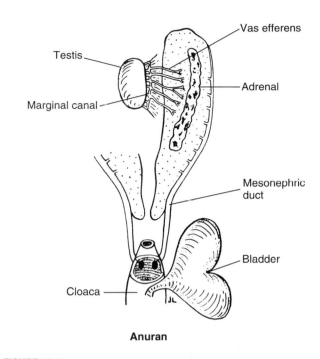

Anuran

FIGURE 15.17

Urogenital system and adrenal of a male frog, ventral view.

is called an **allantois.** The ureters of placental mammals empty into the urinary bladder. The bladder is drained by the urethra.

The distal tip of the urinary bladder of placental mammals is connected to the umbilicus (navel) by a **middle umbilical ligament,** or **urachus.** This is a fibrous remnant of the part of the embryonic allantois that persists within the coelom distal to the bladder after birth (fig. 14.38). The umbilicus and the urachus are more prominent in some mammals (primates, for instance) than in others (cats, for example). The urachus occupies the anterior border of the ventral mesentery of the bladder along with the obliterated umbilical arteries. The bladders of amphibians may also exhibit a persistent ventral mesentery. This mesentery and the ventral mesentery of the liver (falciform ligament) are the sole remnants in adult vertebrates of the ventral mesentery that, in embryos, extended the entire length of the coelom.

Active water reabsorption from the bladder, a water-conservation adaptation, is evoked by **antidiuretic hormone.** This is a neurosecretory hormone from the hypothalamus that is released from the posterior lobe of the pituitary when dehydration threatens (fig. 18.2). In one species or another, stored urine is put to an additional use. Some freshwater turtles have voluminous bladders and, sometimes, accessory bladders that are used by females to carry water for softening and moistening the soil when a nest is being prepared for eggs (figs. 15.27 and 15.41). Urinary bladders in mammals make it possible for pheromones in the urine to be deposited at sites where they can serve as a signaling device to other members of the species, male or female. (Fire hydrants seem popular in cities.) In some lower vertebrates, essential ions that are scarce in the specific environment are also reclaimed from stored urine.

The smooth muscle layers in the bladder wall are disposed in a reverse pattern to that in the digestive tract, having inner longitudinal and outer circular layers. Much elastic tissue is interspersed among the muscle fibers, and the cells of the epithelial lining slide over one another, thinning out as the bladder fills and heaping as it empties. Consequently, the size of the lumen is at all times just sufficient to accommodate the contained urine, whatever the quantity.

Organs called urinary bladders in fishes, when present, arise from nephrogenic mesoderm rather than from the cloaca. They are not homologues of tetrapod bladders. Some ray-finned fishes have an unpaired, sometimes bicornuate sinuslike enlargement at the caudal junction of the two mesonephric ducts. In some species it develops into an elongated sac termed a **tubal bladder** (fig. 15.18). Lobe-finned fishes have a small diverticulum from the dorsal wall of the cloaca that has been called a urinary bladder. Cyclostomes and cartilaginous fishes have no such structures.

GENITAL ORGANS

Gonadal Primordia

Gonads have two major functions: (1) to produce gametes and (2) to synthesize steroidal hormones. The hormones are essential for the differentiation, growth, and maintenance of **accessory sex organs** (reproductive ducts and their glands), **secondary sex characteristics** (sex-specific plumage in birds, for example), and sexual behavior.

Gonads arise as a pair of embryonic **genital ridges,** which are thickenings of the coelomic mesothelium just

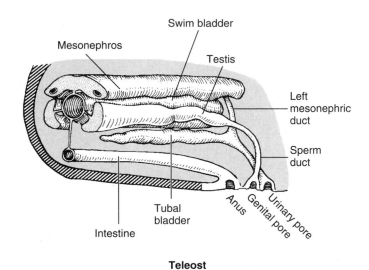

Teleost

FIGURE 15.18

Caudal end of urogenital system of a male teleost (pike), left lateral view. The unpaired urinary bladder arises as a bud off the conjoined caudal ends of the two mesonephric ducts. Note absence of cloaca.

Source: Data from E. S. Goodrich, *Studies on the Structure and Development of Vertebrates,* 1930, Macmillan and Company, Ltd., London.

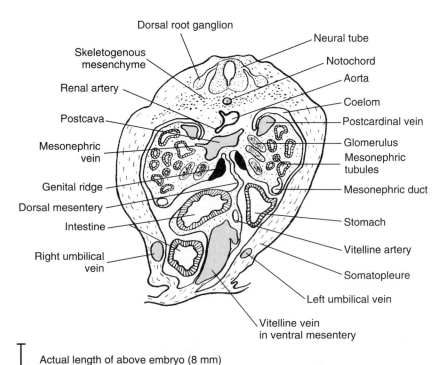

FIGURE 15.19

Opossum embryo (*Didelphis*) 6-1/2 days after cleavage, cross section. The gonadal primordia are shown in *black*.

Actual length of above embryo (8 mm)

Actual length of opossum at birth (10 mm)

medial to the mesonephroi (fig. 15.19, black). The ridges are longer than the resulting mature gonads, which suggests that at one time gonads may have extended the length of the pleuroperitoneal cavity, as they do in living cyclostomes. **Primordial germ cells** (progenitors of sperm and eggs) migrate into the genital ridges and establish themselves just internal to the peritoneum (fig. 15.20). This region of a gonadal primordium thereby becomes the **germinal epithelium.** Although primordial germ cells are first observed in the germinal epithelium, their progenitors are in the endoderm of the yolk sac and elsewhere before genital ridges have developed. From these locations they migrate to the genital ridges.

During early differentiation, the gonads have the potential to become either ovaries or testes. Under the influence of genes, hormones, and factors not yet fully identified, these bisexual gonads develop into the gonads of a specific sex. As gonads enlarge, they generally acquire a dorsal mesentery, the **mesorchium** in males, the **mesovarium** in females.

In gonadal primordia that are becoming testes, **primary sex cords** invade the deeper region of the genital ridge and become **seminiferous tubules** lined with germinal epithelium (figs. 15.20a–d, 15.21, and 15.22). These tubules produce sperm. Primary sex cords form also in females, but they are not the source of ova. Oocytes form from **secondary sex cords** that also arise from the germinal epithelium (fig. 15.20b, e, and f). These cords do not grow deep into the medulla. Consequently, with the exception of some teleosts, eggs form just internal to the surface of the ovary and are shed to the exterior when mature (fig. 15.35). The primary sex cords of females regress, but remnants remain deep in the ovarian medulla, where they are called medullary cords.

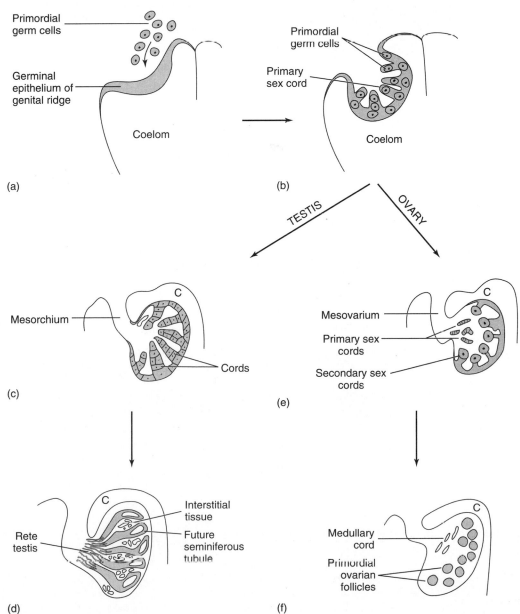

FIGURE 15.20

Early development of mammalian gonads. (a) Germinal epithelium on genital ridge of sexually indifferent primordium. (b) Primary sex cords of bisexual stage. (c)–(f) Differentiating testis and ovary. **c,** coelom.

Spermatozoa

Primary spermatocyte

Spermatogonia

Seminiferous tubule

FIGURE 15.21

Section of mammalian testis. Portions of four seminiferous tubules are shown. Interstitial cells (*light red*) are source of male sex hormones (chapter 18).

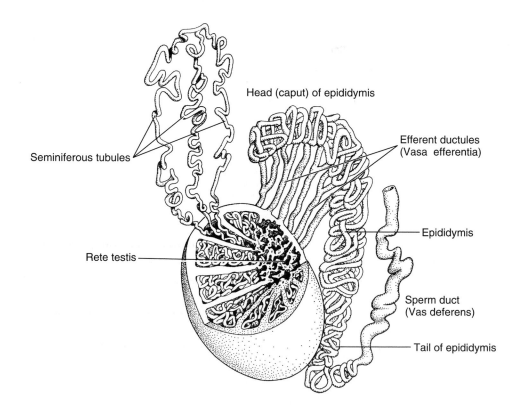

Head (caput) of epididymis

Efferent ductules (Vasa efferentia)

Seminiferous tubules

Epididymis

Rete testis

Sperm duct (Vas deferens)

Tail of epididymis

FIGURE 15.22

Mammalian testis. The efferent ductules (*vasa efferentia*) are modified mesonephric tubules.

A few adult vertebrates have a single gonad because of fusion of the two ridges across the midline, as in lampreys and a few teleosts, or because one of the juvenile gonads, most commonly an ovary, fails to differentiate. Among vertebrates with a single functional ovary are hagfishes, some viviparous elasmobranchs, a few reptiles, and most species of birds. A few female mammals, including the platypus and some bats, have only one ovary.

Instances of natural or experimental sex reversal can be demonstrated in many vertebrates. The embryonic testis of anurans consists of an anterior segment, **Bidder's organ,** which in frogs usually disappears before sexual maturity, and a more caudal portion that becomes the functional testis. Bidder's organ persists in adult male toads (fig. 15.23) and contains large undifferentiated cells resembling immature ova. If the testes of a toad are removed experimentally and Bidder's organs are allowed to remain with a blood supply, the latter develop into functional ovaries and the rudimentary female duct system enlarges under the influence of female sex hormones produced by the new ovaries. Hens that have not been subjected to any treatment have been known to cease laying eggs, crow, and develop other male characteristics. This comes about when the left ovary atrophies and the right ovary, which is rudimentary in birds, enlarges and produces male hormones. However, these birds are not known to have produced sperm.

In all vertebrates, mature *eggs are released into the coelom* or a compartment of it before being swept into an oviduct. *Sperm,* on the contrary, except in cyclostomes, *are conducted in a closed system of vessels* that conveys them from the testis to the exterior without having entered the coelom. All cyclostome gametes are shed into the coelom.

Testes and Male Genital Ducts

Vertebrate testes are essentially the same with few exceptions. Germinal epithelium lines the seminiferous tubules where sperm are formed. Mature sperm are microscopic and abundant (3.5 billion/cc in roosters). They separate from the lining of the tubules and, propelled by flagellumlike tails, swim the length of the tubule to reach the vasa efferentia, which leads to the sperm duct. In mammals, sperm are first collected in a network of fine channels, the **rete testis,** before entering the vasa efferentia (fig. 15.22). Vasa efferentia are mesonephric tubules that invaded the developing testis instead of becoming associated with glomeruli (fig. 15.24). The vasa efferentia (usually called **efferent ductules** in mammals) conduct sperm to the spermatic duct. Efferent ductules number a dozen, more or less, in humans. In most vertebrates with **mesonephric kidneys,** the mesonephric ducts carry *both urine and sperm* (fig. 15.25a and d).

Hagfishes, lampreys, some jawed fishes, and urodeles lack seminiferous tubules. In these vertebrates, the germinal epithelium may be on the surface of the testis or deep within, and primordial germ cells migrate from the germinal epithelium into cystlike **seminiferous ampullae,** where they mature before being discharged. At the end of the spawning season, after hundreds of thousands of spermatozoa have been shed from each ampulla, the cysts collapse.

In some fishes and urodeles, there has been a tendency to form a new sperm duct to replace the mesonephric duct as a carrier of sperm (figs. 15.10a and b, **marginal canal;** and 15.25d and e). In teleosts this has culminated in a mesonephric duct that carries no sperm whatsoever (fig. 15.25f). *In all other fishes and amphibians, and from reptiles to humans, the embryonic mesonephric duct remains in adult males to carry sperm.* A duct that carries only sperm is termed a **vas deferens (ductus deferens).**

Spermatic ducts generally empty into the cloaca or a derivative thereof in vertebrates below placental mammals (figs. 15.26, 15.27, 15.28, 15.29). In placental mammals, they empty into the urethra at the prostate gland (fig. 15.30). This development is a result of the complete separation of the embryonic cloaca into a urogenital sinus and rectum (fig. 15.49e and f, duct in black).

As a result of caudal migration of the testes in mammals during late fetal life, the spermatic ducts become "caught" or "hung up" on the ureters, so that thereafter they loop over the latter en route to the urethra (figs. 15.30 and 15.31).

A number of multicellular glands produce secretions that contribute to seminal fluid in amniotes. These are most abundant in mammals, in which one or more open into the genital tract at or near the junction of the

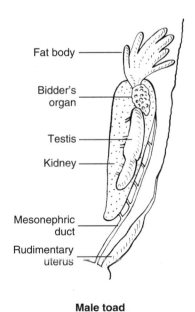

Fat body

Bidder's organ

Testis

Kidney

Mesonephric duct

Rudimentary uterus

Male toad

FIGURE 15.23

Bidder's organ and the rudimentary female reproductive tract of a young male *Bufo,* ventral view. Only the left organs are illustrated. The mesonephric duct and rudimentary uterus empty into the cloaca.

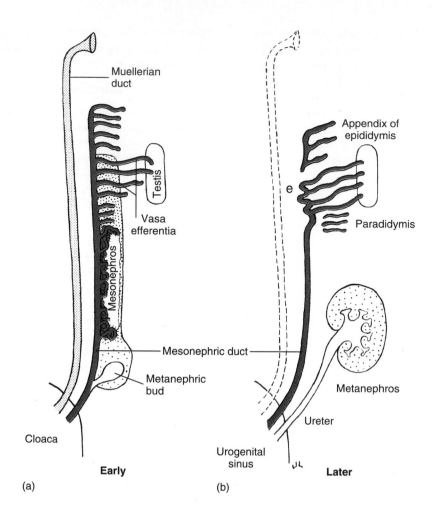

FIGURE 15.24

Developmental changes in the urogenital system of male mammals. In the early (*bisexual*) stage, there is a rudimentary female duct (*muellerian duct*) and a mesonephros. Some of the mesonephric tubules have invaded the genital ridge (*testis*) to become vasa efferentia. Later (*right*), the muellerian duct has regressed (*broken lines*), the mesonephros has regressed except for remnants (*appendix of epididymis, paradidymis*), and the mesonephric duct remains to carry sperm. **e,** epididymal portion of mesonephric duct.

spermatic ducts and the urethra (fig. 15.31). Among these are **ampullary glands,** several median and paired **prostate glands, seminal vesicles, coagulating glands,** and **bulbourethral (Cowper's) glands.** Not all are present in every mammalian order. Seminal vesicles have only a glandular role in mammals, although they serve as storage sites for sperm in lower vertebrates. Coagulating glands cause semen to coagulate in the vagina. This results in a **copulation plug** that precludes another mating for several hours or days. Other glands produce mucus, nutritive secretions, or fluids that neutralize the slight acidity of the vaginal contents, thereby providing a more suitable milieu for motile sperm.

The urethra in male mammals is termed the **prostatic urethra** where the prostate glands empty, **membranous urethra** from the prostate glands to the base of the penis, and **spongy urethra** within the penis (fig. 15.30).

Intromittent Organs

In species in which fertilization is internal, males with few exceptions develop **intromittent (copulatory) organs** for introducing sperm into the female reproductive tract. These are present in fishes with internal fertilization, in the anuran family Ascaphidae (frogs lacking ear drums), in reptiles, in a few birds, and in all mammals. In urodeles with internal fertilization and in most birds, the cloaca is eversible, facilitating sperm transfer. Internal and external fertilization are discussed in chapter 5.

The intromittent organs of elasmobranchs are grooved fingerlike **claspers,** which are modified pelvic fins (fig. 10.8c). Embedded in each fin at the base of some claspers is a muscular **siphon sac** that adds copious quantities of an energy-rich mucopolysaccharide to the semen. In many teleosts, a modified anal fin, or **gonopodium,** serves for sperm transfer. In the frog

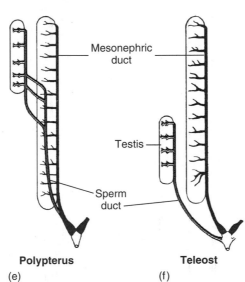

FIGURE 15.25

The mesonephric duct (*black*) as a carrier of sperm and urine. (*a*) Carrying both sperm and urine. (*b*) Carrying urine from the anterior part of the kidney only; chiefly a spermatic duct. (*c*) Carrying sperm only. (*d*) to (*f*) Trend toward a separate sperm duct (*red*) in fishes, the mesonephric duct ultimately carrying only urine. Note the reverse in amniotes.

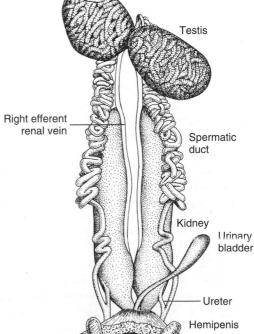

FIGURE 15.26

Urogenital organs of the male lizard *Anolis carolinensis,* ventral view. The sperm duct is the persistent mesonephric duct. The hemipenes are seen in the everted (erected) position.

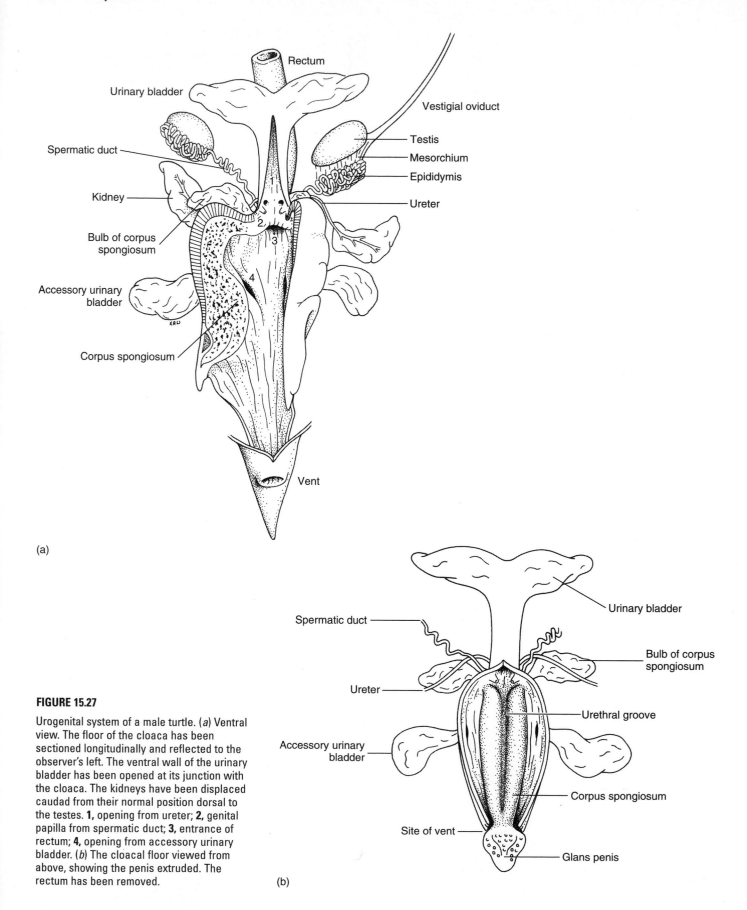

(a)

(b)

FIGURE 15.27

Urogenital system of a male turtle. (*a*) Ventral view. The floor of the cloaca has been sectioned longitudinally and reflected to the observer's left. The ventral wall of the urinary bladder has been opened at its junction with the cloaca. The kidneys have been displaced caudad from their normal position dorsal to the testes. **1,** opening from ureter; **2,** genital papilla from spermatic duct; **3,** entrance of rectum; **4,** opening from accessory urinary bladder. (*b*) The cloacal floor viewed from above, showing the penis extruded. The rectum has been removed.

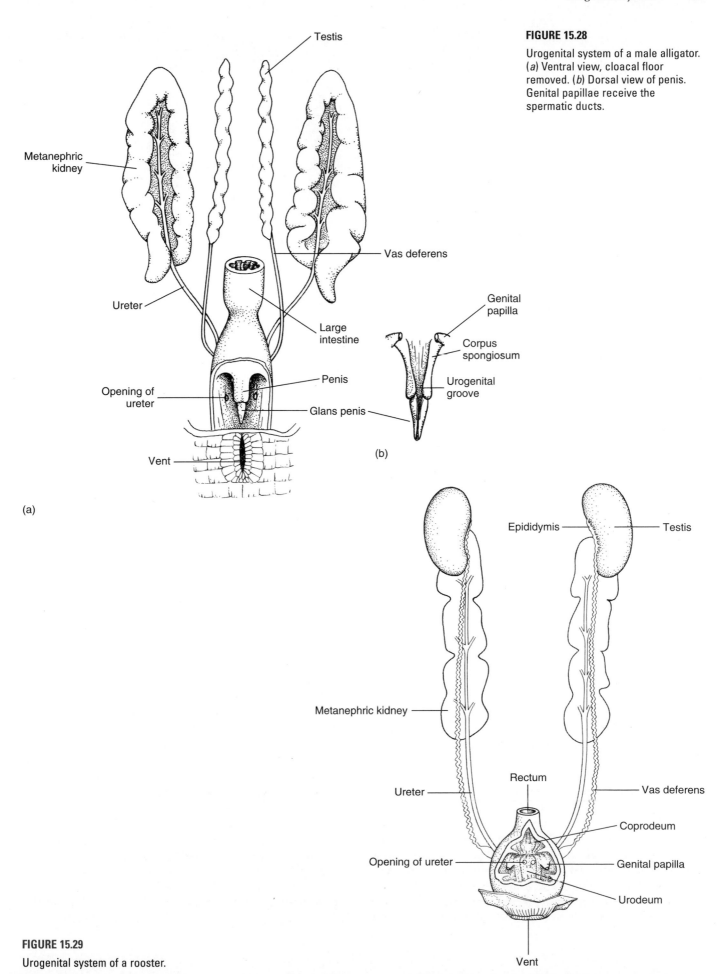

Testis

Metanephric kidney

Vas deferens

Ureter

Large intestine

Opening of ureter

Penis

Glans penis

Vent

(a)

FIGURE 15.28

Urogenital system of a male alligator. (*a*) Ventral view, cloacal floor removed. (*b*) Dorsal view of penis. Genital papillae receive the spermatic ducts.

Genital papilla

Corpus spongiosum

Urogenital groove

(b)

Epididymis

Testis

Metanephric kidney

Rectum

Ureter

Vas deferens

Coprodeum

Opening of ureter

Genital papilla

Urodeum

Vent

FIGURE 15.29

Urogenital system of a rooster.

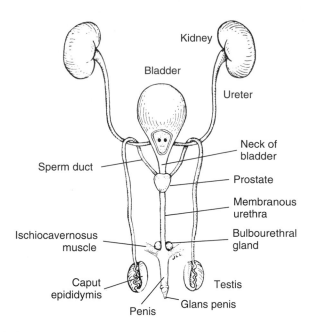

FIGURE 15.30
Urogenital system of a male cat, ventral view.

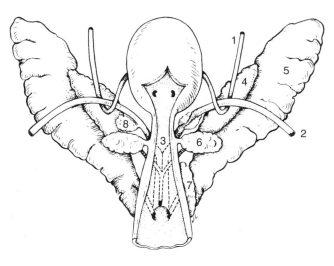

FIGURE 15.31
Secondary sex glands of a male hamster, ventral view. The bladder and urethra have been opened to show entrances of ducts. **1**, ureter; **2**, spermatic duct; **3**, prostatic urethra; **4**, coagulating gland; **5**, seminal vesicle; **6**, cranial prostate; **7**, caudal prostate; **8**, ampullary gland. A bulbourethral gland enters the urethra farther caudad.

Ascaphus, the intromittent organ is a permanent tubular, tail-like extension of the cloaca.

Male snakes and lizards have hemipenes, a pair of protrusible saclike diverticula stored in an inverted position in pockets under the skin at the entrance to the cloaca (fig. 15.26). They are held in the pocket by retractor muscles. Each hemipenis has a deep spiral groove on its surface. During copulation the retractor muscles relax, the sacs turn inside out and protrude on the surface, and

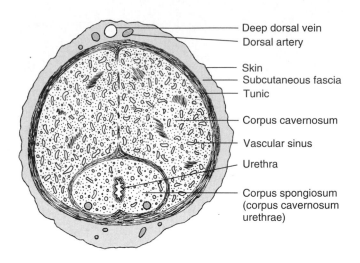

FIGURE 15.32
Proximal cross section of a fetal primate penis.

semen flows along the groove and into the cloaca of the female. Rudimentary hemipenes are present in females.

Male turtles, crocodilians, a few birds (ostriches, ducks, others) and monotremes exhibit an unpaired erectile **penis.** In its simplest form, as seen in reptiles, the penis is a thickening of the floor of the cloaca consisting chiefly of spongy erectile tissue, the **corpus spongiosum,** which has a **urethral groove** on its surface and ends in a **glans penis** (figs. 15.27 and 15.28). The spongy tissue consists of cavernous blood sinuses that, when engorged, cause the glans penis to be extruded through the cloacal aperture. The glans is richly supplied with sensory endings that reflexly stimulate ejaculation. The urethral groove channels sperm toward the cloaca. The penis of monotremes is similar to that of reptiles, being in the cloacal floor; but the groove on the dorsal surface has become folded into the corpus spongiosum to form a tube, and it carries only sperm, urine being conveyed to the cloaca by a separate passageway.

The penis in placental mammals develops from an embryonic **genital tubercle** and a pair of **genital folds** that border the urogenital sinus in the embryos of both sexes (fig. 15.48b). The folds grow toward one another in the midline to form a groove and, eventually, a closed tube within which is the urogenital sinus. This tube, surrounded by the corpus spongiosum, becomes the spongy urethra within the penis. It eventually carries urine and sperm. The glans penis, the expanded distal end of the corpus spongiosum, is ensheathed by a fold of skin, the **prepuce,** except during erection. Two additional erectile masses, the **corpora cavernosa,** develop within the penis of placental mammals (fig. 15.32).

The genital tubercle of females generally develops into a spongy, penile prominence, the **clitoris,** embedded in the floor of the urogenital sinus (vestibule of the vulva in higher primates). Like the penis, the clitoris is erectile, sensitive to tactile stimulation, and ensheathed by a prepuce.

The fate of the clitoris of female rodents and hyenas is slightly different. In rodents, it develops into a closed tube, as in males, and becomes a urinary papilla (fig. 15.48c). Like the penis, it encloses the terminal segment of the urethra and conducts urine to the exterior. In female hyenas, the clitoris also becomes a tube; but in these mammals it encloses an extension of the urogenital sinus that, in most mammals, is a common terminal passageway for both the urinary and genital systems (fig. 15.49c). Therefore, in hyenas, the clitoris carries urine, enables copulation, and serves as a birth canal. Yet it looks precisely like a male penis. Perhaps this is what hyenas are laughing about.

Ovaries

Although ovaries are usually thought of as being compact organs like those of sharks and placental mammals, this is not the case in a large number of female vertebrates. The ovaries of many teleosts have a central cavity (fig. 15.33). Those of anurans and urodeles are folded thin-walled sacs, and those of some reptiles and of birds and monotremes exhibit numerous internal, irregular, fluid-filled lacunae. On the other hand, the ovaries of fishes other than teleosts are generally compact, as are those of turtles, crocodilians, and placental mammals.

The hollow ovary of teleosts results from entrapment of a small part of the coelomic cavity within the developing ovary (fig. 15.34). Because of this, the cavity—an isolated pocket of coelom—is lined by the germinal epithelium; and the small eggs or, in viviparous teleosts, the young, are discharged into that cavity. The latter may be directly continuous with an oviduct (fig. 15.33) or partially surrounded by an oviducal funnel composed of peritoneal folds that have become so arranged as to form a tube. The ovary of some other teleosts is also hollow but for a different reason: it results from a secondary hollowing out of the interior of the ovary at each ovulation. However, the eggs are discharged in a similar manner. The germinal epithelium of most other vertebrate ovaries, whether compact, saccular, or lacunate, is in a

Teleost

FIGURE 15.33

Female reproductive system of a teleost. Ova are shed into the ovarian cavity. In some teleosts, the ovary reaches almost to the genital pore.

cortex immediately internal to the visceral peritoneum, and the eggs are discharged into the coelom through eroded areas in the visceral peritoneum. At the end of each reproductive season, the ovaries of most vertebrates below placental mammals regress to a state resembling those of juveniles.

The ovaries of placental mammals (fig. 15.35) are compact, the only cavitation being the fluid-filled **antra** within the egg follicles (**graafian follicles**) as ovulation approaches. The number of ovulated ova is not large in mammals—1 to 3, generally—although the number in some carnivores and small rodents is larger. The South

FIGURE 15.34

Two methods of entrapment of coelom to form a permanently hollow ovary in teleosts. The genital ridges are shown in cross section.

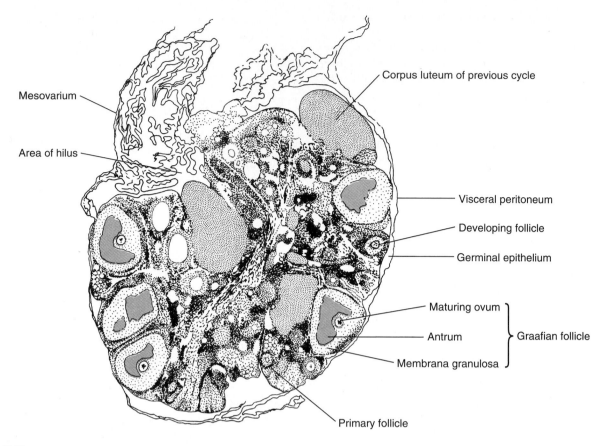

Mesovarium

Area of hilus

Corpus luteum of previous cycle

Visceral peritoneum

Developing follicle

Germinal epithelium

Maturing ovum ⎤
Antrum ⎬ Graafian follicle
Membrana granulosa ⎦

Primary follicle

FIGURE 15.35

Section of hamster ovary at approach of estrus (heat) and ovulation, which occurs every fourth day unless pregnancy intervenes. The follicles are in the germinal epithelium.

African elephant shrew is atypical. It ovulates 50 or more eggs, although there is room for only 2 to implant and develop to term. The average number of mature eggs ovulated in most species is slightly larger than the average number that implant, and the average number that implant exceeds the average number born. Thus, nature compensates for possible prenatal loss just as in teleost fishes, in which millions of eggs may be extruded, but far fewer are fertilized.

The fully mature mammalian graafian follicle bulges on the surface of the ovary. Its wall becomes very thin in the hours immediately preceding ovulation, and it finally ruptures. The ovum, surrounded by a corona of follicular cells, then oozes through a narrow slit in the wall and into the coelomic fluid. The escape into the coelomic fluid is the process of **ovulation.** Following ovulation, the many cells that remain in the ovulated follicle are altered histologically and physiologically under the stimulus of luteinizing hormone from the pituitary, and they organize a **corpus luteum** (fig. 15.35). Thereafter, the corpus luteum is a major source of the progesterone required for the maintenance of pregnancy. Before ovulation, estrogen is the predominant secretion of these cells.

Corpora lutea are not confined to mammals. They have been identified in every vertebrate class having taxa that retain ovulated eggs or embryos in the female tract for a period of time. The endocrine role of gonads is discussed in chapter 18.

Translocation of Mammalian Gonads

The caudal pole of each embryonic gonad is connected, by a ligament, to the coelomic floor of one of the **genital swellings,** the latter being shallow evaginations of the embryonic coelom (fig. 15.48*b*). These swellings evaginate further to become **scrotal sacs** in males and **labia majora** (the major lips of the **vulva,** or external genitalia) in females (fig. 15.36). Partly because elongation of the ligaments does not keep pace with elongation of the fetal trunk and partly as a result of actual shortening of the ligaments during late fetal development, the gonads, anchored to the caudal floor of the coelom, are displaced caudad toward the scrotal sacs in males and toward the labia majora in females. Testes are displaced farther caudad than ovaries, generally entering the scrotal sacs on a temporary or permanent basis. The scrotal sacs are lined by coelomic peritoneum, which, in this location, is the **internal spermatic fascia** (fig. 15.36, male).

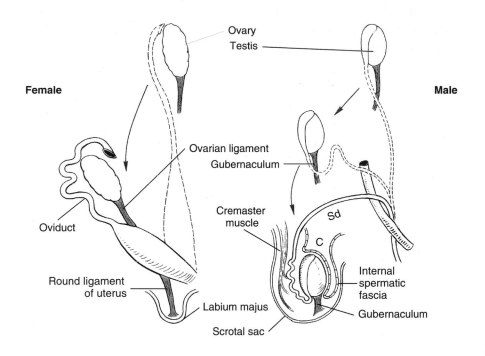

FIGURE 15.36

Caudal displacement of mammalian gonads, ventral view. The ovarian ligament and round ligament of the uterus collectively are homologous with the male gubernaculum. *Arrows* indicate the route of translocation of the right gonads. **C,** scrotal recess of coelom; **Sd,** spermatic duct arching over the ureter. The internal spermatic fascia is part of the coelomic peritoneum.

In males the ligament connecting the testis with the floor of a scrotal sac is the **gubernaculum.** In females it is the **ovarian ligament** between the ovary and the uterus, and the **round ligament of the uterus** between the uterus and the floor of the labium majus (fig. 15.36). The gubernacula appear to play a role in the descent of the testes, although their removal does not always prevent descent.

Testes are housed permanently in the scrotal sacs in many mammals, including ungulates, carnivores, some marsupials, and higher primates. In other mammals, the testes can be lowered into the scrotal sacs and retracted at will (rabbits, bats, a few rodents, some lower primates, and others). The passageway between the abdominal cavity and the scrotal cavity is the **inguinal canal.** The entrance to the canal is the **inguinal ring** (fig. 15.37). In species that retract the testes, the canal remains broadly open to the coelom. In species in which the testes are permanently confined to the **scrotum** (the two scrotal sacs), the inguinal ring eventually becomes only wide enough to accommodate the spermatic cord. An abnormally wide ring may become the site of an inguinal hernia.

The **spermatic cord** is a composite structure composed of the spermatic duct and the spermatic arteries, veins, lymphatics, and nerves, all wrapped in the internal spermatic fascia and all dragged into the scrotal cavity along with the testis. Scrotal sacs do not develop in monotremes, elephants, whales, or a few other mammals. In these, the testes remain permanently in the

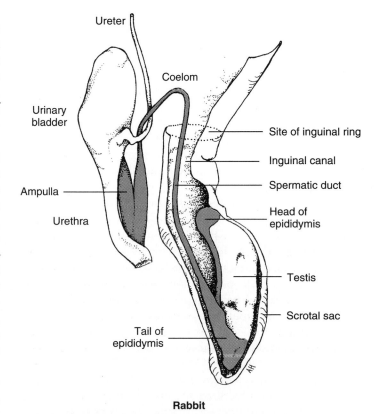

Rabbit

FIGURE 15.37

Rabbit testis in scrotal cavity. The inguinal canals are broadly open to the main coelom at the inguinal ring; therefore the testes are retractable. Sperm pathway from testis to urethra is shown in *red.*

abdomen in a retroperitoneal location; that is, between the parietal peritoneum and the body-wall muscle. Testes that descend do so by sliding behind the peritoneum (internal spermatic fascia, fig. 15.36).

Scrotal sacs are adaptations that maintain mammalian sperm at a temperature cooler than that within the abdominal cavity in which the sperm of most mammals cannot survive. A further adaptation is seen in the relationship between the internal spermatic artery and vein as they approach and leave the testis. These vessels are intimately associated in a rete, the **pampiniform plexus,** where heat is transferred from arterial to venous blood. As a result, blood entering the testis has been cooled in the plexus, and blood returning to the trunk has been warmed. The plexus also conserves body energy in the form of heat.

Female Genital Tracts below Placental Mammals

With the exception of agnathans, which have no genital ducts, and ray-finned fishes, whose female genital tracts are aberrant, the female duct systems of vertebrates below placental mammals all reflect a generalized pattern of structure. It consists of paired muscular and glandular **oviducts** that commence at an **ostium,** an opening to the coelom, leading to the neck of an **oviducal funnel (infundibulum)** (fig. 15.38). Oviducts end in the cloaca or a subdivision of it. Any differences along the length of the tracts are in the degree of regional specializations that support species-specific functions such as collecting eggs from the coelom; coating them with protective or nutrient substances; temporarily housing eggs, embryos, or larvae and maintaining them until the eggs are shed or young are delivered; expelling the eggs or the young; receiving the male intromittent organ, if any; and temporarily housing viable sperm in species in which eggs normally are not mature at the time of mating.

In vertebrates other than ray-finned fishes, the female duct system differentiates from a pair of embryonic **muellerian ducts** (fig. 15.13). These develop in both sexes, but do not persist in males. The muellerian ducts of elasmobranchs and amphibians arise by longitudinal splitting of the pronephric ducts, and their ostia develop from one or several anterior pronephric nephrostomes. In other vertebrates, each muellerian duct arises as a longitudinal groove in the coelomic epithelium paralleling the mesonephric duct. The groove subsequently becomes a tube except at the anterior end where the ostium at least partially surrounds the ovary. Eventually, it separates from the mesonephros but remains attached to the body wall by a dorsal mesentery, the **mesotubarium.**

The muellerian ducts of female sharks give rise to an oviduct that develops a **shell gland** and **uterus** (fig. 15.39). The cephalic half of the shell gland secretes albumen. The caudal half secretes a shell that is very thin in viviparous species, leathery-tough in oviparous

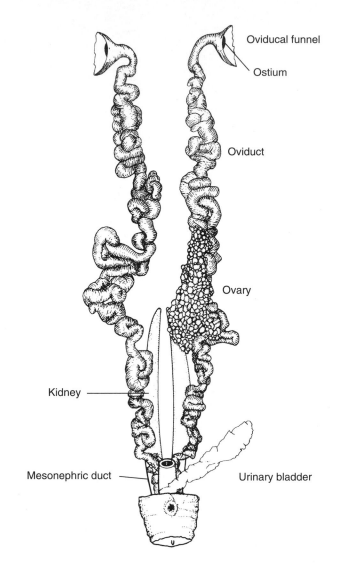

FIGURE 15.38

Generalized female duct system (*Necturus*).

species, and often with tendrils that attach the egg case to seaweed. Two embryonic ostia unite to form a single ostium at the neck of an oviducal funnel. The funnel in sharks is supported by the falciform ligament.

After observing the large size of a mature shark egg, laboratory students often ask how such a huge egg can enter the small ostium and traverse the narrow upper region of the oviduct. We know the answer, because we know how this happens in chickens. At the time the egg is released from the ovary it is a shapeless, almost flowing mass of yolk—no albumin—contained in a nonrigid cell membrane like the yolk of a fresh chicken egg. Under the influence of ovulatory hormones, the fringe (**fimbria**) of the oviducal funnel waves about in a gentle undulating movement. When it comes in contact with the shapeless mass, whether the yolk is still in the ovary or separated from it, the fimbria embraces the yolk, gently at first and then more firmly, drawing the yolk into

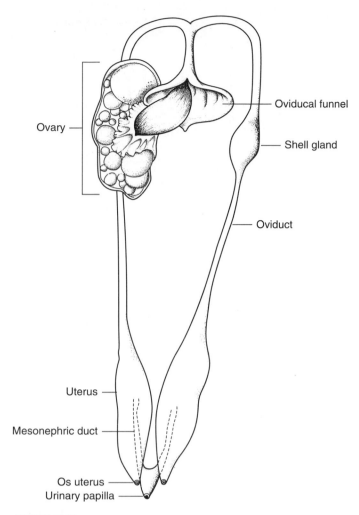

FIGURE 15.39

Reproductive system of female *Squalus,* ventral view. The left ovary has been removed. The right ovary shows an ovulating ovum being drawn into the oviducal funnel. An adult ovary typically exhibits follicles at varying stages of maturation. A urinary sinus precedes the urinary papilla.

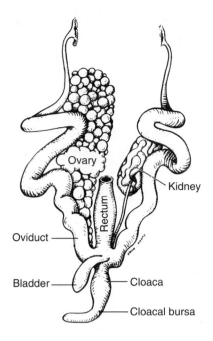

FIGURE 15.40

Urogenital system of female aquatic turtle, *Trionyx euphraticus,* ventral view. The left ovary has been removed.

Source: Courtesy of Mohamad S. Salih, University of Baghdad.

the funnel until the egg is entirely engulfed (fig. 15.39). Muscular contraction of the funnel then squirts the shapeless mass through the ostium and into the oviduct. Thereafter, peristalsis of the oviducal wall moves the yolk caudad. Cilia play an insignificant role in retrieval of macrolecithal eggs from the coelom. However, in placental mammals, cilia along with undulations of the oviducal funnel *are* important since the eggs are microlecithal.

The oviducts of teleosts are not homologous with those of other vertebrates. They arise as caudally directed folds of the coelomic peritoneum created by the genital ridges as the latter entrap the coelom to form a hollow ovary (fig. 15.34). Consequently, the oviducts are usually continuous with the ovarian cavity throughout life (fig. 15.33). These oviducts are short funnels that open into paired or unpaired external genital papillae. The unpaired papillae of some teleosts elongate to form tubelike **ovipositors** through which the eggs are

extruded. The continuity of the oviducts with the ovarian cavity is lost secondarily in some species, and in these species the eggs are extruded into the coelom.

The linings of amphibian oviducts are richly supplied with glands that secrete several layers of protective jelly envelopes around each egg as it moves down the tube. The caudal portions in anurans become voluminous thin-walled **ovisacs** where ovulated eggs accumulate until they are shed. In terrestrial anurans, this occurs after a heavy rain has provided appropriate pools for egg-laying during amplexus (mating). Similar sacs in ovoviviparous urodeles shelter unborn young. The oviducts of dipnoans are similar to those of oviparous urodeles.

Female tracts of reptiles, birds, and monotreme mammals conform closely to the basic pattern (figs. 15.40 to 15.43). However, only one of the two embryonic muellerian ducts may differentiate, as in crocodilians and most birds. The rudimentary right muellerian duct of birds is seen in figure 15.42. Albumen glands line part of the oviduct, and there is usually a prominent shell gland just anterior to the cloaca (figs. 15.42 and 15.43). The shell remains leathery in species whose eggs develop in moist environments; it becomes brittle in eggs that develop in dry air. The difference depends on the nature of the secretions. Ovoviviparous reptiles as well as oviparous ones cover their eggs with a shell.

Although the female tracts of birds are similar to those of reptiles, some of the anatomic terminology differs.

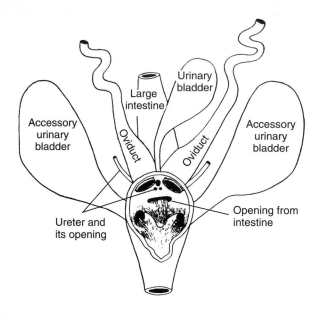

FIGURE 15.41

Cloaca and associated structures of a female terrestrial turtle, ventral view. Clitoris has been removed.

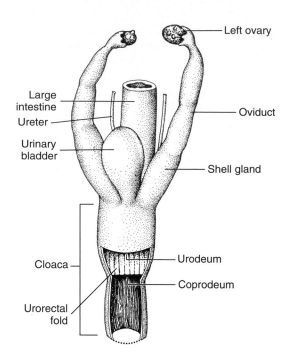

FIGURE 15.43

Genital tract and cloaca of a female monotreme, ventral view. The right ovary is usually the smaller, or it is rudimentary. The cephalic half of the cloaca is partitioned by the urorectal fold into a urodeum and coprodeum. An exterior view of the vent is seen in fig. 15.48*a*.

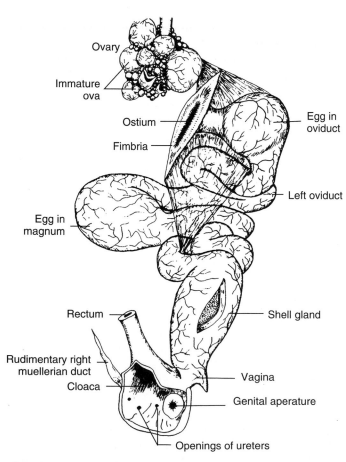

FIGURE 15.42

Genital tract of a hen. The presence of two eggs in the oviduct is unusual.

The albumen-secreting area in fowl is termed the **magnum,** the shell gland is a **uterus,** and the short glandular and muscular segment at the entrance to the cloaca is a **vagina.** The vagina secretes mucus that seals the pores of the shell, preventing the loss of vital water vapor from within the shell without affecting the supply of oxygen to the chorioallantoic membrane (fig. 5.13). The vagina then expels the egg to the exterior.

Spermathecae, which are dorsal diverticula of the female cloaca or tubular crypts in the oviducal lining, are present in many urodeles, lizards, snakes, and domestic fowl. They store sperm received during mating. This ensures viable sperm at the time of next ovulation, which, in some species, is weeks or months after mating.

Female Tracts of Placental Mammals

As in lower vertebrates, paired muellerian ducts in mammals give rise to the female genital tract commencing with the ostium of the oviduct. However, the muellerian ducts of mammals above marsupials unite at their caudal ends to one degree or another. The extent of the union varies with the mammalian order. The tract of true placentals (eutherians), therefore, consists of two **oviducts,** a **uterus** with or without horns **(cornu),** and a **vagina.**

Oviducts

Oviducts, or **Fallopian tubes** as they are also called in mammals, are relatively short, small in diameter (accommodating microlecithal eggs), more or less convoluted, and lined with cilia. The oviducal funnel has a fringed border. In many mammals, a membranous fold of peritoneum, the **ovarian bursa,** envelopes the ovary and the oviducal funnel, entrapping them and a small coelomic cul-de-sac. The bursa increases the probability that ovulated eggs will enter the oviduct rather than be lost in coelomic fluid, perhaps to implant somewhere in the coelomic cavity. The bursa may be broadly open to the main coelom, as in cats and rabbits; it may communicate with the coelom by a mere slit, as in most carnivores and rats; or it may be completely closed off from the coelom, as in hamsters. There is no ovarian bursa in humans. Ectopic pregnancies with fetuses implanted in the ventral abdominal wall are occasionally found in cats and rabbits in anatomy laboratories, and they occur in humans. The fetuses, of course, cannot be delivered.

Uteri

The muellerian ducts of marsupials are paired all the way to the urogenital sinus, a subdivision of the embryonic cloaca (fig. 15.44). Marsupials therefore have two completely separate uteri (a **duplex uterus**) and two vaginas. The uteri of true placental mammals exhibit varying degrees of fusion distally, resulting most often in a **bicornuate uterus**—a uterus with two horns in which the young develop (figs. 15.45 and 15.47b). A uterus with two horns may have two totally separate passageways

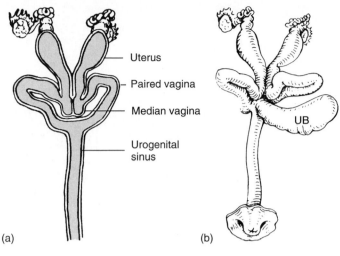

Duplex uteri and vaginae
Marsupial

FIGURE 15.44

Reproductive tract of a female opossum. **UB,** urinary bladder. (a) Section showing internal structure, urinary bladder is absent. (b) Surface view.

within the body of the uterus although this is not discernible from external view (figs. 15.46, rabbit, and 15.47a). When this is the case, the uterus is more precisely termed **bipartite.** In some mammals, one uterine horn is much larger than the other, and blastocysts implant in that horn—the right in impalas—even though both ovaries produce viable eggs.

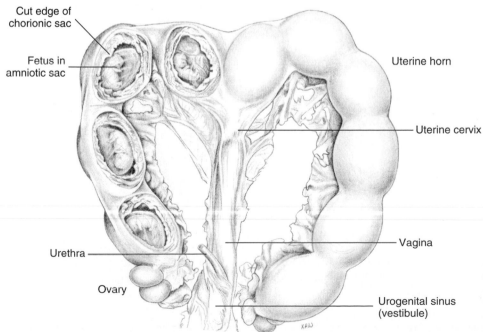

FIGURE 15.45

Gravid bicornuate uterus of a dog. The chorionic sacs have been partially cut away to expose the fetuses within the amniotic sacs. The body of the uterus is insignificant in the pregnant dog.

FIGURE 15.46

Bipartite and simplex uteri. **1**, ovary; **2**, oviduct; **3**, uterine horn; **4**, body of uterus; **5**, vagina; **6**, urogenital sinus; **7**, urinary bladder; **8**, urethra; **9**, vestibule of vulva. There are two complete passageways within the rabbit uterus all the way to the vagina.

Maximal fusion of the muellerian ducts occurs in armadillos, some bats, and in monkeys, apes, and humans, in which fusion commences at the ends of short oviducts and there are no uterine horns. This is a **simplex** uterus (figs. 15.46, monkey, and 15.47*c*). Blastocysts implant in the body of the uterus, and there is usually only one fetus per pregnancy. Armadillos are exceptional; they always give birth to identical quadruplets.

The narrow neck, or **cervix,** of the uterus of placental mammals projects into the vagina as the **lips of the cervix.** These surround the opening, or **os uteri,** of the uterus into the vagina (fig. 15.47*c*). Sperm deposited in the vagina pass between the lips of the cervix en route to the upper end of the oviduct, where fertilization takes place. The uterine lining **(endometrium)** of placental mammals becomes highly vascular under the stimulus of hormones before implantation of blastocysts. The thick muscular layer of the uterine wall **(myometrium)** assists in ejection of the young at birth, provided it, too, has been hormonally prepared. Before delivery, the cervix must dilate to provide a passageway large enough for the young to pass through. Dilation takes place under the influence of the hormone **relaxin** from the ovary. For review of the action of relaxin on the pubic symphysis, see chapter 10, Pelvic Girdles.

Vaginas

The vagina of placentals is the unpaired terminal portion of the embryonic muellerian ducts (fig. 15.47*b*). The cloaca contributes to it in some mammals (fig. 15.47*a* and *c*, red). Except in rodents and higher primates, the vagina empties into the **urogenital sinus,** a derivative of the embryonic cloaca (figs. 15.44, 15.45, 15.46, rabbit, and 15.47*b*). In many rodents, the vagina opens directly to the exterior (fig. 15.48*c*). In apes and humans, it opens into a shallow **vestibule of the vulva,** which also receives the urethra (fig. 15.46, monkey). The vestibule is bordered by two fleshy folds, the **labia minora,** derived from the genital folds (fig. 15.48*b*). It is overlaid by two additional folds, the **labia majora,** derived from the genital swellings. The latter are homologues of the scrotal sacs of males. The vagina is lined by a cornified epithelium for reception of the penis.

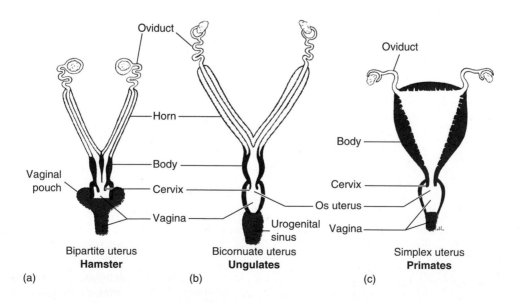

FIGURE 15.47

Varieties of mammalian uteri. *Blackened* regions are the fused caudal ends of the muellerian ducts; *red* is a derivative of the cloaca. Note the two lumens in the body of the bipartite uterus. The urogenital sinus in ungulates, as in most female mammals, receives the vagina and urethra.

(a)

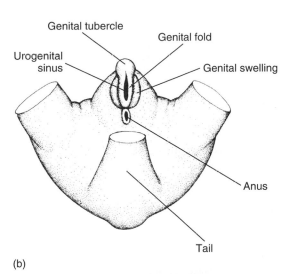

Genital tubercle

Genital fold

Urogenital sinus

Genital swelling

Anus

Tail

(b)

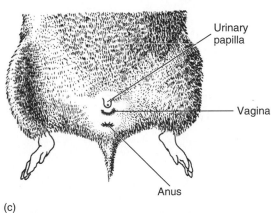

Urinary papilla

Vagina

Anus

(c)

FIGURE 15.48

External openings from mammalian cloaca or its derivatives. (*a*) Cloacal vent of the monotreme *Echidna*. (*b*) External genitalia of bisexual stage of 75 mm human embryo (about 12 weeks of age). The cloaca at this stage has been divided into urogenital sinus and rectum. For adult derivatives of these genitalia see table 15.1. (*c*) Perineal region of female hamster. At the tip of the urinary papilla is the urethral aperture.

The two pouchlike components of the **median vagina** of marsupials (fig. 15.44) lie against the urogenital sinus, separated from the latter by thin septa. At birth, the fetus is usually forced through the septa directly into the urogenital sinus, and the passageway thus established may remain throughout life. As an adaptation to paired vaginas, some male marsupials have a bifurcated penis.

Muellerian Duct Remnants in Adult Males

Although muellerian ducts do not mature in males, vestiges are common. In sharks, a pair of rudimentary oviducts commence at an ostium in the falciform ligament and curve around the anterior end of the liver on the surface of the coronary ligament, after which they disappear in connective tissue. Paired sperm sacs of male sharks (fig. 15.9) are caudal remnants of the muellerian ducts. A more or less complete, but rudimentary, muellerian duct persists in many male amphibians (fig. 15.23). In anurans, removal of the testes, the source of male sex hormones, has been shown to induce the rudimentary muellerian ducts to develop into functional uteri and oviducts. Vestiges are also common in other male vertebrates. Remnants in male mammals include an **appendix testis** (not to be confused with appendix epididymis), a small cystlike nodule with an epithelia-lined lumen, a homologue of the oviducal funnel; and a **prostatic sinus (vagina masculina),** an unpaired saccular remnant of the fused caudal ends of the muellerian ducts homologous with the vagina and opening into the male urethra near the prostate gland.

CLOACA

Urinary and genital ducts empty into the cloaca in amphibians and in fishes other than those in which an adult cloaca is shallow or absent. In the latter—predominantly teleosts—genital and urinary ducts open directly to the exterior (figs. 15.18 and 15.33). They also empty into the cloaca in some reptiles. In other reptiles, and in birds and monotremes, a horizontal **urorectal fold** partitions the cephalic end of the cloaca into two channels—a **urodeum** that receives the urogenital ducts and a **coprodeum** that receives the alimentary canal (figs. 15.29 and 15.43). The terminal (unpartitioned) portion of the cloaca is sometimes called a **proctodeum,** but it is not homologous with the ectodermal proctodeum of embryos.

Placental Mammals

In mammals that deliver live fetuses—that is, marsupials and true placentals—the urorectal fold grows caudad until it reaches the cloacal membrane separating the cloaca from the exterior. This modification completely divides the cloaca into a urogenital sinus and a **rectum** (fig. 15.49*a, b,* and *c;* and *a, b,* and *e,* male). Rupture of the embryonic cloacal membrane at two sites provides two openings to the exterior, an **anus** and a **urogenital aperture** (fig. 15.48*b*). The embryonic urogenital sinus of both sexes

Bisexual stages

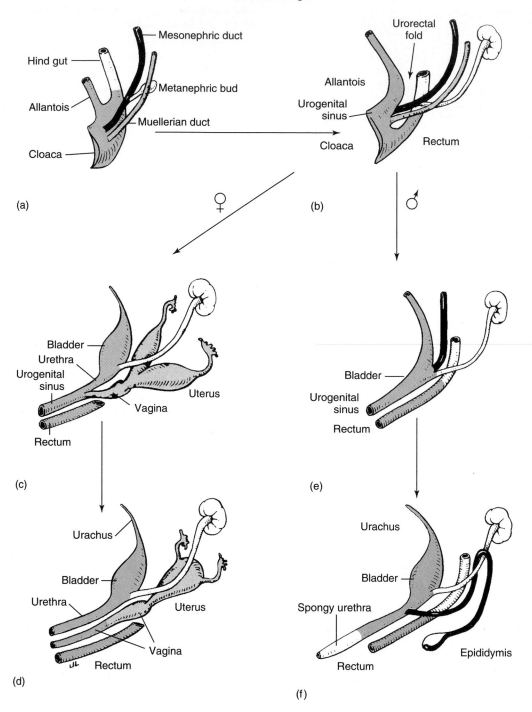

FIGURE 15.49

Fate of the cloaca and allantois (*red*), muellerian ducts (*gray*), and mesonephric duct (*black*) in placental mammals. (*a*) and (*b*) Bisexual stages. Only the left muellerian and mesonephric ducts are shown. In (*b*) the cloaca is becoming subdivided by the urorectal fold into a urogenital sinus ventrally and a rectum dorsally. (*c*) Typical adult female mammal. (*d*) Female primate. In (*c*) and (*d*) the contributions of both the left and right muellerian ducts are shown. (*e*) Developing male, intermediate stage in reorientation of mesonephric and metanephric ducts. (*f*) Adult male.

TABLE 15.1   Some Homologous Urogenital Structures in Male and Female Mammals

Indifferent Structure	Mature Male	Mature Female
Mesonephric duct	Ductus deferens (vas deferens)	Gartner's duct
	Epididymis	
Mesonephric tubules	Vasa efferentia	
	Appendix of epididymis*	Epoophoron*
	Paradidymis*	Paroophoron*
Muellerian duct	Appendix testis*	Oviduct
		Uterus
	Prostatic sinus (vagina masculina*)	Vagina
Gonadal ridge	Testis	Ovary
	Rete testis	Rete ovarii*
Gubernaculum	Gubernaculum	Ovarian ligament
		Round ligament of uterus
Genital swellings	Scrotal sacs	Labia majora
Genital tubercle	Penis	Clitoris
Genital folds	Contribute to penis	Labia minora
Urogenital sinus	Urethra, prostatic, and membranous portions	Urethra
		Urogenital sinus
		Lower vagina in rodents and primates

*Vestigial

receives the mesonephric ducts, muellerian ducts, and the future urinary bladder (allantois, fig. 15.49b).

As development progresses in males, the muellerian ducts regress and the urogenital sinus and rectum elongate (compare fig. 15.49b and e). The urogenital sinus eventually becomes continuous with the spongy urethra that has developed in the penis (fig. 15.49f). The terms "urogenital sinus" and "urethra" have now become synonyms when applied to males. By differential growth, the ureters become reoriented to open into the bladder, and the mesonephric ducts (now sperm ducts) continue to empty into the urogenital sinus (figs. 15.30 and 15.49f). This is the condition in all males.

As development progresses in females, the mesonephric ducts regress and the muellerian ducts unite at their caudal ends to form the body of the uterus and the vagina (fig. 15.49c, gray). The part of the urogenital sinus between the bladder and the entrance of the vagina is now known as the **urethra.** As a result of these developments, *most adult female mammals have two caudal openings to the exterior, a urogenital aperture and an anus.*

Female apes, monkeys, humans, and some female rodents develop an additional partition in the embryonic cloaca, this one in the urogenital sinus. It divides the urogenital sinus into a urethra and a vagina (fig. 15.49d). As a result, the embryonic female cloaca in these mammals becomes subdivided into a *urethra, vagina,* and *rectum,* each with its own exit. In monkeys and rodents *all three exit directly* to the exterior (fig. 15.48c). In apes and humans, the vagina and urethra open separately into the shallow vestibule of the vulva, noted earlier, and the anus opens directly to the exterior. The urogenital systems of these females have therefore evolved farther than those of males. The vagina in these females has a dual origin: the cephalic part is derived from the muellerian ducts, and the terminal portion is cloacal (fig. 15.47a and c).

From the discussions in this chapter it may have become apparent that all vertebrate embryos are bisexual; that is, they exhibit embryonic structures competent to become the urogenital organs of either sex. Under the initial influence of genes and, perhaps other unknown factors, and the subsequent stimuli of gonadal hormones, the indifferent embryonic rudiments differentiate to become the adult genital structures of one sex, or they remain rudimentary, become vestigial, or disappear. Abnormal, incomplete development of either or both sets of organs is a result of chromosomal aberrations and/or hormonal imbalances during fetal life.

True **hermaphroditism,*** the production of viable eggs and sperm by the same individual, is common in cyclostomes and some species of bony fishes, but it is rare among other lower vertebrates and absent in higher ones. These species lack sex chromosomes. Sexual differentiation then is determined by non-genetic factors. In hermaphroditic teleosts studied, the social environment, such as the disappearance of a dominant male or female, commonly directs an animal's sexual development.

Parthenogenesis, or asexual cloning, occurs in some lizards. In whiptail lizards, for example, there are no *genetic* males, although individuals engage in mating behavior, alternating between sexually typical male and female behaviors. David Crews has shown that sexual interactions between parthenogenetic animals yield more eggs laid for each than if they had bred alone.

Table 15.1 summarizes the fate in adult males and females of the chief sexually indifferent urogenital rudiments of mammalian embryos.

---

*Hermaphroditus, of Greek mythology, was the beautiful mythical son of Hermes and Aphrodite. While bathing in the mythical fountain of Salmacis, he became united in one body with the nymph in the fountain.

SUMMARY

1. The earliest role of kidneys was osmoregulation in an aquatic environment. Elimination of nitrogenous wastes is an added excretory role that has become significant in tetrapods. Kidneys replaced gills in this function.

2. The basic functional components of kidneys are glomeruli, renal tubules surrounded by peritubular capillaries, and a pair of longitudinal excretory ducts that empty into the cloaca or a derivative thereof.

3. Glomeruli are arterial retia that filter water and certain solutes from arterial blood. External glomeruli secrete into coelomic fluid, which is then swept by nephrostomes into the tubules. Internal glomeruli are in the body wall and filter into Bowman's capsules, which are extensions of tubules. Several tubules generally empty into collecting tubules that terminate in the longitudinal duct.

4. A glomerulus and its associated capsule constitute a renal corpuscle. A renal corpuscle and its tubule constitute a nephron.

5. Osmoregulation is affected by the tubules, which add to or subtract from glomerular filtrate in accordance with the metabolic needs of the organism.

6. Kidney tissue arises from a ribbon of intermediate (nephrogenic) mesoderm. Primitive pronephric tubules form at the anterior end of the ribbon, and thereafter a wave of differentiation sweeps along the ribbon, giving rise to tubules of increased complexity. As tubules form farther caudad, many that appear earlier regress.

7. An archinephros is a hypothetical archaic kidney with external glomeruli, nephrostomes, and simple tubules arranged metamerically the length of the trunk. The holonephros of larval hagfishes most nearly resembles the archinephros.

8. The anteriormost renal tubules constitute a temporary pronephric kidney in all vertebrate embryos and then regress. A mesonephros develops caudal and subsequent to the pronephros. The ducts of the pronephros continue to serve as two longitudinal mesonephric ducts.

9. The mesonephros is the functional kidney of adult fishes and amphibians and a temporary functional kidney in embryonic amniotes. In adult anamniotes, it is also called an opisthonephros.

10. Some of the more anterior mesonephric tubules invade the developing testis and become vasa efferentia that conduct sperm to the mesonephric duct. This duct in generalized fishes and amphibians carries both sperm and urine. It persists in amniotes to serve as the spermatic duct (vas deferens).

11. The caudal end of the nephrogenic ribbon is displaced laterad and cephalad to form the metanephric kidney of adult amniotes. Thereafter, the mesonephros regresses.

12. The metanephros consists of cortex, medulla, and pelvis. Renal corpuscles pack the cortex; long loops of tubules (loops of Henle) and their peritubular capillaries comprise much of the medulla. The tubules empty into the renal pelvis, which is drained by a ureter.

13. Vertebrates that imbibe excessive quantities of salt (marine vertebrates, birds whose food comes from the sea, reptiles that inhabit arid environments) excrete salt extrarenally via gills or rectal, nasal, orbital, oral, or integumentary salt glands.

14. Bladders of most fishes are derived from the conjoined ends of mesonephric ducts. They are absent in cyclostomes and elasmobranchs. Those of dipnoans arise as dorsal evaginations of the cloaca.

15. Tetrapod bladders arise from the floor of the cloaca and, except in placental mammals, empty directly into it. Bladders of placental mammals develop from the base of the allantois, a cloacal derivative, and empty into a urogenital sinus or, in higher female primates, to the exterior. They are drained by a urethra.

16. The primary role of urinary bladders in tetrapods other than placental mammals is to serve as a reservoir for urine, the water of which is recycled when needed by body tissues.

17. Gonads arise from paired genital ridges adjacent to developing kidneys. Germinal epithelium arises on the surface of the ridges and may remain there or be transposed to a deeper location.

18. Mature eggs are released into the coelom before entering an oviduct. Sperm are transported in a closed system of ducts that begins within the testes.

19. The germinal epithelium of most male vertebrates lines seminiferous tubules that empty into a rete testis, drained by vasa efferentia. These conduct sperm to the spermatic duct.

20. The testes remain in the abdominal cavity in a few mammals, descend permanently in some, and are retractile in others.

21. Except in a few lower vertebrates, the mesonephric duct is the spermatic duct of adult males. Below placental mammals, spermatic ducts open into the cloaca. In placentals, they open into the urethra.

22. Male intromittent organs of fishes are modifications of pelvic fins (claspers) or anal fins (gonopodia). Snakes and lizards have paired hemipenes. Turtles, crocodilians, some birds, and monotremes have an

unpaired penis in the cloacal floor containing a corpus spongiosum. Placental mammals have an external penis with a corpus spongiosum and two corpora cavernosa.

23. In females of reptilian species in which there is a male penis and in female mammals, a clitoris develops. It is the female homologue of the penis.

24. The ovaries of teleosts have a central cavity lined by the germinal epithelium. The cavity may be continuous with the cavity of the oviduct. Anuran and urodele ovaries are thin-walled sacs; those of some reptiles and of birds and monotremes have irregular fluid-filled lacunae. Other ovaries are generally solid, with the germinal epithelium on the surface. Following ovulation, the ovarian (graafian) follicles of mammals organize a corpus luteum that becomes a source of the progesterone that assists in the maintenance of pregnancy.

25. Mammalian external genitalia of both sexes differentiate from a median genital tubercle and paired genital swellings and folds. The tubercle gives rise to the penis or clitoris. Genital swellings give rise to scrotal sacs or labia majora. Genital folds contribute to the penis or become labia minora.

26. Ligaments in mammals connect the embryonic testes to the floor of the scrotal sacs and the embryonic ovaries to the labia majora. Shortening of the male ligaments (gubernacula) assists caudal migration of the testes toward or into the scrotum. Shortening of the female ligaments (ovarian ligaments and round ligaments of the uterus) causes lesser displacement of ovaries.

27. Female genital tracts from ostium to cloaca or its derivatives differentiate, except in teleosts, from muellerian ducts present in embryos of both sexes. Muellerian ducts in males regress. Teleost female tracts arise from peritoneal folds rather than from muellerian ducts.

28. A generalized female genital tract consists of an oviduct within a funnel-like ostium and terminates in the cloaca. Segments of the generalized oviduct are specialized for species-specific functions such as secretion of egg envelopes, temporary storage of ova or developing young, and expulsion of eggs or offspring.

29. The female tracts of placental mammals consist of Fallopian tubes (oviducts); duplex, bipartite, bicornuate, or simplex uteri; a uterine cervix; and a vagina. The latter empties into a urogenital sinus in most female mammals, into the vestibule of the vulva in monkeys, apes, and humans.

30. The ovaries of some placental mammals occupy an ovarian bursa, a cul-de-sac of the coelom that may be partly or completely closed off from the main coelom. It ensures that ovulated eggs will not become implanted in an ectopic location.

31. An adult cloaca receives the large intestine, urinary bladder when present, and urinary and genital ducts. A cloaca is shallow or absent in many fishes. A single aperture opens to the exterior.

32. In reptiles, birds, and monotremes, a urorectal fold divides the anterior portion of the embryonic cloaca into a urodeum that receives the urinary and genital ducts, and a coprodeum that receives the rectum. The caudal portion of the cloaca remains undivided.

33. In female placental mammals except monkeys, apes, and humans, the embryonic cloaca becomes completely divided into a urogenital sinus and rectum and two apertures open to the exterior.

34. In some female rodents and in female monkeys, apes, and humans, the embryonic cloaca is divided into a urethra that drains the bladder, a vagina, and a rectum, and each opens to the exterior independently.

SELECTED READINGS

Crews, D.: Animal sexuality, Scientific American **271**:108–114, 1994.

Dawley, R. M.: An introduction to unisexual vertebrates. In Dawley, R. M., and Bogart, J. P.: Evolution and ecology of unisexual vertebrates, Albany, New York, 1989, Bulletin 466, New York State Museum. An introduction to parthenogenesis and other forms of reproductive unisexuality in all-female fishes, amphibians, and reptiles. Knowledge of genetics recommended.

Fox, H.: The amphibian pronephros, Quarterly Review of Biology **38**:1–25, 1963.

Fox, H.: The urinogenital system of reptiles. In Gans, C., and Parsons, T. S., editors: Biology of the reptilia, vol. 6, New York, 1977, Academic Press.

Moffat, D. B.: The mammalian kidney, New York, 1975, Cambridge University Press.

Mossman, H. W., and Duke, K. L.: Comparative morphology of the mammalian ovary, Madison, 1973, The University of Wisconsin Press. Informative, clearly written, richly illustrated.

Pang, P. K. T., Griffith, R. W., and Atz, J. W.: Osmoregulation in clasmobranchs, American Zoologist **17**:365, 1977.

Peaker, M., and Linzell, J. L.: Salt glands in birds and reptiles, Monographs of the Physiological Society, New York, 1975, Cambridge University Press.

Prosser, C. L., editor: Comparative animal physiology, ed. 4, vol. 2, Philadelphia, 1990, W. B. Saunders Co.

Rankin, J. C., and Davenport, J: Animal osmoregulation, New York, 1981, John Wiley and Sons.

Shapiro, D. Y.: Differentiation and evolution of sex change in fishes, BioScience **37**:490–497, 1987.

Stahl, B. J.: Vertebrate history: problems in evolution, New York, 1974, McGraw-Hill Book Co.

Vincent, A.: A seahorse father makes a good mother, Natural History, December, p. 34, 1990. Courtship, mating, and true pregnancy in male seahorses, and a rare photograph of their elaborate filamentous dermal appendages.

Wourms, J. P., and Callard, I. P.: Evolution of viviparity in vertebrates, American Zoologist **32**:251, 1992.

CHAPTER SIXTEEN

NERVOUS SYSTEM

In this chapter we will examine the nervous system and its parts. We will see how the components are assembled and what they do to ensure survival. Much of our attention will be focused on neurons, the living cells that perform the actual conductive and secretory activities of the system.

OUTLINE

The vertebrate nervous system plays three basic roles. It acquaints the organism with its external environment and stimulates the organism to orient itself favorably in that environment; it participates in regulation of the internal environment; and it serves as a storage site for information. These functions are accomplished by the nerves, spinal cord, and brain in association with **receptors** (sense organs) and **effectors** (chiefly muscles and glands).

The organism must constantly monitor the external environment. Information is supplied by **afferent (sensory) nerves** commencing in sense organs. The response (body movement) is initiated by nerve impulses over **efferent (motor) nerves** that stimulate the skeletal muscles of the body and thus cause the fish to swim or the tetrapod to crawl, run, or fly. Information from the external environment is also employed in the regulation of internal secretions, such as seasonal release of reproductive hormones (fig. 18.5).

The organism also has an internal environment that must be continually monitored and controlled. Afferent nerves from visceral receptors carry information in the form of nerve impulses to the central nervous system; efferent nerves carry impulses from the center to visceral effectors, chiefly smooth and cardiac muscles and glands.

Memory (information storage) is a function of the nervous system. Without information storage and recall, no animal could modify its behavior in accordance with experience, and every situation would be faced as if it were the first time. As experiences multiply, information accumulates and the penalty of past errors and the rewards of successes modify behavior accordingly.

The nervous system is subdivided for convenience into central and peripheral nervous systems. The **central nervous system (CNS)** consists of the brain and spinal cord. The **peripheral nervous system** consists of cranial, spinal, and autonomic nerves. Cranial nerves emanate from the brain and spinal nerves from the spinal cord.

NEURON

To understand the anatomy of the nervous system, one must be acquainted with the **neuron**—the living nerve cell. The neuron is to the nervous system what a muscle cell is to the muscle system: it performs the specific function of the system. In this instance, it is to transmit nerve impulses.

Neurons exhibit many shapes, but all have a cell body containing **Nissl material** (sites of high protein synthesis) and one or more processes (figs. 16.1 and 16.2). The longest process, distinguished by the absence of Nissl material, is an **axon.** It transmits nerve impulses to other neurons or to an effector. Axons are surrounded by a layer of fatty **myelin** that varies in thickness from heavy to ultramicroscopic, depending on the role of the axon

(fig. 16.3). The axon and its myelin constitute a **nerve fiber.** On the surface of nerve fibers outside the CNS is a sheath of living cells, the **neurilemma.** The neurilemma produces and maintains the underlying myelin.

Nerve fibers within the CNS extend for short or long distances up and down the spinal cord or within the brain, aggregated into functional bundles called **fiber tracts** (T, in fig. 16.4a and b). Nerve fibers in the peripheral nervous system are in nerves (fig. 16.3). In fact, a **nerve** is one or more bundles of nerve fibers outside the CNS, wrapped in a fibrous sheath **(epineurium)** and supplied by blood vessels **(vasa vasorum).**

The other processes of neurons are **dendrites** (fig. 16.2). They are short extensions of the cell body that provide additional surface areas for receipt of incoming impulses from other neurons. Their cytoplasm, like that of the cell body, contains Nissl material. To observe how the neuron fits into the peripheral nervous system, we will examine a sensory nerve, a motor nerve, and two mixed nerves.

A typical sensory nerve is diagrammed in figure 16.4a. Its fibers commence in a sense organ (in this instance, the membranous labyrinth) and terminate in the brain (this example) or the spinal cord. The cell bodies of sensory nerve fibers, with few exceptions, are found in a **sensory ganglion** on the pathway of the nerve. A **ganglion** is a group of cell bodies outside the central nervous system. A sensory ganglion contains sensory cell bodies. In lower vertebrates, some sensory cell bodies are scattered along the nerve.

A typical motor nerve is diagrammed in figure 16.4b. The cell bodies of motor nerve fibers in cranial and spinal nerves are inside the CNS in a **motor nucleus.** Neurologically speaking, a **nucleus** is a group of cell bodies within the brain or cord. Motor nuclei contain the cell bodies of motor nerve fibers. The motor fibers of cranial nerve XII terminate peripherally in striated muscle. There are few purely motor nerves in vertebrates, since most nerves supplying somatic muscles have sensory fibers for proprioception from the muscle (fig. 17.26).

Mixed nerves contain both sensory and motor fibers (fig. 16.5).Their sensory cell bodies are in sensory ganglia and, excepting some nerves of the autonomic system, their motor cell bodies are in motor nuclei. The different fibers in a mixed nerve are simultaneously carrying impulses in opposite directions and at different speeds. Most vertebrate nerves are mixed.

The site where a nerve impulse is transferred from one neuron to another is a **synapse.** As an axon approaches a synapse, it sprays into a multitude of fine branches, or **telodendria,** each of which ends in a **synaptic knob (terminal button)** that is in contact with the cell body, dendrite, or axon of the next neuron (fig. 16.2). Nerve impulses are propagated across the synapse by short-lived secretions, chiefly amino acids and amines such as norepinephrine, acetylcholine, serotonin, melatonin, and others, which are

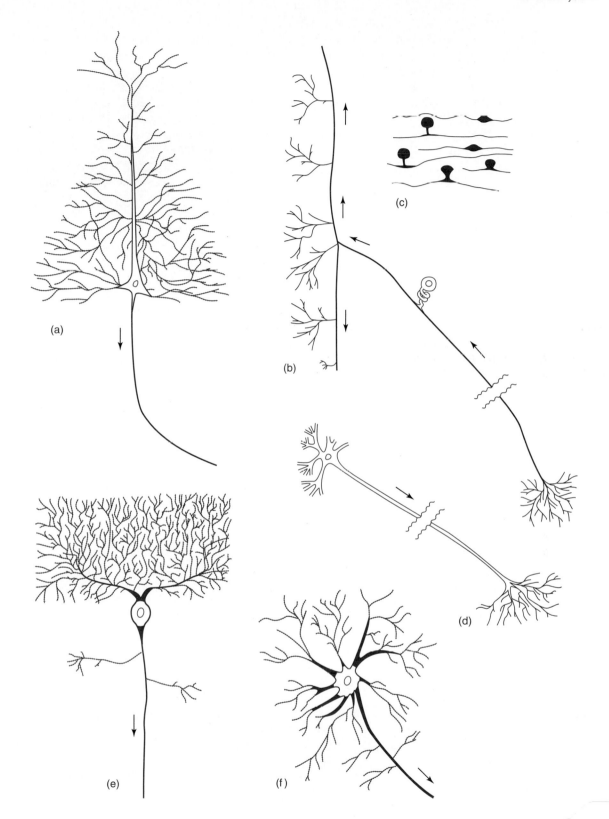

FIGURE 16.1

Several morphological types of neurons. (*a*) Pyramidal cell from motor cortex. (*b*) Dorsal root ganglion cell, unipolar. The single process is an axon. Ascending and descending branches are in fiber tracts in the spinal cord. (*c*) Embryonic dorsal root ganglion cells in transition from bipolar to unipolar. (*d*) Motor cell body from cord or brain stem. The axon ends on a motor end plate (fig. 11.2). (*e*) Purkinje cell from cerebellum. (*f*) Motor cell from sympathetic ganglion. Arrows indicate direction of impulses.

FIGURE 16.2

Synaptic endings on a motor neuron. **A,** synapse between the synaptic knob and a cell body; **B,** between synaptic knob and a dendrite; **C,** between synaptic knob and another axon. Dark splotches in dendrites and cell body are Nissl material. The knobs contain neurotransmitters.

FIGURE 16.3

Cross section of a portion of a nerve. The axons, (*black dots*) are surrounded by myelin sheaths (*white*) of varying thickness. The dark ring around the myelin is the neurilemma.

FIGURE 16.4

Typical locations of cell bodies (*black*) of sensory and motor fibers. (*a*) Sensory nerve with sensory cell bodies in a sensory ganglion. Upon entering the brain the fibers branch and pass toward synapses in several directions. (*b*) Motor nerve (hypoglossal) with motor cell bodies in a motor nucleus in the brain. **T,** descending fiber tract (corticospinal); **X,** decussating fibers. *Arrows* indicate direction of nerve impulses. Note the exceptional bipolar nature of the cell bodies in the vestibular ganglion. The cell bodies in most sensory ganglia are unipolar as in fig. 16.5.

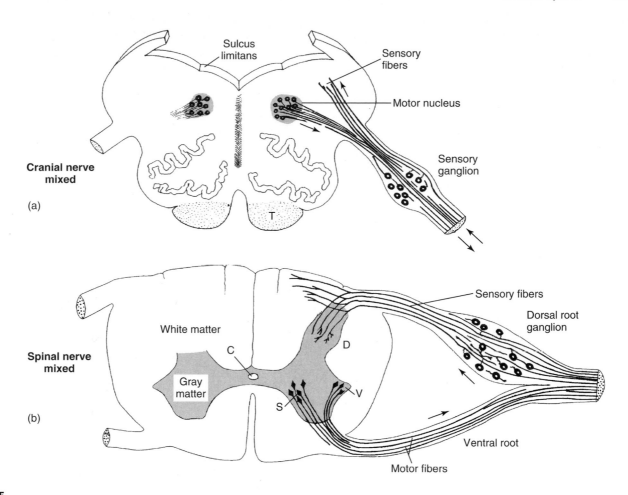

FIGURE 16.5

Locations of cell bodies of mixed nerves. (*a*) Cranial nerve with sensory cell bodies in a ganglion and motor cell bodies in a motor nucleus in the brain. **T**, descending fiber tract. (*b*) Spinal nerve with sensory cell bodies in a ganglion and motor cell bodies in gray matter of ventral horn of cord. **C**, central canal; **D**, dorsal horn of gray matter; **S**, somatic motor nucleus in ventral horn; **V**, visceral motor nucleus in lateral horn. *Arrows* indicate direction of nerve impulses.

released from the synaptic knobs when an electrical nerve impulse arrives. These amino acids and amines are **neurotransmitters.** Neurotransmitters are also released from axon terminals in contact with effectors, causing the effector (muscle, gland, pigment cell) to respond (fig. 11.2). The short life of neurotransmitters ensures against a protracted response from a single nerve impulse.

Most long axons within the central nervous system give off **collateral branches** along their pathway, thereby distributing a single impulse to sometimes thousands of neurons. Collateral branches are seen on the ascending and descending branches of the sensory neuron illustrated in figure 16.1*b*.

Some neurons **(neurosecretory neurons)** with cell bodies in the central nervous system secrete small polypeptides from their axon terminals. These polypeptides are **neurohormones** (neurosecretions). Instead of terminating in synapses or at effectors, most neurosecretory fibers terminate at sinusoidal vascular channels into which they release their secretions (fig. 18.1). The anti-diuretic hormone, which is released in the posterior lobe of the pituitary gland, is such a secretion (fig. 18.2, neurosecretory fiber).

GROWTH AND DIFFERENTIATION OF THE NERVOUS SYSTEM

To gain additional insight into the architecture of the nervous system, it is important to understand how the components of the nervous system develop. **Neurulation,** the initial stages in the formation of the nervous system, and the morphogens that regulate its induction, are discussed in chapter 5. (You may be interested in hedgehog genes, which are the source of these morphogens.) Neurulation results in formation of a dorsal, hollow, neural tube. The central cavity is the **neurocoel** (fig. 16.7*a*). The next discussions describe maturation of the neural tube and the embryonic sources of motor and sensory neurons.

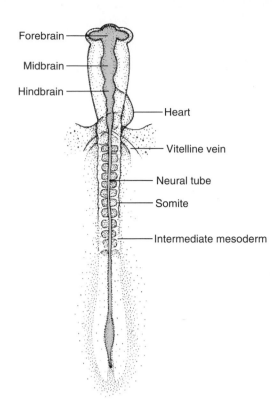

FIGURE 16.6

Neural tube (*red*) of a 33-hour chick embryo viewed through tissue rendered transparent. The optic vesicles are beginning to evaginate from the forebrain.

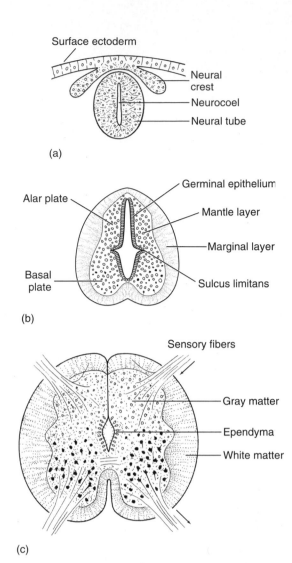

FIGURE 16.7

Differentiation of neural tube at spinal cord level. (*a*) Cross section of neural tube caudal to brain. (*b*) and (*c*) Later developmental stages. In (*c*), sensory fibers from a dorsal root ganglion are seen entering alar plate to synapse with second-order sensory neurons, and motor fibers are emerging from basal plate neuroblasts to innervate striated muscles.

Neural Tube

An early stage in neurulation is seen in figure 5.9*a–c*. A later stage is illustrated in figure 16.6. The cephalic end of the neural tube becomes the brain. The rest is future spinal cord. A cross section of the embryonic spinal cord at this stage is seen in figure 16.7*a*. Shortly thereafter (fig. 16.7*b*), it displays three zones, a **germinal layer** of actively mitotic cells, a **mantle layer** of cells proliferated from the germinal layer, and a **marginal layer** that is practically devoid of nuclei. The germinal layer is the source of the cells in the mantle layer. Some of the latter are **neuroblasts.** These sprout dendrites and an axon, and become neurons. The rest are **spongioblasts.** These become **neuroglia,** the supportive cells of the brain and cord. As neuroblasts and neuroglia differentiate, they contribute to the volume of the mantle layer, and axons from differentiating neuroblasts grow into the marginal layer, contributing to it. Because axons eventually become coated by fatty **myelin,** the marginal layer appears white, hence is **white matter.** The preponderance of cytoplasm in the mantle layer, which is composed chiefly of cell bodies, causes this zone to appear gray—hence the name **gray matter.**

The embryonic cord, hindbrain, and midbrain consist of **alar and basal plates** located, respectively, above and below a **sulcus limitans** (figs. 16.7*b* and 16.34). The cell bodies that develop in the alar plate receive sensory impulses from cranial and spinal nerves and distribute them elsewhere in the cord or brain. Those in the basal plate become the cell bodies of motor neurons (fig. 16.7*c*).

Eventually, the cells of the germinal layer cease to divide, and thereafter additional neurons are rarely formed. The undifferentiated cells that remain adjacent to the neurocoel then become **ependymal cells.** The **ependyma** is the nonnervous lining of the adult neurocoel (figs. 16.7*c* and 16.8*b*, 5).

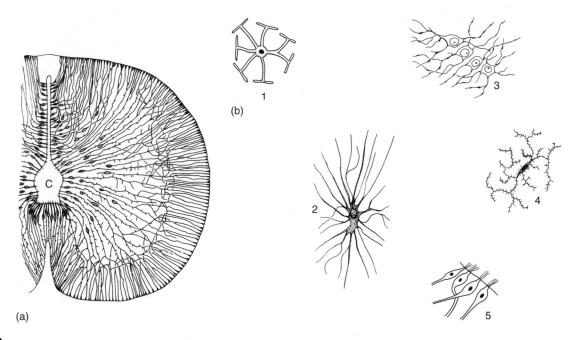

FIGURE 16.8

Neuroglia. (*a*) Neural tube of a 10-week human fetus stained to show spongioblasts in the germinal, mantle, and marginal layers. Those that remain in the germinal layer at birth will have become ependymal cells. (*b*) Differentiated neuroglia. **1,** protoplasmic astrocyte; **2,** fibrous astrocyte; **3,** oligodendroglia; **4,** microglia; **5,** ependymal cells. **C,** neurocoel.

(*a*) Source: From Cajal, *Histologie du Systeme Nerveux de L'Homme et des Vertebres,* Vol. 1, 1909, A. Maloine, Paris.

Development of Motor Components of Nerves

Most of the axons that sprout from neuroblasts in the basal plate grow out of, and away from, the neural tube to make contact with striated muscle cells. These axons become motor fibers of the cranial and spinal nerves. Because they sprout from neuroblasts within the central nervous system, the cell bodies of these motor fibers are within the adult brain or cord (figs. 16.4*b* and 16.5).

Other fibers that sprout from neuroblasts in the basal plate grow out of the neural tube to make contact with neuroblasts in autonomic ganglia (fig. 16.36, sympathetic ganglion). Neuroblasts in autonomic ganglia sprout **postganglionic fibers** (fig. 16.36, *POST*) that grow toward and innervate smooth muscles and glands. Thus *all motor neurons have their cell bodies in the cord or brain with one exception—postganglionic neurons of the autonomic system have their cell bodies in autonomic ganglia.*

Most neuroblasts of autonomic ganglia are migrants from neural crests (next paragraph). A few, in amphibians at least, migrate outward from the basal plate of the neural tube.

Development of Sensory Components of Nerves

At the time the neural groove is closing to form a tube, a longitudinal ribbon of ectoderm separates from the developing neural tube dorsolaterally on each side and soon segments to form a metameric series of **neural crests,** one pair in each body segment (fig. 6.7*a*). Some neural crest cells become neuroblasts that give rise to sensory neurons in spinal and cranial nerves. In doing so they pass through a bipolar stage (fig. 16.1*c*) in which one process grows into the alar plate of the cord or brain and the other grows to a sense organ. Thus there is established a neuronal connection between sense organ and central nervous system. Since neural crests give rise to a large number of sensory cell bodies, the result is a swelling, the sensory ganglion, on the nerve close to the cord or brain. Neurons with cell bodies in sensory ganglia are **first-order sensory neurons.** They conduct impulses from a sense organ to the central nervous system. Within the alar plate they synapse with **second-order sensory neurons (association neurons)** that distribute the impulses elsewhere in the central nervous system. Because the cell bodies of first-order sensory neurons arise from neural crests, we can make the following generalization: *the cell bodies of sensory neurons of cranial and spinal nerves are generally in sensory ganglia on the pathway of the nerves.*

There are three major exceptions to the rule that cell bodies of first-order sensory neurons are in ganglia on the pathway of nerves: (1) olfactory nerve fibers sprout from neuroblasts in the embryonic olfactory epithelium, and their long processes grow into the nearest part of the brain, which is the olfactory bulb. Therefore, *the cell bodies of olfactory nerves are in the olfactory epithelia* (figs. 16.28*a* and 17.21). (2) Neuroblasts that give rise to sensory fibers of the optic nerves differentiate in the nervous layer of the embryonic retinas, and long processes grow brainward along the optic stalk (fig. 17.6*b*). Therefore *the cell bodies of optic nerve fibers are in the retina.* The retina is actually part of the brain, since it arises as an evagination from the diencephalon and never separates from it (fig. 17.6*b*). (3) The cell bodies of **proprioceptive fibers** (sensory fibers that monitor the activity of skeletal muscles) in the first ten cranial nerves are not in ganglia. Neuroblasts within the alar plate of the embryonic midbrain sprout long processes that grow out to the muscles and tendons innervated by these nerves, providing them with proprioceptive innervation (fig. 17.26). Therefore *the cell bodies of proprioceptive fibers in the first ten cranial nerves are in the mesencephalic nucleus of the midbrain.* (The cell bodies of other proprioceptive fibers are in sensory ganglia.)

Olfactory neurons and the rods and cones of the retina, all of which serve as first-order sensory neurons, are termed **neurosensory cells** because they receive the stimulus directly instead of via an intervening epithelial cell that usually serves as a transducer of sensory stimuli. An axonic process of the neurosensory cell transmits the impulse to a synapse. Neurosensory cells are the only afferent neurons in many diploblastic invertebrates.

The primitive location of sensory cell bodies appears to have been in sensory epithelia rather than ganglia. Their gradual withdrawal from epithelia and their aggregation in ganglia appear to have been a later phylogenetic development. Sensory cell bodies in the lowest invertebrates are in epithelia or scattered within the nerve, rather than in ganglia. The phylogeny of neurons and their gradual organization into complex nervous systems has been described by Bullock and Horridge (1965).

Some of the neuroblasts that contribute to sensory ganglia of the head arise from **epibranchial placodes** of embryonic cranial ectoderm (chapter 5). The extent of the contribution of neural crests and epibranchial placodes to sensory ganglia of the head varies with the species.

NEUROGLIA AND NEURILEMMA

Not all the undifferentiated cells of the embryonic neural tube become neuroblasts. Nearly half the bulk of the brain and cord consists of a variety of interstitial cells known as **neuroglia**, which develop from **spongioblasts** (fig. 16.8*a*). They are generally smaller than most neurons, have nonnervous dendritic processes, and fill all the interstices in the brain and cord that are not occupied by neurons or blood vessels.

The oldest neuroglial elements phylogenetically are **ependymal cells.** These line the neurocoel of the cord and brain. Each has cilia that project into the fluid of the neurocoel and a long process that extends radially, sometimes to the circumference of the neural tube (fig. 16.8*a*). Ependymal cells are the sole neuroglial elements of

amphioxus and agnathans, and they are a stage in the differentiation of neuroglial cells of higher vertebrates. Blood vessels do not enter the spinal cord of an amphioxus; the long processes of the ependymal cells are thought to supply nutrients from the neurocoel to the neurons.

A progressive increase in the variety of glial cells is seen from cyclostomes to teleosts and from amphibians to amniotes. Some ependymal cells lose their connection with the periphery of the neural tube, become isolated among nerve cell bodies and their processes, and become functionally specialized. Other glial cells form a **glial membrane** on the surface of the cord and brain. Among the varieties of glial cells are astrocytes, microglia, and oligodendroglia.

Astrocytes, so named because they have the shape of a radiant star, are interposed between blood capillaries and the cell bodies of adjacent neurons. They obtain nutrients from the bloodstream and transfer these to the neurons (fig. 16.8*b*). **Microglia** are phagocytic, engulfing bacteria and cellular debris, which they digest. Unlike other glial cells, they are of mesodermal origin. The cytoplasmic processes of **oligodendroglia** wrap around naked axons in the brain and cord and elaborate myelin. When a nerve impulse sweeps along an axon, there is a change in electric potential on the surface of the axon. The fatty sheath insulates axons against action potentials in adjacent fibers and speeds conduction of the nerve impulse within the axon.

In the early stages of formation of nerves, when motor axons are growing out of the basal plate and sensory axons are arriving in the alar plate (fig. 16.7*c*), spongioblasts from the developing cord and brain migrate outward along the developing nerve root, mature, embrace the naked axons, and deposit myelin on them. These oligodendrogliale cells persist throughout life as **neurilemmal cells** (originally named **Schwann cells** after their discoverer). Heavily myelinated fibers conduct nerve impulses faster than lightly myelinated ones. The thickest sheaths are on fibers from encapsulated nerve endings for touch (chapter 17), on proprioceptive fibers, and on motor fibers that supply skeletal muscles. They are thinnest on visceral fibers. Life in the raw is seldom mild, as the saying goes, and a speedy reaction to environmental threats, as opposed to visceral ones, increases the chances of survival.

SPINAL CORD

The spinal cord occupies the bony vertebral canal formed by the centra of the vertebrae and the successive neural arches. It is flattened in agnathans but tends to be rounded or quadralateral in higher tetrapods. Generally, the neurocoel is relatively large in lower vertebrates, narrower or even obstructed at some levels in mammals.

Closely applied to the outer glial membrane of the cord and brain in most fishes is a delicate connective tissue membrane, the **meninx primitiva.** A similar meninx appears during embryonic development in higher vertebrates below mammals, but it eventually differentiates into an inner vascular **leptomeninx** and a dense outer membrane, the **dura mater.** Mammals have three meninges—an inner **pia mater,** a filmy weblike **arachnoid,** and a tough fibrous dura mater. The latter loosely surrounds the other meninges. The pia mater adheres closely to the glial membrane of the cord, and it is attached to the arachnoid by a network of fibrous strands that traverse a subarachnoid space filled with cerebrospinal fluid. A similar subdural space separates the arachnoid from the dura mater, and a peridural space filled with adipose tissue separates the cord and its meninges from the bony vertebral canal. The cerebrospinal fluid and the cushion of fat buffer the mammalian cord against mechanical trauma. (The same meninges surround the brain but there is no peridural fat, the dura mater having a somewhat different relationship to the periosteal lining of the skull.)

The spinal cord begins at the foramen magnum, but there is no abrupt transition between cord and brain. Instead, there is a gradual rearrangement of the gray matter in the transitional area. In the cord, the gray matter is confined to the area surrounding the neural canal. In the brain stem (brain without the cerebral hemispheres and cerebellum), it is dispersed into many separate gray masses, or nuclei, separated by fiber tracts. The transition from cord to brain in mammals is completed over the length of about one body segment (fig. 16.10, compare *a* with *b*). In lower vertebrates, the transitional region is longer and the nerves in the transitional region (occipitospinal nerves, fig. 16.30) are atypical.

The adult cord is coextensive with the vertebral column in vertebrates with abundant tail musculature, as in some urodeles. In these species, each spinal nerve emerges from the vertebral canal via an **intervertebral foramen** at the same level at which it comes off the cord. Generally, however, the embryonic vertebral column elongates more rapidly than the cord, with the result that the adult cord is shorter than the column. Consequently, the spinal nerves of the lower trunk must pass caudad within the vertebral canal before exiting via their foramen. In doing so, they constitute a bundle, the **cauda equina,** or "horse's tail" (fig. 16.9). The ependyma and meninges of the cord surround the bundle, enclosing it in a connective tissue sheath. After the last nerves have exited, the sheath alone continues for a distance as a threadlike **filum terminale.** It, too, eventually terminates. In humans the adult cord ends at the upper border of the second lumbar vertebra, in cats and rabbits at about the middle of the sacrum, and in frogs anterior to the urostyle. In a few bony fishes, the spinal cord is actually shorter than the brain.

The spinal cord exhibits cervical and lumbar enlargements at the levels of the anterior and posterior appendages. The enlargements result from the large number of cell bodies and fibers required to innervate the appendage. When one pair of appendages is excep-

FIGURE 16.9

Brain, cord, and dorsal rootlets of spinal nerves of a cat. The peridural fat and dura mater have been removed. **C,** first cervical spinal nerve; **T,** thoracic; **L,** lumbar; **S,** sacral; **Ca,** caudal spinal nerve.

tionally muscular, as were the hind limbs of massive dinosaurs, the corresponding enlargement of the cord is pronounced. In dinosaurs it was often larger than the brain. Conversely, the spinal cord of turtles is very slender in the trunk because much of the body wall musculature internal to the shell is vestigial. The cord of many fishes exhibits a neuroendocrine swelling, the **urophysis,** near its caudal end (fig. 18.4).

FIGURE 16.10

Selected cross sections of the human spinal cord. (*a*) Transitional zone between human medulla and cord. Descending fibers in the pyramidal tracts are decussating to continue caudad on the opposite side as the lateral corticospinal tract. The spinal nucleus of the trigeminal nerve is a caudal continuation of the nucleus of that nerve. The ventral horn of the cord coming into view. In (*b*), two ascending and two descending fiber tracts (*arrows*) are identified. In (*c*), six nuclei of the gray matter are highlighted in *red*. (*a* is a composite drawing of selected features of several successive stained microsections.)

Cross sections of a cord, appropriately stained, reveal gray matter, white matter, nuclei, fiber tracts, and other features (fig. 16.10). Nuclei are responsible for gray matter. Nerve fibers occupy the periphery of the cord and, with neuroglia, constitute the white matter. Ascending and descending fiber tracts (arrows in 16.10b) conduct impulses with a similar function up or down the cord and to or from the brain. The **fasciculus gracilis,** for example, consists of heavily myelinated fibers for touch and proprioception. These fibers synapse in the **nucleus gracilis.** From there, the impulse is carried in other fibers until it arrives in the **somesthetic cortex** (fig. 16.26), where it evokes a specific sensation. **Corticospinal tracts** conduct motor impulses for voluntary muscle contraction from the voluntary motor cortex. There are many tracts in a mammalian cord. The agnathan cord has relatively few, since their sensory and motor systems are relatively simple. The tracts and nuclei in the low cervical region of humans are large, reflecting, in part, the large number of sensory and motor fibers that innervate the primate hand.

SPINAL NERVES

Roots and Ganglia

Spinal nerves emerge from the spinal cord by an uninterrupted series of dorsal and ventral **rootlets** in each body segment (figs. 16.9 and 16.32). The dorsal rootlets unite while still within the vertebral canal to form a **dorsal root** just proximal to the dorsal root ganglion, and the ventral rootlets unite to form a **ventral root.** Except in lampreys, the roots then unite to form a spinal nerve.

Dorsal roots are predominantly or wholly sensory, and ventral roots are purely motor. However, it is evident that in the earliest vertebrates (1) dorsal and ventral roots did not unite; (2) dorsal roots contained visceral motor fibers as well as sensory fibers; (3) there were no dorsal root ganglia, bipolar sensory cell bodies being scattered within the nerve along its course; and (4) when sensory cell bodies first aggregated in ganglia they were still bipolar.

As vertebrate nervous systems evolved (1) dorsal and ventral roots united, (2) the dorsal roots became predominantly or totally sensory, (3) sensory cell bodies aggregated in the dorsal root to form ganglia, and (4) dorsal root ganglion cell bodies became increasingly unipolar.

Dorsal and ventral roots alternate in lampreys and do not unite; in hagfishes, they unite in the trunk but not in the tail. Some sensory cell bodies are in ganglia, some are scattered along the nerve, and most are bipolar. The dorsal roots contain visceral motor fibers as well as sensory fibers. Ventral roots are purely motor.

Above agnathans, dorsal and ventral roots unite. The dorsal roots of many bony fishes contain visceral motor fibers, but in cartilaginous fishes and tetrapods, most or all of the visceral motor fibers have been lost from the dorsal root. The cell bodies of first-order sensory neurons are bipolar in cartilaginous fishes; bipolar, intermediate, and unipolar in bony fishes; chiefly unipolar in amphibians; and almost entirely unipolar in amniotes.

In an amphioxus, spinal nerves have only one typical root, this being the equivalent of the mixed dorsal root of lower vertebrates. It arises from the cord at the level of each myoseptum and enters the myoseptum. It is then distributed to the integument (sensory) and viscera (visceral motor). The cell bodies for the visceral motor fibers are in the cord, and those for the sensory fibers are either in the cord or scattered along the course of the nerve. There is no dorsal root ganglion. Sensory cell bodies are bipolar.

What grossly resembles a ventral root in amphioxus emerges from the cord between two myosepta and enters a myomere. Electron microscopy reveals that these ventral "roots" contain no nerve fibers. Instead, they consist of filamentous extensions of the striated muscle fibers from the myomere. These myofilaments enter the cord, where they receive their motor innervation directly from neurons within the central nervous system. Thus, the somatic muscles of amphioxus actually "come to the cord" for their stimuli. A similar condition is seen in echinoderms and some other invertebrates.

Occipitospinal Nerves

In many fishes and amphibians, one or more pairs of **occipitospinal nerves** arise between the last cranial nerve and the first pair of typical spinal nerves (fig. 16.30). They generally lack dorsal roots and supply the hypobranchial musculature, including tongue muscles when present. Embryonic frogs have an occipitospinal nerve between the last cranial nerve and the first spinal nerve, the latter supplying the tongue; but the occipitospinal nerve is suppressed during later development. Cranial nerves XI and XII of amniotes lack dorsal roots and appear to be derived in part from occipitospinal nerves.

Spinal Nerve Metamerism

With the exception of the lampreys and hagfishes, a spinal nerve arises from each segment of the cord except near the end of the tail. The fibers in these nerves, other than fibers to the viscera, are distributed metamerically to the skin and muscles of the body wall and tail. Where fins or limbs develop, nerves at those levels innervate the appendages (fig. 16.11). The segmental distribution of spinal nerves is best observed in fishes, since the metamerism of their body wall muscles is so evident. It is this metamerism, inherited with adaptive modifications by tetrapods from fishes to human beings, that enables swimming by lateral undulation (fig. 11.8). Tadpoles swim like fishes and they have as many as 40 spinal nerves. However, they lose all but the first 10 when, at metamorphosis, the tail is resorbed and tetrapod appendages develop.

FIGURE 16.11

Metameric innervation of the skin of a mammalian forelimb. **C,** cervical, and **T,** thoracic somites and the skin areas supplied by their associated spinal nerves.

Rami and Plexuses

Shortly after emerging from the vertebral canal, spinal nerves divide into two or more branches (fig. 16.12). A **dorsal ramus** supplies the muscles and skin of the back, and a larger **ventral ramus** enters the lateral body wall and supplies the hypaxial muscles and skin to the midventral raphe. In the thoracic and lumbar region of mammals, **white and gray rami communicantes** (sing., ramus communicans) connect the spinal nerves with the ganglia of the sympathetic trunk (figs. 16.12 and 16.35). These rami conduct impulses that have visceral functions.

The ventral rami of successive spinal nerves of vertebrates with fins or limbs unite to form **spinal nerve plexuses**—sites where two or more nerves or branches of nerves converge and their fibers commingle in one or more common trunks prior to being distributed. These plexuses are relatively simple in fishes, but become increasingly complex in tetrapods (compare shark and mammal, fig. 11.17). The chief spinal nerve plexuses are the **brachial** and **lumbar** (**cervicobrachial** and **lumbosacral** in tetrapods that have cervical, thoracic, lumbar, and sacral vertebrae). Nerves emerge from spinal nerve plexuses to supply the fins or limbs. Highly complex spinal nerve plexuses with fibers that also supply pelvic viscera are present in the pelvis of mammals, including humans.

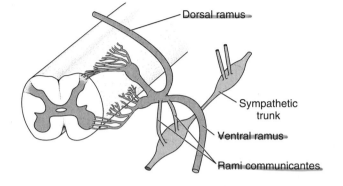

FIGURE 16.12

Primary branches of thoracic and lumbar nerves in mammals.

Functional Components of Spinal Nerves

The fibers in a typical spinal nerve are either sensory (afferent) or motor (efferent) and they are either somatic or visceral, in accordance with the structures they innervate. They represent, therefore, four functional varieties of nerve fibers: somatic afferent (SA), visceral afferent (VA), somatic efferent (SE), and visceral efferent (VE). These fibers are widely distributed throughout the body and are present in some cranial nerves. They are referred to as *general* fibers to differentiate them from *special* fibers that are found only in cranial nerves. Table 16.1 lists the kinds of structures they innervate.

TABLE 16.1   Fiber Components of Typical Spinal Nerves[1]

Components	Innervation
Sensory	
General somatic afferent fibers (GSA)	General cutaneous receptors (touch, pain, temperature, and pressure)
	Receptors in striated muscle, tendons, and bursae (proprioceptive)
General visceral afferent fibers (GVA)	Viscera, including general receptors in endoderm
Motor	
Somatic efferent fibers (SE)[2]	Myotomal muscle
General visceral efferent fibers (GVE)	Smooth and cardiac muscle, and glands (via autonomic ganglia)

[1]These components are found also in some of the cranial nerves.

[2]The fibers to myotomal muscle are designated simply as SE, rather than as GSE (general somatic efferent), because there are no special somatic efferent fibers.

BRAIN

The cranial end of the early embryonic neural tube of all vertebrates from fishes to humans exhibits three primary brain vesicles, a **prosencephalon** (future forebrain), **mesencephalon** (future midbrain), and **rhombencephalon** (future hindbrain, fig. 16.13a). The rhombencephalon was named for its shape. The primary vesicles soon differentiate to form the anlagen of the five major subdivisions of the adult brain: **telencephalon, diencephalon, mesencephalon, metencephalon,** and **myelencephalon** (fig. 16.13b).

Differentiation of the primary vesicles to form an adult brain is accomplished by (1) localized thickenings of the lateral walls and floor of some vesicles, and (2) dorsal, lateral, or ventral evaginations, median or paired, in others. The surface structures that develop in all vertebrates are readily seen to be homologues (fig. 16.14). Because the huge cerebral hemispheres and the cerebellum of mammals overlie the other structures, it is necessary to remove the hemispheres to reveal the underlying structures from a dorsal view. When they have been removed, what remains is the **brain stem** (fig. 16.15).

The embryonic prosencephalon differs from the rest of the neural tube in not being divided into alar and basal plates. Early in development, except in ray-finned fishes, it exhibits two pairs of evaginations, the **telencephalic vesicles,** the walls of which become the cerebral hemispheres, and the **optic vesicles**, which become the retinas of the eyeballs (figs. 16.13c and 17.6). The optic vesicles become associated with the diencephalon. The mesencephalon does not undergo subdivision. The rhombencephalon becomes the metencephalon and myelencephalon.

The adult brain is surrounded by the meninges described earlier. The major components of the brain are listed in table 16.2. We will begin study with the least specialized region, the hindbrain.

Metencephalon and Myelencephalon: The Hindbrain

The myelencephalon is represented chiefly by the **medulla oblongata,** which merges imperceptibly with the spinal cord. The area of transition is characterized internally by gradual relocation of the fiber tracts, which are the white matter (fig. 16.10a). As a result, whereas in the cord the gray matter is centrally located, in the brain it is dispersed into small and large nuclear masses of gray matter interspersed among fiber tracts.

A conspicuous feature of the hindbrain is the **cerebellum,** a dorsal evagination of the metencephalon (figs. 16.14, 16.23, and 16.30). In mammals it is connected to the brain stem by three pairs of **cerebellar peduncles** (stalks) composed of afferent and efferent fiber tracts (fig. 16.15, 3–5). The cerebellum coordinates the response of skeletal muscles to input from the membranous labyrinth, lateral-line canals of fishes, proprioceptors in muscles, joints, and tendons, and from reflex and voluntary motor centers in the brain stem and forebrain. Its size is correlated with the complexity of the activities

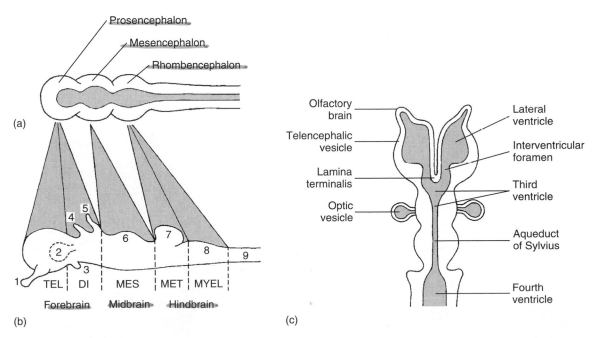

FIGURE 16.13

Primary brain vesicles and initial stages of differentiation, generalized scheme. (a) Three-vesicle stage, sagittal view. (b) Early differentiation, lateral view. **1,** olfactory bulb; **2,** future site of optic vesicle; **3,** neural lobe of pituitary; **4,** parapineal; **5,** pineal; **6,** optic lobe; **7,** cerebellum; **8,** medulla; **9,** postcranial neural tube. (c) Initial formation of cerebral hemispheres, except in actinopterygians, frontal section.

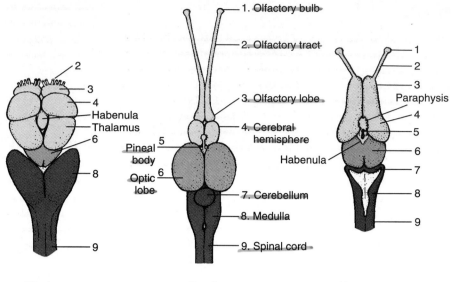

1. Olfactory bulb
2. Olfactory tract
3. Olfactory lobe
4. Cerebral hemisphere
5. Pineal body
6. Optic lobe
7. Cerebellum
8. Medulla
9. Spinal cord

Habenula
Thalamus

Paraphysis
Habenula

Myxine   **Perch**   **Necturus**

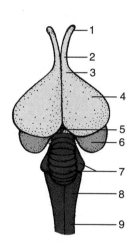

Thalamus

Frog   **Snake**   **Chicken**

FIGURE 16.14

Vertebrate brains, dorsal view. The roof of the fourth ventricle has been removed in necturus, frog, and snake. The forebrain, midbrain, and hindbrain are differentially colored. The habenuli, thalami, and pineal bodies are part of the diencephalon. **1** to **9**, see perch for key; **10**, auditory lobe of amniotes. Mammalian brains are shown in fig. 16.23.

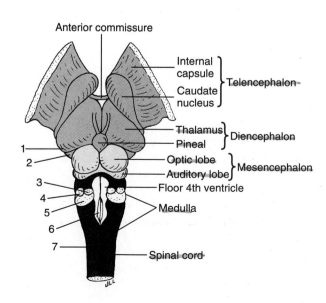

Anterior commissure
Internal capsule
Caudate nucleus
Telencephalon
Thalamus
Pineal
Diencephalon
Optic lobe
Auditory lobe
Mesencephalon
Floor 4th ventricle
Medulla
Spinal cord

FIGURE 16.15

Brain stem of a sheep. The cerebral hemispheres and cerebellum have been cut away to reveal the underlying structures. **1,** location of lateral geniculate body of thalamus; **2,** medial geniculate body; **3 to 5,** anterior, middle, and posterior cerebellar peduncles. The peduncles had to be cut to remove the cerebellum. **6,** hypoglossal trigone in floor of the fourth ventricle (location of the hypoglossal nuclei); **7,** posterior funiculus (ascending fibers for proprioception). Optic and auditory lobes constitute the corpora quadrigemina. The habenulae lie under cover of the pineal. A paired fiber tract, the stria medullaris, is seen passing toward the habenulae on the surface of the thalamus. Three shades of *red* indicate forebrain, midbrain, and hindbrain.

TABLE 16.2 Subdivisions of the Brain and Their Major Components

Subdivisions		Chief Components
Prosencephalon (forebrain)	Telencephalon	Cerebral hemispheres Lateral ventricles
	Diencephalon	Epithalamus Thalamus Hypothalamus Third ventricle
Mesencephalon (midbrain)		Tectum Tegmentum Cerebral aqueduct
Rhombencephalon (hindbrain)	Metencephalon	Cerebellum Tegmentum Fourth ventricle
	Myelencephalon	Medulla oblongata Fourth ventricle

Note: For a list of all components discussed in this chapter, see chapter summary.

of the striated musculature. It is larger in fishes than in amphibians since swimming, which involves schooling, vertical movements, adjusting to water currents, and keeping the dorsal part of the body from tipping over, requires more synergistic muscle activity than dragging the belly along the ground or squatting on a lily pad. Lacking a large cerebellum, aquatic urodeles rely to a large extent on spinal cord reflexes and primitive nuclei in the hindbrain for muscle coordination when swimming. The cerebellum is largest in birds and mammals, which require a large computerlike neural center to coordinate the muscles of the head, neck, trunk, and appendages in such diverse activities as flying, running, climbing, balancing, or, in the case of humans, playing a piano. The cell bodies of the cerebellum are on the surface. In cyclostomes, the cerebellum is not well developed and does not cause a bulge on the brain.

Other topographical features of the hindbrain include various swellings caused by underlying nuclei, and elevated ridges or transverse bands from superficial fiber tracts. These topographical markings are most prominent in mammals. One sensory nucleus (**nucleus solitarius**) in the alar plate becomes enormous in fishes that have taste buds over the entire surface of the body. The resulting swelling, or **vagal lobe** (fig. 16.16), is the termination of the many incoming sensory fibers for taste. The nucleus contains second-order sensory neurons whose fibers are projected to reflex and relay centers elsewhere in the brain. Among ventral ridges on the mammalian hindbrain are the **pyramids** (figs. 16.10*a* and 16.17), which contain the **corticospinal tracts** that carry voluntary motor impulses from the cerebral cortex to the cord; and the **ventral decussation of the pons** in those mammals that have relatively large cerebellums (fig. 16.17). The hindbrain also houses a reflex vasomotor center that maintains normal blood pressure and a

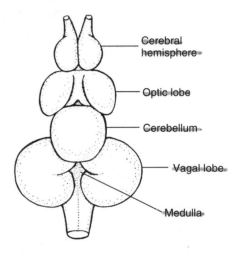

Modification for bottom-feeding

FIGURE 16.16

Brain of the buffalo fish *Carpiodes velifer.* Note unusual bulge (vagal lobe) on the alar plate of the medulla. Here terminate the many incoming taste fibers characteristic of bottom feeders.

reflex respiratory center that is vital in maintaining pulmonary respiration.

The cavity of the hindbrain is the fourth ventricle. The cerebellum is part of its roof (figs. 16.18 and 16.24). The remainder of the roof anterior and posterior to the cerebellum is a thin membrane that includes the ependyma of the neural tube and, posteriorly, a layer of highly vascular pia mater from which vascular tufts hang into the cavity. The tufts constitute the **choroid plexus of the fourth ventricle.**

Mesencephalon: The Midbrain

The **tectum,** or roof, of the mesencephalon displays a pair of prominent **optic lobes** in all vertebrates (figs. 16.14 and 16.15). These bulging gray masses serve partly as reflex

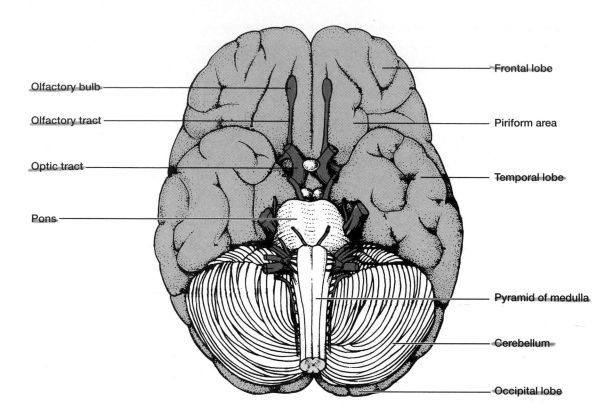

FIGURE 16.17

Human brain, ventral view.

From David Shier, et al., *Hole's Human Anatomy and Physiology,* 7th edition. Copyright © 1996 Times Mirror Higher Education Group, Inc., Dubuque, Iowa. All Rights Reserved. Reprinted by permission.

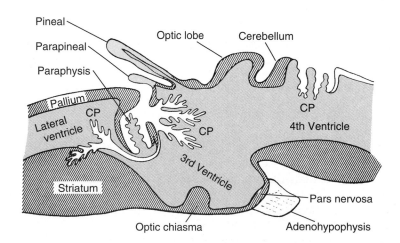

FIGURE 16.18

Diencephalon and adjacent areas of a larval anuran brain, sagittal section, anterior end to the left. **CP,** choroid plexus of lateral, third, and fourth ventricles. The choroid plexus of the lateral ventricle enters the latter via an interventricular foramen. *Red,* brain ventricles.

and relay centers that receive impulses originating in the retina. They are especially large in birds, which have large eyes and rely on visual stimuli for much information about the environment. A pair of **auditory lobes** lies caudal to the optic lobes in the tectum commencing with reptiles, and the four bodies (optic and auditory lobes) constitute the **corpora quadrigemina** (fig. 16.15). Fishes have auditory nuclei in this location, but they are not large enough to bulge on the surface. The auditory lobes receive input from the part of the membranous labyrinth that is sensitive to vibratory stimuli, and also from other sources. Phylogenetically, they enlarge as the receptors for sound expand. The **mesencephalic nucleus of the trigeminal nerve** is a prominent gray mass in the alar plate. It contains the cell bodies of the proprioceptive fibers of most, if not all, of the cranial nerves that have them.

The basal plate of the mesencephalon, the **tegmentum,** is greatly thickened by nuclei and by aggregates of fiber tracts that connect higher levels of the brain with the hindbrain and spinal cord. In mammals, some of these nuclei and tracts are massive. One of these is the **red nucleus,** a large mass of gray matter that, in fresh brains, has a pink tinge. Most of its afferent fibers come from the cerebellum; some come from the cerebral cortex via a massive pair of fiber tracts, the **cerebral peduncles.** Efferent impulses from the red nucleus influence, directly or via relays, motor functions of the medulla and spinal cord, thus participating in control of the striated musculature.

The ventricle of the midbrain is quite large in fishes and amphibians, and extends dorsad into the hollow optic lobes. The optic lobes of higher vertebrates are not hollow, and the midbrain ventricle is constricted to a narrow canal, the **cerebral aqueduct,** also known as the **aqueduct of Sylvius** and the **iter** (figs. 16.13*c* and 16.22, human).

Diencephalon

The diencephalon consists of three major components, **epithalamus, thalamus,** and **hypothalamus** (fig. 16.19). Its neurocoel is the third ventricle.

Epithalamus

The epithalamus is the dorsalmost component of the diencephalon and, as such, constitutes the roof of the third ventricle. It consists of a **pineal** or **parapineal** organ, or both, a choroid plexus like that of the fourth ventricle, and a pair of elevated thickenings (median in *Myxine*), the **habenulae** (fig. 16.19).

The pineal is a club-shaped or knoblike organ, sometimes threadlike or saccular, projecting above the diencephalon (fig. 16.18). It is connected to the diencephalon by a stalk that frequently contains an extension of the third ventricle. In lampreys, the pineal is a photoreceptor. In gnathostomes, it functions as an endocrine organ that is stimulated in part by light that first enters the body through the retina. The neural pathway from retina to pineal, and the endocrine role of the pineal are described under "Pineal Organ" in chapter 18. The pineal is vestigial or absent in a few vertebrates, including hagfishes, crocodilians, and a few permanently aquatic mammals. It is relatively large in primates and sheep (fig. 16.15).

The parapineal was a constant feature of bony fishes of the Devonian and of ancestral amphibians and reptiles, in which it served as a **parietal eye.** It is now present in only a few bony fishes, lampreys, larval and a few adult anurans, and some lizards (fig. 17.19, **median eye**). When functional, it lies under a patch of translucent skin. The structure and neural connections of the parietal eye are described under "Median Eyes" in chapter 17.

The habenulae are elevations of a pair of underlying habenular nuclei that perform an ancient role associated with olfaction. They receive fibers from olfactory nuclei, the hypothalamus, and other forebrain nuclei, and discharge impulses to the thalamus and midbrain, eliciting reflex responses to olfactory stimuli. One incoming tract, the **stria medullaris,** is seen in figure 16.15 emerging from the thalamus and passing caudad to disappear beneath the pineal, which overlies the habenulae in sheep. The habenulae are largest in species such as sharks and bloodhounds, which depend heavily on olfaction for locating food. They are inconspicuous in most birds, many of which have a poorly developed olfactory sense. They are poorly developed in permanently aquatic mammals, which do not use their olfactory epithelium in seeking food, and for a good reason—they would drown! The habenula of *Myxine* is a median elevation (fig. 16.14), but the underlying nuclei are paired.

Thalamus

The thalamus is the largest structure of the diencephalon. It is a paired mass of many nuclei in the lateral walls of the third ventricle and it bulges dorsally just behind the cerebral hemispheres (fig. 16.14, *Myxine* and frog). In amniotes, it is hidden by the caudal poles of the cerebral hemispheres, and must be observed on a brain stem preparation (fig. 16.15).

All *sensory pathways* ascending to the telencephalon from the spinal cord, hindbrain, or midbrain, *synapse in a thalamic nucleus* before continuing to the telencephalon. With development of the neocortex in mammals, the

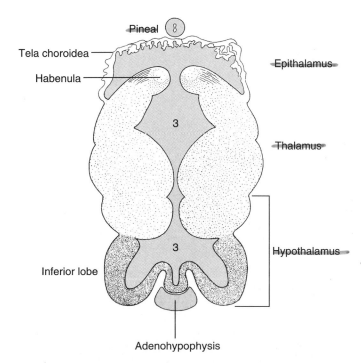

FIGURE 16.19

Major regions of diencephalon of a shark, cross section. Third ventricle, **3,** in *red.*

thalamic nuclei have become so numerous that the left and right thalami bulge into the third ventricle and meet at one site to form an oval bridge of gray matter, the **massa intermedia,** or false commissure (fig. 16.24). The term "false commissure" reflects the fact that it does not consist of decussating fiber tracts as do true commissures.

Hypothalamus and Associated Structures

A ventral view of the brain reveals the optic chiasma, inferior lobes or other components of the hypothalamus, a saccus vasculosus in fishes only, and the pituitary (figs. 16.17 and 16.20a).

The **optic chiasma** is the cephalic boundary of the diencephalon, ventrally. This is where the optic nerves reach the brain, and where some or all of its fibers cross to the opposite side. Caudal to the optic chiasma are the components of the hypothalamus and, at the caudal boundary of the diencephalon, the pituitary.

The floor of the diencephalon exhibits a midventral evagination within which is an **infundibular recess** of the third ventricle (fig. 16.21, 3). An elongated evagination results in an **infundibular stalk** (fig. 16.21c). The diencephalic floor at the end of the infundibular recess or stalk becomes the **posterior lobe (neural lobe, pars nervosa)** of the pituitary. Intimately associated with it is the adenohypophysis, which arises embryonically from the roof of the stomodeum (fig. 18.6).

The **hypothalamus** is the floor and ventrolateral walls of the third ventricle (figs. 16.19 and 16.21c, hypothalamic nuclei). It is prominent on the ventral surface of the diencephalon in cartilaginous fishes because of the extraordi-

nary size of the **inferior lobes,** which have connections with the cerebellum (fig. 16.20a). In all vertebrates, the hypothalamus is an important neural center for homeostasis, a visceral function. Hypothalamic nuclei exert major reflex control over the autonomic nervous system, produce neurohormones that regulate the pituitary and gonads, and monitor the sodium chloride and glucose content of the blood. There is an appetite regulating center and, in endotherms, a temperature regulating center. In general, the anterior nuclei are associated with parasympathetic functions, the caudal nuclei with sympathetic ones. Hypothalamic nuclei also have connections via fiber tracts with the thalamus, with one or more nuclei of the basal ganglia (fig. 16.25f), and with the **hippocampus,** an ancient olfactory cortex tucked under the temporal lobes of the cerebral hemispheres in mammals. These interconnections (the **limbic system**) are involved in evoking emotional responses in primates, an indication of the close relationship between visceral functioning and the emotions.

The **saccus vasculosus** is a highly vascularized, thin-walled ventral evagination of the diencephalic floor of elasmobranch and ray-finned fishes, located just behind the pituitary (fig. 16.20). Within it is a fluid-filled recess of the third ventricle. The saccus vasculosus, a sense organ, is discussed in the next chapter.

Third Ventricle

The third ventricle is continuous caudad with the cerebral aqueduct of the midbrain and cranially with the ventricle in each cerebral hemisphere via a left and right **interventricular foramen** (figs. 16.13c, and 16.22,

(a)

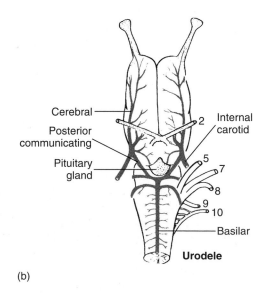

(b)

FIGURE 16.20

Brains of *Squalus,* (*a*) and *Necturus* (*b*), ventral views. Numerals in (*b*) identify cranial nerve roots.

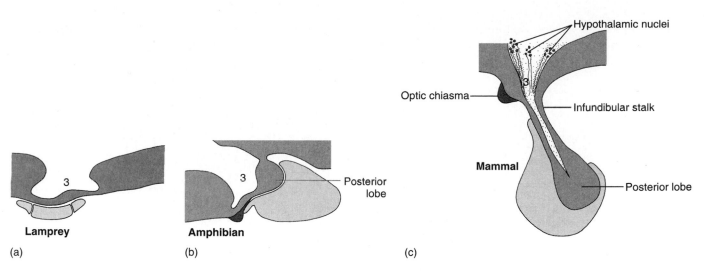

FIGURE 16.21

Infundibular recess, **3,** of third ventricle in three vertebrates. *Light gray,* adenohypophysis.

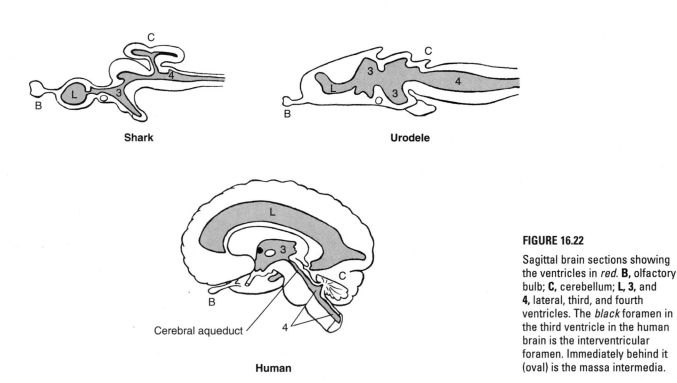

FIGURE 16.22

Sagittal brain sections showing the ventricles in *red.* **B,** olfactory bulb; **C,** cerebellum; **L, 3,** and **4,** lateral, third, and fourth ventricles. The *black* foramen in the third ventricle in the human brain is the interventricular foramen. Immediately behind it (oval) is the massa intermedia.

human). The third ventricle becomes compressed laterally with expansion of the thalami toward the midline. An optic recess extends toward the optic chiasma, the infundibular recess extends into the infundibular stalk, and a narrow canal extends upward into the stalk of the pineal. A choroid plexus is present in the ventricular roof.

Telencephalon

The telencephalon in all vertebrates other than ray-finned fishes is the portion of the brain that develops in the walls, floor, and some regions of the roof of the paired embryonic telencephalic vesicles (fig. 16.13c). The typical vertebrate telencephalon therefore consists

of paired left and right hemispheres, each with a ventricle. These hemispheres, which anatomically are mirror images, are referred to collectively as the **cerebrum.** Extending forward from each hemisphere is an **olfactory tract and bulb** (fig. 16.14). In mammals these are hidden from dorsal view by the enormous overgrowth of the hemispheres, a result of millions of years of hemispheric expansion (figs. 16.23 and 16.24). The olfactory bulbs lie in contact with the ethmoid cartilages or bones and receive axonlike processes from the olfactory cells in the olfactory epithelium. The foramina through which the processes reach the olfactory bulbs in mammals are seen in figs. 9.4 and 9.6.

FIGURE 16.23

Brain of a monotreme and a sheep.

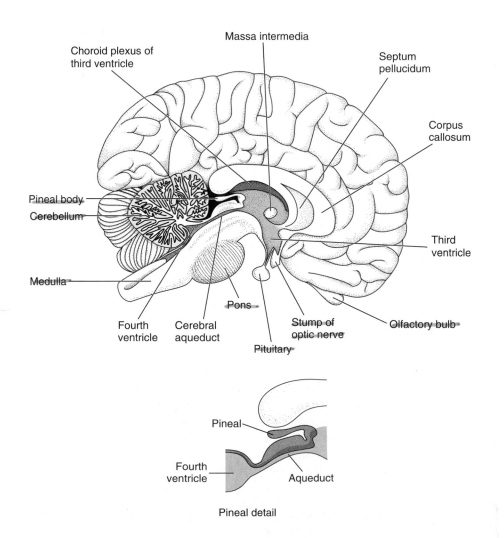

FIGURE 16.24

Human brain, sagittal section.

The regions of the cerebrum devoted to olfaction—olfactory bulbs, tracts, lobes, and nuclei that process olfactory information—are prominent in those vertebrates that depend on olfactory stimuli for survival. Sharks, for example, have olfactory lobes that are larger than the rest of the cerebrum (fig. 16.30). On the other hand, olfactory components remain rudimentary or are vestigial in many marine mammals, and they are poorly developed in birds other than those that feed on carrion. Olfaction is also a relatively minor source of environmental information in bats and primates, including humans.

At the boundary between the diencephalon and telencephalon, the thin roof of the embryonic neurocoel evaginates dorsad in some members of every vertebrate class to form a wrinkled thin-walled sac, the **paraphysis** (fig. 16.18). However, in amniotes other than *Sphenodon*, it is limited to embryos. It is seldom observed during dissection of sharks because it is delicate and readily torn. Little is known of its function. It resembles a choroid plexus, but seems to have a different role. It is assignable arbitrarily to either the telencephalon or diencephalon.

Because of its origin from paired embryonic vesicles, the neurocoel of the cerebrum in all vertebrates other than ray-finned fishes consists of paired lateral ventricles. Each ventricle is continuous with the third ventricle via an **interventricular foramen,** which marks the site of the initial embryonic evagination of the cerebrum. In mammals, a perpendicular wall, the **lamina terminalis,** separates the two foramina and marks the original cephalic boundary of the embryonic neural tube (fig. 16.13*c*). Ray-finned fishes employ a slightly different means of producing a cerebrum. The telencephalic neurocoel remains unpaired, hence there is a single cerebral ventricle and no interventricular septum. As in other vertebrates, paired nuclei form in the walls of the ventricle, but, in proliferating, the dorsal and lateral walls turn outward (that is, they are "everted").

Fishes

The cerebrum of fishes consists of (1) a primitive sensory and association area (**pallium,** fig. 16.25*a* and *b*) that receives and processes input from the olfactory epithelium and, to a lesser degree, from the thalamus; and (2) a motor area, the subpallial **globus pallidus,** which receives projections from the pallium and from the thalamus and projects efferent fibers into descending tracts that innervate motor nuclei of cranial and spinal nerves. Efferent impulses from the globus pallidus evoke reflex, survival-oriented responses of the muscles of locomotion and feeding. (The globus pallidus is one of a group of striated nuclear masses known as the **striatum** in higher vertebrates. The arrangement of fiber tracts passing through the region is responsible for the striations.) The nuclei of the pallium and the globus pallidus

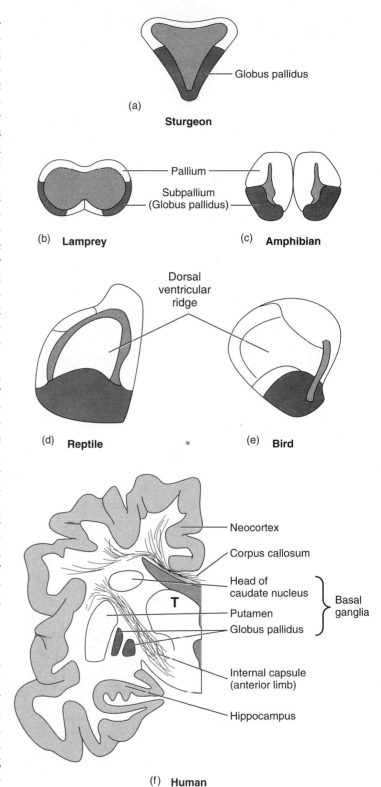

FIGURE 16.25

The globus pallidus from fishes to humans. (*d*)–(*f*), only the left cerebral hemispheres are shown. Ventricles are *red*. **T,** thalamus. Not to scale.

persist in tetrapods, although eventually becoming subsidiary to more recently evolved nuclei.

Amphibians

The primitive pallium and globus pallidus that function in fishes are still prominent in amphibians, and they function in a similar relationship (fig. 16.25c). Additional nuclei that have evolved in the amphibian subpallium enable the cerebrum to participate more extensively in coordinating incoming sensory information and in directing responses to appropriate somatic musculature, which now includes an appendicular musculature that is more complex than that of fishes.

Reptiles

The cerebrum of reptiles has become huge compared to that of amphibians. The hemispheres now bulge laterally, dorsally, and backward over the diencephalon so that the thalamus is no longer visible from above (fig. 16.14, snake and chicken). One of the factors responsible is a massive new area of association neurons, the **dorsal ventricular ridge,** in the lateral wall of the hemispheres. It increases considerably the bulk of the cerebrum and bulges into the lateral ventricle, converting the latter into a narrow slit (fig. 16.25d). The dorsal ridge receives visual, auditory, and general afferent somatic stimuli (table 16.1) that have been relayed from the thalamus. The ridge processes this information and projects fibers to the globus pallidus and to other nuclei in the subpallium that constitute, collectively, the striatum. From there descending fiber tracts conduct efferent impulses to motor nuclei in the brain stem and cord, evoking somatic motor activity appropriate to the exigencies of the environment.

Contributing also to the size of the reptilian hemispheres are the additional upper-level neurons necessitated by the musculature of the neck, a reptilian innovation, and especially of the limbs. Since the days of the cotylosaurs, amniote limbs have supported the rest of the body above the ground; and, excepting limbless offshoots, many reptiles have been capable of locomotion equal, or superior, to that of many mammals.

Birds

The cerebrum of birds is basically reptilian. The dorsal ridge has been inherited from reptiles, but another stratum of neurons has been added, capping the ridge (fig. 16.25e). The source of those neurons is unknown. The eyes of birds, and especially birds of prey such as hawks and eagles, have unique specializations (chapter 17, Visual Organs), and the sensory input via the optic nerves is highly complex. The avian ridge collates and processes this information. Any other competence the avian ridge may have is speculative. It has been suggested that it may have a role in migration, homing, and stereotyped behavior such as nest building. It must be noted, however, that lampreys and many bony fishes exhibit equally remarkable behavior with the very simple brain that they have inherited.

Mammals

The walls and roof of the lateral ventricles of mammals have become expanded in all available directions with the formation of the mammalian **neocortex** consisting, in humans, of the cell bodies of an estimated 13 billion neurons. For convenience in discussing neocortical locations, the neocortex has been divided topographically—and arbitrarily—into **frontal, parietal, temporal,** and **occipital** lobes (fig. 16.26). In most mammals, the cortex is so voluminous that it is folded into ridges (**gyri**) and grooves (**sulci**) that increase the working area of the hemisphere. These folds are lacking in monotremes (fig. 16.23), a few marsupials, and in many rodents. Formation of the neocortex resulted in a massive band of white matter, the **internal capsule,** connecting the cortex with the brain stem (figs. 16.15 and 16.25). It is the sole pathway for (1) afferent fibers radiating to the cortex of the frontal lobe from the lateral nucleus of the thalamus, and (2) the great efferent tracts that descend from the cortex to the red nucleus, to other nuclei in the brain stem, and to motor nuclei in the spinal cord. Among the efferent tracts is the corticospinal, mentioned earlier. These efferent tracts enter the tegmentum of the mesencephalon just anterior to the pons. In the roof of the ventricles, except in monotremes and marsupials, a broad transverse sheet of decussating fibers, the **corpus callosum,** connects the neocortex of the two hemispheres (figs. 16.24 and 16.25f). Separating the lateral ventricles is a thin, double-walled membrane, the **septum pellucidum.**

An ancient olfactory pallial area that has persisted in mammals, the **hippocampus,** has become tucked under the temporal lobe of the cerebral hemispheres as a result of overgrowth of the neocortex. It was mentioned earlier as part of the mammalian limbic system. Memory storage has been attributed to the mammalian hippocampus, but ongoing neuroresearch has not confirmed the hypothesis. Modern techniques for identifying fleeting activity within the brain while it is occurring promise to provide uncontroversial evidence for the localization of many cortical functions.

The development of the many additional components of the cerebrum of mammals has displaced the globus pallidus to a position deep within the hemispheres (fig. 16.25f). It has been joined by adjacent striated nuclei, a **caudate nucleus** (fig. 16.15), (so called because it has a long tail), a **putamen** (fig. 16.25f), and an **amygdaloid nucleus** (not illustrated). Early homologues of the associated nuclei are present in submammalian amniotes. Collectively, these nuclei constitute what has been known as the **basal ganglia.** The globus pallidus, as in fishes, is a source of descending motor fibers that end in synapse with cell bodies of somatic motor fibers of cranial and

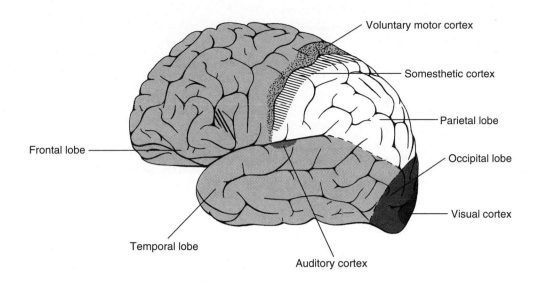

FIGURE 16.26

Topographical lobes of the human cerebral hemisphere and cortical loci for voluntary motor control, hearing, sight, and somesthetic sensibilities. Most of the visual and auditory cortices are on infolded gyri not visible on the surface. Cortical centers for taste and olfaction are less localized.

spinal nerves. This role is shared by the other nuclei, whose total function has not yet been ascertained. However, the motor control exerted by the globus pallidus has been largely preempted by the red nucleus and other recently evolved nuclei in the brain stem, and by motor stimuli from the voluntary motor cortex in the frontal lobe (fig. 16.26). Pathologic changes in the basal ganglia result in certain motor disfunctions (dyskinesias) such as the rhythmical tremor seen in aged victims of Parkinson's disease (paralysis agitans). Expansion of the hemispheres in mammals has hidden the basal ganglia, thalamus, and midbrain from dorsal view. The large cerebellum overlies almost all the remainder of the brain stem (fig. 16.23). However, viewing the brain stem from above should convince the most dubious observer that the mammalian brain, including that of humans, is built in accordance with the same architectural pattern as are the brains of other vertebrates.

The mammalian neocortex has three fundamental roles: (1) it is the center to which sensory impulses from special and general receptors are projected for evoking an awareness of the environment through physical sensation. General receptors evoke sensations of a discriminative (**epicritic**) nature, such as perceiving differences in the temperature, texture, or weight of objects held in the hand. (2) The voluntary motor cortex is the site where voluntary motor activity is initiated. (3) Having received sensory information from the environment, the neocortex collates the information and then shunts it elsewhere in the cerebrum for storage as memory. "Elsewhere" is currently the subject of active investigation. The prefrontal cortex, phylogenetically the newest, retrieves bits of pertinent stored information and employs them as *working memory,* such as recalling a

telephone number. The process of decision making is much more complicated; but the manner in which the neocortex is employed in the solution of society's problems will determine the future of civilization insofar as it is under human control.

Blood Supply to Brain

The internal carotids and the anterior (ventral) spinal artery supply the brains of most vertebrates, the spinal artery continuing onto the ventral surface of the brain as the basilar artery (fig. 16.27). In many reptiles and mammals, the vertebral arteries become a major supply, emptying into the basilar just anterior to the foramen magnum. Communicating arteries connect the basilar and internal carotid streams in mammals and some other vertebrates, providing an arterial **circle of Willis** around the base of the diencephalon (fig. 16.27). The internal carotids supply principally the cerebral hemispheres, whereas the basilar supplies principally the brain stem. There is considerable variation in the specific configuration of the circle of Willis, even among humans.

Brains are drained of blood by venous sinuses and smaller veins that empty chiefly into the internal jugular vein. The cavernous venous sinuses of mammals, of which the **superior sagittal sinus** is the largest, are located between two layers of the mammalian dura mater. There is no lymphatic drainage of either the brain or the cord.

Choroid Plexuses and Cerebrospinal Fluid

The cavities of the brain, the neural canal of the cord, and the subarachnoid space, when present, are filled with **cerebrospinal fluid.** This is a clear, watery liquid of low specific gravity that is removed from the blood by

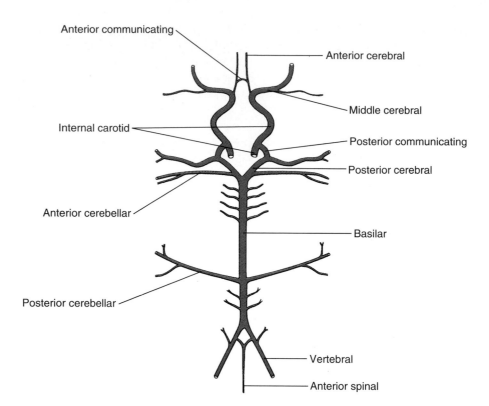

FIGURE 16.27

Arterial supply to brain of domestic cat. The anterior and posterior communicating arteries link the basilar and internal carotid streams, forming a circle of Willis around the base of the diencephalon.

diffusion and secretion. Most of it is secreted into the ventricles by the choroid plexuses; in mammals, at least, a smaller amount comes from the bloodstream directly. **Choroid plexuses** consist of tufts of a highly vascularized pia mater and ependyma that dangle into the third and fourth ventricles. The plexus in the third ventricle extends forward into the lateral ventricles via the interventricular foramina (fig. 16.18). Obstruction of these foramina results in accumulation of fluid in the lateral ventricles. When this condition is present in fetal or neonatal mammals, the fontanels of the immature skull allow the head to enlarge, a condition known as hydrocephalus. The plexus in the fourth ventricle is in the roof of the medulla.

Cerebrospinal fluid moves sluggishly caudad from the fourth ventricle into the neural canal of the cord. In mammals it also enters the subarachnoid spaces via a median and two lateral apertures in the roof of the fourth ventricle under cover of the cerebellum. These spaces surrounding the cord of mammals exhibit several cisterns where fluid normally accumulates. One of these in humans surrounds the cauda equina in the lumbosacral region, from which a *spinal tap* may aspirate some of the fluid for study of its contents or diagnosis of disease.

From the subarachnoid spaces, the cerebrospinal fluid travels along narrow perivascular spaces that accompany blood vessels entering or leaving the substance of the cord and brain; it follows the rootlets of the spinal and cranial nerves where they pierce the subarachnoid spaces and dura mater to emerge from the cord; and it seeps

into the cavity surrounding the membranous labyrinth where it contributes to the perilymph.

Cerebrospinal fluid of the brain and cord of lower vertebrates is removed by diffusion into adjacent venous channels. In mammals, clusters of **arachnoid granulations,** which are tufts of the pia-arachnoid membrane, project into large subdural venous sinuses characteristic of higher vertebrates, and the fluid diffuses out of the granulations into the venous blood of the sinuses.

In addition to cushioning the brain and cord against mechanical trauma, cerebrospinal fluid selectively exchanges metabolites with the tissues it bathes.

CRANIAL NERVES

The first ten cranial nerves of vertebrates are distributed in accordance with a basic pattern. On a functional basis, these may be grouped as follows: purely or predominantly sensory nerves (I, II, and VIII), extrinsic eyeball muscles nerves (III, IV, VI), and branchiomeric nerves (V, VII, IX, and X). Amniotes have two additional cranial nerves (XI and XII) that are purely or predominantly motor. Nerves II and VIII, although functionally sensory, contain a number of efferent fibers from the brain to the deep layer of the retina (II) or to the vestibular hair cells (VIII). These fibers apparently modulate the discharge of sensory impulses from the receptor.

The site where a cranial nerve emerges from the surface of the brain, or enters it, is its **superficial origin** as distinguished from the deep origins of its fibers, which arise (motor fibers) or terminate (sensory fibers) in a number of nuclei located at distances from one another within the brain stem. Cranial nerves III, VI, and XII emerge from the basal plate and resemble spinal nerves that have ventral roots only. Nerve IV is unique in that, although its nucleus is in the basal plate, it emerges from the roof of the brain. Branchiomeric nerves have lateral roots only (fig. 16.32, nerve XI). The three purely sensory nerves have superficial origins that will be discussed shortly.

Predominantly Sensory Cranial Nerves

Nerve I (Olfactory)
The cell bodies of olfactory nerve fibers are located in olfactory epithelium, and the fibers terminate in the olfactory bulb (fig. 16.28a). In *Squalus,* the olfactory epithelium lies so close to the bulb that an olfactory nerve cannot be distinguished as an anatomical entity. In *Scoliodon,* another shark, and in some teleosts, the sac containing the olfactory epithelium is sufficiently distant from the olfactory bulb that an olfactory nerve is demonstrable (fig. 16.28b). In most vertebrates, a number of short, discrete bundles of olfactory nerve fibers, the **filia olfactoria,** extend between the olfactory epithelium and the bulb, separated only by the olfactory capsule of the skull. These filia collectively constitute the olfactory nerve in these species.

In mammals, the olfactory epithelium is in the upper part of the nasal passage, separated from the olfactory bulb by the cribriform plate of the ethmoid bone. The foramina in the plate (figs. 9.4 and 9.6) transmit the filia olfactoria. When the brain is lifted from the cranial cavity, the olfactory nerve bundles are torn, and only stumps remain attached to the brain. The number of olfactory fibers is small in birds which, as explained earlier, have poorly developed olfactory systems. It is also small in the platypus, which spends most of its time in creeks or rivers, or in a burrow, hence has few natural enemies on land. Tactile organs and electroreceptors on the beaks of the platypus substitute for olfaction as they rout for food in the water bottom. In many marine mammals, the olfactory nerve is vestigial.

In tetrapods that have a vomeronasal organ, the olfactory nerve has a separate subdivision, the **vomeronasal nerve,** and an accessory olfactory bulb. The vomeronasal organ, a chemoreceptor, is discussed in chapter 17.

A **terminal nerve,** distinct from the olfactory nerve, lies close to the olfactory bulb and tract in representatives of most vertebrate classes. It arises from the ventral surface of the forebrain and consists of general sensory and, at least in some instances, vasomotor fibers that supply a restricted area of the nasal mucosa. Anatomic relationships suggest that it may be a vestige of an ancient branchiomeric nerve anterior to the trigeminal. The nerve conventionally has been without a numerical designation. It is seldom seen in human anatomy laboratories.

Nerve II (Optic)
The cell bodies of the optic nerve fibers are in the retina. The nerve emerges from the rear of the eyeball and extends to the optic chiasma. Beyond that it is termed the **optic tract.** Except in mammals, all or nearly all optic nerve fibers decussate in the chiasma and enter the opposite side of the brain (fig. 16.29a). In higher primates, the eyes are directed forward, and only the fibers from the nasal side of the retina decussate (fig. 16.29b). These correlated primate features result in overlap of the visual field and concomitant depth perception (binocular vision).

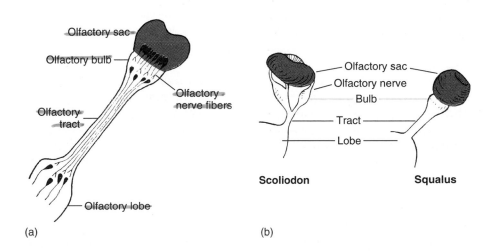

FIGURE 16.28

(*a*) Olfactory sac (*red,* containing the olfactory epithelium), bulb, and tract of *Squalus.* The cell bodies of the olfactory nerve fibers are in the olfactory epithelium. These fibers synapse in the olfactory bulb with second-order sensory neurons whose axons constitute an olfactory tract. The latter terminates in the gray matter of the olfactory lobe. (*b*) Contrasting spatial relationship of the olfactory sacs to the bulbs of two species of sharks. Note the discrete nerve in *Scoliodon* only.

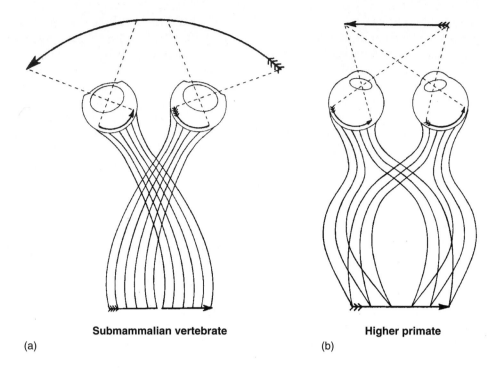

FIGURE 16.29

Decussation of optic nerve fibers in optic chiasma in vertebrates below mammals contrasted with higher primates. Below mammals, only half of the arrow is projected onto the opposite retina. In higher primates, the entire arrow is projected onto each retina, and from a slightly different angle of vision, resulting in depth perception.

(a)   **Submammalian vertebrate**   **Higher primate**   (b)

The term optic *nerve* is, from an embryological viewpoint, a misnomer since the retina arises from paired optic vesicles that never separate from the forebrain (fig. 16.13). The term *tract* (a bundle of nerve fibers within the central nervous system) is appropriate for the pathway from the optic chiasma to the tectum of the mesencephalon, where many of the fibers within the tract terminate.

Nerve VIII (Vestibulocochlear)

The eighth nerve in all vertebrates comes off the medulla very close to the roots of nerves V and VII and innervates the membranous labyrinth. It consists of two major primary branches in fishes, one innervating the ampullae on the anterior vertical and horizontal semicircular ducts, and the utriculus; the other innervating the ampulla on the posterior vertical duct, and the sacculus and lagena (fig. 17.4, shark).

In tetrapods the lagena elongates to eventually become the cochlear duct (fig. 17.4), and the branch supplying this region thereafter is known as the **cochlear nerve.** The other branch is then the **vestibular nerve.** The nerves have a **cochlear** or **vestibular ganglion** on their pathway. The cell bodies in these ganglia are exceptional for first-order sensory neurons in that they remain in a primitive bipolar state. Since the cochlear ganglion spirals within the cochlea of mammals, it is also called the **spiral ganglion.**

In some mammals (rat and mouse but not cat or bat), large unipolar sensory cell bodies are distributed within the eighth nerves along their entire length. These are cell bodies of second-order sensory neurons. They are stimulated by collateral (side) branches of incoming fibers; and because their axons end on other second-order neurons in the cochlear nuclei of the medulla, they reinforce (augment) stimuli coming from the receptor.

Eyeball Muscle Nerves

Nerves III, IV, and VI (Oculomotor, Trochlear, and Abducens)

The third, fourth, and sixth nerves supply the superior oblique (IV), external rectus (VI), the four remaining extrinsic eyeball muscles (III), and certain other myotomal muscles of the eyes (table 11.3). These eyeball-muscle nerves resemble spinal nerves that have lost their dorsal roots. In addition to somatic motor fibers, the nerves contain sensory fibers for proprioception from the muscles innervated.

Nerve III arises ventrally from the mesencephalon. Nerve IV is the only nerve arising dorsally from the brain (anterior roof of the fourth ventricle) and one of the few nerves with motor fibers that decussate before emerging. Nerve VI emerges ventrally at the anterior end of the hindbrain. Nerves IV and VI are the smallest of the cranial nerves, having fewest fibers. The cell bodies of all fibers in these nerves, both motor and proprioceptive, are in nuclei within the central nervous system. Therefore the nerves have no sensory ganglia.

Nerve III also contains visceral motor fibers that end in the ciliary ganglion of the autonomic nervous system (fig. 16.35). From the ganglion, postganglionic fibers pass to the muscles of the iris diaphragm that constrict the pupil (sphincter pupillae muscles), and to the ciliary muscles of the eye (fig. 17.18). These govern the position or thickness of the lens for visual accommodation, depending on the species.

Branchiomeric Nerves

Pharyngeal arches were the unique solution arrived at by chordates for getting oxygen into the bloodstream. The arches required muscles to operate them, and a series of similar nerves—V, VII, IX, and X—perform this function. These are the branchiomeric nerves of vertebrates.

The primitive pattern of distribution of branchiomeric nerves is readily demonstrable in a dissection that reveals cranial nerves IX and X of a modern shark (fig. 16.30). Nerve IX innervates two successive pharyngeal arches via two major branches, a **pretrematic nerve** that enters the arch in front of the first gill slit (Pre, in fig. 16.30), and a **posttrematic nerve** (Post, in fig. 16.30) that enters the arch behind it. The posttrematic branch contains the motor fibers that operate the muscles of the arch that it enters. It also has sensory fibers that monitor the condition of the gill filaments in the posterior wall of the first gill chamber. It is the only branch that contains motor fibers to striated muscles. The pretrematic branch

monitors the gill filaments in the anterior wall of the chamber. A third branch, the **pharyngeal**, extends onto the palate to monitor the pharyngeal mucosa at that body level, including any endodermal taste buds that may be present. Nerve X exhibits the identical pattern, four times over. Nerves V and VII, which provide motor innervation to the mandibular and hyoid arches, will be discussed shortly.

In addition to innervating the pharyngeal arches of fishes or their derivatives in tetrapods, branchiomeric nerves have functions that are unrelated to respiration. With the exception of nerve V, they have visceral motor fibers that are components of the autonomic nervous system and sensory fibers that innervate several varieties of sense organs. Each motor branch has proprioceptive fibers. The distribution of branchiomeric nerves in *Squalus* reveals the primitive pattern of distribution to the pharyngeal arches. Alterations in tetrapods result chiefly from elimination of gills and from other adaptations to terrestrial life.

FIGURE 16.30

Brain and cranial nerves 2 to 10 (6 not visible) of *Squalus acanthias,* dorsal view. Branchiomeric nerves are in *red.* **Do,** deep ophthalmic; **G,** petrosal ganglion; **Hy,** hyomandibular; **In,** infraorbital; **Md,** mandibular; **Pre,** pretrematic; **Post,** posttrematic; **So,** superficial ophthalmic; **Rl,** ramus lateralis (lateral line nerve) of vagus; **Rv,** ramus visceralis of vagus; **T$_3$** and **T$_4$** tela choroidea of third and fourth ventricles. The pineal body has been removed. The three eyeball muscles shown are, commencing anteriorly, superior oblique, superior rectus, and lateral rectus.

With awareness of the basic pattern of distribution of nerves IX and X, nerves V and VII in sharks become easier to understand. And since tetrapods have inherited all these nerves with modifications, the distribution and roles of the branchiomeric nerves of tetrapods will be more comprehensible.

Nerve V (Trigeminal)

The fifth nerve in all vertebrates arises from the anterior end of the hindbrain and exhibits three divisions: **ophthalmic, maxillary (infraorbital** of fishes), and **mandibular.** All branches contain sensory fibers. Only the mandibular, a posttrematic nerve in fishes, contains motor fibers.

Via one or another of the three divisions, cranial nerve V is sensory for general sensation to the skin of the entire head (fig. 16.31), including the conjunctiva of the eye, the lining of the outer ear canal, and the ectodermal surface of the eardrum. It also supplies the jaw teeth, oral mucosa, nasal passageway, and anterior surface of the tongue for general sensation. The cell bodies for all sensory fibers in the nerve, except those for proprioception, are in the trigeminal ganglion unless the ophthalmic division has its own, as in some lower vertebrates.

The mandibular nerve is motor to the muscles derived from the mandibular arch. These are chiefly the muscles of the jaws. It also innervates the tensor tympani muscle

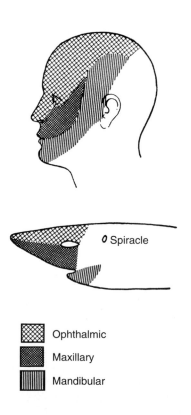

Ophthalmic

Maxillary

Mandibular

FIGURE 16.31

Cutaneous distribution of the trigeminal nerve in a human and a shark.

that is attached to the malleus, an ear ossicle. This should not be surprising, since the malleus is the displaced posterior tip of the embryonic lower jaw (fig. 9.40). Table 11.6 lists the muscles supplied by the trigeminal nerve in *Squalus* and tetrapods. The fiber components of the nerve are listed in table 16.4.

Nerve VII (Facial)

The seventh nerve arises from the anterior end of the hindbrain in close association with the fifth nerve. In sharks, the fifth and seventh nerves have a common trigeminofacial root.

Nerve VII is sensory to the neuromast organs on the head of fishes and aquatic amphibians and to taste buds in the ectoderm of fishes, which can extend all the way to the tip of the tail. With adaptation to land, those specific branches were lost. A **palatine (pharyngeal)** branch in fishes supplies taste buds in the mucosa of the pharynx at the level of the second pharyngeal arch. In tetrapods, the taste buds on the anterior portion of the tongue are also derived from the second arch, and they continue to be innervated by the facial nerve. Nerve VII of fishes also contains sensory fibers for general sensation from the endoderm of the second arch. The cell bodies for sensory fibers in nerve VII, except those for proprioception, are in the facial ganglion. Another name for this ganglion is geniculate, so named because it is at a kneelike bend in the facial canal within the temporal bone of mammals.

The facial nerve is motor to the muscles of the hyoid arch in all vertebrates. A list of these muscles will be found in table 11.6. They include the mimetic muscles of mammals and the stapedius muscle, which is attached to the stapes, an ear ossicle derived from the hyomandibular cartilage of the embryonic hyoid arch.

The facial nerve in mammals also contains visceral motor fibers to the submandibular and sphenopalatine (pterygopalatine) ganglia of the autonomic nervous system (figs. 16.33 and 16.35). These ganglia contain the cell bodies for postganglionic fibers that innervate the submandibular and sublingual salivary glands, the lacrimal glands, and the mucous membranes of the nose.

Nerve IX (Glossopharyngeal)

The ninth nerve arises from the medulla. In sharks it has three major branches, **pretrematic, posttrematic,** and **pharyngeal.** The role of these branches was discussed in the introductory paragraphs to branchiomeric nerves. A small **lateral-line branch** supplies a short segment of the lateral-line canal at the junction of the head and trunk. The nerve in all vertebrates is motor to the muscles derived from the third visceral arch (table 11.6). Of these, only a stylopharyngeus remains in mammals. The muscular branches contain proprioceptive fibers.

With loss of gills and neuromast organs in tetrapods, the ninth nerve lost many fibers. It supplies taste buds on the posterior surface of the tongue in mammals, and general receptors (touch, pain, temperature) on that same mucosa.

Preganglionic fibers of the autonomic nervous system are present in nerve IX. They have not been fully explored in all vertebrates, but in mammals they synapse in the otic ganglion, from which postganglionic fibers supply the parotid salivary gland.

The sensory cell bodies for nerve IX, other than those for proprioception, are found in the **petrosal ganglion** of lower vertebrates, and in the **superior (petrosal)** and **inferior glossopharyngeal ganglia** of birds and mammals. The superior ganglion, chiefly somatic, is derived from a neural crest. The inferior ganglion, chiefly visceral, is derived largely from an ectodermal placode. The lateral-line branch of fishes has its own ganglion.

Nerve X (Vagus)

The vagus arises from the medulla by lateral rootlets. In *Squalus* (fig. 16.30), it consists of four **branchial trunks** (more in sharks with more gill chambers) that subserve branchiomeric functions, plus a **ramus lateralis** to the lateral-line canal system, and a **ramus visceralis** to the coelomic viscera.

Each branchiomeric trunk in *Squalus* has a pretrematic, posttrematic, and pharyngeal branch. The posttrematic branches are motor to the fourth and remaining pharyngeal arches, and are therefore the chief respiratory nerves of fishes. They are also sensory to the posterior walls of the last four gill chambers. The pretrematics are sensory to the demibranchs in the anterior walls of those chambers. The pharyngeal branches are sensory for general sensation and taste on the palate between the fourth arch and the esophagus.

The ramus lateralis of *Squalus* is sensory to the lateral-line canal all the way to the tip of the tail; the ramus visceralis and its branches carry preganglionic fibers of the autonomic nervous system to autonomic ganglia in the coelom. The ganglia innervate coelomic viscera. The same branches contain visceral afferent fibers.

With the rise of land vertebrates, the vagus lost the sensory fibers from the gill chambers and the lateral-line nerve. The remainder of the vagus continues to function as in fishes. The *ramus visceralis* is now the major branch of the vagus, supplying the heart and most of the viscera via autonomic ganglia in the thorax and abdomen (fig. 16.35). Visceral sensory fibers accompany the motor fibers.

The vagus of tetrapods continues to supply muscles derived from the fourth and successive pharyngeal arches. In amniotes, these are chiefly the cricothyroid, cricoarytenoid, thyroarytenoid, and several small but important muscles of the soft palate and pharyngeal walls that are active in swallowing (palatoglossus, for example, which is seldom seen in routine mammalian dissection).

Finally, as in fishes, the tetrapod vagus carries fibers for general sensation and any fibers for taste from the pharyngeal mucosa—this being, in tetrapods, the mucosa of the oral pharynx.

The cell bodies of all sensory fibers, except those for proprioception, are in sensory ganglia. In some elasmobranchs, each of the four or more branches to the gills has its own epibranchial ganglion. The vagus of birds and mammals, like nerve IX, has two sensory ganglia on its root, a **superior vagal ganglion** that is chiefly somatic, and an **inferior vagal ganglion** that is chiefly visceral. The superior ganglia are derived from neural crests, the inferior ganglia from epibranchial placodes.

The innervation of the mucosa of the oral cavity and pharynx by nerves V, VII, IX, and X *in that sequence from front to rear, in all vertebrates from fish to humans,* is evidence of the genetic ties between fishes and tetrapods and of the negligible effect of a terrestrial habitat on the sensory innervation of constantly moist membranes.

The pretrematic, posttrematic, and pharyngeal branches of nerves IX and X of Squalus *have their counterparts in nerves V and VII. The ophthalmic and mandibular branches of nerve V are functionally equivalent, respectively, to pretrematic and posttrematic branches farther caudad. Nerve V has no pharyngeal branch. The pretrematic branch of VII has joined sensory fibers from nerve V to form an infraorbital trunk; the hyomandibular nerve, behind the spiracle and supplying the second arch, is the posttrematic branch of VII; and the palatine is the pharyngeal branch. Some deviation from the basic pattern is to be expected in the first two arches, since they are specialized for procuring and handling food.*

Accessory and Hypoglossal Nerves

Nerve XI (Accessory)

The spinal accessory nerve constitutes an eleventh cranial nerve in amniotes. It is purely motor. The details of assembly and distribution of its components vary among reptilian, avian, and mammalian orders.

In mammals, rootlets of the accessory nerve emerge from the medulla immediately behind, but not in line with, the roots of the vagus (fig. 16.32). A small number of the fibers in those rootlets innervate small muscles of the pharynx, such as those in the caudal border (velum) of the soft palate. The majority of the fibers in those rootlets join the vagus to be distributed to autonomic ganglia in the coelom. In mammalian terminology, the latter fibers constitute the **internal ramus of the accessory nerve.**

In mammals, additional rootlets of the accessory nerve emerge from the spinal cord along with ventral rootlets of a variable number of cervical spinal nerves. The spinal rootlets of the accessory nerve unite to form a nerve trunk that passes cephalad close to the cord and enters the cranial cavity via the foramen magnum.

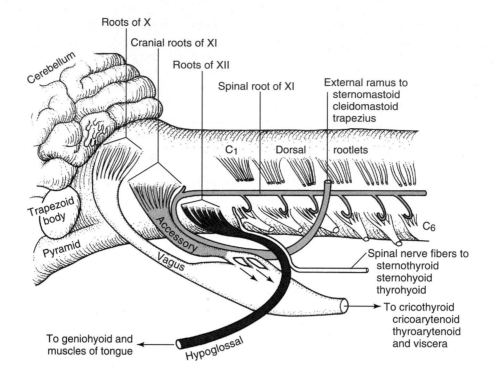

FIGURE 16.32

Superficial origins of the vagus, accessory (*light red*), and hypoglossal nerves (*dark red*) of a generalized mammal. Details vary among mammalian orders. The rootlets of XII are in series with the ventral rootlets of the spinal nerves. **C₁**, dorsal rootlets of the first cervical spinal nerve; **C₆** ventral rootlets of the sixth cervical spinal nerve. Nerve XI contributes an internal ramus (*arrows*) to the vagus.

Within the braincase the trunk joins the cranial root of XI for a short distance, after which most of the spinal fibers leave the accessory nerve to become its **external ramus.** This ramus exits the cranial cavity via the jugular foramen along with nerves IX and X. It is composed largely of fibers that innervate the trapezius and sternocleidomastoid complex of muscles. A corresponding nerve in reptiles and birds arises from medullary roots, rather than from the spinal cord.

What is the history of the eleventh nerve of amniotes? The components of the nerve are not new; only the way they are assembled is new. If we look at the location of the cell bodies of fibers that arise from the spinal cord in reptiles and birds, we find they are in neither the SE column, which supplies myotomal muscles, nor in the SVE column, which supplies branchiomeric muscles (fig. 16.34). The location varies, but it is very close to the SVE column that supplies branchiomeric fibers to nerve X. In mammals, the SVE column does not extend into the cord, and the cell bodies are close to the SE column, but clearly not part of it. Originally, they could have been part of one or the other of these columns.

As mentioned earlier, there is a series of occipitospinal nerves between the vagus and the first cervical spinal nerve of lower vertebrates. These nerves are either missing in amniotes, or their fibers are now part of nerve XI. On the basis of available facts, which indeed are slight, it is reasonable to believe that the fibers of the eleventh nerve of amniotes are contributions from the posteriormost branchiomeric nerves of fishes and from one or more ancestral occipitospinal nerves.

Nerve XII (Hypoglossal)

The twelfth nerve of amniotes is motor except for proprioceptive fibers. It arises from the **hypoglossal nucleus** in the medulla by a series of ventral rootlets and passes into the base of the tongue from which location it sends branches to the hyoglossus, styloglossus, genioglossus, and lingualis muscles (figs. 11.6, 16.4b and 16.32). On emerging from the hypoglossal foramen in mammals, the nerve is joined by fibers from one or more anterior cervical spinal nerves, the number depending on the species. Some of these spinal nerve fibers are distributed to the geniohyoid muscle. The rest of the spinal fibers leave nerve XII to join a loose plexus of cervical spinal nerves supplying the hypobranchial muscles of the neck (sternothyroid, sternohyoid, thyrohyoid, or their homologues).

That the hypoglossal is a cranial rather than a spinal nerve is dictated solely by the location of the foramen magnum. *Actually,* it is a spinal nerve that became "locked up" in the braincase. Like spinal nerves, the hypoglossal nerve of some mammals develops an embryonic dorsal root and dorsal root ganglion (**Froriep's ganglion**). The root and ganglion later disappear.

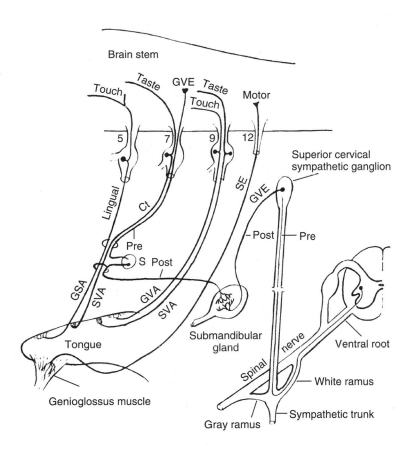

FIGURE 16.33

Innervation of the tongue and submandibular gland of a mammal (based on cat and human). **5, 7, 9,** and **12,** cranial nerves; **Ct,** chorda tympani; **Pre** and **Post,** preganglionic and postganglionic fibers of the autonomic nervous system; **S,** submandibular ganglion of the autonomic system. A key to the fiber components is given in tables 16.1 and 16.3.

The twelfth nerve is a derivative of the occipitospinal series of fishes and amphibians, as may be deduced from the following facts: (1) Occipitospinal nerves have been reduced in number above fishes, whereas the number of cranial nerves has increased. (2) The twelfth nerve, like many occipitospinal nerves, lacks a dorsal root. (3) The occipitospinal nerves of lower vertebrates supply hypobranchial muscles, and the tongue is hypobranchial muscle. (4) Whereas tongue muscles are supplied by the last cranial nerve in amniotes, the tongue is supplied by the first spinal nerve in amphibians.

Innervation of the Mammalian Tongue: An Anatomic Legacy

The innervation of the mammalian tongue illustrates how a single organ may be served by numerous nerves, depending on the ontogenetic and phylogenetic history of its parts (fig. 16.33). The mucosa of the anterior part of the tongue is first arch endoderm and is therefore innervated by nerve V for general sensations. The taste buds on this part of the tongue are innervated by nerve VII, which supplies taste buds on the first and second arches in fishes. The mucosa on the posterior part of the tongue is innervated by nerve IX for both general sensation and taste because of the origin of this mucosa from the third pharyngeal arch. The muscles of the tongue are myotomal and are innervated by nerve XII.

Although four cranial nerves innervate the tongue, only three can be traced into it, since the branch of nerve VII conducting taste fibers (chorda tympani, fig. 16.33) unites with the lingual branch of nerve V just before the latter reaches the tongue.

Functional Components of Cranial Nerves

As noted earlier, fibers in spinal nerves may be classified in four functional categories: GSA, GVA, SE, and GVE. Each supplies general receptors or effectors (table 16.1). One or more of these components may be found also in cranial nerves. In addition to general fibers, cranial nerves contain *special* fibers, of which there are three categories: SSA, SVA, and SVE or EB (table 16.3). Each

TABLE 16.3   Special Fiber Components of Cranial Nerves

Components	Innervation and Nerve
Special somatic afferent fibers (SSA)	Special somatic receptors Retina (II) Membranous labyrinth (VIII) Neuromast organs (VII, IX, and X)
Special visceral afferent fibers (SVA)	Special visceral receptors Olfactory epithelium (I) Taste buds (VII, IX, and X)
Efferent branchial (EB)*	Branchiomeric muscle (V, VII, IX, X, and XI)

Note: In addition to these special components, cranial nerves other than I, II, and VIII have one or more of the general components listed in table 16.1.
*Arbitrarily classifiable also as special visceral efferent fibers (SVE).

functional category—general and special—commences or terminates in the central nervous system in a continuous, or, more often, discontinuous column of gray matter composed of the nuclei subserving that specific function, motor columns being in the basal plate, sensory columns in the alar plate (fig. 16.34). Table 16.4 summarizes the major distribution of the cranial nerves in vertebrates in general.

AUTONOMIC NERVOUS SYSTEM

The autonomic nervous system innervates glands and smooth and cardiac muscle. It consists chiefly of autonomic nerves, plexuses, and ganglia (fig. 16.35). It does not, however, constitute an anatomical entity; that is, it cannot be completely dissected away from the rest of the nervous system, since its components begin inside the central nervous system and emerge via cranial or spinal nerves. It is entirely a visceral *motor* system. However, sensory fibers from the viscera use autonomic pathways to reach cranial and spinal nerves.

The autonomic nervous system is composed of (1) a **sympathetic system** (**thoracicolumbar system** in mammals) that emerges from the cord via most of the spinal nerves of the trunk, and (2) a **parasympathetic (craniosacral) system** that emerges from the brain and from the sacral region of the spinal cord. Most visceral effectors except those in the skin are innervated by fibers from

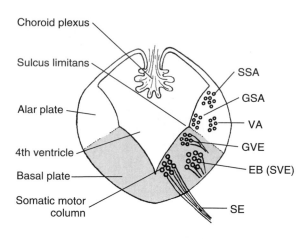

FIGURE 16.34

Cross section of amniote medulla, generalized, showing location of the columns of sensory and motor nuclei of cranial nerves. Sensory nuclei are in the alar plate; motor nuclei, in the basal plate. **SE,** somatic motor fibers with cell bodies in the nucleus in the somatic motor column. **VA,** column for all general and special visceral afferent nuclei except those for smell. The remaining nuclei are identified in tables 16.1 and 16.3.

both systems (fig. 16.35, iris diaphragm, heart, stomach, urinary bladder). The stimulatory effects of one system modulate the inhibitory effects of the other to bring about an appropriate response. Impulses from the sympathetic and parasympathetic systems that innervate the intestines are delivered to delicate neural plexuses (**Meissner's** and

---

TABLE 16.4 The Cranial Nerves and Their Major Distribution in a Generalized Vertebrate

	Predominant Distribution and Fiber Type[1]		Other Significant Distributions
Predominantly Sensory			
I Olfactory	Olfactory epithelium	SVA	Vomeronasal organ
II Optic	Retina	SSA	
VIII Vestibulocochlear	Membranous labyrinth	SSA	
Predominatly Motor			
III Oculomotor			Constrictors of iris diaphragm
IV Trochlear	Extrinsic eyeball muscles	SE[2]	
VI Abducens			
Branchiomeric Mixed			
V Trigeminal	Arch I	EB[2]	Head ectoderm
VII Facial	Arch II	EB	Taste buds, oral glands, neuromasts
IX Glossopharyngeal	Arch III	EB	Taste buds, oral glands, neuromasts
X Vagus	Arches IV–VII	EB	Taste buds, neuromasts
	Coelomic viscera	GVE[3]	
Amniotes			
XI Accessory	Branchiomeric muscles	EB[2]	
	Coelomic viscera	GVE[3]	
XII Hypoglossal	Tongue muscles	SE[2]	

[1]Fiber types are as follows: SVA = special visceral afferent; SSA = special somatic afferent; SE = somatic efferent; EB = efferent branchial; GVE = general visceral efferent.
[2]Proprioceptive fibers accompany SE and EB fibers.
[3]GVA fibers accompany most GVE fibers.

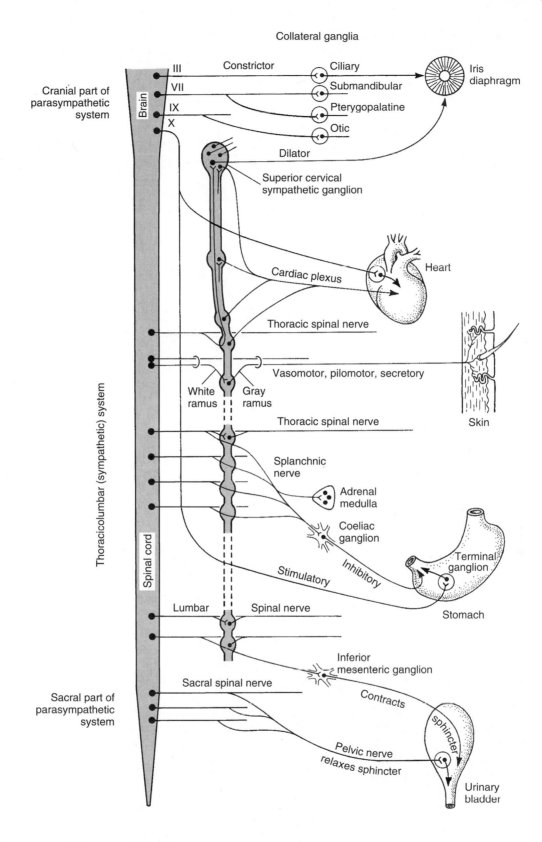

FIGURE 16.35

Representative components of the mammalian autonomic nervous system, sympathetic trunk in *gray*. Autonomic fibers are not confined to the spinal nerves shown. Arrows emphasize dual control of visceral effectors by craniosacral and thoracicolumbar systems. The cell bodies (*black dots*) of preganglionic fibers are in the brain or cord; those of postganglionic fibers are in a sympathetic, collateral, or terminal ganglion. *Red* indicates parasympathetic innervation (lacking in skin and adrenal).

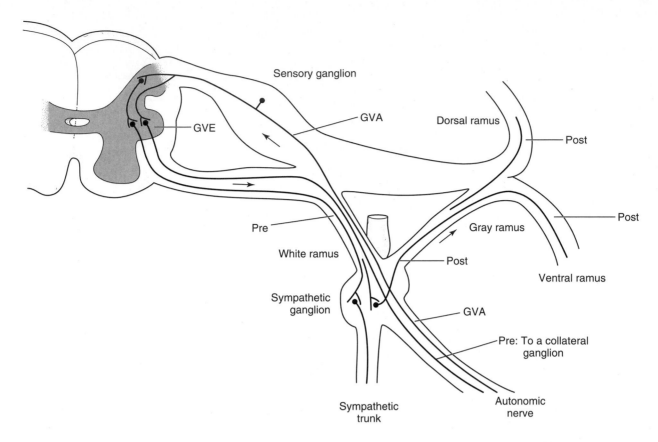

FIGURE 16.36

Motor fibers (*red*) of autonomic system in a thoracic spinal nerve. Fibers in the dorsal and ventral rami are vasomotor, pilomotor, or secretory. The red fiber in the autonomic nerve distal to the ganglion will synapse in a collateral ganglion such as the coeliac (see splanchnic nerve, fig. 16.35). **GVA,** general visceral afferent fiber; **GVE,** visceral motor cell body in lateral horn; **Post, Pre,** postganglionnic and preganglionic fiber.

Auerbach's) located in the external muscle layer of the gut (fig. 12.16). This **enteric plexus,** consisting of enteric neurons, is responsible for peristalsis.

Two finely myelinated neurons in series conduct the impulse from the brain or cord to a visceral organ (fig. 16.36). The cell body of the first neuron (preganglionic neuron) is in the GVE column of the brain or cord and its preganglionic fiber terminates in an autonomic ganglion. The cell body of the second neuron (postganglionic neuron) is in an autonomic ganglion. Its fiber extends to the effector. Autonomic ganglia are classified in three categories, **sympathetic (paravertebral), collateral,** and **terminal.**

Sympathetic ganglia are interconnected by autonomic fibers that form a **sympathetic trunk** close to the vertebral column (figs. 16.35 and 16.37). There is generally one ganglion for each spinal nerve except in the cervical region, where there are fewer, and in the sacrum, where there are none. The sympathetic ganglia in the trunk are connected to a spinal nerve by **white and gray rami communicantes** (fig. 16.36). White rami contain preganglionic fibers from the cord and afferent visceral fibers with cell bodies in the dorsal root ganglion. Gray rami contain postganglionic fibers to the skin.

In general, fibers with cell bodies in the cervical ganglia of the sympathetic trunk supply visceral organs in the head, neck, and thorax. Fibers with cell bodies in other sympathetic ganglia supply vasomotor, pilomotor, and secretory fibers to the skin via gray rami and spinal nerves. (**Vasomotor fibers** supply cutaneous arterioles, **pilomotor fibers** supply muscles that move hairs, and **secretory fibers** supply skin glands.) In cold-blooded vertebrates, the cutaneous fibers also supply some types of chromatophores, and in birds they supply **plumomotor fibers** to muscles that move feathers. A considerable majority of preganglionic fibers that enter the sympathetic trunk pass through sympathetic ganglia without synapsing, enter an **autonomic nerve** such as the splanchnic, and terminate in the coeliac and other collateral ganglia in the abdominal cavity.

Postganglionic fibers with cell bodies in collateral ganglia in the head—**ciliary, submandibular, sphenopalatine (pterygopalatine)** and **otic**—innervate visceral effectors in the head. Their neural connections are given in table 16.5. Postganglionic fibers with cell bodies in collateral ganglia in the coelom, such as the coeliac and inferior mesenteric, innervate organs in the abdomen. The collateral ganglia in the abdomen are part of auto-

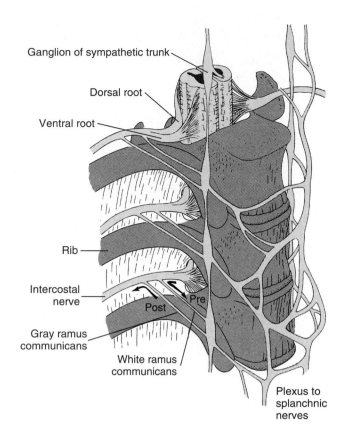

Ganglion of sympathetic trunk

Dorsal root

Ventral root

Rib

Intercostal nerve

Post   Pre

Gray ramus communicans

White ramus communicans

Plexus to splanchnic nerves

FIGURE 16.37

Sympathetic (paravertebral) ganglia and their interconnections in thoracic region of a cat, oblique ventral view. Arrows indicate direction of preganglionic **(Pre)** and postganglionic **(Post)** impulses.

TABLE 16.5 Innervation and Peripheral Distribution of Autonomic Ganglia of the Head of Mammals

Ganglion	Receives Fibers From:	Projects Fibers To:
Ciliary	Oculomotor nerve	Ciliary muscle of eye Constrictor muscles of iris
Submandibular	Facial nerve	Submandibular gland Sublingual gland
Sphenopalatine (pterygopalatine)	Facial nerve	Lacrimal gland Glands of nose and pharyngeal mucosa
Otic	Glossopharyngeal nerve	Parotid gland

nomic plexuses on the major arterial branches of the abdominal aorta (fig. 16.38).

Terminal ganglia are embedded in the walls of visceral organs within the coelom. The cell bodies in these ganglia send very short postganglionic fibers to the organs they innervate.

After passing through the sympathetic trunk, most autonomic fibers join **autonomic plexuses** en route to their destination. These plexuses are entwined around blood vessels, the vertebral column, and other available structures including, in the neck, the esophagus (fig. 16.37).

The neurotransmitter released at the endings of all preganglionic fibers, both sympathetic and parasympathetic, and the one released at postganglionic endings of the parasympathetic system is chiefly **acetylcholine.** These fibers are therefore said to be **cholinergic.** (Acetylcholine also mediates transfer of impulses from somatic motor fibers to skeletal muscle cells.) The neurotransmitters released at postganglionic endings of the sympathetic system are chiefly **norepinephrine (noradrenaline)** and small amounts of **epinephrine (adrenaline).** These fibers are said to be **adrenergic.**

The **adrenal medulla** (fig. 16.35) is functionally equivalent to a sympathetic ganglion. The cells of the medulla, like those of sympathetic ganglia, arise from neural crests, are innervated by preganglionic fibers, and synthesize and release adrenaline and noradrenaline. They differ in that the adrenal medulla does not sprout axons. Consequently their secretions are carried away by the bloodstream.

Although the autonomic nervous system is involuntary (you cannot prevent blushing), some voluntary control can be exerted by biofeedback. Autonomic systems are present in all vertebrates; anurans have the basic components of amniote systems.

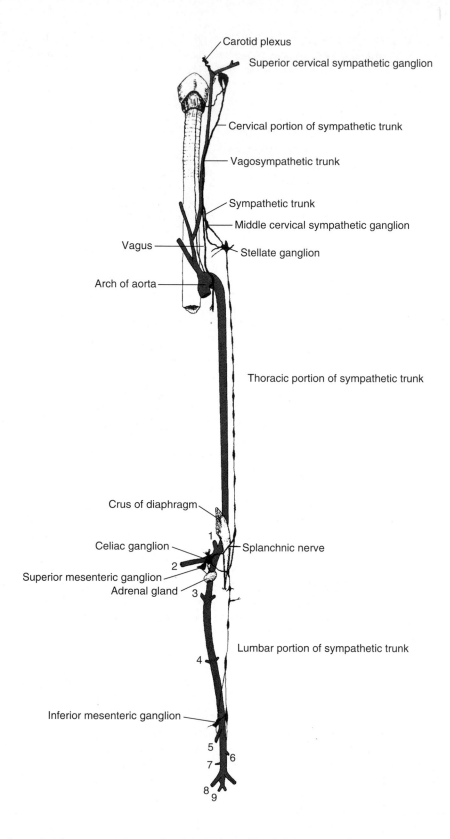

FIGURE 16.38

Left sympathetic trunk and associated structures of a cat. **1** to **9**, major branches of abdominal aorta: **1**, celiac; **2**, superior mesenteric; **3**, renal; **4**, spermatic or ovarian; **5**, inferior mesenteric; **6** and **7**, left and right iliolumbars; **8**, external iliac; **9**, internal iliac. In cats, as in some other mammals, the vagus nerve and sympathetic trunk in the neck are in a common sheath to form a vagosympathetic trunk. Fibers in the carotid plexus have postganglionic cell bodies in the superior cervical sympathetic ganglion.

SUMMARY

1. The neuron is the functional unit of the nervous system. It consists of a cell body and one or more processes. The long process is an axon, and short processes are dendrites.
2. A nerve fiber is an axon and its myelin sheath.
3. A nerve is a bundle of nerve fibers outside the central nervous system (CNS). A tract is a bundle of nerve fibers within the CNS.
4. The CNS consists of the brain and spinal cord. The peripheral nervous system includes all neurons and their processes outside the CNS.
5. A nucleus is a group of similarly functioning cell bodies inside the CNS. Nuclei constitute the gray matter of the brain and cord.
6. A ganglion is a group of cell bodies outside the CNS. Ganglia are sensory except those of the autonomic nervous system, which are motor.
7. Neurotransmitters are amines that facilitate or inhibit nerve impulses across synapses. Neurosecretions are polypeptide hormones. They are secreted by neurosecretory neurons and carried in the blood.
8. The cell bodies of motor neurons are inside the cord or brain with one exception. The cell bodies of postganglionic fibers of the autonomic nervous system are in autonomic ganglia.
9. The cell bodies of first-order sensory neurons are in sensory ganglia on cranial and spinal nerves with three major exceptions: the cell bodies of olfactory nerve fibers are in the olfactory epithelium, those of optic nerve fibers are in the retina, and those of proprioceptive fibers of cranial nerves are in the mesencephalon.
10. Sensory ganglia arise chiefly from neural crests. A few in the head arise from ectodermal placodes.
11. Olfactory neurons and the rods and cones of the retina are sometimes referred to as neurosensory cells.
12. The embryonic neural tube caudal to the forebrain consists of alar and basal plates that are sensory and motor, respectively. The alar plate gives rise to second-order sensory neurons. The basal plate gives rise to motor neurons other than postganglionic.
13. Neuroglia in the CNS consist of ependyma, oligodendroglia, microglia, and astrocytes.
14. Schwann cells are oligodendroglia that surround nerve fibers in the periphery. They form a living membrane, the neurilemma. It secretes the myelin sheath.
15. The spinal cord consists of nuclei surrounding the central canal, and of fiber tracts that ascend, descend, or cross the cord from one side to the other. It is enclosed within one or more meninges.
16. A meninx primitiva surrounds the brain and cord of fishes. A dura mater and leptomeninx are present in lower tetrapods. The latter differentiates into a pia mater and arachnoid membrane in some birds and in mammals.
17. The spinal cord has cervical and lumbar enlargements at the level of the paired appendages. When shorter than the column, the cord terminates in a filum terminale accompanied by a cauda equina. Fishes exhibit an unpaired neuroendocrine swelling of the cord, the urophysis at the base of the tail. It houses neurosecretory cells.
18. Spinal nerves are metameric in origin and distribution. Most spinal nerves exhibit dorsal and ventral rootlets and roots, and dorsal, ventral, and communicating rami. Ventral rami often unite to form simple or complicated plexuses.
19. Most spinal nerves have sensory ganglia on their dorsal roots. The following cranial nerves also have sensory ganglia: V, VII, VIII, IX, and X.
20. Occipitospinal nerves lack sensory roots and supply hypobranchial muscles. They arise between the tenth cranial and the first spinal nerves. They are common in lower vertebrates. They are represented in amniotes by cranial nerves XI and XII.
21. Spinal nerves contain the following fiber components: GSA, SE, GVA, and GVE. Cranial nerves may contain one or more of the preceding and one or more of the following: SSA, SVA, and EB (SVE).
22. Anamniotes have 10 cranial nerves, amniotes have 12. Nerves I, II, and VIII are purely sensory to the olfactory epithelium, retina, and membranous labyrinth. Nerves III, IV, and VI supply myotomal muscles of the eyeball. Nerves V, VII, IX, and X supply the pharyngeal arches with motor and sensory fibers.
23. Cranial nerve V is the chief nerve for cutaneous sensation on the head and the ectodermal part of the oral cavity. Nerves VII, IX, and X supply neuromast organs and taste buds.
24. Nerve XI has medullar and spinal roots. It is derived from the caudal end of the branchiomeric series and from occipitospinal nerves. The internal ramus contains GVE fibers that are distributed with the vagus. The external ramus innervates the trapezius and sternocleidomastoid muscles.
25. Nerve XII represents one or more occipitospinal nerves. It supplies the muscles of the tongue.
26. A small anteriormost cranial nerve, apparently vestigial, arises from the ventral surface of the forebrain and supplies a limited region of the nasal mucosa with general sensory and vasomotor fibers. It is called the terminal nerve.
27. The autonomic nervous system innervates smooth and cardiac muscles and glands. Preganglionic fibers of the craniosacral (parasympathetic) division emerge from the brain via cranial nerves III, VII, IX, X, and XI, and from the sacral region of the cord via

sacral spinal nerves. Preganglionic fibers of the thoracicolumbar (sympathetic) division emerge from the cord via thoracic and lumbar spinal nerves.

28. Autonomic ganglia are paravertebral (sympathetic chain), collateral (in head or close to abdominal aorta), and terminal (in trunk adjacent to or within the organ innervated).

29. The autonomic ganglia of the head and their associated nerves are the ciliary (III), sphenopalatine (pterygopalatine) (VII), submandibular (VII), and otic (IX). These are parasympathetic ganglia.

30. Most viscera are supplied by sympathetic and parasympathetic fibers. Visceral effectors in the skin receive only sympathetic innervation.

31. Cerebrospinal fluid is secreted by choroid plexuses in the lateral, third, and fourth ventricles. It fills the brain ventricles and central canal of the cord and escapes to submeningeal spaces via foramina in the roof of the fourth ventricle. It is retrieved directly by venous channels of the brain and cord, and by arachnoid villi that return it to major venous sinuses of the brain.

32. The brain has three major subdivisions: prosencephalon (forebrain), mesencephalon (midbrain), and rhombencephalon (hindbrain). The more prominent components of these subdivisions in mammals are listed below.

Prosencephalon (Forebrain)

Telencephalon
- Olfactory bulbs
- Olfactory tracts
- Cerebral hemispheres
  - Neocortex
  - Basal ganglia
  - Corpus callosum
  - Internal capsule
  - Lamina terminalis
- Lateral ventricles

Diencephalon
- Epithalamus
  - Habenulae
  - Pineal
- Thalamus
  - Massa intermedia
- Hypothalamus
- Optic chiasma
- Infundibular stalk and pituitary
- Third ventricle
- Circle of Willis

Rhombencephalon (Hindbrain)

Metencephalon
- Cerebellum
- Pons
- Fourth ventricle

Myelencephalon
- Medulla oblongata
- Vagal lobes
- Pyramidal tracts
- Fourth ventricle

Mesencephalon (Midbrain)
- Optic lobes } Tectum
- Auditory lobes }
- Tegmentum
  - Cerebral peduncles
  - Red nucleus
- Cerebral aqueduct

SELECTED READINGS

Ariens Kappers, C. U., Huber, G. C., and Crosby, E. C.: The comparative anatomy of the nervous system of vertebrates, including man, 2 vols., New York, 1936, The Macmillan Co. (Republished by Hafner Publishing Co., 1960, New York.)

Bullock, T. H., and Horridge, G. A.: Structure and function in the nervous systems of invertebrates, 2 vols., San Francisco, 1965, W. H. Freeman and Co., Publishers.

Finley, B. L., and Darlington, R. B.: Linked regularities in the development and evolution of mammalian brains, Science **268**:1578, 1995.

Hopkins, W. G., and Brown, M. C.: Development of nerve cells and their connections, New York, 1984, Cambridge University Press.

Jacobson, M.: Developmental neurobiology, second edition, New York, 1978, Plenum Press.

Morell, P., and Norton, W. T.: Myelin, Scientific American **242**(5):88, 1980.

Myers, P. Z.: Spinal motorneurones of the larval zebrafish, Journal of Comparative Neurology **236**:555, 1985.

Nichol, J. A. C.: Autonomic nervous system in lower chordates, Biological Reviews **27**:1, 1952.

Norris, H. W., and Hughes, S. P.: The cranial, occipital, and anterior spinal nerves of the dogfish, *Squalus acanthias,* Journal of Comparative Neurology **31**:293, 1920.

Pearson, R., and Pearson, L.: The vertebrate brain, New York, 1976, Academic Press.

Rovainen, C. M.: Neurobiology of lampreys, Physiological Reviews **59**:1007, 1979.

Shuangshoti, S., and Netsky, M. G.: Choroid plexus and paraphysis in lower vertebrates, Journal of Morphology **120**:157, 1966.

Szekely, G. The morphology of motor neurones and dorsal root fibers in the frog's spinal cord, Brain Research **103**:275, 1976.

CHAPTER SEVENTEEN

SENSE ORGANS

In this chapter we will study the organs that enable vertebrates to monitor two environments, the one surrounding
them and the one within. Since the former presents more
challenges to moment-by-moment well-being, proportionately more time will be devoted to monitors of the ambient
environment, the exteroceptors. Along the way, we will look
at the sensory feedback mechanism that enables skeletal
muscles to respond to any challenge.

OUTLINE

ime has resulted in the evolution of a large variety of simple and complex sense organs for monitoring the external and internal environment. Sense organs (receptors) are transducers of specific forms of kinetic energy. They change mechanical, electrical, thermal, chemical, or radiant energy into nerve impulses in sensory neurons.

Vertebrates arose with an array of essential sense organs and the necessary central nervous system pathways for processing the information: a mechanically stimulated lateral-line canal system, electroreceptors, a mechanoreceptive vestibular system that may have had some auditory competence, light receptors (lateral and median eyes) and their central pathways, an olfactory system, chemoreceptors other than olfactory, a general somatic sensory system, and a general visceral sensory system. Because many of the receptors are, and have been, protected by skeletal capsules that fossilize readily, or because they leave indelible impressions in fossilized dermal armor, or because of telltale foramina, it can be said with confidence that all these systems were present in the oldest known vertebrates, the ostracoderms. During subsequent vertebrate evolution, receptors have undergone mutation—some more than others. Some, such as the lateral-line system, disappeared among terrestrial vertebrates. Meanwhile, new systems such as thermal receptors in the infrared range have made an appearance and play a role in survival on land.

Somatic receptors provide information about the external environment **(exteroceptors)** and the activity of the skeletomuscular system **(proprioceptors).** Their sensory fibers terminate in nuclei in the *somatic sensory column* of the alar plate of the brain and cord (fig. 16.34, GSA, SSA). In reacting to this information, animals make appropriate skeletal muscle responses. For example, stimuli from an approaching potential enemy initiate locomotion or a change in posture, and input from surrounding fishes in a school of fishes reflexly maintains the individual's orientation in the dynamic mass. The category, somatic receptors, does not include chemical receptors such as those for taste and smell.

Visceral receptors provide information about the environment within the animal **(enteroceptors)** and olfactory and gustatory information from the environment. Their sensory fibers, except those for smell, end in nuclei in the *visceral sensory column* of the brain and cord (fig. 16.34, VA). As a result of information from visceral receptors, an appropriate internal environment is reflexly maintained. Olfactory stimuli evoke both somatic and visceral responses, particularly in higher tetrapods, and they are more or less arbitrarily categorized as visceral.

Although the peripheral endings of many sensory nerve fibers, especially visceral, are stimulated directly, **intermediary nonnervous receptor cells,** or batteries of them, often serve as transducers of the energy actuated by the stimulus, and the receptor cells pass on a stimulus to the sensory nerve fiber. **Hair cells** are one variety of receptor cells (fig. 17.1). They are elongated or bulbous *epithelial* cells with a variable number of short stereocilia and usually a single long kinocilium projecting into the fluid that bathes the sensory epithelium. The cilia are sometimes embedded in a gelatinous mass, the **cupula.** Receptor cells with cytoplasmic apical processes instead of cilia are present in taste buds (fig. 17.23).

For convenience of discussion, sense organs can be classified as general or special receptors. **General receptors** are widely distributed in the skin and within the body. **Special receptors** are those with a limited distribution. They are confined to the head except in fishes and aquatic amphibians. In the remainder of this chapter, we will examine some of the general and special receptors.

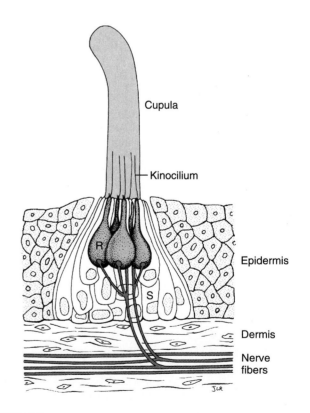

FIGURE 17.1

A neuromast organ in the epidermis of a necturus. **R,** hair cell (receptor cell) with a single kinocilium and three or four short delicate stereocilia (at base of kinocilium) extending into the cupula. Contacting the base of the hair cells is a heavily myelinated sensory nerve fiber. **S,** supporting cell.

SPECIAL SOMATIC RECEPTORS

Neuromast Organs of Fishes and Aquatic Amphibians

Neuromasts are epithelial receptors that consist of aggregations of hair cells and, generally, a cupula that projects into a fluid (fig. 17.1). Associated with the hair cells are **supportive (sustentacular) cells.** Neuromasts are disposed singly or in groups, depending on their location and the species. In fishes and aquatic amphibians, they monitor principally mechanical stimuli in the pond or seawater, although some serve as electrical, thermal, or chemoreceptors. The simplest neuromasts occupy shallow pits or grooves in the epidermis, with the cupulae projecting into the pond or seawater. These **external neuromasts** are characteristic of agnathans, larval amphibians, and aquatic urodeles. More specialized are the **internal neuromasts** in the membranous labyrinth (inner ear) of vertebrates from fishes to humans.

Neuromasts in some jawed fishes occupy fluid-filled pits that lie beneath the epidermis and open to the surface via a pore. These are **pit organs.** Other neuromasts occupy vesicles or ampullae that have a long duct that connects them with the surface. Among these are the **vesicles of Savi** in electric rays and the **ampullae of Lorenzini** in sharks (fig. 17.2a). The jellylike fluid in the vesicles and ampullae is secreted by glandular cells that line the organ. *Amia*, a primitive bony fish, has as many as 3,700 ampullae on the head alone. Pit organs, vesicles, and ampullae are found chiefly, but not exclusively, on the head.

The most common systems of neuromast organs are the fluid-filled **lateral-line canal** and the **cephalic canal system** (fig. 17.2). In their simplest manifestation, these are linear series of neuromasts located in numerous shallow canal-like grooves on the surface of the head and in

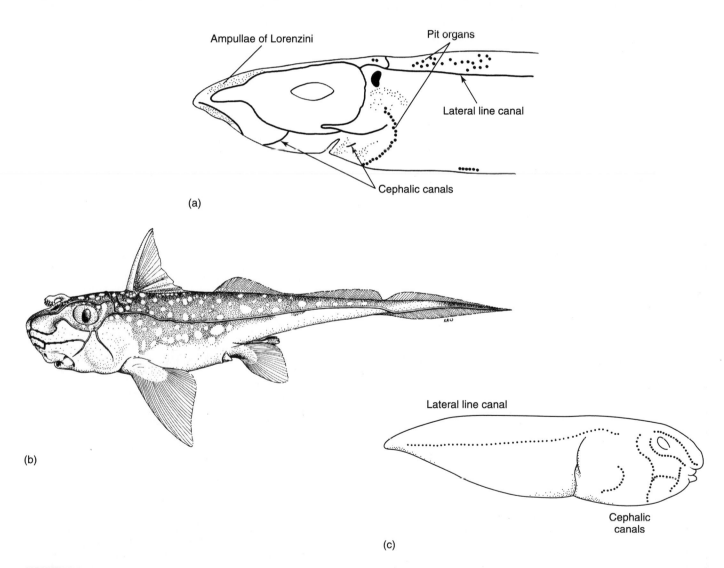

(a)

(b)

(c)

FIGURE 17.2

Distribution of neuromast organs in selected aquatic vertebrates. (*a*) External openings of several varieties in *Squalus.* (*b*) The elevated canal systems of *Chimaera.* (*c*) An anuran tadpole.

a groove that may extend along the lateral line of the body all the way to the tip of the tail. Identical grooves in the dermal armor of fossil acanthodians and placoderms reveal that, primitively, the system was a network of superficial canals covering the entire body. The network became reduced in later fishes by the loss of many branches, and by interruption of the canals at one or more locations. In sharks, the canals have sunk into the skin of the head and trunk but remain as open grooves on the tail. In chimaeras the canals form elevated ridges on the surface where they are readily visible (fig. 17.2*b*). When neuromasts are in enclosed canals, pores may open to the surface at regular intervals. In modern bony fishes, the canals become even more isolated from the external environment, often being embedded in dermal bone underlying the integument (fig. 17.3). In some teleosts, the canals are restricted to the head.

Free-living amphibian larvae, particularly anuran tadpoles, ordinarily lose the lateral-line and cephalic canal systems at metamorphosis. Among amphibian larvae that do not are those of the urodele *Notophthalmus.* When these larvae metamorphose into red efts and migrate to land, the neuromast organs become buried under the proliferating stratum corneum. Later—several years later in some localities—when the eft returns to water as a sexually mature newt, the stratum corneum is shed and the canal systems are again exposed.

The major portion of the canal systems of fishes responds to mechanical stimuli such as compression waves of low frequency in the water, water currents, and, perhaps, hydrostatic pressure. Some portions respond to low electric potentials in certain species. Other regions possibly respond to thermal stimuli. Water currents include, for instance, those generated in flowing streams, those caused by locomotor movements of nearby fishes, and waves reflecting from stationary objects. Response to compression waves of low frequency was an early basis for the belief that fishes can hear sounds via the inner ear. (As will be seen shortly, at least some fishes *can* hear.) Input from mechanoreceptors enables fishes to orient appropriately in flowing streams (**rheotaxis),** to participate in schooling, and to avoid potential enemies. Blind carnivorous fishes inhabiting dark caves locate some of their food with this input. The roles of the canal systems vary with the habitat of the species (swiftly flowing mountain streams, for example, as opposed to ocean

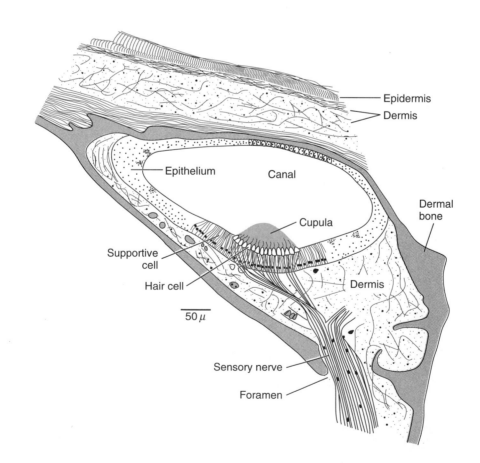

FIGURE 17.3

Neuromast organ in lateral-line canal of a bony fish, cross section. The canal is embedded in bone (*red*). The sensory nerve penetrates the bone via a foramen.

depths), with feeding habits of the species (predators as opposed to vegetarians), with schooling proclivities, and with other environmental and behavioral differences.

Electroreception has been demonstrated in ampullary neuromast organs of elasmobranchs and in the cephalic or lateral-line systems of most bony fishes. The mere contraction of respiratory and locomotor muscles in a single fish produces sufficient electric potential to make its presence detectable at short range to other fishes via the electroreceptive neuromasts. Electroreception evidently arose independently several times in fishes. Nerve fibers from electroreceptors terminate in the cord and brain in nuclei that are distinct from those receiving mechanoreceptive input. Also, the nuclei of teleosts do not occupy the same location as those in other fishes. Electroreceptors are present also in larval urodeles, some aquatic apodans, and in aquatic monotremes, who use them to detect underwater quarry.

Neuromast organs of the lateral-line system arise from a linear series of embryonic **ectodermal placodes** that extends from the head to the tip of the tail (chapter 5). One pair of these placodes lateral to the hindbrain sinks into the head to become the membranous labyrinth of the inner ear, which is a complex of neuromast organs. The membranous labyrinths and the lateral-line system of fishes constitute, collectively, an **acousticolateralis (octavolateralis) system** of receptors innervated by cranial nerves VII, VIII, IX, and X. In postlarval amphibians and in amniotes, the only neuromast organs are those innervated by nerve VIII.

Membranous Labyrinth

All vertebrates have a pair of fluid-filled membranous labyrinths (**inner ears**) located within a similarly shaped cartilaginous or bony labyrinth that is part of the otic capsule of the neurocranium (fig. 17.4). The membranous labyrinth is a highly specialized complex of neuromast mechanoreceptors. In lower vertebrates, membranous labyrinths function chiefly in equilibration; in higher vertebrates, they are also receptors for sound.

The membranous labyrinth is separated from the walls of the skeletal labyrinth by a **perilymphatic space** filled with **perilymph** (fig. 17.11, *light red*). This fluid cushions the membranous labyrinths in their cartilaginous or bony canals. Extending across the space between the skeletal and the membranous labyrinth are fine strands of connective tissue that stabilize the membranous labyrinth and anchor it in place. Perilymph consists of cerebrospinal fluid that seeps into the space by one route or another. The fluid *within* the membranous labyrinths is **endolymph.** It resembles interstitial fluids elsewhere in the body. The endolymph and perilymph are not confluent at any point.

Each membranous labyrinth of lower vertebrates other than lampreys and hagfishes consists of three **semicircular ducts** and two membranous sacs, the **utriculus** and **sacculus.** In the floor or wall of each sacculus in fishes and amphibians, there is a small outpocketing, the **lagena** (fig. 17.4, *shark*). The lagena has expanded to become a **cochlea** in mammals. The lumina of the semicircular ducts are continuous with the cavity within the utriculus, and each duct has a dilation, or **ampulla,** near the junction. The lumen of the utriculus is continuous with that of the sacculus (fig. 17.4). The anterior and posterior vertical semicircular ducts are in vertical planes that are perpendicular to each other. The third, or lateral, duct is in a horizontal plane. Thus there is a semicircular canal in each of the three planes of space. In lampreys, each labyrinth has only two semicircular ducts, an anterior and posterior vertical. Hagfishes have only the posterior vertical duct with, uniquely, an ampulla at each end. Also, in some agnathan species there is little or no distinction between sacculus and utriculus. One can only speculate whether their labyrinths bespeak a modified vertebrate condition rather than a primitive one.

Emerging dorsally from approximately the site of confluence of the sacculus and utriculus, except in a few teleosts, is an **endolymphatic duct** that usually terminates in a blind **endolymphatic sac** (figs. 17.4 and 17.11). Uniquely, in elasmobranchs the ducts open to the

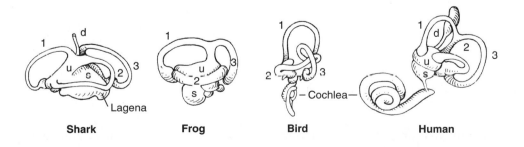

Shark   **Frog**   **Bird**   — Cochlea —   **Human**

FIGURE 17.4

Left membranous labyrinths of representative vertebrates. **1,** anterior semicircular duct with ampulla at base; **2,** horizontal semicircular duct; **3,** posterior semicircular duct. **d,** endolymphatic duct in shark, duct and endolymphatic sac in human. In sharks, the endolymphatic duct opens to the exterior. **s,** sacculus; **u,** utriculus.

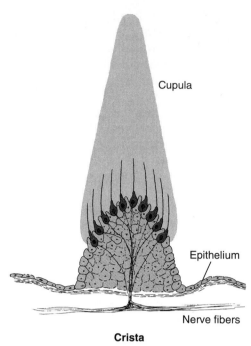

FIGURE 17.5

Neuromast organs of membranous labyrinth of lower vertebrates, generalized. (*a*) Macula with otolithic membrane. (*b*) Crista, found only in ampullae of semicircular ducts. *Dark red,* hair cells.

exterior through two endolymphatic pores on top of the head, rather than ending in blind sacs. In mammals, the endolymphatic sacs lie in the subarachnoid space of the brain beneath the temporal bones. A pair of **perilymphatic ducts** emerge from perilymphatic spaces in most vertebrates, and they, too, usually terminate in a small sac nearby. In sharks, they parallel the endolymphatic ducts and terminate in a small sac in the endolymphatic fossa (a depression on the middorsal aspect of the neurocranium between the two otic capsules). In mammals, the perilymphatic ducts open directly into the subarachnoid space overlying the brain, providing a channel connecting the cerebrospinal fluid and the perilymph. Little is known of the role of the perilymphatic ducts in most vertebrates.

The membranous labyrinth is a connective tissue complex lined by a simple squamous epithelium. In this epithelium, within each of the components of the labyrinth—sacculus, utriculus, ampullae, and in the lagena or cochlea of birds and mammals—are one or more neuromast sites. Each site contains large numbers of hair cells innervated by sensory fibers of the eighth cranial nerve. A neuromast site within an ampulla is a crista. Those elsewhere are maculae.

Cristae and maculae differ in structural details (fig. 17.5). A **crista** is a papilla of epithelial cells surmounted by hair cells. The kinocilia of the hair cells are embedded in a tall, gelatinous cupula that projects far out into the endolymph. The cilia are mechanically displaced as the inert endolymph responds to rotatory movements of the head. A **macula** in most anamniotes is a mound with a flattened cupula that usually has on its surface a layer of otoliths that constitutes an **otolithic membrane.** The maculae of most reptiles and of birds and mammals lack a cupula, and overlying the hair cells is a tectorial membrane (fig. 17.10, 2).

Otoliths (otoconia), found only in lower vertebrates, are crystals of calcium carbonate combined with a protein. They grow by accretion. In many fishes, a single, large otolith nearly fills the sacculus. In others, otoliths are microscopic, abundant, and constitute an amorphous mass. The crystals spill out of the sacculus of a shark whenever the sacculus is ruptured during dissection. They can be readily seen with a low power lens, and their crystalline nature can be observed in transmitted light. Species differences in otolith morphology make it possible to assign an otolith to a species without seeing the fish from which it came. Otoliths are generally confined to otolithic membranes in tetrapods.

Membranous labyrinths in all vertebrates arise during embryogenesis as a pair of ectodermal **otic placodes** on the side of the head lateral to the hindbrain. In fishes, the placodes are in series with those that give rise to the neuromasts of the lateral-line canal system. The otic placodes sink into the embryonic head to become a pair of fluid-filled vesicles, or **otocysts** (fig. 17.6*a* and *b*). Eventually, an otic capsule is deposited around the developing labyrinth.

The sacculus, utriculus, and semicircular ducts are primarily organs of equilibration (balance, in a broad sense). However, in at least some fishes, maculae function in hearing to one degree or another.

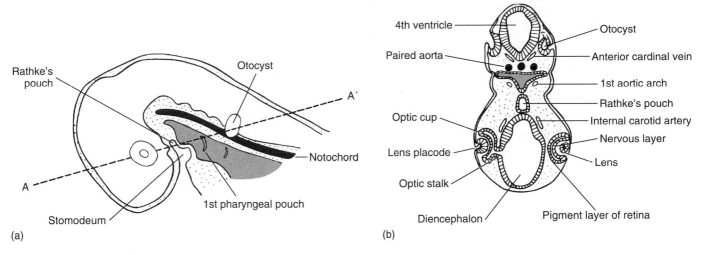

FIGURE 17.6

Anlagen of amniote inner ear (otocyst), retina (optic cup), and lens (lens placode). (*a*) Embryonic head, sagittal plane. (*b*) Cross section of head at level A–A′. In (*b*) the structures are shown as slightly earlier in development on the left. *Red* is the foregut.

Equilibratory Role of the Labyrinth

The sacculus, utriculus, and semicircular ducts monitor the position of the head in space, whether at rest (in *static* equilibrium) or in motion (in *dynamic* equilibrium). Information from these organs is employed reflexly by skeletal muscles to maintain a moment-to-moment posture of the body that favors survival. But what is the role of each of these in equilibration?

Answering this question with reference to the sacculus and utriculus in vertebrates as a group is possible only in the most general terms, largely for lack of sufficient experimental evidence from the several vertebrate classes. There are facts we are sure of. The maculae of the sacculus and utriculus lie in different planes; not all kinocilia in a single neuromast site are oriented in the same direction; the orientation of the kinocilia is affected by movements of the endolymph and of either an otolithic or tectorial membrane, the latter having no otoliths; and displacement of kinocilia is translated into nerve impulses that are transmitted across synapses in the brain. Thus, different maculae and different hair cells in the same macula are stimulated differentially as the endolymph is displaced by movements of the head. Therefore, the sacculus and utriculus respond to changes in linear velocity of a fish; that is, appropriately oriented hair cells discharge when stimulated as a result of acceleration of the head when the fish is swimming in a straight line. One explanation proposes that linear acceleration tends to carry along the kinocilia but that the otolithic membrane, because of inertia, initially resists acceleration. As a result, certain kinocilia that are in contact with the overlying membrane are bent in a direction opposite that of acceleration, and this stimulates the sensory nerve endings at the base of hair cells. It is also thought that a tilted position of the head differentially stimulates one macula as opposed to another.

Stimulation by *rotation* of the head in higher vertebrates seems to be the province of the semicircular ducts. The cupula of a crista projects far out into the endolymph, where it is subject to displacement by the shearing forces generated in the endolymph as the head turns in one plane or another. Consequent bending of the enclosed kinocilia is translated into a nerve impulse.

The sensory impulses generated in the semicircular ducts, sacculus, and utriculus are transmitted to nuclei in the brain stem and cerebellum where, as one result, *reflex movements of the eyeballs* are initiated. As a result, the eyes are always looking in a direction compatible with the position of the head. Voluntary effort is required to turn your head swiftly to the side while continuing to gaze straight ahead. (Try it!) The eyeball reflex impulses travel over the vestibular branch of cranial nerve VIII to the anterior end of the medulla, where they synapse in a vestibular nucleus. Second-order sensory fibers from this nucleus enter an ascending tract that distributes them to the motor nuclei of nerves III, IV, and VI. From these nuclei, motor fibers innervate the extrinsic eyeball muscles. This is one of several *vestibular reflexes* initiated by stimulation of the labyrinth.

That the labyrinth controls reflexive eyeball movements can be demonstrated by spinning someone in a revolving chair rapidly a number of times and then stopping the chair abruptly. The individual will feel dizzy, and an observer will note that the eyeballs continue to exhibit rapid, jerky, lateral oscillation (side-to-side movements; *nystagmus*), which stops as the inert endolymph ceases to apply shearing forces on the cristae of the semicircular ducts. The nystagmus is the cause of the dizzy feeling. Children, in fun, often turn round and round until, for the same reason, they become dizzy.

Input from the equilibratory components of the labyrinth in amniotes also results in *reflex motor activity that alters the position of the head and forelimbs,* keeping these parts oriented with respect to the rest of the body and to gravity. For example, a cat that is turned upside down and dropped from a height will land on its feet. This is the result of a vestibulocollic reflex that terminates in the sternomastoid and other muscles of the neck, and of a reflex pathway that utilizes the vestibulospinal tract to terminate in motor nuclei that supply the brachial plexus. Maintenance of equilibrium is more readily demonstrable in tetrapods, since they have a larger number of independently movable members of the body and they are not suspended in a dense medium, but it is equally essential for survival of a fish in water. In its role of maintaining balance, the labyrinth is assisted by proprioceptors, to be discussed shortly.

Auditory Role of the Labyrinth and Evolution of the Cochlea

The labyrinth has an auditory function in some fishes in which maculae respond to longitudinal sinusoidal waves of the amplitude and frequency of sound (low amplitude, high frequency). In Cypriniformes, a group of mostly freshwater teleosts that includes catfish, goldfish, and carp, sound waves in water evoke waves of similar frequency in the gas in the turgid swim bladder, and these are transmitted to the labyrinth by **weberian ossicles** (fig. 17.7). These are a series of modified transverse processes of the first three (occasionally four or five) trunk vertebrae that extend between the swim bladder and the **sinus impar,** an extension of the perilymphatic space. These fish can hear, although the mechanism presumably provides no information as to the direction from which sounds are coming. The specific maculae that detect these sounds are not known. It is known, however, that the maculae that respond to sound in *lower tetrapods* are in the sacculus near the lagena. In Clupeiformes, an order of herringlike teleosts, an anterior extension of the swim bladder comes in direct contact with the labyrinth, but whether this is a route for transmitting sound waves is not known. Without doubt, there are other teleosts that have maculae with low enough thresholds to detect sound waves. The extent of the phenomenon of hearing in fishes and the receptor sites involved is not yet known.

An account of the evolution of the amniote cochlea begins with amphibians. In addition to the usual maculae associated with the sacculus and lagena of fishes, amphibians have two receptor sites that have a **tectorial membrane** instead of a cupula. (A mammalian tectorial membrane is illustrated in fig. 17.10.) One amphibian receptor site is the **amphibian papilla** of the sacculus, so named because it is not found in any other vertebrate

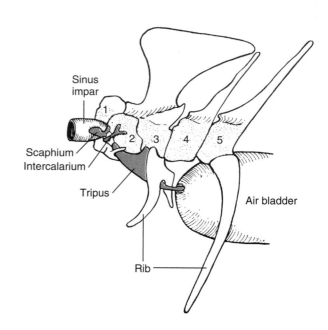

FIGURE 17.7

Weberian ossicles (*red*) of a cypriniform teleost. **1** to **5**, centra of the first five vertebrae. The sinus impar is a caudal extension of the perilymphatic space. The ossicles are interconnected by modified intervertebral ligaments.

(fig. 17.8*a*). The other is the **basilar papilla** in a recess in the wall of the sacculus. These two papillae are the receptors for sound in amphibians.

In most reptiles and in birds, the lagena has become a prominent sac suspended from the sacculus (fig. 17.8*b*). It still houses the lagenar macula, the function of which is unknown. The basilar papilla has become incorporated in its epithelial lining to become, in crocodilians, birds, and monotremes, the **organ of Corti,** the sole receptor for sound (fig. 17.8*b*). Elongation of the lagena was accompanied by a corresponding extension of the surrounding perilymphatic space.

In placental mammals, the lagena elongated still more to become the **cochlear duct,** or **scala media** (fig. 17.8*c*). In its epithelial floor is the organ of Corti. The cochlear duct spirals around a bony pillar of the petrosal bone, the **modiolus** (fig. 17.8*c*), accompanied by perilymphatic spaces named, in this location, the **scala vestibuli** and **scala tympani** (fig. 17.9). The scala media, scala vestibuli, and scala tympani collectively constitute the mammalian **cochlea,** so named because it resembles a snail's spiral shell. The spiral exhibits from one to five turns, according to the species. Embedded in the modiolus is the spiral ganglion of the cochlear nerve.

The scala vestibuli ("ladder of the vestibule") begins at the vestibule (fig. 17.9), a chamber of perilymphatic fluid where sound waves from the stapes enter the cochlea. The scala vestibuli spirals *upward* to the apex of the cochlea where it is continuous with the scala tympani via a small aperture, the **helicotrema.** The scala

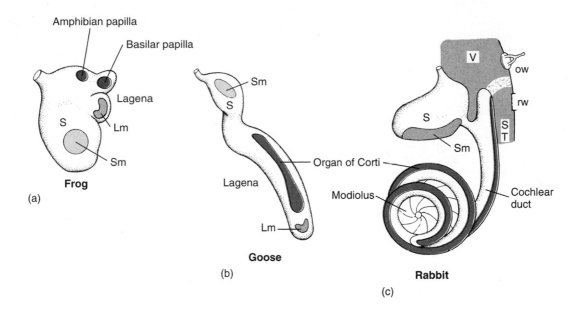

FIGURE 17.8

Receptor sites for sound (*red*) and evolution of cochlear duct in selected tetrapods. **Lm,** lagenar macula; **ow,** oval window; **rw,** round window; **S,** sacculus; **Sm,** saccular macula; **ST,** base of scala tympani; **V,** vestibule of cochlea. The organ of Corti of birds and mammals is derived from the amphibian basilar papilla. For orientation of the cochlear duct within the cochlea, see figs. 17.9 and 17.10, **1** (not to scale).

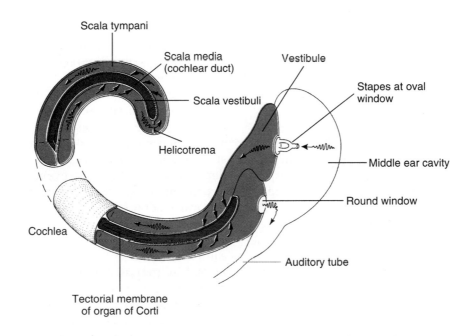

FIGURE 17.9

Pathway of sound waves (*arrows*) from the middle ear cavity, up the scala vestibuli, down the scala tympani, and back to the middle ear cavity in a mammalian cochlea.

tympani spirals *down* the helix and ends at the membrane stretched across the **round window (fenestra rotunda)** in the lateral wall of the otic capsule.

The mammalian organ of Corti is a long, maculalike neuromast in the epithelium of the floor of the cochlear duct (fig. 17.10). It rests on a **basilar membrane** that is attached to a spiraling, bony shelflike projection of the modiolus, and it projects into the endolymph. A continuation of the basilar membrane separates the scala vestibuli from the scala tympani. Overlying the organ of Corti and suspended in the endolymph is a **tectorial membrane,** which is attached at one end to the inner

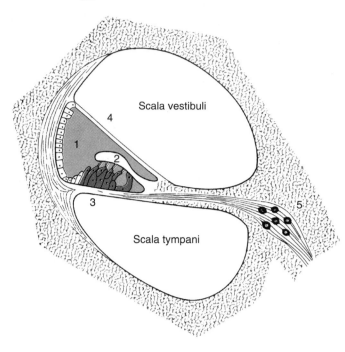

FIGURE 17.10

Cross section of one turn of a mammalian cochlea. **1,** cochlear duct (scala media) containing endolymph (*red*); **2,** tectorial membrane; **3,** basilar membrane (attached at left to the spiral shelf of the modiolus), and organ of Corti (*dark red*) resting on the basilar membrane; **4,** Reissner's membrane; **5,** spiral ganglion and cochlear nerve fibers. The scala tympani and scala vestibuli are perilymphatic spaces. The diameter of the human cochlea is less than 1 mm.

lining of the cochlear duct. Each hair cell of the neuromast has several rows of stereocilia of differing lengths that are embedded in the tectorial membrane. The wall of the cochlear duct opposite the organ of Corti is reduced to a thin epithelium, **Reissner's membrane.** It separates the endolymph in the scala media from the perilymph in the scala vestibuli.

Sound waves induced in the perilymph of the vestibule by vibrations of the footplate of the stapes are transmitted up the scala vestibuli (fig. 17.9), down the scala tympani, and, via Reissner's and the basilar membranes, into and through the endolymph of the scala media and back into the perilymph at all levels of the spiral. Expansion and compression waves induced in the endolymph displace the organ of Corti toward the tectorial membrane at the same frequency, and the hair cells transduce the mechanical stimuli into volleys of nerve impulses in the sensory fibers that innervate the organ. It has been ascertained that hair cells in the first spiral respond maximally to vibrations of higher frequencies, hence higher pitches; those near the apex of the scala media respond only to low frequencies, and those in between exhibit graduated responses. Vibrations in the perilymph are instantaneously discharged back into the middle ear cavity at the round window. Lower tetrapods,

as well as mammals, have oval and round windows in their otic capsules.

With expansion of the basilar papilla to become an organ of Corti, the relay nuclei on auditory pathways in the brain became increasingly larger. Reptiles were the earliest vertebrates in which the auditory nuclei in the tectum of the mesencephalon bulged above the surface and became auditory lobes (figs. 16.14, snake, and 16.15). Differentiation of an auditory cortex in the temporal lobes of the cerebral hemispheres (fig. 16.26) was accompanied by a further increase in complexity of the central pathways that mediate sound.

Middle Ear Cavity

The usual route for conduction of sound waves from the **tympanic membrane** (eardrum) to the inner ear in tetrapods is via one or more ossicles that transmit sound waves across a middle ear cavity (**tympanic cavity; cavum tympanum**) to a **secondary tympanic membrane** in the **oval window (fenestra ovale)** of the otic capsule (fig. 17.11). To perform this function, or to assist in it, all tetrapods have a **columella,** better known as a **stapes** in mammals. The columella in amphibians, reptiles, and birds is part of a cartilaginous or bony complex that includes an **extracolumella** attached to the tympanic membrane. The columella, or stapes in mammals, is derived from the dorsalmost segment of the embryonic hyoid arch, and is, therefore, a homologue of the hyomandibula of fishes.

A stage in the evolution of a columella is found in the urodele *Ranodon* (fig. 9.12*b*), whose middle ear cavity, like that of other urodeles, is rudimentary and who lacks a tympanic membrane. In this urodele, the columella still occupies its primitive and embryonic location as the dorsalmost segment of the hyoid arch skeleton, abutting against the prootic bone (otic capsule) at the site where the oval window forms in other tetrapods. The columella of anurans, on the other hand, is a strong, rodlike replacement bone that is firmly attached to the large eardrum and traverses a broadly open middle ear cavity to end at the membrane in the oval window of the prootic bone.

Mammals have two additional middle ear ossicles, a **malleus** and **incus,** derived from the articular and quadrate bones, respectively, of early reptiles (fig. 17.12). The history of the three ear ossicles of mammals is related in more detail, along with the evidence, in chapter 9 under "Ear ossicles from the hyomandibula and jaws." Because the surface area of the eardrum of tetrapods approaches 20 times that of the membrane in the oval window, vibratory stimuli impinging on the eardrum have been amplified many times before they are delivered to the vestibule of the cochlea. The leverage achieved by the series of nonlinear articulations between the ear ossicles of mammals intensifies the effect.

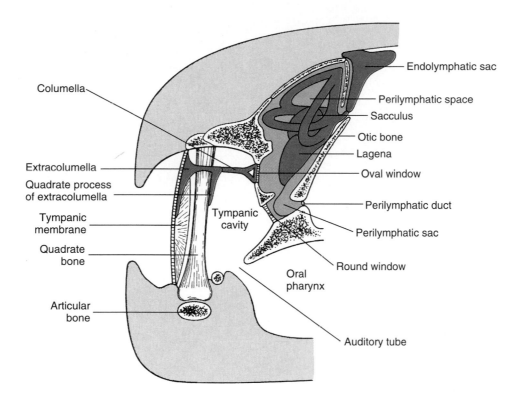

FIGURE 17.11

Columellar complex, perilymphatic space, and membranous labyrinth of a lizard, schematic.

Source: Data from J. S. Kingsley, *Outline of Comparative Anatomy of Vertebrates,* 1920, The Blakiston Company, Philadelphia, after Versluys.

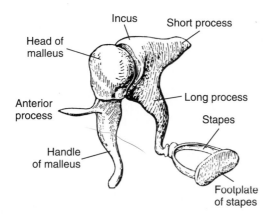

FIGURE 17.12

Human middle ear ossicles. The handle of the malleus is attached by ligaments to the eardrum, the anterior process is anchored by a ligament to the petrous portion of the temporal bone, and the footplate of the stapes rests against the membrane at the oval window.

From B. A. Schottelius and D. D. Schottelius, *Textbook of Physiology,* 17th edition. Copyright © 1973 Mosby-Year Book, Inc. Reprinted by permission of D. Schottelius.

Two small muscles are present in the middle ear cavity of mammals. A **tensor tympani,** derived from the first pharyngeal arch and innervated by the fifth cranial nerve, inserts on the malleus (once a lower jaw bone). It regulates the tension in the tympanic membrane. A **stapedial muscle,** derived from the second pharyngeal arch and innervated by the seventh cranial nerve, inserts on the stapes. It restrains the movement of this ossicle. Very loud sounds cause the tensor tympani and stapedial muscles to contract, which dampens, to a degree, the vibrations of the ossicles and protects the hair cells of the organ of Corti. However, it does not protect against repeated loud noises that are not normal in a natural environment.

The middle ear cavity arises ontogenetically as an evagination of the first pharyngeal (spiracular) pouch, which grows toward the developing ear ossicle or ossicles and partially surrounds them, isolating them from other tissues. The cavitation process is completed by erosion of any remaining mesenchyme surrounding these bones. The middle ear cavity remains in communication with the pharynx throughout life via an **auditory tube (eustachian tube),** which ensures that the air pressure on both sides of the eardrum will be equal. Anurans and some lizards have a wide, short tube that opens into the buccopharyngeal cavity (figs. 12.3a and 17.11). That of crocodilians, birds, and mammals is longer, of small diameter, and opens above the secondary palate (figs. 13.13a and 17.13). The mammalian auditory tube remains closed at its nasopharyngeal exit except during the act of swallowing. Anyone who has suffered from increased pressure in the middle ear as a result of chronic blockage of the auditory tube will recall the sensation and its effect on hearing. Frequent swallowing relieves

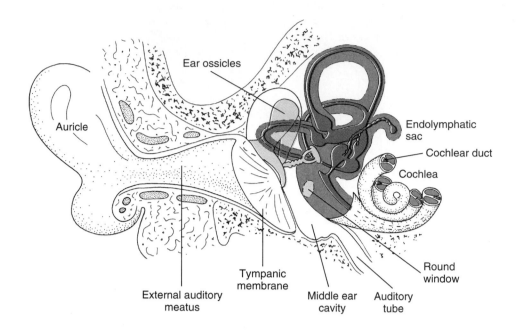

FIGURE 17.13

Mammalian ear complex. *Dark red,* membranous labyrinth; *light red,* middle ear ossicles; *gray,* perilymphatic space within bony labyrinth. The footplate of the stapes rests against the oval window.

the symptoms by opening the entrance to the tube, thereby equalizing the pressure on the two sides of the eardrum.

Outer Ear

The tympanic membrane (eardrum) is the usual portal to the ear ossicles of tetrapods. In anurans, turtles, and primitive lizards, the eardrum is more or less flush with the surface of the head (fig. 17.14). In more specialized lizards, crocodilians, birds, and mammals, the eardrum lies at the end of an outer ear canal (**external auditory meatus**) that may be short or long and has an entrance on the side of the head behind the angle of the jaws (figs. 6.16*a* and 17.13). In terrestrial mammals a fibrocartilaginous appendage, the **auricle (pinna)** collects sound waves and directs them into the outer ear canal. Cetaceans, pinnipeds, sirenians, and some moles have a very small auricle or, like earless seals, none at all. The outer ear canal of whales is of very small diameter, but whales have excellent hearing, which is used to receive ultrasonic intraspecies communications and for echolocation.

Tetrapod Hearing in the Absence of an Eardrum

Although a tympanic membrane and middle ear cavity are the usual routes to the auditory receptors of tetrapods, they are not always present. Urodeles, apodans, a few anurans (*Ascaphus,* for instance), *Sphenodon,* and most limbless reptiles have no eardrum, the middle ear cavities and auditory tubes are vestigial, and, except

FIGURE 17.14

Head of an iguanid lizard showing slightly depressed eardrum just behind angle of jaws.

© Thomas Gula/Visuals Unlimited.

in snakes and amphisbaenians, the columella is rudimentary. Yet these tetrapods are sensitive either to airborne or substrate (seismic) sound waves, or both.

Anurans and many urodeles have an aberrant **opercular bone,** a flat plate that is in contact with the membrane in the oval window. An **opercular muscle** extends between the scapula and the opercular bone. Seismic vibrations detected by the forelimb and scapula are transmitted via the opercular muscle to the opercular bone. This evokes sound waves in the endolymph. The opercular complex of amphibians bears no relationship to the opercular structures of fishes.

In amphisbaenians (squamates that live in burrows in the ground), a very long, slim extracolumella extends forward immediately under the skin of the lower jaw on each side of the head. It vibrates in response to seismic waves in the soil of the burrow and also to airborne sounds. In some species, it ends anteriorly in a broad, flat subcutaneous plate that increases the surface area of exposure to vibrations. The extracolumella articulates with an unusually large columella that ends at an oval window. The hearing complex of amphisbaenians appears to be an adaptation to a subterranean habitat.

Seismic signals are received from the substrate via the lower jaw in many lizards and snakes. In boas, which lack a tympanic membrane and middle-ear cavity, the columellar complex connects the quadrate bone and the oval window. The quadrate receives vibrations via the articular bone of the lower jaw (fig. 9.17c). An articulation of the extracolumella with the quadrate in a lizard is seen in figure 17.11. Many hearing aids for humans employ bone conduction to bypass immobilized middle ear ossicles.

Echolocation

The ability to emit vocal sounds of ultrasonic frequencies in a specific pattern of clicks or structured pulses and to detect the reflection of these from distant objects as small as a flying insect is a sonarlike phenomenon. It is exhibited by mammalian species whose natural environment is such that other sources of sensory input, particularly olfaction and vision, cannot provide sufficient monitoring. Without echolocation, insectivorous bats would starve. A large larynx, grotesque outer ear pinnas, and large and extremely sensitive cochleas combine to provide this capability in bats. Information gleaned at close range includes the size of the insect, its shape, its distance, and escape velocity (Novacek, 1988). Echolocation is less well developed in bats that feed on fruit or nectar, its chief role in these species being obstacle-avoidance. Echolocation is also used by shrews and a few other mammals. Whales employ it for navigation, and they also use the sound-emitting competence for intraspecies communication. A few birds also employ echolocation.

Saccus Vasculosus

Suspended from the floor of the diencephalon just caudal to the pituitary in jawed fishes other than dipnoans is a highly vascular, thin-walled ventral evagination of the third ventricle, the saccus vasculosus (fig. 16.20a). In some species the sac underlies much of the brain stem. It is lined with hair cells, the cilia of which project

into the cerebrospinal fluid. Its sensory nerve fibers pass to the hypothalamus and other brain centers. Its function is unknown. It may be that the organ monitors the pressure of the cerebrospinal fluid, which varies with the depth of the fish, and that the information is used to regulate the volume of gas in the swim bladder. However, elasmobranchs have the organ and no swim bladder. The sac is largest in deep sea fishes, not well developed in shallow freshwater species, rudimentary in cyclostomes, and absent in lungfishes.

Visual Organs: Lateral Eyes

Many cold-blooded vertebrates have two varieties of photoreceptors, paired (lateral) eyes and unpaired (median) eyes. Lateral eyes focus light from objects in the environment onto a photosensitive epithelium, the retina, where an image is formed. The most primitive response elicited by the information is reorientation of the body with respect to the objects viewed, a necessity for moment-to-moment survival of the individual. Median eyes form no image; captured light stimulates neuroendocrine reflex arcs that maintain biologic rhythms (fig. 18.5).

Retina

The retina is the receptor site of the lateral eye. It is a membrane rich in nervous tissue and located at the rear of a fluid filled **vitreous chamber** of the eyeball (fig. 17.15). Retinas arise as lateral evaginations, or **optic vesicles,** from the embryonic forebrain (fig. 16.13b). The vesicles soon become double-walled **optic cups** (fig. 17.6b). The optic cups retain an attachment to the brain via **optic stalks.** The invaginated layer of the optic cup that light will first strike becomes the nervous layer of the retina (fig. 17.16). Some of the retinal cells become

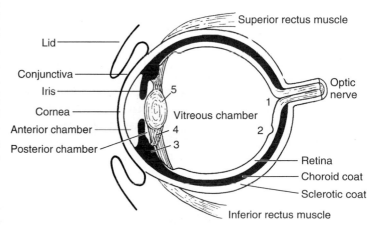

FIGURE 17.15

Generalized spherical eyeball in sagittal section. **1,** blind spot; **2,** fovea; **3,** ciliary body; **4,** suspensory ligaments; **5,** lens. *Red* represents the bulbar conjunctiva (on the cornea) and the palpebral conjunctiva (on the under surface of the lids).

Direction of light

Optic nerve fibers

Ganglion cells

Bipolar cells

Rods and cones

Pigmented epithelium

(a)

(b)

FIGURE 17.16

The retina. (*a*) Areas of synapse, diagrammatic. (*b*) A photomicrograph. Light passes through all layers until it reaches the rods and cones, the cellular photoreceptors. The pigmented epithelium forms a dark choroid coat.

(*b*) SEM From *Tissues and Organs: A Text Atlas of Scanning Electron Microscopy* by Richard G. Kessel and Randy H. Kardon.

rods and cones, which are the photoreceptors. Others become bipolar association neurons, and still others become cell bodies of the optic nerve fibers. These neurons (ganglion cells) sprout long processes that grow along a groove **(choroid fissure)** in the optic stalk and establish synapses within the brain. The processes constitute the optic nerve. From the discussion, it can be seen that the retina arises from the brain and never becomes completely detached from it.

Rods and cones are neurosensory cells, so called because their cell bodies lie in an epithelium rather than in a sensory ganglion. Rods are insensitive to color because the pigment within them, **rhodopsin** (visual purple), absorbs light over the entire range of the visual spectrum. They give only black and white impressions and are most effective in relatively low intensities of light, in the shade, at dusk, or at night. The pigment is bleached when exposed to light, and in this respect the part of the retina containing rods is somewhat analogous to a photographic film. There are three types of cones, and the pigments in each absorb light in only one part of the visible spectrum—blue, green, or red. The cones

function only when there is ample radiation from a source that emits light of these wavelengths.

Not all vertebrates, nor all mammals, have color vision. Cones are generally few in number or lacking in fishes that inhabit depths where light rays become increasingly weaker. The same is true of many terrestrial vertebrates that live in dark caves, as do some salamanders, in burrows, like moles and some rodents, or that are nocturnal, like bats and some birds. On the other hand, the retina of many diurnal birds has a preponderance of cones or, in some instances, no rods at all. The environment in which they compete for food is well illuminated by the sun, and rods have little survival value. Other diurnal vertebrates have a mixture of rods and cones, with the cones sometimes concentrated in a **fovea** (fig. 17.15).

Raptorial birds (owls, hawks, eagles, other birds of prey) and some lizards have two foveas in each eye. The grand champion of foveas is probably that of hawks, which have a man-sized eyeball with approximately one million cones per square millimeter in each fovea. The few other birds with two foveas in each eye are diurnal birds that catch flying insects.

The greatest concentration of rods is peripheral to the fovea. This can be verified by looking at a very faint star at night. It will not be visible if looked at directly, since very little light is falling on the fovea. But by shifting one's gaze to one side of the star, the light will strike peripheral to the fovea, the rods will be stimulated, and the star will come into view. Nocturnal animals have a larger proportion of rods than diurnal ones and may lack a fovea.

Toward the medial quadrant of the retina there is a blind spot, the **optic disc,** the epithelium of which lacks both rods and cones. This is where the optic nerve fibers leave the retina and blood vessels enter and leave (fig. 17.15).

Choroid and Sclerotic Coats and the Pupil

The embryonic mesenchyme surrounding the optic cup gives rise to a pigmented, highly vascular **choroid coat** and a tough fibrous **sclerotic coat (sclera)** that are external to the retina (figs. 17.15 and 17.18). The vessels of the choroid coat provide supplementary nutrients and oxygen to the retina, and the pigment assures a pitch-black background for the retinal "screen." The choroid coat is perforated anteriorly by the **pupil,** which appears black because there is no light in the interior of the eyeball. The pigmented ring surrounding the pupil is the **iris diaphragm,** a modified portion of the choroid coat.

The size of the pupil is fixed in most fishes, but in sharks, some teleosts, and tetrapods, the diaphragm has dilator and constrictor muscles that reflexly alter the size of the pupil, chiefly in response to changes in light intensity. These **intrinsic eyeball muscles** are smooth in amphibians and mammals, striated in most reptiles and birds. The effect of light may be demonstrated by shining a flashlight into a dark-adapted human eye and observing the pupil constrict.

Although the pupil is more or less circular, it can be constricted to become *vertically* slitlike in snakes and in mammals such as cats that are active day and night. It is *horizontally rectangular* in a few mammals, including whales. The **dilator fibers** are radially disposed in all types and are innervated by the sympathetic nervous system. The **constrictor** fibers are disposed circumferentially around a circular pupil and parallel to slits that are vertical or horizontal. They are innervated by the oculomotor nerve (fig. 16.35).

The sclerotic coat, which is the "white" of the eye, is so named because it is often sclerotized more or less by cartilage or bone. In reptiles and birds, the bones take the form of **sclerotic plates** that assist in maintaining the shape of the eyeball (fig. 17.17). In front of the iris diaphragm and pupil, the sclerotic coat is transparent and more convex. This portion of the sclerotic coat is the **cornea.** It is through the cornea that one sees the iris diaphragm and the pupil. Inserting on the outer surface of the sclera are the extrinsic muscles that rotate the eyeball within the orbit (fig. 11.16). The extrinsic muscles are basically similar from fish to humans except that in birds they are so poorly developed that they are practically useless. As a result, a bird must move its entire head to change the direction of its gaze.

Lens

The crystalline **lens** consists of many thin strata of epithelial cells layered on one another in a manner similar to the layers of an onion (fig. 17.18). It is mildly elastic and resilient, being capable of regaining its resting shape after having been deformed. It is suspended in the fluid interior of the eye behind the pupil by a system of fine fibers that form the **suspensory ligament.** The ligament is attached to the ciliary body, which completely encircles the periphery of the cornea. The role of the lens is to focus light rays onto the retina.

The lens arises in embryos as an ectodermal **lens placode** that sinks into position in front of the optic cup (fig. 17.6b). Formation of a lens placode is induced in part by inductors from the developing retina. This has been demonstrated in amphibian embryos by exchanging undifferentiated ectoderm from the region of the future thigh with that at the site where the lens will form. A lens placode is induced in the ectoderm transplanted to the head but not in the potential lens ectoderm that was removed from the influence of the retina. Removal of the optic cup results in no lens placode being formed.

Vitreous and Aqueous Chambers

The large vitreous chamber of the eyeball (fig. 17.18) is filled with a jellylike viscous refracting mass, the **vitreous** (glassy) **humor.** Two smaller chambers are filled with a watery lymphlike **aqueous humor.** These, the **anterior** and **posterior aqueous chambers,** one in front of the iris diaphragm and the other immediately behind it (fig. 17.18), are confluent at the site where the iris diaphragm lies against the lens.

Sclerotic plates

(a)   (b)

FIGURE 17.17

Ossicles of the sclerotic coat of the eye. (*a*) Owl's eye, showing sclerotic plates in place. (*b*) Scleral ring of overlapping ossicles from a lizard's eye.

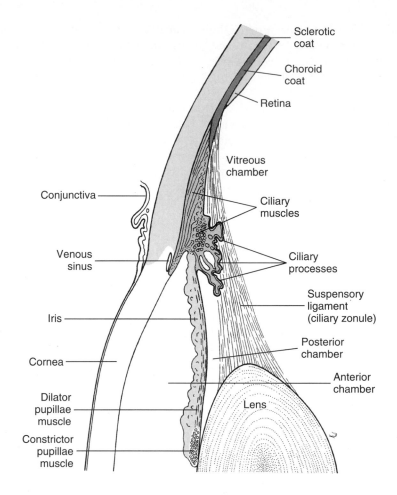

FIGURE 17.18

Ciliary body (*red*) and associated structures of the human eye.

In teleosts, squamates, and birds, a somewhat pigmented and highly vascular fold of the choroid coat protrudes into the vitreous chamber from a site near the blind spot. In fishes, it is a **falciform** (sickle-shaped) **process.** It is a cone-shaped papilla **(papillary cone)** in reptiles, and a **pecten** (shaped like a comb) in birds. Their intense vascularity enables them to participate in metabolic exchange between the bloodstream and the vitreous humor, providing nutrients and oxygen to the retina and removing metabolites. The processes in reptiles and birds extend nearly to the lens, and it is thought they may alert the animal to nearby moving objects by casting a moving shadow on the retina.

Ciliary Bodies: The Source and Drainage of the Aqueous Humor

Within the eyeball, at the periphery of the iris diaphragm, is the **ciliary body,** a low, wedge-shaped ring of muscle with a richly vascular fimbria (fringe) that projects into the posterior chamber (fig. 17.18, *red*). Extending between the ciliary body and the lens is the suspensory ligament described earlier. The ciliary muscles and suspensory ligament in amniotes other than snakes play a major role in regulating the curvature of the lens for near and far vision.

The fimbria of the ciliary body consists of delicate **ciliary processes** that project into the posterior chamber and secrete the aqueous humor. The fluid seeps around the border of the iris where it lies against the lens and into the anterior chamber, where it nourishes the cornea. The cornea is devoid of any other source of nutrients and oxygen because it lacks blood vessels. If present, blood vessels would distort or interrupt the visual light rays en route to the retina. The rate of secretion of aqueous humor is controlled reflexly, maintaining an appropriate pressure within the chambers. It drains slowly out of the anterior chamber via tiny canals (of Schlemm), and from these it diffuses into venous sinuses in the sclerotic coat. Obstruction of the drainage in humans is a major cause of *glaucoma,* a pathologic increase in intraocular pressure.

Shape of the Eyeball

The shape of an eyeball depends on the size, shape, and spatial relations of the parts within, particularly the retina and lens. In general, eyeballs are spherical, like those of mammals and many other vertebrates; tubular, like those of abyssal fishes and predatory birds (fig. 17.17, owl); or ovoid, as in most diurnal birds. The shape is a product of natural selection and evinces evolutionary convergence in species with similar habitats and habit. Variability in eyeball shape illustrates adaptations superimposed on an ancestral organ.

Accommodation

Light passes from a medium of a given refractive index (water or air) into a medium of a higher refractive index (the interior of the eyeball). This reduces the speed of the light and deflects the rays. Regardless of whether the light is coming from a distant or near object, it must be redirected to come to a sharp focus on the retina. The combined refractory effects of the lens and vitreous humor, and also of the cornea (except in saltwater fishes), accomplishes this necessary revision of the direction of the rays.

The mechanism of accommodation, which is the *reflex adjustment of the eye for viewing objects at different distances, and in different intensities of light,* varies among the classes and orders. Accommodation for near or distant vision is a natural application of the physical law of refraction of light taught in elementary physics. These are the same principles, of course, that dictate that the lens of a camera must be moved farther from the film for near objects and closer to it for distant objects.

Light from distant objects under water becomes increasingly reduced at increasing depths. However, near vision at any depth usually suffices, especially for fishes whose food, generally algae or smaller fishes, is always in the immediate vicinity. It is not surprising, then, that the lenses of most aquatic vertebrates are highly concave (spherical), close to the cornea, hence more or less remote from the retina, thereby providing a focus for near objects (nearsighted eyes). The eyes of teleosts at rest are focused in that mode, and a **retractor muscle** *pulls the lens backward* toward the retina for distant vision. The eyes of elasmobranchs and amphibians are focused at infinity (farsighted eyes), and a **protractor muscle** *pulls the lens forward* for near vision. Moving the lens, rather than changing its curvature, is the usual method of accommodation in anamniotes.

Lampreys, which are nearsighted, use the cornea to *push the lens backward.* This is accomplished by contracting an extrinsic **corneal muscle** that arises in the skin and inserts on the cornea just behind the conjunctiva. When this muscle contracts it pulls on the cornea from the side, causing it to push backward on the lens, with which it is almost in contact at all times. This pushes the lens closer to the retina, causing light from distant objects to be properly focused. When the corneal muscle relaxes, the lens resumes its earlier nearsighted position.

The cornea of marine fishes plays only a minor role in refraction since its refractive index is not very different from that of seawater. As a correlate, many fish lenses are more convex—more spherical—than those of most terrestrial vertebrates, in which the cornea participates in refraction. If you have examined the lens of a shark in the laboratory, you may have been surprised at how spherical it is.

In snakes, unlike in other amniotes, increased pressure on the vitreous humor generated by muscles near the iris *pushes the lens forward* for near accommodation. As in lampreys, the lens returns passively to the resting position when the force that displaced it is removed.

The process of accommodation is quite different in other amniotes. The resting eyes of reptiles, birds, and mammals are focused at infinity, and, except in snakes, the *curvature of the lens* is altered by the action of ciliary muscles. We shall examine accommodation for near objects in mammals, whose resting lens is less convex than that of reptiles and birds.

When the mammalian eye is at rest there is tension within the suspensory ligament stretched, as it is, between the lens and the ciliary body and choroid coat (fig. 17.18). This causes the lens to be anteroposteriorly flattened, hence focused at infinity. (Infinity in humans is a distance of 20 feet or more from the retina.) Accommodation for near vision is accomplished by contraction of the ciliary muscles, which draws the choroid coat forward and downward, hence closer to the lens. This reduces the tension within the suspensory ligament, allowing the lens to become more spherical. This increase in convexity increases its refractive power, and near objects are then focused onto the retina.

The ciliary muscles of reptiles and birds are striated. Those of amphibians and mammals are not. For this reason, reptiles and birds can change from infinity to near vision, and reverse, more rapidly than can you and I and other mammals. As in most vertebrates, accommodation for light intensity is brought about by reflex operation of the constrictor and dilator muscles of the pupil.

Because of the shape of the eyeball, cornea, and lens, the number, ratio, and disposition of rods and cones, the presence of two foveas, the striated nature of the ciliary muscles, and other adaptations, the resolving power of a hawk's eye is said to be at least eight times that of the human eye. Such an eye might be useful in watching a game from the top of a stadium!

Conjunctiva

The surface of the eyeball that you can touch when the lids are open is covered with transparent skin, the **bulbar conjunctiva.** This is continuous with the **palpebral conjunctiva** on the inner surface of the eyelid (fig. 17.15). In snakes and some lizards, the lids are permanently closed, but they are transparent, constituting a **spectacle.** Each time the snake or lizard molts, the epidermal surface of the spectacle is shed along with the rest of the skin, and this carries away any scratches.

Moisturizing and Lubricating Glands

Epidermal glands in terrestrial vertebrates bathe and lubricate the cornea. They are unnecessary in fishes. The duct of a prominent **lacrimal gland** located behind the upper lid opens into the lateral corner of the eye and secretes a slightly salty fluid (**lacrima;** tears when produced in quantity in humans only). The gland is poorly developed in some reptiles and birds but enormous in marine turtles, in which it functions as an excretory salt gland. In many mammals, the duct of a **Harderian gland** opens into the medial corner of the eye. It secretes a more viscous fluid that lubricates the **nictitating membrane** (third eyelid), which is more prominent in some mammals than in others. The gland is absent in primates and some other mammals, especially permanently aquatic ones. Some mammals also have an **infraorbital (zygomatic) gland** that opens into the base of the **palpebral fissure** (the slit between the margins of the eyelids, or **palpebrae**). The fluids secreted into the palpebral fissure generally drain into a **nasolacrimal duct** (tear duct) that leads to the nasal cavity.

Absence of Functional Eyeballs

Species that live in caves or other dark recesses (e.g., some fishes, cave salamanders, caecilians, moles) are frequently blind, and the eyeballs may be vestigial. Often, the lids do not open. In hagfishes no eyeball whatsoever differentiates. In these species information necessary for survival is obtained from the environment via other receptors.

Median Eyes

Many vertebrates below birds have a functional third eye on the top of the head (fig. 17.19). Among these are lampreys, ganoid fishes, a few teleosts (especially larvae), anuran larvae, some adult anurans, *Sphenodon,* and some lizards. The median eye is an evagination of the roof of the diencephalon. Primitively, it was one of two evaginations constituting an **epiphyseal complex** (fig. 17.20). The more anterior evagination is the **parapineal,** the posterior one is the **pineal.** It is usually the parapineal that is photosensitive, but in lampreys both are.

In lampreys (fig. 17.20*a*), the pineal ends as a hollow knob beneath the cornea, an area of skin devoid of pigment between the lateral eyes. The upper wall of the knob consists of several layers of cells that form a lens. The lower wall contains photosensitive cells and, beneath these, ganglion cells with long processes that pass down the stalk to sensory nuclei in the *right* side of the diencephalon. The parapineal of lampreys is similar, but its fibers terminate on the *left.*

The parapineal of lizards, or **parietal eye,** lies in the parietal foramen immediately under a single, translucent epidermal scale in the midline of the head (fig. 17.20*c*). It consists of a cornea, lens, retina with photoreceptive cells resembling those of the vertebrate retina, ganglion

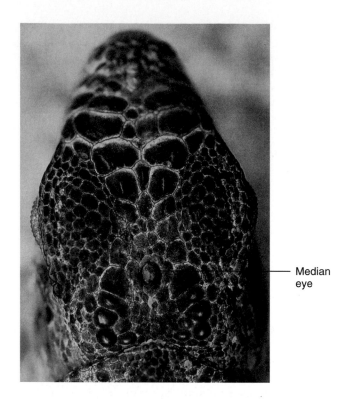

— Median eye

FIGURE 17.19

Parapineal eye of an iguana.

© Times Mirror Higher Education Group, Inc./Bob Coyle, photographer.

cells, and a sensory fiber tract that extends down the epiphyseal stalk and enters the roof of the diencephalon.

In larval frogs, the third eye is called a **frontal organ (stirnorgan).** Authorities are uncertain whether this is the pineal or parapineal part of the epiphyseal complex. At metamorphosis, the photoreceptive part of this organ regresses, leaving only the glandular component that produces the hormone melatonin. In at least one tree frog, however, a functional third eye persists throughout life. The role of melatonin is discussed under "Pineal Organ" in chapter 18.

Median eyes, unlike lateral eyes, form no retinal image. Instead, they monitor the *duration* of environmental photoperiods, and the input affects internal biologic rhythms such as daily spontaneous motor activity and seasonal changes in the gonads. They probably also monitor the *intensity* of solar radiation. The metabolic rate is correlated with body temperature, which, in turn, depends on how much solar energy has been received. Third eyes were present in all major groups of Devonian fishes and in the earliest amphibians and reptiles before being lost in most later vertebrates.

Infrared Receptors of Snakes

Pit vipers (venomous snakes in the family Crotalidae, such as rattlesnakes, copperheads, and water moccasins) have an infrared receptor in a deep pit on each side of the

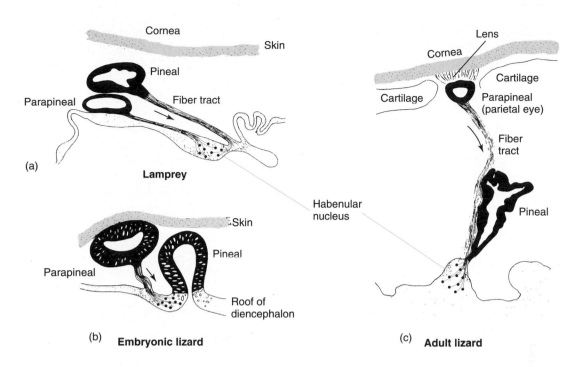

FIGURE 17.20

Epiphyseal complex of lamprey and embryonic and adult lizard. The lizard parapineal occupies the parietal foramen of the skull.

Source: From M. Novikoff, "Untersuchungen uber den Bau, die Entwicklung und die Debeutung des Parietalauges von Saurien" in *Zeitschrift fur eissenschaftliche Zoologie,* 96:118, 1910.

head between the eye and the nostril. The pits are lined with an epithelium that houses hair cells that are sensitive to infrared (thermal) radiation. The tissue that underlies the epithelium is highly vascular. Because of their location in the **lore** (the region in front of the eye in reptiles and birds), these are called **loreal pits.** They are readily visible, being several millimeters wide, twice as deep, and directed forward.

Physiologic and behavioral studies have shown that loreal pits can detect temperature changes of as little as 0.001° C at a distance of several feet. This enables pit vipers to detect the presence of warm-blooded vertebrates such as mice or small birds at night or in daytime when the prey is hidden under the substrate or in a nest or burrow, so long as the environmental temperature is lower than that of the prey. The receptors also enable the viper to strike accurately in the dark. They can also detect objects that are colder than the surrounding environment.

Pythons (family Pythonidae) and some boas (family Boidae) have a series of similar but smaller and less sensitive **labial pits** with slitlike openings between the scales surrounding the mouth. Some pythons have up to 30 of these infrared receptors. All thermal receptors on the head of snakes are innervated by sensory fibers from the fifth cranial nerve. The fibers terminate in a cranial nucleus that is separate from, but closely associated with, the trigeminal nucleus that receives mechanoreceptive input.

Evolution of reptilian pit organs, which lie above the maxillary bones, entailed restructuring of the maxillae to accommodate the pit organs while at the same time con-

tinuing to support and move the fangs. Pit organs supplement vision, mechanoreception, and chemoreception in the search for food.

SPECIAL VISCERAL RECEPTORS

There are two varieties of special visceral receptors, **olfactory** (for smell) and **gustatory** (for taste). Both are *chemoreceptors* that are sensitive to certain amino acids. Olfactory receptors respond to less specific chemical configurations than taste receptors. Differences in sensitivity apparently are related to side chains of the acids. Taste and smell are related phenomena, as is evident to anyone who has been unable to savor food because of a head cold.

Special visceral receptors evoke both somatic and visceral responses. Somatic responses consist of striated muscle activity that reorients the organism in the environment. Unconditioned visceral responses to stimulation of taste buds of teleosts have been demonstrated by applying arginine, an amino acid common in foodstuffs, to the buds, and recording the impulses evoked in branches of the vagus nerve that innervate digestive organs. Unconditioned and conditioned salivation and secretion of digestive juices following olfactory and gustatory stimulation occurs also in mammals.

Olfactory Organs

The olfactory organs of gnathostomes arise as a pair of ectodermal **olfactory placodes** that form just anterior to the embryonic stomodeum. The placodes sink into the head to form a pair of **nasal pits** (fig. 13.7*b*), the epithelial lining of which differentiates into olfactory, supportive, and mucous cells. Olfactory cells, like rods and cones, are neurosensory cells that differentiate in an epithelium and sprout processes that grow into the brain (fig. 17.21). These processes are the fibers of the olfactory nerve. There are an estimated 50 million olfactory cells in humans. For a substance to act as an odorant, it must have a chemical configuration compatible with that of a binding substance on the membrane of the olfactory cell, and it must be in solution.

In fishes other than lobe-fins, the differentiating olfactory epithelium becomes surrounded by mesenchyme that forms a blind olfactory sac. A current of water into and out of the sac is ensured because each external naris is partitioned into incurrent and excurrent apertures so situated that the forward motion of the fish propels a stream of water into one aperture and out the other. The mucosa containing the olfactory epithelium may exhibit folds that increase the surface area. The olfactory cells monitor the water stream and are stimulated by odorants that may have their source in potential food, mates, or enemies. The most primitive somatic response to olfac-

FIGURE 17.21

Section of olfactory epithelium from the channel catfish, *Ictalurus punctatus.* The axons of three olfactory cells are readily identifiable. A cilium at the apex of the olfactory cells projects into the moisture on the surface of the epithelium. Height of the section was approximately 20 microns.

tory stimuli is reflex contraction of locomotor muscles, which propel the fish closer to, or farther from, the source of the odorant.

In lungfishes and tetrapods the olfactory pits push deep into the head to acquire an opening into the oral cavity or pharynx. The openings are internal nares. When this occurs, the olfactory epithelium is confined to a portion of the lining of the nasal canal, so that it is appropriate to distinguish the *olfactory* epithelium from the *nasal* epithelium of the rest of the nasal passageway. The olfactory epithelium of tetrapods contains olfactory cell bodies just as in fishes, but in tetrapods, it monitors an airstream instead of a water stream. Compound tubular mucous glands (**Bowman's glands,** fig. 6.10*d*) keep the epithelium moist. Odorants in the airstream dissolve on the moist olfactory epithelium and stimulate the olfactory cells.

An olfactory epithelium is well developed in fishes, least developed in birds, which therefore have a poor sense of smell. The olfactory nerves of some species of whales disappear during embryonic life. However, a mammal trying to inhale under water would drown, so the evolutionary loss of the olfactory nerves could not have been a disadvantage to whales.

Vomeronasal Organs

In many tetrapods, a ventral section of the olfactory epithelium becomes more or less isolated from the nasal passageway to serve as an accessory olfactory chemoreceptor. This vomeronasal organ, named for its proximity to the vomer bone, receives sensory input via an accessory olfactory bulb and a vomeronasal nerve. The size of the bulb and nerve is proportional to that of the organ.

In urodeles, the organs are a pair of deep grooves in the ventromedial floor of the nasal canals. In anurans and apodans, they are blind sacs that open into the nasal canals. In squamates, they lose their connection with the nasal canals and open into the anterior roof of the oral cavity as two moist pockets (fig. 17.22). These receive the forked tips of the tongue each time it darts in and out of the mouth, monitoring the environment. Because these organs are clearly chemoreceptors, it was long assumed that their role is chemical detection of appropriate foodstuffs. However, removing the forked tip of the tongue of test snakes prevents them from following pheromone trails left by other members of the species. This, then, is one of the roles of the organs in snakes, at least. Garter snakes have no tongue, but autoradiographic studies show that chemicals on the lips, derived from potential foodstuffs, reach the vomeronasal epithelium by diffusion. Vomeronasal organs appear in embryonic crocodilians and birds and then regress—mute evidence of an ancestral function. They have not been found in turtles.

Most mammals have vomeronasal organs just above the hard palate. They open either in the floor of the nasal

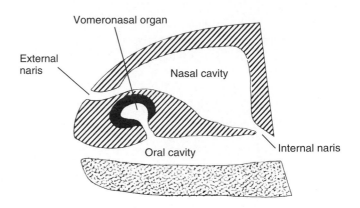

FIGURE 17.22

Location of the vomeronasal organ in a generalized lizard. Only the left organ is shown. Diagonal lines designate a parasagittal section.

canals, as in many rodents, or onto the hard palate via nasopalatine ducts that pass through incisive foramina, as in cats. They are especially well developed in monotremes, marsupials, generalized insectivores, and many carnivores. They are absent in cetaceans (where they would be as useless as other olfactory epithelia), some bats, and adult higher primates. They and their nerves develop in human embryos, reach maximum size about the fifth month of gestation, and then regress.

Organs of Taste

Taste buds are barrel-shaped clusters of elongated **taste cells** (nonnervous receptor cells) and **supportive cells** (fig. 17.23). The apices of the receptor cells have prominent apical processes that protrude into a **taste pore** in the moist epithelium. The apical processes are covered with microvilli. Intimately associated with the base of each receptor cell are the sensory nerve endings. The functional life of a taste receptor cell is only about 10 days, at which time it dies from "wear and tear." They are replaced by supportive cells, which are reserve receptor cells. The supply of supportive cells is continually being replenished from the basal layer of the epithelium.

In fishes, taste buds are widely distributed in the roof, side walls, and floor of the pharynx, where they monitor the incoming stream of respiratory water. In bottom feeders or scavengers, such as catfish, carp, and suckers, taste buds are distributed over the entire surface of the head and body to the tip of the tail. They are abundant on the barbels ("whiskers") of catfish. The exaggerated size of the sensory nucleus that receives incoming taste fibers in such fishes is illustrated in figure 16.16.

In most tetrapods, taste buds are restricted to the tongue, posterior palate, and oral pharynx, which are moist. There are fewer on the tongue in reptiles and birds than in mammals, and human embryos have more taste buds than children seven years old since some that are lost are not replaced.

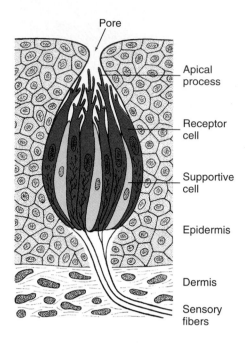

FIGURE 17.23

Receptor cells and their innervation in a taste bud.

Taste buds from fish to humans are supplied by the seventh, ninth, and tenth cranial nerves *in that sequence* from mouth to the end of the oral pharynx. Therefore, taste buds on the anterior surface of the human tongue are supplied by nerve VII, those on the posterior surface of the tongue by nerve IX, and those in the vicinity of the glottis by nerve X. Nerve VII also supplies all taste buds in the skin of fishes all the way to the tip of the tail.

GENERAL SOMATIC RECEPTORS

There are two categories of general somatic receptors: (1) **cutaneous receptors** for light touch, pain, temperature, and pressure (touch sufficient to indent the surface) and (2) **proprioceptors**, found in striated muscles, joints, and tendons.

Cutaneous Receptors

Naked Endings
The skin of all vertebrates contains free, or naked, endings of sensory nerve fibers that ramify among the epidermal cells and are stimulated by contact (fig. 17.24, naked endings in epidermis). These are the most primi-

FIGURE 17.24

Some general receptors. All are somatic except the naked endings in the mucosa and any pacinian corpuscles in visceral locations. Corpuscles are encapsulated receptors.

tive cutaneous endings in vertebrates, and in agnathans they are the only ones. They give rise to what has been called **protopathic** sensation. This is a crude, poorly localized, phylogenetically ancient sensation. It is purely protective. It is not necessary that a fish know the texture of whatever touches it or whether the object is warm or cool. The mere fact of contact indicates possible danger. The impulses reach the thalamus, where reflex motor activity is initiated to reorient the animal so that it avoids the stimulus. In mammals, naked endings give rise to vague, poorly localized, and sometimes unpleasant sensations including pain, as in toothache or when the skin is pricked. Free nerve endings for touch entwine around the base of each mammalian hair. Displace a single hair on your arm and note the sensation. Localization of the stimulus is a function of the cerebral cortex.

Within the epidermis of some mammals, unique meniscuslike terminals of sensory nerve fibers are applied to the underside of a single concave epidermal cell (fig. 17.24, tactile discs). These **tactile discs** are the simplest receptors in which an epithelial cell combines with the terminal of a nerve fiber to form a functional receptive unit. They are found in clusters in the stratum malpighii of the sensitive snout of moles and pigs, in the epidermal sheaths surrounding the base of the vibrissae of rodents, and in other sensitive sites.

Encapsulated Endings

In addition to naked endings that ramify in the skin, tetrapods have acquired bulbous encapsulated endings that add to the repertoire of receptors that detect localized integumentary stimuli (fig. 17.24). These consist of nerve endings in association with epithelialike cells wrapped in a connective tissue capsule, and, with one exception, located within dermal papillae at the dermal-epidermal interface (fig. 6.12, touch corpuscle). **Grandry's corpuscles** along the margins of the beaks of many shore birds and freshwater fowl consist of a nerve ending and *two* epithelial cells surrounded by a thin capsule. **Corpuscles of Herbst,** on the beak, tongue, and palate of water birds, have a thick lamellar capsule and a central core of epithelialike cells. **Genital corpuscles** are distributed to the external genitalia of mammals, especially the glans penis and clitoris; and, in humans, on other erogenous areas. **End bulb (corpuscles) of Kraus** and **Ruffini corpuscles** of mammals may be thermal receptors. **Meissner's corpuscles,** found only in primates and chiefly on hairless areas—toes, soles, palms, and especially fingertips—are probably tactile receptors. Still other corpuscles have been described in avian and mammalian skin of one species or another. The sensation of touch as opposed to temperature perception (heat or cold) cannot be ascribed to every known encapsulated ending with certainty. All the foregoing are in dermal papillae.

Pacinian corpuscles, the largest encapsulated receptors, are found deep in the dermis or in subdermal connective tissue (fig. 6.12) where they are stimulated by compression or, when an object is handled, where they participate in stereognosis (see next paragraph). They are not, however, restricted to the category of cutaneous receptors. They are also present in the bursas of diarthroses where they participate in proprioception, and in coelomic mesenteries and peritonea, including the pericardial peritoneum, where they serve an undefined visceral function. In the latter locations, they are not known to evoke conscious sensations.

Encapsulated exteroceptive endings in mammalian skin are engaged in *epicritic* sensation: those that enable one to differentiate between small differences in warmth or coolness of objects, to distinguish textures, to perceive as separate sites two pinpoints on the surface of the skin that are very close together and are being touched simultaneously, and to perceive the shape and weight of an object being handled, the latter being **stereognosis.** After synapsing in the thalamus, epicritic stimuli are projected to the **somesthetic cortex** of the mammalian brain (fig. 16.26), the cortical center for epicritic sensibilities.

A few simple exteroceptive corpuscles are found in amphibians and reptiles, but birds and mammals have the largest number and variety.

Miscellaneous Cutaneous Receptors

Lizards have mechanoreceptors and thermal receptors that are widely distributed over the surface of the body. The most numerous of these occupy **apical pits,** distributed in groups of one to seven at the apices—the posterior free borders—of the epidermal scales. In another variety of cutaneous receptors of lizards, a filamentous hairlike bristle, or **protothrix,** projects to the surface between scales and transmits tactile stimuli to a receptor cell in a pit. Receptors such as these provide sites for input of exteroceptive stimuli in a scale-covered epidermis that is otherwise difficult to penetrate by usual tactile or thermal stimuli.

Some snakes, black snakes for example, have mechanoreceptors located between scales over the entire trunk and tail except on the venter. The majority of these receptors have a high threshold and are sensitive to vibrational stimuli in the same frequency range (up to 800 Hz) that induces cochlear potentials in those species.

Proprioceptors

Skeletal muscles, their tendons, and the bursas of joints are supplied with sensory endings that monitor continuously the activity status of muscles and joints. These proprioceptors are distinguished from exteroceptors, which monitor the external environment, and enteroceptors, which monitor visceral effectors. Proprioceptors maintain tonus reflexly in a resting muscle and assure synergy in the activity of functional groups of muscles when the latter are working against a load.

The sensory endings in most mammalian skeletal muscles, especially appendicular, are found in **muscle spindles** (fig. 17.25). In mammals, these are tiny fusiform bundles of 3 to 20 specialized striated muscle fibers that are wrapped in a connective tissue sheath and interspersed near the extremities of the muscle among and parallel to extrafusal fibers not in the spindle. Each modified muscle fiber (**intrafusal fiber**) is supplied with at least one **annulospiral** proprioceptive ending (fig. 17.26). Intrafusal fibers also exhibit **flower spray** endings, which ramify along the fiber. They have a motor innervation, being supplied by **gamma motor neurons,** a term employed to distinguish them from **alpha motor neurons** that innervate striated muscle fibers that are not in a spindle. Differentiation of mature intrafusal muscle fibers requires both sensory and motor innervation. Spindles are **stretch receptors,** being extremely sensitive to very small changes in the length of a muscle, hence to muscle activity.

Some mammalian muscles have fewer spindles than others. Appendicular muscles have the largest number per unit volume, and the extrinsic eyeball muscles have none. In the latter, annulospiral endings are found on regular striated muscle fibers.

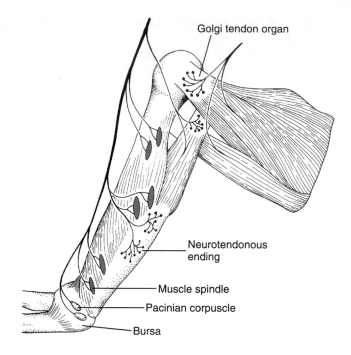

FIGURE 17.25

Representative proprioceptor sites (exaggerated) occurring in muscle, tendon, and bursa.

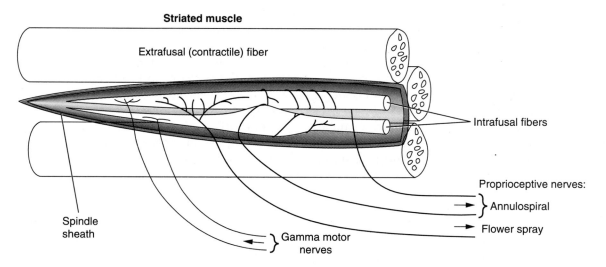

FIGURE 17.26

Muscle spindle (*red*) in section revealing two inner intrafusal fibers, which are modified muscle fibers. Two types of proprioceptive (sensory) nerves are shown, which monitor muscle tension. These nerves synapse in the central nervous system with neurons that stimulate extrafusal muscle fibers.

Muscle spindles are not common in amphibians and have not been described in fishes. Proprioceptive endings in the muscles of these lower vertebrates are on regular muscle fibers and are relatively simple dendritic sprays of one variety or another.

In maintaining muscle tonus, which is a dynamic state, muscle spindles function as follows. The connective tissue envelope of the spindle is continuous with the endomysium (strands of perimysium, fig. 11.3) that surrounds extrafusal fibers beyond the spindle so that any change in the length of a muscle affects the spindle. When a muscle is at rest it is elongated, the spindle is slightly stretched, and proprioceptive nerve fibers are firing. Since proprioceptive fibers synapse in the central nervous system with alpha motor neurons, discharge from the spindle triggers contraction of a limited group of extrafusal fibers, and this increases the tonus of the muscle, shortening the muscle slightly. Shortening relieves some of the tension within the spindle, and as a result, firing of the proprioceptive fibers slows. Thereupon, the muscle tends to elongate once again, and this triggers new discharges from the spindle. Thus, feedback from muscles to central nervous system reflexly maintains the muscle in a state of dynamic tonicity while the muscle is at rest.

The role of spindles when muscles are working against a load is more complicated. Alpha and gamma motor neurons in motor nuclei of the spinal cord are being stimulated by motor nerve fibers whose cell bodies are in the brain. Stimulation of large numbers of gamma motor neurons results in contraction of intrafusal fibers in more than one muscle and consequent discharge of many proprioceptive impulses at a high frequency. The resulting steady reflexive firing of alpha motor neurons maintains a steady contraction of large numbers of extrafusal muscle fibers, and this continues, subject to brain control, whether the vertebrate is locomoting or playing a piano.

Proprioceptive endings on tendons (neurotendonous endings) are flower spray endings that respond to forces generated in the tendon when its muscle contracts. Proprioceptive receptors in mammalian bursas are chiefly pacinian corpuscles (fig. 17.25). Naked sensory endings in myosepta, presumably for proprioceptive input, have been described in lampreys.

Most proprioceptive stimuli do not reach centers of consciousness but are shunted into the cerebellum and elsewhere, where they make reflex connections. To demonstrate conscious proprioception, first close your eyes, then extend your arm or leg. Your awareness of the change in position is an example of conscious proprioception, also called *kinesthesia* and *muscle sensibility.* That the act was performed smoothly is a result of reflex responses to volleys of proprioceptive stimuli that did not reach the level of consciousness.

GENERAL VISCERAL RECEPTORS

General visceral receptors are mostly naked endings in the mucosa of the tubes, vessels, and organs of the body, in cardiac muscle, in smooth muscle including that of the blood vessels, and in the capsules, mesenteries, and meninges of the viscera (fig. 17.24, naked endings in mucosa). Pacinian corpuscles serve a visceral function in mesenteries, in the connective tissue stroma of some viscera, and in the peritonea of the coelom.

General visceral receptors are chiefly stretch and chemoreceptors, but they include a few baroreceptors and osmoreceptors. Among other roles, chemoreceptors monitor the pH of the blood (including oxygen and carbon dioxide content), which affects cardiorespiratory functions; and they monitor the pH of the contents of the stomach and proximal intestine, which affects digestive functions. Baroreceptors monitor blood pressure, and osmoreceptors in the hypothalamus and elsewhere monitor the concentration of certain solutes in the bloodstream. General visceral receptors are stimulated by tactile and thermal stimuli in the pharynx but not beyond. Such input would have no survival value. Most visceral pain from healthy hollow visceral organs results from distension (stretch) and is detected by naked endings.

Three grossly observable vascular monitors are present in amniotes. These are the **carotid bodies,** the **aortic bodies,** and the **carotid sinuses.** The paired carotid bodies lie close to, or are embedded in, the walls of the common carotid or internal carotid arteries; the aortic body, present in mammals only, is on the aortic arch. The carotid and aortic bodies are chemoreceptors. They are richly supplied with sensory endings, the carotid bodies by the ninth nerve and the aortic bodies by the vagus. They are stimulated by a deficiency of oxygen or by an excess of carbon dioxide in the blood. The stimuli evoke discharge of sensory impulses that reflexly increase the depth and rate of respiration. The respiratory center is in the medulla.

The carotid sinuses are slight dilations of the internal carotid arteries at their origin from the common carotid. They are baroreceptors that respond to distension of their walls resulting from an elevation in blood pressure. The sensory nerve impulses travel over the ninth cranial nerve and eventually reach the cardioinhibitory and vasoconstrictor centers in the medulla. From there autonomic fibers in the vagal nerve (fig. 16.35) reflexly slow the heart and dilate the peripheral arteries, reducing the blood pressure.

Most of the sensory input from general visceral receptors gives rise to no conscious sensation, although the reflex response may do so. Some of the same kinds of monitoring of the viscera are necessary in all vertebrates whether in water or on land, so visceral receptors are generally similar from fishes to human beings.

SUMMARY

1. Receptors are transducers of mechanical, electrical, thermal, chemical, or radiant energy. General receptors are widely distributed in the body. Special receptors have a limited distribution and mostly are paired.

2. Somatic receptors provide information about the external environment (exteroceptors) and about skeletal muscle activity (proprioceptors). Visceral receptors monitor the internal environment (enteroceptors).

3. Neuromast organs consist of hair cells, sensory endings, and supporting cells. They are most abundant in fishes and aquatic amphibians and occupy shallow pits, grooves, ampullae, or canals. They function principally in mechanoreception, less commonly in electroreception, thermoreception, and chemoreception. The lateral-line and cephalic canals are neuromast systems.

4. The membranous labyrinth (inner ear) is a neuromast system consisting of semicircular ducts, sacculus, utriculus, endolymphatic and perilymphatic ducts and sacs, and either a lagena or a cochlea. It is filled with endolymph and surrounded by perilymph.

5. The primitive function of the labyrinth was equilibratory. An auditory function employing a cochlea was a later development.

6. Cristae and maculae are the receptor sites in a membranous labyrinth. Otoliths are generally associated with maculae of anamniotes. The organ of Corti of birds and mammals is the most specialized macular complex.

7. Some fishes detect sound waves via air bladders and weberian ossicles. Cephalic and lateral-line canals subserve an analogous function in most fishes and larval amphibians.

8. A middle-ear cavity (cavum tympanum) containing a columellar complex characterizes tetrapods below mammals. Mammals have two additional ear ossicles, a malleus and an incus. The columella of mammals is called a stapes. Striated tensor tympani and stapedial muscles attach to the malleus and stapes.

9. Airborne sounds in tetrapods are generally detected by a tympanic membrane (eardrum) and conducted to the membranous labyrinth by the ear ossicles. Seismic waves are normally conducted by the skeleton of the forelimbs or jaws in some amphibians and reptiles.

10. Tympanic membranes are flush with the surface of the head in anurans. They are at the end of a shallow or long outer ear canal (external auditory meatus) in amniotes. Eardrums and middle ear cavities are rudimentary or lacking in urodeles, apodans, and most limbless reptiles.

11. A sound-collecting auricle (pinna) is found on the head of most mammals.

12. A saccus vasculosus of uncertain function evaginates from the floor of the third ventricle of fishes. It is lined by hair cells that project into cerebrospinal fluid.

13. The receptor site for lateral eyes is the retina. It is nourished and protected by a vascular and light-absorbing choroid coat. A fibrous and sometimes bony sclerotic coat (sclera) is the outer covering of the eyeball.

14. Accommodation for near or far vision generally involves altering the distance between the lens and retina in anamniotes, and altering the shape of the lens in amniotes. A ciliary body in amniotes participates in accommodation for distance.

15. Accommodation for intensity of light is the role of constrictor and dilator muscles of the iris diaphragm.

16. Median (parietal) eyes are photoreceptors but form no retinal image. They are unpaired and include a pineal and/or parapineal organ. They were present in the earliest fishes and are still present in agnathans, some bony fishes, larval and a few adult anurans, and some lizards.

17. Olfactory and vomeronasal organs are special visceral chemoreceptors that arise from embryonic nasal pits. Neurosensory cells receive the stimuli.

18. Taste buds are special visceral chemoreceptors. In fishes they may be distributed over much of the surface of the body and also in the entire buccopharyngeal mucosa. They are confined to the buccopharyngeal cavity in tetrapods and to the tongue and oral pharynx of mammals.

19. Infrared receptors (pit organs) are found in pit vipers, pythons, and some boas. They detect thermal stimuli.

20. Cutaneous receptors are naked or encapsulated nerve endings distributed universally in skin. They detect touch, temperature, and pressure on the surface, and they subserve protopathic and epicritic sensibilities. Apical pits are cutaneous mechanoreceptors confined to lizards.

21. Proprioceptors are found in striated muscles, tendons, and bursas of joints. They monitor the activity of skeletal muscles.

22. General visceral receptors are mostly naked endings for chemoreception in the digestive tract and for monitoring smooth-muscle activity everywhere. Carotid bodies and aortic bodies house chemoreceptors that monitor the oxygen content of blood. Carotid sinuses are mechanoreceptors that monitor blood pressure. Osmoreceptors monitor the concentration of certain solutes in the circulation.

SELECTED READINGS

Adams, W. E.: The comparative morphology of the carotid body and carotid sinus, Springfield, Ill., 1958, Charles C. Thomas, Publisher.

Ariëns Kappers, C. U., Huber, G. C., and Crosby, E. C.: The comparative anatomy of the nervous system of vertebrates, including man, New York, 1936, The Macmillan Company. (Republished by Hafner Publishing Co., 1960.)

Berger P. J.: The reptilian baroreceptor and its role in cardio-vascular control, American Zoologist **27**:111, 1987.

Branson, B. A., and Moore, G. A.: The lateralis components of the acoustico-lateralis system in the sunfish family Centrarchidae, Copia, no. 1, p. 1, 1962.

Bullock, T. H., Northcutt, R. G., and Bodznick, D. A.: Evolution of electroreception, Trends in Neurosciences **5**:50, 1982.

Buning, T. d'C.: Thermal sensitivity as a specialization for prey capture and feeding in snakes, American Zoologist **23**:363, 1983.

Finnell, R. B., editor: Symphony beneath the sea, Natural History, March, 1991, p. 36. Eleven articles on echolocation and communication sounds of marine mammals, including illustrations of the ear complexes. Fascinating reading.

Gaskell, W. H.: On the structure, distribution, and function of the nerves which innervate the visceral and vascular systems, Journal of Physiology (London) **7**:1, 1986.

Gilbertson, T. A.: The physiology of vertebrate taste reception. Current Opinion in Neurobiology **3**:532, 1993.

Gregory, E.: Tuned-in, turned-on platypus, Natural History, May, 1991.

Jones, D. R., and Milsom, W. K.: Peripheral receptors affecting breathing and cardiovascular function in non-mammalian vertebrates, Journal of Experimental Biology **100**:59, 1982.

Levine, J. S.: The vertebrate eye. In M. Hildebrand, and others, editors: Functional vertebrate morphology, Cambridge, Massachusetts, 1985, Harvard University Press.

McLaughlin, S., and Margolskee, R. F.: The sense of taste, American Scientist **82**:538, 1994.

Novacek, M. J.: Navigators of the night, Natural History, October, 1988. FM listening in bats.

Parker, D. E.: The vestibular apparatus, Scientific American **243**:98, 1980.

Purves, E., and Pilleri, G. E.: Echolocation in whales and dolphins, New York, 1983, Academic Press.

Tavolga, W. N., Popper, A. N., and Fay, R. R., editors: Hearing and sound communication in fishes, New York, 1981, Springer-Verlag.

Ulinski, P. S.: Design features in vertebrate sensory systems, American Zoologist **24**:717, 1984.

Weaver, E. G.: The amphibian ear, Princeton, New Jersey, 1985, Princeton University Press.

CHAPTER EIGHTEEN

ENDOCRINE ORGANS

In this chapter we will look at some of the organs whose hormonal secretions adapt vertebrates to daily and seasonal fluctuations in the external environment, or which maintain some of the major metabolic activities at levels optimal for survival. We will find that the brain is one of these organs, translating cyclical environmental input into hormonal messages. We will learn that vertebrates from fishes to humans have the same array of such organs, that they form from the same embryonic precursors, and that they synthesize essentially the same molecules. We will also note some of the anatomical and molecular mutations that have occurred during phylogeny. Along the way we will have glimpses of the endocrine basis for biological rhythms.

OUTLINE

ndocrine organs synthesize hormones. Hormones are organic molecules that have a regulatory effect on other cells, often referred to as "target" cells. Target cells may be right next to the source of the hormone, but usually they are remote and the hormone arrives via the bloodstream.

The term "target cell" might lead to the mistaken notion that a hormone passes only to the "target." This is not the case. Most hormones are transported in the bloodstream, and the fluid constituents of the stream escape from capillaries everywhere in the body and bathe every living cell. Only those cells having an appropriate molecular "receptor site" on their membranes can be affected by the hormone. For example, thyrotropic hormone from the pituitary gland bathes the cells of the big toe in the same concentration that it bathes the cells of the thyroid gland, which is its target; but only certain cells of the thyroid gland are able to respond. A few hormones, such as insulin and thyroxine, have receptor sites on every cell of the organism. These are **general metabolic hormones.**

The earliest hormones probably had only local action, affecting nearby cells as they diffused from their source, rather than being transported by a closed circulatory system to distant organs. This is a necessity in coelenterates (Hydra, for example) because they lack a circulatory system; but hormones that act locally by diffusion are not uncommon in vertebrates. Certain gonadal hormones, for example, have both local and remote effects.

Most hormones are combinations of amino acids—chiefly polypeptides, proteins, and glycoproteins—or they are steroids, a type of lipid. A few are amines. All these compounds are products of biochemical synthesis not only among vertebrates but also among many invertebrates and plants. Steroids, for example, are found in yeast, in many green plants, and in invertebrates; and thyroxine, a product of the thyroid gland, is found also in seaweed, annelids, molluscs, and other invertebrates.

The synthesis of hormones in vertebrates is not limited to endocrine organs. There are endocrine cells in the liver, kidney, stomach, small intestine, and many other locations, including the brain. In this chapter we will discuss only organs that are solely endocrine, or that produce hormones as one of their major functions. We will begin with the endocrine roles of the nervous system. Thereafter, we will discuss briefly the endocrine tissues derived from ectoderm, mesoderm, and endoderm, in that order.

ENDOCRINE ROLE OF THE NERVOUS SYSTEM

The brain in all vertebrates, and the posterior end of the spinal cord in fishes, produces hormones that are synthesized by the cell bodies of **neurosecretory neurons.**

These **neurosecretions (neurohormones)** are small polypeptides, and should not be confused with neurotransmitters, which induce other neurons to fire. Neurosecretions are found in the animal kingdom commencing with coelenterates.

Neurosecretory neurons have cell bodies in **neurosecretory nuclei** in the central nervous system (fig. 18.1). The secretions move along the axon **(neurosecretory fiber)**, bound chemically to proteins **(neurophysins)**, and accumulate in the axon terminals until released reflexly into blood sinusoids. The axon terminals plus the sinusoids constitute a **neurohemal organ.** The major neurohemal organs of vertebrates are the **posterior lobe of the pituitary gland** (fig. 18.2), the **median eminence** of the diencephalic floor immediately caudal to the optic chiasma (figs. 18.2 and 18.3), and the **urophysis** at the end of the spinal cord in elasmobranchs, chondrosteans, and teleosts (fig. 18.4).

Neuron cell bodies with stainable droplets were seen in the caudal region of the cord of fishes early in the twentieth century, but their significance was unknown. Later, similar cell bodies were reported in certain hypothalamic nuclei. It was theorized that these might be secretory cells, but it was not until suitable staining techniques were developed that this was confirmed. Thus, study of neurons in the spinal cord of a fish led to one of the most important discoveries in neurobiology—the endocrine role of the hypothalamus.

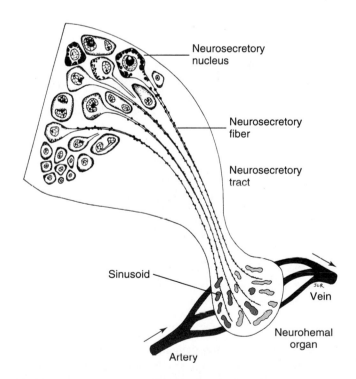

FIGURE 18.1

Neurosecretory neurons in a functional neurosecretory unit. The neurohemal organ consists of vascular sinusoids into which the neurosecretion is released.

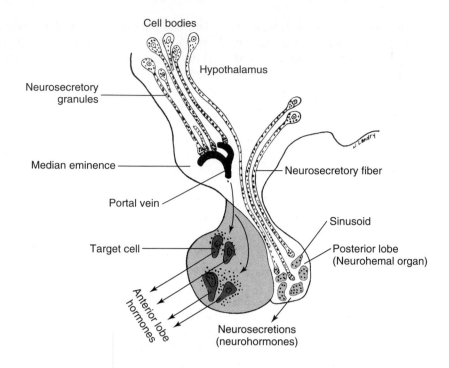

FIGURE 18.2

Hypothalamic neurosecretory neurons. Cell bodies in hypothalamic nuclei synthesize neurosecretions (black granules) that flow along the axons (neurosecretory fibers) and are discharged into the hypophyseal portal vein or into sinusoids in the posterior lobe of the pituitary. Neurosecretions released into the portal vein help regulate the hormone-producing cells (dark red) of the anterior lobe. Arrows show direction of blood flow. Those released in the posterior lobe affect tissue remote from the pituitary.

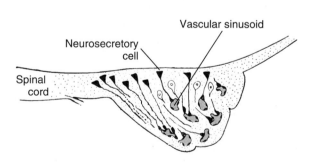

FIGURE 18.3

Schema of the hypophyseal portal system (black) of mammals. Arrows indicate direction of blood flow.

FIGURE 18.4

Urophysis (caudal neurohemal organ) of a carp. In some species this organ is more pendulous.

The majority of hypothalamic neurosecretions are released into blood sinusoids in the median eminence and transported by the **hypophyseal portal system** to the anterior lobe, where the target cells are located (fig. 18.3). There, a neurosecretion of a specific chemical configuration (thyrotropin-releasing hormone, for example) causes the release of a specific anterior lobe hormone (thyrotropin, in this example). Other neurosecretions in the portal blood (somostatin, for example), inhibit the secretion of a specific anterior lobe hormone (growth-promoting hormone, in this example). Therefore, hypothalamic neurosecretions in the portal stream are either

releasing hormones (releasing factors), or **inhibiting hormones** (inhibiting factors). Neural stimuli from elsewhere in the brain, delivered to a neurosecretory neuron in the hypothalamus, trigger a nerve impulse that travels down the neurosecretory fiber and releases the neurosecretion into the portal system.

Synthesis and release of these releasing and inhibiting hormones is regulated in part by the cyclic external environment and partly by feedback from the remote endocrine glands affected. The obvious environmental influences include, among other variables, the circadian cycles of light and darkness, and seasonal change in temperature, day length, rainfall, and, in an aquatic environment, salinity. These are monitored by sense organs. Among the results of this input is promotion of gametogenesis and appropriate reproductive behavior (territorial defense, mating behavior, nest building, care of eggs and young) at the precise time of year when environmental conditions are most suitable for survival of offspring (fig. 18.5). Such adaptive **neuroendocrine reflex arcs** (receptor—sensory nerve—association tracts within the brain—hypothalamus—releasing hormone—median eminence—hypophyseal portal system—anterior lobe hormone—effector) are the result of natural selection.

Neurosecretions from other hypothalamic nuclei are released directly into sinusoids in the posterior lobe (fig. 18.2). These so-called "posterior lobe hormones" are transported in the systemic bloodstream to their target organs. The role of these neurohormones will be discussed shortly.

Fishes have neurosecretory cells in the spinal cord at the base of the tail (fig. 18.4). Their axons terminate in the **urophysis,** a neurohemal organ that varies from an inconspicuous swelling of the spinal cord to a pendulous organ, depending on the species. The neurosecretions of this neurosecretory organ raise the blood pressure of teleosts, hence have been termed **urotensins.** The precise target for urotensin is unknown. Experimental evidence suggests that the observed effect of urotensin may be attributable to a role in osmoregulation.

ENDOCRINE ORGANS DERIVED FROM ECTODERM

Pituitary Gland

The pituitary gland, or hypophysis, is suspended beneath the diencephalon, cradled in gnathostomes in the sella turcica. The pituitary consists of two major components with very different embryonic origins, a nonglandular **neurohypophysis** that arises from the floor of the diencephalon, and a glandular **adenohypophysis** that arises from the roof of the embryonic stomodeum (fig. 18.6). The components of these lobes are as follows:

> Neurohypophysis
>   Median eminence
>   Infundibular stalk
>   Pars nervosa (posterior lobe; neural lobe)
> Adenohypophysis
>   Pars intermedia
>   Pars distalis
>   Pars tuberalis

Neurohypophysis

The neurohypophysis forms from the floor of the diencephalon (fig. 18.6, dark gray). It contains a recess of the third ventricle that is shallow in lower vertebrates, deepest in mammals (figs. 18.6 and 18.7). In mammals, the diencephalic floor is drawn out into a long **infundibular stalk** with the **posterior lobe (pars nervosa),** a neurohemal organ, at the end (figs. 16.21 and 18.7, cat). Just behind the optic chiasma, the neurohypophysis has another neurohemal organ, the **median eminence** (fig. 18.2).

FIGURE 18.5

Regulatory effects of the environment on reproduction. Vascular transport of hormones is indicated by red lines.

FIGURE 18.6

Embryogenesis of amniote pituitary. (a) Rathke's pouch stage. (b) Isolation of adenohypophyseal anlage (light gray) in contact with the floor of the diencephalon. (c) Young pituitary consisting of adenohypophysis (light gray) and neurohypophysis (dark gray). **1,** median eminence; **2,** infundibular stalk; **3,** pars nervosa (posterior lobe). The subdivisions of the adenohypophysis are labeled.

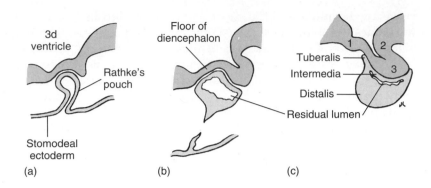

FIGURE 18.7

Pituitary glands of representative vertebrates, sagittal sections, anterior to the left. **3,** third ventricle. Gray, neurohypophysis and contiguous brainstem; red, pars intermedia; stippled areas, par distalis (anterior lobe); black, pars tuberalis. The pars distalis of teleosts, as exemplified by the trout, exhibits only two cytological regions, a rostal part (coarse stipple) and a proximal part (fine stipple).

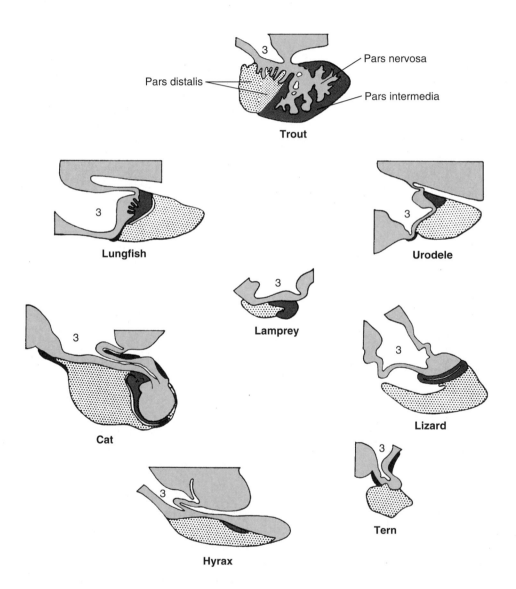

The posterior lobe *synthesizes* no known hormones. Hypothalamic neurosecretions *released* there are carried to the heart for distribution throughout the body. The most ubiquitous, and probably the oldest, hormone released from the posterior lobe is **arginine vasotocin,** found in cyclostomes and all higher vertebrate classes, although in mammals it is secreted only during fetal life. It is the only neurohormone found in cyclostomes. Its function in fishes other than lungfishes is unknown, but in many terrestrial vertebrates it is an antidiuretic hormone that prevents dehydration by causing the kidney to reclaim water from glomerular filtrate and by inducing absorption of water from the urinary bladder. In estivating lungfishes and in amphibians that seek moist locations during arid conditions, it induces absorption of water from the soil through the skin.

A number of chemical mutants of arginine vasotocin have evolved during vertebrate phylogeny by the substitution of other amino acids at one or more positions on the peptide molecule (fig. 18.8). **Human antidiuretic hormone** (arginine vasopressin), which regulates water and salt excretion in other mammals as well, is one of these mutants. **Oxytocin** is another. Oxytocin causes oviducal contractions in turtles, induces strong uterine contractions during birth in humans, and causes letdown of milk into nipples by stimulating the smooth muscle fibers surrounding the milk-secreting alveoli of a lactating mammary gland.

Hypothalamic neurosecretions released into the median eminence enter hypophyseal portal vessels that transport them to the adenohypophysis. There they stimulate or inhibit the release of hormones synthesized in the anterior lobe. The portal vessels in tetrapods and some fishes are discrete portal veins (fig. 18.3). In teleosts, many capillarylike vessels form a plexus that does not organize into a discrete portal vein, yet performs the same function by emptying into the capillary beds of the anterior lobe. This vascular arrangement is probably correlated with the unusually intimate anatomical relationship between the neurohypophysis and adenohypophysis in teleosts (fig. 18.7, trout).

Adenohypophysis

The adenohypophysis arises as a bud of ectodermal cells from the roof of the stomodeum. In amniotes, sharks, and some lower bony fishes, the bud is hollow and is known as **Rathke's pouch** (fig. 18.6*a*). In other fishes and in amphibians, the bud is solid. When the anlage of the adenohypophysis has made intimate contact over a broad area with the floor of the brain, the connection between the stomodeum and the bud usually disappears (fig. 18.6*b*). However, it remains as an open ciliated duct leading to the oropharyngeal cavity in *Calamoichthys, Polypterus,* and some primitive teleosts. Rathke's pouch may remain in the adult gland as a residual lumen (fig. 18.6*c*).

The origin of the adenohypophysis from the stomodeum suggests that primitively the adenohypophysis may have secreted into the oropharyngeal cavity. It was once thought that the pituitary (*pitua,* phlegm) was the source of phlegm that falls into the throat. Although they were wrong about the source of the phlegm, they may have been correct about the primitive site of secretion of the pituitary!

The adenohypophysis exhibits two chief regions, the **pars intermedia** and **pars distalis** (fig. 18.7). A pair of thin, narrow extensions, the **partes tuberales** (singular, **pars tuberalis**) develops in a few fishes and in most tetrapods other than squamates. They extend upward on the surface of the infundibular stalk or, in amphibians, forward on the ventral surface of the diencephalon, from which they can be readily removed. Their role, if any, is unknown.

The Pars Intermedia   The pars intermedia is in intimate contact with the neurohypophysis (fig. 18.7, red) It synthesizes and secretes **melanophore stimulating hormone (MSH),** known also as **intermedin.** The MSH causes pigment granules in some chromatophores of ectotherms to disperse, thus causing the skin to darken. No discrete pars intermedia is present in birds, cetaceans, and some sirenians, but MSH-producing cells have been seen among the cells of the pars distalis. In mammals, MSH is essential for melanogenesis to occur in the hair follicles. It should be mentioned that an entire 13-peptide molecule of alpha-MSH (one of several species of MSH) is incorporated in a molecule of adrenocorticotropic hormone produced in the pars distalis.

The Pars Distalis   The pars distalis is commonly known as the anterior lobe because of its position in

S————————————S
| |
Cys · Tyr · Ileu · Gln · Asn · Cys · Pro · Leu · Gly   **Human oxytocin**

S————————————S
| |
Cys · Tyr · Ileu · Gln · Asn · Cys · Pro · Arg · Gly   **Cyclostome vasotocin**
 1   2   3   4   5   6   7   8   9

S————————————S
| |
Cys · Tyr · Phe · Gln · Asn · Cys · Pro · Arg · Gly   **Human antidiuretic hormone**

FIGURE 18.8

Two of many evolutionary mutants of arginine vasotocin (cyclostome vasotocin). Gray bars designate the only differences between cyclostome and human posterior lobe neurosecretions. The human antidiuretic hormone is arginine vasopressin.

mammals. However, the term is not always appropriate (fig. 18.7). The anterior lobe synthesizes the following hormones: somatotropin (growth hormone) STH (GH); thyrotropin (thyroid stimulating hormone) TSH; adrenocorticotropin (ACTH); gonadotropins—follicle stimulating hormone (FSH), luteinizing hormone (LH) or interstitial cell stimulating hormone (ICSH); and prolactin (luteotropin) PRL (LTH).

Somatotropin is a general metabolic hormone. It promotes growth by stimulating the synthesis of proteins throughout the body. When administered to youths before the epiphyseal plate has ossified (fig. 7.5), it increases the height of the recipient. **Thyrotropin** stimulates the thyroid to accumulate iodine, to synthesize thyroid hormone, and to release it into the circulation. **Adrenocorticotropin** regulates the zones of the mammalian adrenal cortex that secrete glucocorticoids. **Follicle stimulating hormone** stimulates the growth of ovarian follicles and their secretion of estrogen. In males, it promotes spermatogenesis. **Luteinizing hormone** induces ovulation of mature ova and conversion of a graafian follicle into a corpus luteum. In males, it induces the interstitial cells of the testes to synthesize androgens. It is usually referred to in males as **interstitial cell stimulating hormone.**

Prolactin has a variety of effects from fishes to human beings, at least some of which are evoked only after the responding tissue has been presynergized by corticosteroids or some other synergizing hormone (see "Hormonal Control of Biological Rhythms" at end of chapter). It is a major freshwater-adapting hormone of euryhaline fishes (fishes that adapt to a wide range of salinity), and it stimulates secretion of maternal skin mucus that nourishes hatchlings of some teleosts. It has marked lipogenic effects, inducing cyclical deposit of fat in many vertebrates and it stimulates production of pigeon milk by the avian crop sac and mammalian milk by the mammary gland. The latter effect inspired the name prolactin. Acting on tissue that has been rendered sensitive by synchronizing hormones, it causes some urodeles to migrate to ponds at the approach of sexual maturity; in birds it induces parental behavior such as nest building, turning and incubation of eggs, and protection of nestlings; and it activates the corpora lutea in rats (luteotropic effect). The many and varied roles of prolactin illustrate how new receptors have arisen for essentially the same hormone. Natural selection has thus far maintained those receptors for prolactin in extant vertebrates. These varied effects, when first observed, seemed unrelated. However, there is little doubt that ongoing *basic research* will discover a single, unifying principle that accounts for all these observations. Thereafter, *applied research* will design methods to utilize the knowledge for the benefit of humanity.

Only the most obvious roles of the pars distalis hormones have been mentioned above. Each hormone has additional roles, more subtle, but equally important. These may be found in appropriate textbooks of endocrinology.

Pineal Organ

The pineal organ, so named because in mammals it is shaped like a pine cone, is an evagination from the roof of the diencephalon that synthesizes **melatonin.** The latter, an amine, is synthesized from serotonin, another brain amine, but only during hours of darkness. Experimental evidence shows that light impedes the synthesis of melatonin by inhibiting the activity of the enzyme *serotonin N-acetyltransferase* that catalyzes the rate-limiting step in melatonin synthesis. The effect of light on the pineal body is direct when the pineal is located under translucent skin, as in numerous vertebrates. Otherwise, it is mediated by way of the optic nerves, hypothalamic nuclei, descending fiber tracts in the brain stem and cord, thoracic spinal nerves, superior cervical sympathetic ganglion (fig. 16.35), and the conarial nerves **(nervi conarii),** which pass from the ganglion to the pineal organ.

Mammalian melatonin causes melanin granules in dermal melanophores of cyclostomes, larval amphibians, and many fishes to aggregate, thereby blanching the skin. The effect is opposite that of MSH. Amphibian larvae with intact pineals become pale when grown in darkness. Pinealectomy abolishes this response.

Melatonin has a role in preparing the organs of some mammals for reproductive activity, but the effect appears not to be direct. The *duration* of melatonin secretion (rather than total secretion), probably monitored by the hypothalamus, evidently signals the duration of photoperiods, and these signals are transduced into neurosecretory stimuli that influence reproductive cycles. The details of this regulatory mechanism are still to be elucidated. Other aspects of the pineal gland have been discussed in chapters 16 and 17 and its role in certain biological rhythms is discussed later in this chapter.

Aminogenic Tissue and the Adrenal Medulla

The adrenal glands of tetrapods consist of two entirely different components: (1) an **aminogenic tissue** that produces **norepinephrine (noradrenaline)** and **epinephrine (adrenaline),** which are catecholamines, and (2) a **steroidogenic tissue** that produces steroid hormones. These two tissues are spatially separated in many fishes, but they aggregate and become wrapped in a single capsule in amniotes. We will therefore discuss the adrenal complex as two separate glands. We will examine first the aminogenic component, which constitutes the adrenal medulla of mammals. Because of its staining reaction, aminogenic tissue is also called **chromaffin tissue.** Chromaffin tissue is widely distributed elsewhere in the coelom of most vertebrates, but not all of it is aminogenic.

Nodules occur close to the sympathetic ganglia of the trunk and in the gonads, kidneys, heart, and other viscera.

Fishes and Amphibians

In lampreys and elasmobranchs, aminogenic cells lie in clusters along the length of the postcardinal vein (fig. 18.9, ray), and in lungfishes they are scattered along the dorsal aorta of the trunk. These clusters lie close to sympathetic ganglia, whose cell bodies synthesize the same amines. In teleosts, the aminogenic cells are generally at the anterior end of the kidneys within vestiges of the pronephroi, where they are interspersed with the other component of the adrenal complex (fig. 18.9, teleost). In anurans, the two components are interspersed in a diffuse gland on the ventral surface of the each kidney. In urodeles, they form small, bright flecks and nodules along the postcava and are difficult to locate without a hand lens.

Amniotes

In amniotes, discrete adrenal glands are generally located at or near the cephalic pole of each kidney and the two glandular components are usually interspersed within a capsule, as in crocodilians and birds (fig. 18.9, bird). However, in lizards the aminogenic tissue tends to form an

almost complete capsule around the steroidogenic tissue (fig. 18.9). In some mammals, including humans, the condition is just the reverse, the steroidogenic tissue forming an adrenal **cortex,** and the aminogenic tissue forming a **medulla.** However, there are mammalian species in which the steroidogenic tissue does not form a homogeneous cortex. In sea lions, for example, cortical tissue is scattered also in the medulla, and clumps of medullary cells are scattered throughout the cortex. The adrenal glands are also termed **suprarenal glands** in mammals because of their location in the bipedal human body.

The aminogenic cells of the adrenal and the cell bodies of postganglionic neurons of the sympathetic nervous system are of the same lineage. Both arise from neural crests, both are innervated by preganglionic fibers of the sympathetic system, and both synthesize norepinephrine and epinephrine. In fact, the aminogenic cells of the adrenal gland can be thought of as being postganglionic cell bodies of the sympathetic nervous system that fail to sprout processes.

Epinephrine

Norepinephrine is a precursor of epinephrine. It is converted to epinephrine by addition of a methyl group.

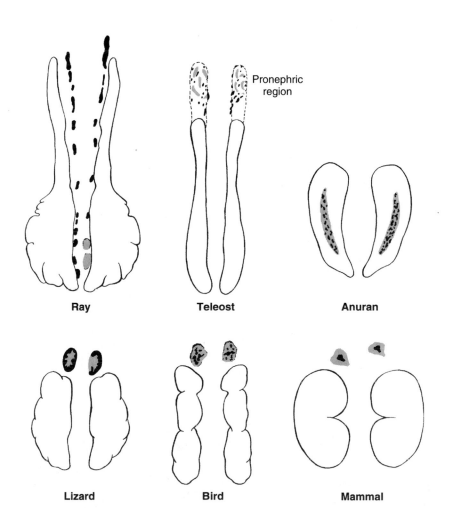

FIGURE 18.9

Adrenal components in selected vertebrates. Aminogenic tissue (medulla in mammals) is shown in black, steroidogenic tissue (cortex in mammals) in gray. The kidneys are shown in outline. Note the reversed location of the two components in lizards and mammals. The two masses of steroidogenic tissue between the caudal ends of the kidneys in the ray are interrenal bodies.

This conversion generally takes place to a greater degree in the adrenal medulla than in postganglionic cell bodies. Consequently, epinephrine is the predominant amine of the adrenal, whereas norepinephrine is the predominant amine of postganglionic neurons of the sympathetic system. The proportions of each in each tissue are fairly constant in any species.

Among numerous roles, epinephrine released at times of exceptional stress stimulates **glycogenolysis**—the breakdown of glycogen into glucose—in the liver. The effect is the immediate release of glucose (glucose-6-phosphate) into the circulation. (Glycogen stored in muscles is not directly convertible to glucose. It must first be converted to lactic acid and then transported to the liver, where it is reconverted into glycogen and then to glucose.) Elevated blood glucose levels satisfy the increased metabolic demand of the heart and skeletal muscles for glucose during stress. Epinephrine and norepinephrine have similar actions at some receptor sites, but their relative potency differs with the number of specific receptors at the site. Because norepinephrine is a potent vasoconstrictor, it has a significant role in the day-by-day management of the peripheral circulation.

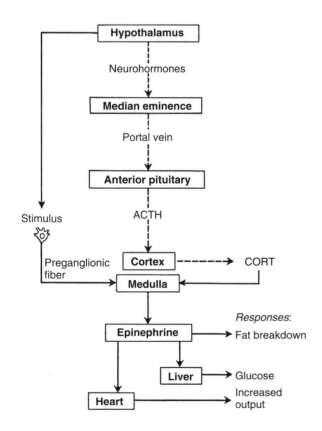

FIGURE 18.10

Some regulatory functions of the adrenal medulla. When presented with a suitable neural stimulus (far left), a preganglionic neuron of the sympathetic nervous system stimulates the medulla to release epinephrine. The latter elicits many responses, three of which are indicated at the right. **CORT,** adrenal corticoids.

Both epinephrine and norepinephrine relax the smooth musculature of the bronchi and trachea, thereby enhancing the delivery of oxygen to the lungs. The preganglionic fibers that supply the adrenal medulla arrive via a nerve plexus supplied in part by a splanchnic nerve (fig. 16.35). Autonomic control centers in the hypothalamus initiate these incoming neural stimuli (fig. 18.10).

ENDOCRINE ORGANS DERIVED FROM MESODERM

Steroidogenic Tissue and the Adrenal Cortex

We have just seen that the adrenal complex is composed of two entirely different components, aminogenic and steroidogenic, which may be intimately associated or spatially separated. The aminogenic component becomes the adrenal **medulla** in mammals, and the steroidogenic component becomes the **cortex.** Whereas aminogenic tissue receives motor innervation (fig. 16.35), steroidogenic tissue is regulated by hormones.

The steroidogenic component is derived from mesodermal cells that arise from the epithelium of the genital ridge and from the underlying nephrogenic mesoderm (fig. 15.19). In sharks and rays the steroidogenic cells usually form one or more compact **interrenal bodies,** so named because they lie between the caudal ends of the kidneys (fig. 18.9, ray). The steroidogenic cells of teleosts generally aggregate in the vestigial pronephric region interspersed among the aminogenic cells. All such steroidogenic masses are homologues of the adrenal cortex of mammals. Although the term "cortical tissue" is appropriate for this tissue in mammals only, it is sometimes used to designate tissues in other vertebrates that produce "corticoids," that is, steroids that resemble those of the mammalian cortex. There are two major categories of corticoids in mammals, **glucocorticoids** and **mineralocorticoids.** Other species of steroids, some with no known effect, are synthesized in the mammalian adrenal cortex, but in much lower quantities.

Glucocorticoids stimulate the conversion of noncarbohydrates into glucose, a process known as **gluconeogenesis.** They also have other roles. Among the glucocorticoids of mammals are **cortisol, corticosterone,** and **cortisone.** Cortisol is the chief corticoid produced by the human adrenal cortex, and by apes, monkeys, some other mammals, and fishes. Corticosterone is the chief one in birds, rabbits, and rats. Hamsters produce cortisol and corticosterone in approximately equal amounts. Cortisone is not a significant mammalian corticoid, but it has important effects in some lower vertebrates. During prolonged stress, one or more of these glucocorticoids stimulates the synthesis of epinephrine in the adrenal medulla. They also have a degree of sodium-regulating competence. The secretion of glucocorticoids is regulated by adrenocorticotropin.

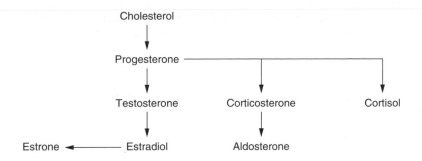

FIGURE 18.11

Synthesis of selected steroid hormones. One or more intermediate molecules occur between most of the hormones shown.

Aldosterone is the most potent mineralocorticoid in tetrapods. Its role is to reclaim sodium from glomerular filtrate in the loop of Henle. The secretion of aldosterone is stimulated by **angiotensin**, a hormonelike compound formed in the bloodstream under the catalyzing effect of the enzyme **renin**, which is produced in the kidneys. (Renin should not be confused with rennin, an enzyme in the stomach of some mammals that curdles milk.)

Only adrenal cortical tissue and gonads produce steroid hormones in vertebrates. The mammalian adrenal cortex produces about 50 different steroids. Some are intermediate products, some are end products, and some have no known survival value. Included are small amounts of male and female sex hormones. A bearded lady is an example of what may happen when the adrenal cortex of a female produces, as end products, excessive quantities of male hormones (androgens).

Gonads as Endocrine Organs

Gonads arise from the coelomic epithelium as a pair of genital ridges medial to the kidneys (fig. 15.19). Ovaries and testes of most vertebrates produce three classes of steroid hormones: **estrogens, androgens**, and **progestogens**.

Estrogens produced by ovarian follicles (fig. 15.35) and androgens produced by the interstitial cells of the testes (fig. 15.21) are primarily responsible for differention, growth, and maintenance of **accessory sex organs** (reproductive ducts and their glands) and for **secondary sex characteristics** (characteristics associated with one sex, such as wattles in roosters, mammary glands in female mammals). They are also responsible for characteristic reproductive behavior from fishes to mammals. Most accessory sex organs and secondary sex characteristics atrophy on ablation of the gonads, but atrophy can be prevented by administration of the appropriate androgen or estrogen. Reproductive behavior also ceases in the absence of gonadal steroids.

Androgens are intermediate stages in the biosynthesis of sex steroids, including estrogens, in all vertebrates from fishes to humans. **Testosterone** is the most potent androgen in mammals, and **17β-estradiol** is the most potent estrogen. As mentioned earlier, androgens are also synthesized in the adrenal cortex, and their overproduction may result in sexual abnormalities.

Progesterone, one of the progestogens, is a precursor in the biosynthesis of androgens, estrogens, and also of corticosteroids throughout the vertebrate classes (fig. 18.11). However, its role below mammals, other than being a step in the biosynthesis of steroids in general, has not been ascertained. In female mammals, small amounts of progesterone from mature graafian follicles are essential for ovulation and preovulatory preparation of the uterus for implantation. Larger amounts from the corpora lutea (fig. 15.35) are essential for final preparation of the uterus for implantation. Progesterone from the corpora lutea and eventually the placenta, depending on the species, maintain the pregnant uterus in a healthy state until term. It is these roles that earned progesterone its name, which means "a steroid promoting gestation." By negative feedback to the hypothalamus, progesterone inhibits the formation of a new wave of ovarian follicles after ovulation "in anticipation" of pregnancy. During pregnancy, it inhibits the maturation of new ova. Pregnancy is the anticipated outcome of mating in mammals.

The mammalian ovary also produces a peptide hormone, **relaxin**. Relaxin produced during pregnancy softens the ligaments of the pubic symphysis and sacroiliac joints of the mother before birth, thus enlarging the birth canal for delivery of the fetus.

Corpuscles of Stannius

Embedded in the posterior part of the mesonephric kidneys or attached to the mesonephric ducts of ray-finned fishes are spherical epithelioid bodies, the corpuscles of Stannius. They are easily mistaken for interrenal bodies, but their embryonic origin is different because they arise as evaginations of the pronephric duct. In most teleosts there are 2, but in *Amia* there are 40 to 50. In large salmon, the corpuscles may reach 0.5 cm in diameter.

Ablation of the corpuscles has been followed by increases in the concentration of calcium in tissue fluids of holosteans and teleosts. The physiological basis for this is not known.

ENDOCRINE ORGANS DERIVED FROM ENDODERM

The endoderm of the embryonic pharynx or pharyngeal pouches of all vertebrates gives rise to the thyroid, parathyroid, and ultimobranchial glands, and to thymus glands, which may or may not produce a hormone but which have a vital role in establishment of the immune system of the vertebrate body. The endoderm immediately caudal to the embryonic pharynx gives rise to endocrine components of the pancreas and to cells that produce gastrointestinal hormones. Except in a few lower vertebrates, the pharyngeal derivatives separate from the pharyngeal lining, sink into the surrounding mesenchyme, and become displaced short or long distances from their origins. We will examine first the endocrine organs that arise from the embryonic pharynx.

Thyroid Gland

Cells in the epithelium of the pharyngeal floor of chordates have the capacity to accumulate iodine and to bind it to molecules of tyrosine, an amino acid, thereby producing an iodinated protein, **thyroxine.** Cells in the hypobranchial groove (endostyle) of protochordates and larval lampreys perform this synthesis. Thyroxine is also found in algae and in various segmented worms, insects, and molluscs. In vertebrates, the synthesis is stimulated by thyroid stimulating hormone from the adenohypophysis. In peripheral tissues thyroxine (T_4) is converted to **triiodothyronine** (T_3), the biologically active form of thyroid hormone. Thyroxine is a general metabolic hormone that regulates the rate of cellular oxidation, stimulates thermogenesis in homeotherms ("warm-blooded" animals), induces metamorphosis of larval amphibians,

and facilitates the action of other hormones on their target tissues.

The mammalian thyroid synthesizes a second hormone, **calcitonin,** which protects calcium in bones and other tissues from depletion when not needed elsewhere. No hormone is required to cause circulating calcium to be stored, at least in tetrapods. Only the release of calcium from storage is a hormonal function (see parathyroid gland). Calcitonin is synthesized in **parafollicular cells** (**C** cells, C for calcium) that lie between the follicles, the C cells having migrated into the mammalian thyroid during embryonic life from developing ultimobranchial glands.

Thyroxine is synthesized in the epithelium of **thyroid follicles** (fig. 18.12). A follicle consists of the epithelium and a colloid-filled cavity that temporarily stores thyroxine as large molecules of **thyroglobulin.** As demanded by metabolic needs, a protease from the epithelium breaks the stored thyroglobulin into smaller molecules that are able to pass through the follicular walls and enter the circulation.

The thyroid gland arises as a median evagination from the pharyngeal floor at the level of the second pharyngeal pouches (fig. 18.13). Typically, the median evagination develops into paired glands in tetrapods (fig. 18.14, necturus, bird, cat). A median thyroid is characteristic of snakes, turtles, a few birds, and the monotreme *Echidna.* Sharks have a median thyroid just caudal to the mandibular symphysis near the insertion of the coracomandibular muscles.

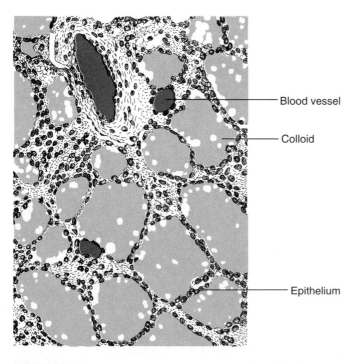

FIGURE 18.12

Actively secreting thyroid follicles. A follicle includes acellular colloid and surrounding epithelial cells.

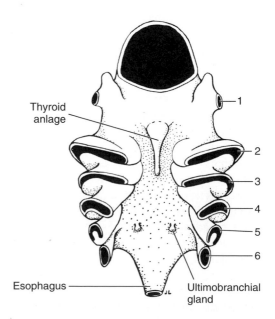

FIGURE 18.13

Pharynx of shark embryo, viewed from below. **1**, spiracle; **2** to **6**, gill slits.

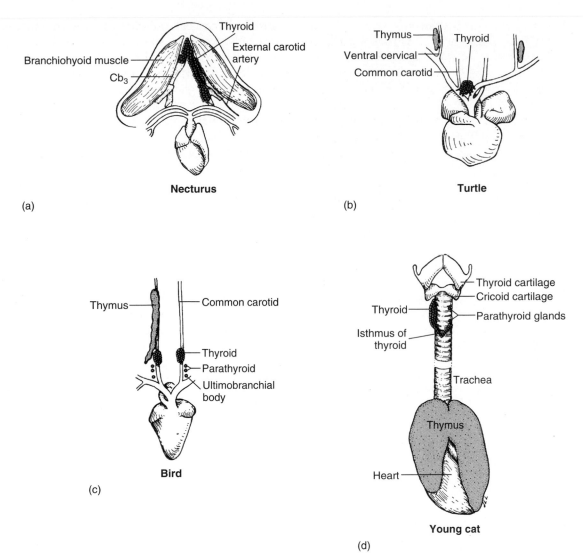

FIGURE 18.14

Thymus (light red), thyroid (dark red), parathyroids, and ultimobranchial bodies in selected vertebrates. **Cb₃**, ceratobranchial cartilage of third pharyngeal arch. The parathyroids of turtles are embedded in the thymus. The thymus of birds and thyroid of mammals are paired. Only one is illustrated.

The larval thyroid of lampreys (endostyle, subpharyngeal gland) opens to the pharyngeal floor until metamorphosis (fig. 18.15). The iodinated protein is secreted into the pharyngeal lumen and absorbed from the digestive tract farther along. After the duct closes at metamorphosis, thyroxine accumulates in follicles and is removed by the bloodstream.

In teleosts, thyroid follicles are usually scattered singly or in small groups along the ventral aorta under the pharyngeal floor. They may also accompany some of the afferent branchial arteries into the gill arches. In a few teleosts, they follow the dorsal aorta caudad and even invade the kidneys. In still others, they form one or two compact masses between the bases of the first gill pouches.

In amphibians, the two glands lie in the floor of the pharynx. In amniotes, they migrate caudad for varying distances, taking a position close to the trachea and to the common carotid arteries from which they receive a rich arterial supply (fig. 18.14b–d). The gland was named for its proximity to the thyroid ("shield-shaped") cartilage in mammals.

After the evaginating organ has reached its final location, thyroid follicles organize and the embryonic stalk connecting the thyroid with the pharyngeal floor usually disappears. However, a thyroid *duct* remains in the shark *Chlamydoselache*. In mammals, a small pit at the root of the tongue, the **foramen cecum,** marks the site where the thyroid evagination took place (foramen cecum, fig. 13.13a). In humans, short sections of the embryonic stalk may persist as one or more thyroglossal *cysts* that occasionally require surgical removal. The cysts may occur anywhere in the neck between the foramen and the gland.

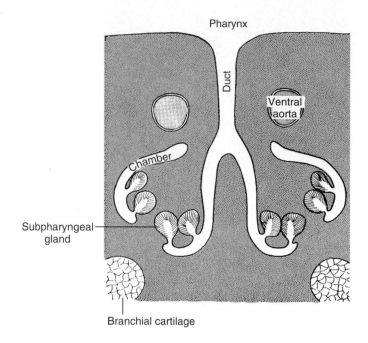

FIGURE 18.15

Cross section of subpharyngeal gland (endostyle) of a larval lamprey. All chambers empty into the duct. At metamorphosis, the duct closes and some of the glandular cells form thyroid follicles.

Parathyroid Glands

Parathyroid glands arise as evaginations from pharyngeal pouches and are so named because they usually lie close to or embedded within the thyroid gland (fig. 18.14c and d). They are not present in fishes or larval or neotenic amphibians. A few reptiles have three pairs which arise as endodermal outgrowths from the second, third, and fourth pharyngeal pouches. However, most tetrapods have only two pairs, since second-pouch anlagen usually fail to mature. When there is a single pair, as in a few urodeles, crocodilians, some domestic fowl, and some mammals, the glands develop from either the third or fourth pouch, depending on the species. Occasionally the single gland on each side has contributions from both pouches.

Parathyroid glands produce **parathyroid hormone.** Levels of serum calcium below a normal range, which varies with the species, evoke release of parathyroid hormone. Parathyroid hormone causes calcium to be released from bone and other storage sites, restoring serum calcium levels to their normal range.

Ultimobranchial Glands

Ultimobranchial glands develop from the epithelium of the last pair of pharyngeal pouches. They synthesize **calcitonin.** They are present as discrete glands in all gnathostomes throughout life, with the exception of adult mammals (figs. 18.13; 18.14c; and 18.16). In elasmobranchs, the left gland alone matures. It lies between the caudal end of the pharynx and the parietal pericardium, not far from its origin. In teleosts, the ultimobranchial glands may be median or paired, depending on the species, and they lie in the septum transversum ventral to the esophagus. In reptiles and birds, they lie close to the thyroid.

Adult mammals have no ultimobranchial glands, with the possible exception of scaly anteaters. The fourth pharyngeal pouches of mammals are unusual in that a small extra pouch evaginates from their caudal walls. These extra pouches are the anlagen of mammalian ultimobranchial glands. Potential calcitonin cells migrate from the thickened lining of these pouches and enter the sub-

Elasmobranch

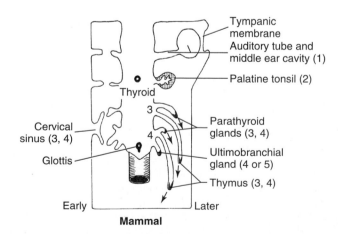

Mammal

FIGURE 18.16

Pharyngeal derivatives of sharks and mammals (diagrammatic frontal sections looking down onto pharyngeal floor). Left sides are earlier in ontogeny than the right. Numbers identify ectodermal grooves or pharyngeal pouches. Arrows indicate caudal growth of anlagen. Whether the ultimobranchial gland of mammals is from pouch 4 or from a vestige of pouch 5 is not certain. Endocrine anlagen, except thyroid, are in red.

stance of the developing thyroid gland. There they become clusters of calcitonin-producing cells (parafollicular cells) lying among and between the thyroid follicles. Whether or not these caudalmost pouches are homologues of the fifth pharyngeal pouches of lower vertebrates is not certain. No ultimobranchial glands have been found in cyclostomes, and no calcitonin has been demonstrated in their blood serum.

Thymus

The thymus is a lymphoid organ. In birds and mammals, at least, it functions during fetal and juvenile life only, during which time it is vital in the establishment of an immune system that functions for the rest of the individual's life. Without an immune system, the organism cannot survive.

Primitively, thymus tissue arose as a thickening in the epithelial lining of all the pharyngeal pouches. The role of those thickenings is unknown. As vertebrate evolution proceeded, fewer and fewer pouches were involved. Thymus tissue still develops in the walls of all seven pouches in lampreys; from all except the first pouch in most gnathostome fishes (fig. 18.16); from the first six in apodans; from the third, fourth, and fifth in urodeles; and generally from the second pouch only in anurans. In most amniotes, the third and fourth pouches are the sole source (fig. 18.16), and in many mammals, including humans, the thymus arises from the third pouch only.

There is a tendency for successive anlagen to unite during development. In bony fishes, they all more or less coalesce to form an elongated gland above the gill chambers. In the larvae of at least one elasmobranch (*Heptanchus cinereus*), ducts from the first six lobes open into the pharyngeal cavity. The ducts sometimes persist in young adults.

The epithelium of a differentiating mammalian thymus is converted into a reticulum by invading mesenchyme. Immigrant cells from embryonic bone marrow invade the interstices of the reticulum, where they proliferate. These immigrants are *stem cells* whose descendants differentiate into **T lymphocytes (T cells)** of the developing immune system. Mature T cells enter the general circulation and lodge in lymph nodes, the spleen, and other lymphoid tissues throughout the body. Proliferation continues at an intensive pace in the juvenile thymus, during which time it is large or massive (fig. 18.14*d*). Once the thymus has completed "seeding" other lymphoid tissues with T cells, it undergoes fatty degeneration and can be dispensed with. Thereafter, T cells are produced in the remaining lymphoid tissues. T lymphocytes are destroyed in acquired immunodeficiency syndrome (AIDS), and immunity to disease is thereby lost. T cells are also responsible for rejection of transplanted organs (foreign proteins) such as the heart or liver. Whether or not the thymus produces an endocrine factor is debatable.

Avian Bursa of Fabricius

The role of the thymus in many young birds is supplemented by a **bursa of Fabricius,** a lymphoid organ that arises as a middorsal evagination of the embryonic cloaca and extends into the pelvic cavity, where it is sandwiched between the large intestine and the synsacrum. It resembles the young thymus in structure and regresses completely at sexual maturity. Its role is similar to that of the thymus.

Endocrine Pancreas

The pancreas, like the adrenal, consists of two functionally different components; **acini,** which synthesize digestive enzymes, and **pancreatic islets (islands of Langerhans),** which synthesize the hormones **insulin** and **glucagon** (fig. 12.27). These two components are more or less interspersed in tetrapods. In many jawed fishes, however, the exocrine and endocrine tissues are more or less separate from each other, and in cyclostomes they are remote to the extent that cyclostomes have no organ that can be called a pancreas. In lampreys, the cells that secrete insulin are in the submucosa of the intestinal wall near its cephalic end. In hagfishes, they constitute a small lobe, or islet, at the base of the bile duct or encircling it. The digestive enzyme-secreting cells, on the other hand, are scattered diffusely within the submucosa over a considerable length of the intestine and also to some extent in the tissue of the liver.

In elasmobranchs, the endocrine cells lie within the epithelium of the pancreatic ductules that drain the acini. In most teleosts, they are aggregated into several compact macroscopic nodules, often two or three, but in the same mesenteries that support the diffusely disposed exocrine components. The aggregation of insulin-secreting cells in a few nodules that can be readily excised made teleosts good subjects for early insulin research. Although in tetrapods the exocrine and endocrine components are more intimately associated, the pancreas remains rather diffuse in many species, and aggregations of chiefly endocrine tissue or chiefly exocrine tissue are not uncommon.

The seemingly random location of the endocrine cells may be more comprehensible in the light of their ontogeny in vertebrates as a group. Both the exocrine and endocrine cells arise from the epithelial lining of the foregut. When this lining is disrupted by the formation of pancreatic and liver buds, some epithelial cells from the lining are carried to new locations (fig. 12.26). Whether no bud forms and the cells remain in the wall of the gut (as in agnathans), whether they become part of a bud but fail to separate from the epithelial lining of the ductules (as in elasmobranchs), or whether they lose all connections with the epithelium and become isolated islands may be correlated in part with the stage of ontogeny during which their inductor substances are synthesized. In the final analysis, of course, their adult location is an expression of inherited genetic directions.

Insulin stimulates the *uptake* of glucose by all living cells, regulates the *utilization* of glucose in most tissues, promotes *glycogenesis* (the conversion of glucose into glycogen for storage in the liver and skeletal muscles), and it performs other roles associated with carbohydrate metabolism. Also, it is the most potent of the currently known hormones that have a lipogenic (fat-synthesizing) effect.

Glucagon has a role essentially opposite that of insulin. It elevates blood glucose by promoting *glycogenolysis* (the breakdown of glycogen to glucose) in the liver, but not in the striated muscles, where glycogen is protected from the action of glucagon.

Phylogenetically, insulin and glucagon are among the oldest polypeptide hormones. Mammalian and fish insulin are so similar that the former causes a decrease in blood sugar levels when administered to agnathans and jawed fishes. Sheep, beef, or pork pancreases were the major source of insulin for human diabetes mellitus until insulins identical with human insulin were synthesized in pharmaceutical laboratories. Because insulin is destroyed by stomach acids, it cannot be administered orally.

Synergistic Regulation of Blood Glucose

Insulin and glucagon synergistically maintain a stable day-by-day level of blood glucose, causing it to be stored when in excess and to be released when needed to meet normal tissue requirements. In times of stress, epinephrine invokes a rapid release of glucose from the liver (fig. 18.10). Under conditons of sustained stress, glucocorticoids stimulate gluconeogenesis.

Gastrointestinal Hormones

Several gastrointestinal hormones that affect the activity of the stomach, pancreas, and gallbladder are synthesized in the epithelium of the stomach and small intestine. These include gastrin, secretin, and pancreozymin-cholecystokinin (a single hormone).

Gastrin is synthesized in the pyloric epithelium and released in response to the presence of food in the stomach. It evokes secretion of hydrochloric acid and pepsinogen by the parietal cells and chief cells of the stomach lining (chapter 12). **Secretin** is released by the duodenal epithelium when **chyme,** the macerated, acidified contents of the stomach, enters the duodenum. Secretin stimulates release of water, salts, and bicarbonates into the intestinal lumen, thereby altering the pH of the chyme for optimal functioning of digestive enzymes. **Pancreozymin-cholecystokinin** is secreted by the intestinal epithelium in response to increasing titers of lipids and peptides (partially digested proteins) in the lumen. Carried by the bloodstream, this hormone stimulates the release, and perhaps the production, of digestive enzymes by the pancreas and it stimulates contraction of the gallbladder, producing a flow of bile.

The autonomic nervous system participates in the release of all these digestive secretions. Most of our knowledge of gastrointestinal hormones comes from mammalian studies. The hormones of lower vertebrates have received much less attention. Secretin, however, has been found in all vertebrate classes.

HORMONAL CONTROL OF BIOLOGICAL RHYTHMS

All organisms exhibit rhythms of metabolism and behavior. These vary from annual cycles in which gonads may regress and recrudesce and reproductive behavior waxes and wanes, to approximately 24-hour cycles (**circadian rhythms**) in which metabolism exhibits phases of elevated and depressed activity and behavior is rhythmically altered. The daily photoperoid (period of light) is the principal *environmental* operant of circadian rhythms, but other environmental factors such as temperature and more subtle stimuli may also be involved (fig. 18.5). Environmental photoperiod rhythms are detected by photoreceptors—pineal or parapineal and the retinas of paired eyes. The information is transmitted to hypothalamic nuclei that serve as pacemakers by rhythmically elaborating and releasing neurosecretions that regulate synthesis and release of anterior pituitary hormones (fig. 18.5). Hormones that regulate circadian rhythms at the tissue level are of two kinds, synchronizers and inducers.

Synchronizing hormones set the *phases* of circadian rhythms. Through their synchronizing effects on responding tissues, they determine the onset, duration, and termination of daily cycles. These effects render specific systems "ready," that is, the tissues develop a *competence to respond* to the inducer hormones. Adrenal corticoids are among the most important synchronizing hormones. The corticosteroid rhythm is under direct control of the daily light-darkness cycle in vertebrates exposed to one. A phase of the corticosteroid rhythm determines, for example, the time of day or night that the liver is able to respond to the fat-synthesizing effects of prolactin and insulin. This has been demonstrated in fishes, several species of birds, and in mammals. Fat in Syrian hamsters is synthesized by the liver at the onset of light, but not throughout the entire day. Seasonal changes in the daily photoperiod also result in seasonal differences in fat synthesis.

Melatonin is also a synchronizing hormone under direct control of the daily photoperiod. The blanching of skin in certain anamniotes discussed previously occurs in the daily dark period as an expression of high melatonin levels during those hours. Hence, melatonin sets the phase for a daily rhythm of skin melanocyte response. In house sparrows, pinealectomy abolishes both the daily body temperature rhythm and a "free-running" locomotor activity rhythm, which occurs in constant lighting conditions. Reproductive cycles in vertebrates are commonly timed by the photoperiod. In species with a single annual reproductive season, higher melatonin levels occur during the short days and long nights of winter. Melatonin exerts an antireproductive effect by way of inhibiting gonadotropic hormonal secretion and thus contributes to synchronizing the overall reproductive cycle.

Inducing hormones (not to be confused with inductor substances) act on tissues presensitized by synchronizing hormones. Inducing hormones too, are synthesized, stored, and ultimately released rhythmically. For example, pituitary prolactin levels in male and female golden hamsters in Baton Rouge, Louisiana, are higher at 8:00 P.M. than at 8:00 A.M. in January and August. Their release into the circulatory system induces those metabolic processes that have been rendered inducible by the earlier release of synchronizing hormone. *The time interval between release of corticosteroids and the subsequent release of prolactin is critical,* and is not necessarily the same for each function of prolactin. Prolactin and the adrenal corticoids "get out of sync" seasonally for certain functions. The result is an annual rhythm superimposed on the circadian rhythm. The phenomenon is adaptive. Reproductive hormones and insulin also act in an inducing capacity. Some of the effects of seasonal environmental changes on reproductive structure, physiology, and behavior were noted earlier in the chapter.

The temporal interaction of synchronizing and inducer hormones determines specific physiological conditions such as growth rates, fat storage, and rhythms of reproductive system and migratory readiness. The greatest net effect occurs when the daily peak of the inducer hormone coincides with the daily peak of greatest tissue responsiveness. Any other temporal relationship between synchronizer and inducer produces a gradation of lesser effects.

The effectiveness of synchronizing and inducing hormones is dependent on **general metabolic hormones,** among which are thyroid hormone and insulin. A tissue that has an inadequate cellular respiratory rate because of lowered thyroid hormone or which is incapacitated by an insulin deficiency cannot respond fully to the stimuli of inducing hormones. Deficiencies of general metabolic hormones affect not only peripheral target tissues, but also the neurosecretory cells of the hypothalamus. General metabolic hormones therefore exert a modifying effect on circadian and annual rhythms.

Biological rhythms can be restored or impeded in humans depending on the schedule of administration of daily medication that affects metabolism. This insight offers an opportunity for new approaches to the medical treatment of metabolic diseases, including diabetes mellitus and cancer.

SUMMARY

1. Hormones are products of specific groups of cells that alter the metabolism of nearby or remote cells of a different nature. They are mostly polypeptides, proteins, glycoproteins, or steroids. A few are amines.
2. Neurosecretions are small polypeptides produced by neurosecretory neurons and released into circulatory channels of neurohemal organs.
3. Known neurohemal organs of vertebrates are the median eminence, posterior lobe of pituitary, and urophysis of elasmobranchs, chondrosteans, and teleosts.
4. The urophysis at the caudal end of the spinal cord of fishes secretes urotensins, which have vasopressor and osmoregulatory effects.
5. The hypothalamus is the major source of vertebrate neurosecretions. It is regulated by hormonal feedback and, in part, by the external environment. Its secretions are released into the median eminence and posterior lobe of the pituitary.
6. The pituitary consists of a neurohypophysis derived from the floor of the third ventricle and an adenohypophysis derived from the roof of the stomodeum.
7. The neurohypophysis consists of a median eminence, an infundibular stalk, and a posterior lobe (pars nervosa). Hypothalamic neurosecretions released from the posterior lobe prevent dehydration and stimulate smooth muscles in restricted locations. Those released from the median eminence reach the adenohypophysis via hypophyseal portal vessels.
8. The adenohypophysis consists of pars distalis, pars intermedia, and pars tuberalis. The pars distalis secretes STH, TSH, ACTH, FSH, LH (ICSH), and LTH. The pars intermedia produces MSH.
9. The pineal synthesizes melatonin during darkness. It causes dermal melanophore cells of fishes and larval amphibians to aggregate melanin pigment and signals the photoperiod to the hypothalamus.
10. Aminogenic tissue synthesizes amines, including norepinephrine and epinephrine. In fishes, it tends to be isolated from steroidogenic tissue. It is the medulla of most mammalian adrenals.
11. Steroidogenic tissue arises from the epithelium of the genital ridges and produces adrenal and gonadal steroids.

12. Adrenal steroidogenic tissue includes the interrenal bodies of fishes and the adrenal cortex of mammals. Aldosterone regulates sodium excretion, and glucocorticoids elevate blood sugar levels by stimulating gluconeogenesis.

13. The gonads of both sexes synthesize steroids that include androgens (which predominate in males) and estrogens and progesterone (which predominate in females).

14. The ovary of mammals also produces relaxin, a nonsteroid hormone that induces dilation of the uterine cervix in preparation for delivery of a fetus.

15. Corpuscles of Stannius, derivatives of pronephric ducts, may synthesize a calcium-regulating hormone in ray-finned fishes.

16. Thyroid tissue arises from a median evagination of the pharyngeal floor. The adult gland consists of fluid-filled thyroid follicles. These are generally scattered along the ventral aorta in fishes and aggregated beneath the pharynx or in the neck in tetrapods.

17. Thyroxine and triiodothyronine are thyroid hormones that increase metabolic rates. Parafollicular cells in mammalian thyroids produce calcitonin, which impedes withdrawal of calcium from storage sites.

18. Parathyroid glands, found only in tetrapods, are derivatives of several pharyngeal pouches. Parathyroid hormone elevates serum calcium by withdrawal of calcium from storage sites.

19. Ultimobranchial glands develop from the last pharyngeal pouches and produce calcitonin. Adult glands are lacking in cyclostomes and mammals.

20. The thymus is a lymphoid organ that arises from one or more pharyngeal pouches. It produces T-lymphocyte stem cells whose descendants are carried by the bloodstream to lymph organs where they participate in the immune response.

21. The bursa of Fabricius is a cloacal derivative of birds that supplements the role of the avian thymus.

22. The endocrine pancreas produces insulin and glucagon. Insulin stimulates glucose uptake, glucose utilization, and glycogenesis. Glucagon elevates blood sugar by inducing glycogenolysis. The exocrine and endocrine components of the pancreas are less intimately associated in fishes than in tetrapods.

23. Gastrointestinal hormones are produced in the epithelial lining of the stomach and duodenum. They stimulate release of digestive enzymes in the stomach, small intestine, and pancreas, and the flow of bile from the gallbladder.

24. Circadian and annual rhythms of reproduction, metabolism, and behavior are entrained chiefly by environmental photoperiods that initiate neuroendocrine and subsequent hormonal responses.

25. Predictable daily rhythms of glucocorticoids and melatonin are hormonal expressions of environmental time-givers. These hormones synchronize rhythmic sensitivities of many tissues to neural stimuli or other hormonal factors.

SELECTED READINGS

Binkley, S.: The pineal: endocrine and nonendocrine function, Englewood Cliffs, New Jersey, 1988, Prentice-Hall.

Cincotta, A. H., Wilson, J. M., de Souza, C. J., and Meier, A. H.: Properly timed injections of cortisol and prolactin produce long-term reductions in obesity, hyperinsulinemia, and insulin resistance in the Syrian hamster (*Mesocricetus auratus*), Journal of Endocrinology **120**:385, 1989.

Gorbman, A., and others: Comparative endocrinology, ed. 2, New York, 1987, John Wiley and Sons.

Hadley, M. E.: Endocrinology, 4th ed., Englewood Cliffs, New Jersey, 1995, Prentice-Hall.

Marcus, N. H., and others: Photoperiodism in the marine environment, American Zoologist **26**:386, 1986.

SYNOPTIC CLASSIFICATION OF CHORDATES

This synopsis has been provided so that readers may readily ascertain the postulated kinships of animals they read about in the text. Traditional evolutionary taxonomy is employed for the most part; postulates and terminology derived from phylogenetic systematics have been incorporated when warranted. We believe this approach will best achieve the objectives of the synopsis. Students interested in alternative and more complete classifications should consult comprehensive works in systematics. All generally recognized extant classes are included. *Subtaxa and extinct taxa have been included only if referred to in the text.* Extinct taxa are designated by asterisks.

PHYLUM CHORDATA

Subphylum Urochordata (Tunicata)
  Class Ascidiacea
  Class Larvacea (Appendicularia)
  Class Thaliacea
Subphylum Cephalochordata. *Branchiostoma, Asymmetron,* sole genera
Subphylum Vertebrata (Craniata)
  Superclass Agnatha. Jawless fishes
    Class Myxini. Hagfishes
    ***Class Pteraspidomorphi (Diplorhina)**
      Order Pteraspidiformes (Heterostraci)
      Order Thelodontiformes (Coelolepida)
    Class Cephalaspidomorphi (Monorhina)
      Order Petromyzontiformes. Lampreys
      ***Order Cephalaspidiformes (Osteostraci)**
      ***Order Galeaspidiformes**
      ***Order Anaspidiformes**
  Superclass Gnathostomata. Jawed vertebrates
    ***Class Acanthodii.** Spiny Paleozoic fishes
    ***Class Placodermi.** Armored Paleozoic fishes. Arthrodires, antiarchs
    Class Chondrichthyes. Cartilaginous fishes
      Subclass Elasmobranchii. Naked gill slits
        ***Order Cladoselachiformes.** *Cladoselache*
        Order Squaliformes. Squaluslike sharks
        Order Rajiformes (Batoidea). Rays, skates, sawfishes

      Subclass Holocephali. Gill slits covered by operculum. *Chimaera*
    Class Sarcopterygii (Choanichthyes). Lobe-finned fishes
      Superorder Crossopterygii. Coelacanths, rhipidistians. *Latimeria*
      Superorder Dipnoi. Lungfishes. *Lepidosiren, Neoceratodus, Protopterus*
    Class Actinopterygii. Ray-finned fishes
      Subclass Chondrostei. Chiefly Paleozoic
        ***Order Palaeonisciformes.** Paleozoic ganoids
        Order Polypteriformes. *Polypterus* and *Calamoichthys.* Sole genera
        Order Acipenseriformes. Sturgeons and paddlefishes. *Acipenser; Polyodon, Psephurus*
      Subclass Neopterygii
        Order Semionotiformes (Holostei). Garfishes (*Lepisosteous*) and *Amia*
      Division Teleostei. Modern ray-finned fishes
        Order Clupeiformes. Herring, sardines, others
        Order Cypriniformes. Minnows, carp, goldfish, others
        Order Anguilliformes. Eels
        Order Gadiformes. Codfishes, others
        Order Perciformes. Perchlike teleosts. And more than 30 additional extant orders
  Class Amphibia. Highest anamniotes
    ***Subclass Labyrinthodontia.** Earliest tetrapods
      Order Ichthyostegalia. *Ichthyostega*
      Order Temnospondyli. *Archegosaurus*
      Order Anthracosauria. *Seymouria*
    ***Subclass Lepospondyli**
    Subclass Lissamphibia. Modern amphibians
      ***Order Proanura**
      Order Anura (Salientia). Frogs, toads, tree toads
      Order Urodela (Caudata). Tailed amphibians
      Order Apoda (Gymnophiona). Caecilians

Class Reptilia
    Subclass Anapsida. No temporal fossae
      ***Order Cotylosauria.** Stem reptiles
    Order Chelonia (Testudinata). Turtles
    Subclass Diapsida. Two lateral temporal fossae
      Superorder Lepidosauria
        Order Rhynchocephalia (Sphenodonta). *Sphenodon*
        Order Squamata. Lizards, snakes, amphisbaenians
      ***Superorder Sauropterygia.** Plesiosaurs
      ***Superorder Ichthyosauria.** Aquatic fishlike reptiles
      Superorder Archosauria
        ***Order Thecodontia.** Stem archosaurs
        ***Order Pterosauria.** Winged reptiles. Pterodactyls
        ***Order Saurischia.** Dinosaurs with reptilelike pelvis
        ***Order Ornithischia.** Dinosaurs with birdlike pelvis
        Order Crocodilia. Crocodiles, alligators, caimans, gavials
    Subclass Synapsida. One temporal fossa
      ***Order Pelycosauria.** Early synapsids
      ***Order Therapsida.** Precursors to mammals
Class Aves. Feathered vertebrates
    ***Subclass Archaeornithes.** Earliest known birds. *Protoavis* (?) and *Archaeopteryx*
    Subclass Neornithes. All other birds
      ***Superorder Odontognathae.** Toothed Cretaceous marine birds. *Hesperornis* and *Ichthyornis*
      Superorder Palaeognathae. Ratites. Ostriches, emus, rheas, cassowaries
      Superorder Neognathae. Carinates
        Order Columbiformes. Pigeons and other doves
        Order Pelecaniformes. Pelicans, cormorants
        Order Anseriformes. Ducks, geese, other waterfowl
        Order Galliformes. Quail, peacocks, domestic fowl
        Order Falconiformes. Hawks, eagles, vultures
        Order Psittaciformes. Parrots, paroquets
        Order Passeriformes. Perching birds, up to 64 families, including song birds
      And about 20 other modern orders

Class Mammalia
    Subclass Prototheria. Oviparous mammals
      Order Monotremata. Platypuses and *Echidna*
    Subclass Theria
      Infraclass Metatheria. Yolk sac placentals
        Order Marsupialia. Opossum, kangaroo, others
      Infraclass Eutheria. True (chorioallantoic) placentals
        Order Insectivora. Shrews, moles, hedgehogs
        Order Xenarthra. Armadillos, sloths, South American anteaters
        Order Tubulidentata. Aardvark, an African anteater
        Order Pholidota. Pangolins (scaly anteaters)
        Order Chiroptera. Bats
        Order Primates
          Suborder Prosimii. Lemurs, lorises, tarsiers, tree shrews
          Suborder Anthropoidea
            Infraorder Platyrrhini. Nostrils open to the side. New World monkeys and marmosets.
            Infraorder Catarrhini. Nostrils open downward
            Superfamily Cercopithecoidea. Old World monkeys
            Superfamily Hominoidea.
              Family Hylobatidae. Gibbons
              Family Pongidae. Apes: Orangutans, chimps, gorillas
              Family Hominidae. Extinct and modern man
                ***Ardipithicus**
                ***Australopithecus.** Several species
                ***Homo erectus**
                ***Homo neanderthalensis**
                Homo sapiens
        Order Lagomorpha. Pika, rabbits, hares
        Order Rodentia
          Suborder Sciuromorpha. Squirrels, gophers, woodchucks, others
          Suborder Sciurognathi. Beavers, others
          Suborder Myomorpha. Mouselike rodents, rats
          Suborder Hystricomorpha. Porcupines
          Suborder Caviomorpha. Cavies
        Order Carnivora. Terrestrial carnivores. Canines, hyenas, bears, others
        Order Pinnipedia. Aquatic carnivores. Seals, sea lions, walruses

Family Phocidae. Earless (wriggling) seals

Family Otariidae. Eared (fur) seals and sea lions

Family Odobenidae. Walruses

Order Perissodactyla. Ungulates with mesaxonic foot. Usually odd-toed. Horses, tapirs, rhinos

Order Artiodactyla. Ungulates with paraxonic foot. Usually eventoed

Suborder Suiformes. Pigs, hippos, peccaries. Relatively primitive

Suborder Tylopoda. Camels, llamas

Suborder Ruminantia. Cud chewers with complex stomachs

Family Cervidae. Deer, caribou, reindeer

Family Giraffidae. Giraffes

Family Antilocapridae. Pronghorn antelope

Family Bovidae. Oxen, sheep, goats, true antelopes, others

Family Tragulidae. Chevrotains (mouse deer)

Order Hyracoidea. Hyraxes

Order Proboscidea. Elephants, mastodons

Order Sirenia. Manatees and dugongs (sea cows)

Order Cetacea. Whales, dolphins, porpoises

SELECTED PREFIXES, SUFFIXES, ROOTS, AND STEMS EMPLOYED IN THE TEXT*

Following are some classical components of terms used in the text, with examples. Familiarity with the entries should enable a reader to deduce, within useful limits, the meanings of many additional words, such as *hemangioep-ithelioblastoma*, which otherwise may be a meaningless jumble of letters. The list is no substitute for a general unabridged or standard medical dictionary, but it is sufficiently long (800 entries) and varied to motivate a reader toward habitual use of those standard works. Such a prac-tice will extend the reader's intellectual horizons far beyond the boundaries of comparative anatomy. Pragmatically, it will result in better recognition and recall and also in more accurate spelling of technical terms.

Meanings are those relevant to the subject matter of the text. Headings that are stems (**acanth-**) are usually the smallest combinations of letters that are common to the derivatives. The symbol > means *hence* and separates a classical from a derived meaning.

A

a- lacking, without; as in *acelous, Agnatha.*

ab- from, away from.

abducens a nerve that abducts (q.v.) the eyeball.

abduct to move a part away from the longitudinal axis, as in raising the arm laterally.

acanth- spine, spiny.

acelous *a + coel* (q.v.).

acetabulum a cup for holding vinegar; the socket in the innominate bone.

acinar resembling a cluster of grapes.

acinus a grape.

acr- extremity highest.

acrodont tooth on the summit of the jaw.

acromion process at proximal extremity of the shoulder (*-omo*).

actin- ray.

Actinopterygii ray-finned fish (see *pter-*).

ad- to, toward, upon.

adduct to draw toward.

adrenal a gland on the kidney.

aden- gland.

adenohypophysis glandular part of the pituitary.

adenoid resembling a gland; a nasal tonsil.

-ae nominative plural ending, as in chordae tendineae (tendinous chords); genitive singular ending as in radix aortae (root of the aorta).

aestivate to spend summer in a state of lowered metabolism.

af- same as *ad-* (to, toward); the *ad-* is changed to *af-* to make the word easier to pronounce.

afferent carrying something to or toward something else (see *-ferent*).

Agnatha *a-* (lacking) + *gnath* (jaws).

ala- wing; see also *ali-.*

alar winglike.

alba white.

alecithal lacking yolk.

ali- wing.

alisphenoid wing of the sphenoid.

alveolus a small chamber or sac.

amel- enamel.

ameloblast a cell that produces enamel.

amphi- on both sides; in two ways.

amphiarthrosis a joint with severely limited movement.

amphicelous having concavities at both ends.

amphioxus an animal with both ends pointed (*-oxy*).

ampulla a flask > a small dilation.

an- without.

anamniote an animal lacking an amnion.

anapsid lacking an arch.

anura lacking a tail.

ana- upward; again.

anadromous *ana-* + *-dromos* (running) > able to migrate from the sea up into freshwater streams.

anastomose to unite end to end.

anatomy *ana-* + *-tome* (q.v.).

anch- gill.

andr- male.

androgen a hormone that induces maleness.

angi- vessel.

ankyl- a growing together of parts.

ankylose to fuse in an immovable articulation.

anlage (pl., anlagen) an embryonic rudiment or precursor of a developing structure.

annulus a ring.

annulus tympanicus bony ring to which the eardrum is attached.

ante- before.

antebrachium the forearm; the part before the brachium.

anthrop- refers to human beings.

anti- against, opposite.

antidiuretic inhibiting loss of water (diuresis) via kidneys.

antrum a cavernous space.

Anura *an-* + *uro* (q.v.); tailless amphibians.

apical at the apex.

Apo- from, away from.

Apoda *a + pod-*; without legs.

aponeurosis *apo-* + *neuron* (a tendon); a broad, flat, tendonous sheet.

apophysis an outgrowth or process.

apsid refers to an arch.

arch- first, primary, ancient.

archenteron primitive gut.

---

*This is not a glossary. Terms not included here may be located in the text by referring to the index.

archetype an early model.

archinephros hypothetical primitive kidney.

archipallium first roof of the telencephalon.

arcuate arched.

arrector pili (pl., **arrectores pilorum**) muscle that erects a hair.

arthro- joint, articulation.

arthrodire placoderm with joints, because of dermal plates, in the neck (-*dire*).

artio- an even number.

artiodactyl having an even number of digits.

arytenoid resembling a ladle.

Ascaphus genus of frog lacking a drum (*scapha*).

aspid a shield.

ataxia *a-* (lacking) + *taxia* (order); a disorder of the neuromuscular system.

-ate having the property of, as septate: with septa.

atlas the vertebra that supports the head like the mythical Atlas holds up the earth.

atrium the courtyard of a Roman home > a cavity that has entrances and exits.

auricle an ear or earlike flap.

auto- self.

autostyly a condition in which the upper jaw braces (-*styly*) itself against the skull.

autotomy cutting one's self, as when a lizard breaks off the end of its tail.

axial in the longitudinal axis.

axilla armpit.

azygos *a-* (lacking) + *zyg-* (a yoke) > on one side only.

B

baro- pressure.

baroreceptor a sense organ that monitors pressure.

basi- most ventral; pertaining to a basal location.

basihyal ventral element of hyoid skeleton.

bi- double, twice.

bicornuate having two horns (*cornua*).

bicuspid having two cusps.

bili bile.

bio- life.

bipartite having two parts.

blast- an embryonic precursor; a germ of something.

blastema an embryonic concentration of mesenchyme.

blastocoel cavity of the blastula.

blastocyst the mammalian blastula.

blastula a little (-*ula*) embryo.

bovine pertaining to oxen or cows.

brachi- arm.

brachiocephalic associated with the arm and head.

branchi- gill.

branchiomeric referring to parts of the branchial arches.

bucco- cheek.

bulbus bulb.

bulbus arteriosus muscular swelling on ventral aorta.

bulbus cordis term sometimes applied to conus arteriosus in lungfishes, amphibians, and mammalian embryos; part of the heart.

bulla, a bubble > a bubblelike part such as the tympanic bulla.

buno a hill or mound.

bunodont a tooth with low cusps.

bursa a sac or pouch.

C

caecum a blind pouch.

calamus a stem or reed, the stem of a feather.

calcaneo- pertaining to the heel bone.

canaliculus a little canal.

capitulum a little head.

caput head.

cardi- heart.

cardinal of basic importance; chief.

carina a keel.

carn- flesh.

carnivore a flesh-eating animal.

carotid from a word meaning heavy sleep; compression of the carotid artery cuts off blood to the brain.

carpo- wrist.

cata- down, complete.

catarrhine having a nose (-*rhin*) with nares directed downward.

cauda tail.

caudad toward (-*ad*) the tail.

cecum see *caecum*.

cel- see -*coel*-.

cen- new.

cenogenic of recent origin.

cephal- head.

cephalaspidomorpha having a head shield.

cera wax.

cerat- horn.

ceratohyal horn of the hyoid.

ceratotrich a horny, hairlike fire support.

-cercal tail.

cerumen secretion of mammalian external ear glands; see *cera*.

cervical pertaining to the *cervix* (q.v.).

cervix neck.

Chelys tortoise.

chiasma shaped like the Greek letter chi (χ) > a crossing.

chir-, cheir- hand.

Chiroptera mammals in which the hand is modified as a wing; bats.

choana a funnel-shaped opening.

chole- bile.

chondr- cartilage.

chorion a membrane.

choroid, chorioid resembling a membrane (see *chorion*).

chrom-, chromato- color.

circa, circum around, about.

circadian about a day in length.

clad a branch.

Cladoselachii a branch of cartilaginous fishes.

clava club.

clavo-, cleido- clavicle.

cleidomastoid a neck muscle attached to clavicle.

cleidoic closed, locked up > a reptilian, bird, or monotreme egg with much yolk and a shell.

cloaca a sewer > common terminus for digestive and urinary tracts.

cochlea a snail with a spiral shell > spiral labyrinth of inner ear.

-coel- hollow, a cavity.

coelom body cavity.

collagen gelatinous, gluelike material.

colli of the neck.

colliculus a little hill.

columella a little (-*ella*) pillar or column.

com-, con-, cor- with, together.

conari of the pineal (q.v.).

conch- shell.

concha the pinna of the ear, shaped like a clamshell.

contra- opposite.

contralateral on the opposite side.

conus a cone.

conus arteriosus chamber of heart after the ventricle.

copr- feces.

coprodeum fecal passage derived from cloaca.

cor- see *com-*.

coracoid shaped like a crow's beak.

corn- horn.

cornified changed to horn by keratinization.

cornu (pl., **cornua**) horn of the hyoid.

corona a wreath or crown.

coronary sinus forms a "wreath" around the heart.

corpus (pl., **corpora**) body.

corpora quadrigemina the two pairs of twins (-*gemini*) of the roof of the mesencephalon of amniotes.

corpus luteum a yellow body of the ovary.

corpus spongiosum spongy body.

costa rib.

costal cartilage a cartilage at the ventral end of a rib.

cotyledonary cup shaped.

coxa hip.

cribriform sievelike.

cricoid resembling a ring.

crin- shed.

crista a ridge or crest.

crotalum a rattle > *Crotalus,* a rattlesnake.

crus leg.

cten- comb.

ctenoid resembling a comb.

cucullaris from a word meaning a hood; the two cucullaris (trapezius) muscles resemble collectively a hood or shawl.

cuneiform wedge-shaped.

cupula a structural peak.

cusp a peak or point.

cutaneous referring to skin.

cyclo- circular.

cyclostome agnathan with round mouthlike funnel.

cyno- dog.

cynodont dog-toothed.

cysto- sac, bladder.

cyt-, cyto-, -cyte cell.

D

dactyl- finger, toe.

de- away from.

deferens *de* + *ferent* (q.v.).

decidu- fall off or be shed.

deciduous placenta one in which the uterine lining is partly shed at parturition.

deltoid resembling the Greek letter delta (Δ).

demi- half.

demibranch gill on one face of a gill arch.

dendro- tree.

dendrite neuronal arborization.

dent- tooth.

dentin bone like that in teeth.

derm- skin.

dermatome layer of a somite giving rise to skin.

Dermoptera mammals in which the skin forms a wing membrane (see *-pter*).

dermato- referring to skin.

dermatocranium skull bones phylogenetically derived from skin.

-deum, -daeum a passageway.

deuter- two.

deuterostome an animal that uses the blastopore as an anus and forms a second mouth.

di- twice, double.

diapophysis one of two lateral processes.

diapsid having two arches.

diarthrosis freely movable joint between two bones or cartilages.

Dipnoi fish with two paired breathing apertures (external and internal).

dia- through, apart, completely, between.

diaphragm a separation (*phragma*) between two parts.

diaphysis ossifying shaft of a long bone.

diastema an intervening space.

-didym- twins > the testes.

digiti- fingers or toes.

digitigrade walking (*-grade*) on the digits.

dino- fearful, terrible.

dinosaur a fear-inspiring reptile.

diphy- double.

diphyodont having two successive sets of teeth.

diplo- double, two.

diplospondyly two vertebrae in each body segment.

dis- separation, apart from.

dissect to disassemble.

diverticulum an outpocketing.

dorsum the back.

dorsad toward the back.

duodenum twelve, in Latin; the length of the human duodenum is about the breadth of 12 fingers.

dura tough, hard.

dys- faulty, painful, difficult.

E

e- without.

Edentata an order of mammals lacking teeth.

ect- outer.

ectopterygoid the outer pterygoid bone.

ectotherm an animal whose temperature varies with the environment.

-ectomy *ex-* (out of) + *tome* (cut), as in appendectomy.

ectopic *ex-* (out of) + *topo-* (place).

ectopic pregnancy a pregnancy in which the fetus is implanted elsewhere than in the uterus.

ef- variant of *ex-*.

efferent that which carries away from; efferent branchial arteries carry blood out of the gills.

elasmo- plate.

elasmobranch with gills composed of flat plates.

-ella a diminutive, as in columella.

en- in, into.

encephalon the brain, a structure in the head.

endo- within, inner; see also *ento-*.

endochondral within cartilage.

endolymph the lymph within the membranous labyrinth.

endomysium *endo* + *mys* (q.v.).

endotherm an animal that maintains a relatively constant body temperature regardless of environmental fluctuations.

enteron the gut.

ento- within, inner; see also *endo-*.

entoglossal within the tongue.

ep-, epi- upon, above, over.

epaxial above an axis.

ependyma an outer garment > the membrane covering the part of the central nervous system exposed to the neurocoel.

epididymis (pl., epididymides) a structure lying on the testis (*didym-*).

epiglottis a skeletal flap over the glottis.

epimere dorsal mesoderm.

epimysium *epi* + *mys* (muscle).

epiphysis *epi* + *physis* (q.v.), pineal complex.

epipodium the leg bones immediately above the foot.

epiploic relating to greater omentum (the epiploon).

epithelium a surface layer of cells.

epitrichium *epi* + *trich-* (hair), a temporary layer of epidermal cells overlying the hair of fetal mammals. The mammalian periderm.

erythro- red.

eso- carrier.

esophagus carrier of substances that have been eaten (*phag-*).

estr- female.

ethmoid sievelike.

eu- true.

eury- wide.

euryhaline able to live in waters with a wide range of salinity (*halo-*).

ex-, exo- out, out of, away from, outer.

excurrent pore a pore for the exit of a current of water.

exoskeleton a skeleton in the skin.

extra- beyond, outside of.

extraembryonic outside of the embryo.

F

falciform shaped like a sickle.

fascicle a bundle.

fauces throat.

fenestra a window > an aperture.

-ferent, -ferous carrying, as in *afferent* (q.v.).

fil thread.

fimbria fringe.

foramen a small opening, usually transmits something.

foramen magnum the large foramen in the occipital region.

fore- before, in front.

forearm the part of the arm before the upper arm.

fossa a pit, cavity, depression, vacuity.

fovea a small pit; site of most acute vision in retina.
frenulum a little bridle > the membrane that bridles (ties) the tongue to the floor of the oral cavity.
frug- fruit.
  frugivore a fruit eater.
fundus the bottom of a cavity.
fusiform spindle shaped.

G
gan- bright.
  ganoid a shining bony scale.
ganglion a swelling > a group of cell bodies outside the central nervous system.
gastr- a belly, a stomach; the digastric muscle has two bellies.
  gastralia ventral abdominal ribs.
  gastrula a little stomach > a stomachlike embryo.
gen- origin.
  geniculate ganglion a kneelike bend of facial nerve; see *genu*.
genio- chin.
genotype the specific assortment of genes in an individual.
genu knee.
geo- earth.
glans acorn > the tip of the penis.
glenoid resembling a socket.
glia glue.
glomerulus a little glomus (ball or skein) > a tiny plexus of blood vessels.
gloss- tongue.
gnath- jaw.
gon- seed > generative, as in glucagon (giving rise to glucose).
  gonad the source of gametes.
gracilis slender.
gubernaculum a rudder > a governor, as the ligament that (partly) governs the position of the testis.
gula throat.
  gular fold a fold at the throat of some tetrapods.
gustatory related to gustation (taste).
gymn- naked.

gyrus a ridge between grooves.

H
haem-, hem- blood.
hallux great toe.
hamate having a hook.
hamulus a little hook.
helico- helix, spiral.
hemi- half; equal to *demi-*.
hemopoiesis formation of blood.
hepat- liver.
hept- seven.
herbivore an animal that devours grasses (herbs).
hetero- other, different; opposite to *homo-*.
  heterochrony a time difference.
  heterodont having different kinds of teeth.
hex- six.
hilum a small notch.
hipp- horse.
hist- tissue.
  histogenesis the formation of a tissue.
holo- entire, whole.
  holonephros a kidney extending the length of the coelom.
hom-, homeo-, homo-, homoio- like, similar.
  homeostasis maintenance of a constant internal environment.
  homeotherm an animal that maintains a steady body temperature despite ambient (surrounding) temperature; an endotherm.
homodont teeth all alike.
hyaline clear, glassy.
hyoid shaped like the Greek letter upsilon (γ).
hyostyly see -*styly*.
hyp-, hypo- under, below, less than ordinary.
  hypaxial below a given axis.
  hypophysis a growth under the brain; the pituitary body.
  hypural a bone below the urostyle of fishes.
hyper- above, beyond the ordinary.
hypso- height.

I
ichthy- fish.
-iform having the shape of.
ileo- pertaining to the ileum of the intestine.
ilio- pertaining to the ilium of the pelvis.
impar unpaired.
in- not.
  innominate not named.
incus an anvil.
infra- beneath, under.
infundibulum a little funnel.
inguen the groin.
  inguinal in the region of the groin.
inter- between.
  intercalary plate part of neural arch between neural plates in fishes.
  interrenal steroidogenic tissue between the kidneys.
intra- within.
  intramembranous within a membrane.
  intrasegmental within a segment.
ipsi- the same.
  ipsilateral on the same side.
irid- rainbow.
  irides pl. for irises.
  iridophore a pigment cell containing refractory bodies that result in iridescence.
iris see *irid*.
ischi- hip, pelvis.
iso- equal, alike.
  isolecithal egg an egg having even distribution of yolk.
-issimus a superlative ending; the longissimus dorsi is the longest muscle of the back.
iter Roman road > passageway.
-itis inflammation of.

J
jejunum empty; part of intestine that is often empty at death.
juga- yoke > something that joins; the **jugal bone** in mammals is a yoke uniting maxilla and temporal bones.
jugular pertaining to the neck.

juxta- next to, near.
  juxtaglomerular near glomeruli.

K
kat- down; same as *cat-*.
keratin from a Greek word meaning horn; a relatively insoluble substance in cornified cells.
kinetic capable of moving.

L
labium lip.
labyrinth a maze.
  labyrinthodont an early tetrapod with greatly folded dentin in the teeth.
lac-, lact- milk.
lacrimal pertaining to tears; from lachryma (a teardrop).
lacuna a lake.
lag- hare.
lagena a flask > flask-shaped part of inner ear.
lambdoidal having the shape of the Greek letter lambda (λ).
lamella a thin plate.
lamina a thin sheet, plate, or layer.
laryng- larynx.
latissimus the broadest; see -*issimus*.
lecith- yolk.
lemmo- sheath or envelope.
lemur from a word meaning a nocturnal being or ghost.
lepid-, lepis-, lepo- scale, husk.
  Lepisosteus a ganoid fish with bony scales.
lepto weak, thin, delicate.
leuco-, leuko- white, colorless.
levator that which elevates or raises.
lien- spleen.
lingua the tongue.
lip- fat.
liss- smooth.
lith stone.
longissimus the longest; see -*issimus*.
lopho- ridge, crest.
lumbar pertain to the loins; the region of the vertebral column just cephalad to the sacrum.
lumen light > an opening that light can pass through, the cavity in a tube.

lunar, lunate moon shaped.

luteo- yellow.

M

macula a spot.

magnus, -a, -um large.

malleus a hammer.

mandibula a jaw.

manu- hand.

manubrium a handle.

marsupium a pouch.

mastoid like a breast (*mast-*).

mater mother.

maximus, -a, -um largest.

meatus a canal or passageway.

medulla bone marrow.

    medulla spinalis the marrow of the backbone > spinal cord.

meg- great, very large.

melan- dark, black.

    melanophore bearing (*-phore*) dark pigment.

meninx (pl., **meninges**) a membrane.

mento- chin.

    mental foramen foramen on mandible near chin.

-mer- a segment, a part, one of a series.

mes- middle, midway, intermediate.

    mesaxonic foot one in which the weight-bearing axis passes through the middle toe.

    mesenchyme a tissue (*enchyme*) that is not yet differentiated.

    mesentery associated with the midline of the enteron.

    mesonephros an intermediate kidney.

met- after, succession, change, behind.

    metacarpal a bone distal to a carpal.

    metamere one of a series of segments.

    metamorphosis change in form.

    metanephros hindmost kidney.

mimetic capable of mimicking; having the characteristics of a mime.

mitral refers to a bishop's miter or headdress; the mitral valve is the bicuspid valve of the mammalian heart.

mono- one.

    monotreme a mammal with one caudal opening.

morph- shape, structure, form.

    morphogenesis development of form.

    morphology study of form; anatomy.

morula a mulberry.

myel- marrow.

    myelencephalon the marrow inside the skull > the medulla.

    myelin fatty material.

mylo- from a word meaning a millstone.

    mylohyoid a muscle attached near the grinding teeth.

myo-, mys- muscle.

    myocardium muscle of heart.

    myomere one of a series of muscle segments.

    myotome part of somite giving rise to muscle.

N

neo- new, recent.

nephr- kidney.

    nephrogenic giving rise to kidney.

    nephron a functional kidney unit.

nomen- (pl., **nomina**) name.

noto- back.

    notochord cordlike skeleton of the back.

nuchal refers to the nape of the neck.

O

occiput part of the head surrounding the foramen magnum; in mammals, the back of the head.

octo- eight.

ocul- eye.

odon-, odont- tooth.

    odontognath a bird with teeth on the jaws.

    odontoid resembling a tooth.

-oid like, having a resemblance to, as in hominoid (humanlike).

-ole small, as in arteriole.

olecranon the head (*cranium*) of the ulna (*oléné*) at the elbow joint.

oligo- few, small.

-oma swelling.

omentum a free fold of peritoneum.

omni- all.

    omnivore an animal that eats plants and animals.

omo- shoulder.

omphalo- navel.

ontogenesis *onto* (individual) + *genesis* (origin); the development of an individual.

oö- egg (pronounced oh-oh).

    oöcyte egg cell.

    oöphoron *oö* + *-phore-* (q.v.); ovary.

operculum a cover or lid.

ophthalm- eye.

opisth- at the rear, behind.

    opisthonephros mesonephric kidney of adult anamniotes.

orb- a circle.

    orbit cavity for the eyeball.

ornith- bird.

oro- mouth.

ortho straight.

-orum of the, as in branchiorum (of the gills).

os bone.

    ossicle a small bone.

    ossify to become bony.

os mouth.

    os uteri entrance to uterus from vagina.

osmoregulation electrolyte homeostasis.

oste- bone.

    Osteichthyes bony fish.

    osteon unit of bone in concentric layers.

ostium an entranceway > a mouth.

ostraco- shell.

oto- ear.

    otic refers to the ear.

    otocyst a vesicle that becomes the inner ear.

-ous having the characteristic of.

ovale shaped like a hen's egg; oval.

ovi-, ovo- egg.

    oviparous *ovi* + *parous* (q.v.); egg laying.

    ovipositor a structure for laying eggs.

ovine pertaining to sheep.

-oxy- sharp, acute, acid.

P

pachy- thick.

paed-, ped- child.

    paedomorphosis evolution of species that resemble earlier ontogenetic stages of the ancestor.

palatine referring to the palate.

paleo- old, ancient.

pallium a cloak > a roof.

panniculus a small piece of cloth > a layer of tissue.

papilla a nipple > a nipple-shaped structure.

par-, para- beside, near.

    parasphenoid parallel to the sphenoid.

    parotid near the ear.

parie- (pl., **parietes**) pertaining to the body wall.

-parous bearing, giving birth to.

pars (pl., **partes**) part.

pectoral refers to the chest.

pedicel a slender stalk.

pelvis a basin.

penta- five.

perennibranchiate having permanent gills.

peri- around.

    perilymph fluid surrounding membranous labyrinth.

periss- odd.

peritoneum something that stretches over or around > the coelomic lining.

pes (pl., **pedes**) foot.

petro- stone, rock.

    petrosal bone a bone which, in humans, resembles a rugged rock.

phag- eat.

phalanx (pl., **phalanges**) a line of soldiers > a bone of a digit.

pharyng- pharynx.

pheno- visible.

    phenetic pertaining to phenotypes.

    phenotype a hereditary character or set of characters expressed in the soma as opposed to genotype.

-phil loving > having an affinity for.

-phore- bearing, one that bears, as in photophore (an organ that emits light).

phrenic refers to the diaphragm.

phylo- tribe.

    phylogenesis creation of new taxa.

    phylogeny evolutionary history of a group.

-physis that which grows.

physo- bellows > lung or air bladder.

    physoclistous lacking a duct from air bladder.

    physostome a fish that can get air to the swim bladder via the mouth.

pia tender, kind.

    pia mater the delicate meninx of the brain.

pilo- hair.

pineal resembling a pine cone.

pinna fin, feather.

pisci- fish.

pisiform shaped like a pea.

placo- thick, flat, platelike.

    placode in embryology, an ectodermal thickening that gives rise to something.

    placoderm fish with (bony) plates in the skin.

planta- the sole of the foot.

    plantigrade a flat-footed stance.

platy- flat, wide, broad.

plesio- close (to) > resembling.

    plesiosaur a marine animal that resembles, and is, a reptile.

pleur- rib, side.

    pleural refers to ribs.

    pleurodont teeth set into the side of the jawbone.

plexus a network.

pneumato- air.

pneumo- lung.

pod- foot.

poikilotherm an ectotherm (q.v.).

pollex the thumb.

poly- many, much, variegated.

pons a bridge.

porcine pertaining to swine.

post- after, behind.

    posttrematic behind a trema or slit.

pre- before, in front of, see also *pro-*.

    pretrematic in front of a trema or slit.

prim- first, earliest.

    primate one first in rank.

pro- favoring, on behalf of.

    prolactin hormone necessary for milk production.

pro- in front of, before, preceding.

    procelous with a cavity at the cephalic end.

    prostate a gland standing (*stat-*) at the beginning of the urethra.

proboscis an elongated snout; an elephant's trunk.

procto- anus.

proprio- one's own.

    proprioception reception of stimuli from muscles, joints, tendons.

pros- toward, near.

    prosencephalon the anterior end of the embryonic brain.

proto- early, first.

pseudo- false.

psoas loin.

pter-, pteryg- wing, feather.

    pterosaur winged reptile.

    pterotic wing of the otic complex.

    pterygoid resembling a wing.

pulmo- lung.

pyg- rump.

pyriform pear shaped.

Q

quadrate square, four sided.

quint- five.

R

rachi- referring to an axial support.

    rachis spine of a feather.

    rachitomous specifies a vertebra consisting of several pieces; see *-tome*.

radix (pl., **radices**) a root.

ramus a branch.

rectus, rectum straight.

rete (pl., **retia**) a network.

rete mirabile a remarkable (*mirabile*) network of vessels.

reticulum a little network.

retro- behind, backward.

rheo- current, flow.

rhin- nose.

    rhinencephalon an olfactory part of the brain.

    rhinoceros an animal with a horn (*cerato-*) on the nose.

rhomb- rhomboid.

rhynch- snout.

rostrum a beak or platform.

ruga (pl., **rugae**) a wrinkle.

S

sacculus a little sac.

sagittal from a word meaning an arrow.

sangui- blood.

sarco- flesh.

    Sarcopterygii fish with fleshy lobe at base of fin.

saur- lizard > reptile.

scala a ladder.

scalene a triangle with sides and angles unequal.

scler- hard, skeletal.

    sclerotome part of somite giving rise to skeletal components.

sebum grease, wax.

    sebaceous having an oily secretion.

seco- refers to cutting.

    secodont shearing molariform teeth of carnivorous mammals.

-sect- cut, divide.

seleno- refers to the moon.

    selenodont teeth with enamel disposed in crescentic folds.

sella turcica a seat (*sella*) shaped like a Turkish saddle.

semi- half.

    semilunar shaped like a half-moon.

seminal pertaining to seed > to semen.

seminiferous carrying sperm.

serrate notched or toothed along the edge; the serratus muscle is serrate.

sex- six.

sigmoid S-shaped.

sinus a cavity.

    sinusoid a thin-walled, sinuslike vascular channel.

soma-, somato- body.

    somite a body segment.

sphenoid wedge shaped.

spiracle a breathing hole.

splanchn- viscera.

splen- spleen.

spondyl- vertebra.

squam- scale.

    squamous scalelike, flattened; squamate.

stapes a stirrup.

    stapedial associated with the stapes.

stato- standing, fixed.

steg- refers to covering plates; also to a roof.

stellate star shaped.

stereo- solid > shape; stereognosis is knowledge (*-gnosis*) of form or weight acquired by feeling or lifting.

stom- mouth.

stratum layer.

strept- twisted, curved.

stria a stripe.

    striate having stripes.

-style pillar > a process such as the urostyle.

    styloid having an elongated shape.

-styly braced; in hyostyly the hyoid braces the jaws against the skull.

sub- under, below.

    subclavian under the clavicle.

    subunguis under the nail.

    subungulate not quite an ungulate.

sudor sweat.

sulcus a groove.

super-, supra- over, above, in addition.

    suprarenal a gland above the mammalian kidney.

sur- over, above; equivalent to *super-*.

    surangular bone a bone above the angular.

sym-, syn- together.

    symphysis a growing together; see *-physis*.

    synapse a junction.

    synarthrosis an immovable suturelike joint.

    synsacrum sacrum united with other vertebrae.

T

tarsus ankle; also, a connective tissue plate in the eyelid.

tax- arrangement.

taxon a taxonomic unit such as a phylum or species.

taxonomy the orderly arrangement of taxons; classification.

tectum a roof.

tegmentum a covering.

tel-, teleo-, telo- end, complete.

telencephalon anterior end of the brain.

teleology ascribing natural phenomena to a preordained purpose.

tela a web > a thin weblike membrane.

temporal refers to the temple or the side of the skull behind the eye.

teres round.

tetra- four.

thalamo- a chamber.

theco- a case or sheath.

thecodont having socketed teeth.

therio- an animal with hair, a beast.

thrombo- clot.

thyroid shield shaped.

-tome cut; also, the result of cutting, as a section or sheet.

trabecula a little beam > a strand, ridge, rod, or bundle.

trans- across.

transect to cut across.

trapezoid a four-sided plane with two parallel sides.

trapezius muscle named for its shape in humans.

-trema a slit.

tri- having three parts.

trigeminal from word meaning triplets; a nerve with three primary branches.

trich- hair.

trochlea a pulley; the trochlear nerve of humans passes through a pulley at its attachment.

troph- nourishment.

truncus trunk.

truncus arteriosus ventral aorta.

tuber a swelling or knob > a tuberosity, tubercle, or protuberance.

tuberculum a little tubercle.

tunic a coat or wrap.

tympanum a drum > eardrum.

U

ulna elbow > the bone at the elbow.

ultimobranchial a gland derived from the last (*ultimo-*) branchial pouch.

ultra- beyond.

-ulus, -ula, -ulum diminutive endings denoting tiny.

uncus a hook.

uncinate hooked.

unguis nail, claw, hoof.

ungula hoof.

ungulate hooved.

uro- from Greek *ouro*, meaning tail or urine.

urophysis a growth at the base of the tail.

uropygium the rumplike tail of a bird.

utricle a little sac or vesicle.

V

vagina a sheath.

vagus wandering.

vas (pl., **vasa**) vessel.

vasa vasorum vessels of the blood vessels.

velum a veil > a thin membrane.

venter abdomen; the part opposite the dorsum or back.

ventricle a cavity in an organ.

vermi- worm.

vesica a bladder or vesicle.

vestibule an antechamber or entrance way.

vitelli- yolk.

vitreous having a glassy appearance.

vivi- alive.

volar pertaining to palm or sole.

vomer a plowshare; the mammalian vomer bone resembles a plowshare.

-vorous eating, devouring, as in insectivore.

X

Xanth- yellow.

Xen- strange, foreign.

xiph- sword.

xiphoid process a swordlike process of the sternum opposite the manubrium.

Y

ypsiloid shaped like the Greek letter upsilon (γ).

Z

zyg- yoke > something that links two things.

zygapophysis a vertebral process that articulates with a more anterior or posterior one.

zygoma an arch.

zygote result of union of gametes.